SENSORIMOTOR CONTROL OF GRASPING: PHYSIOLOGY AND PATHOPHYSIOLOGY

The human hand can take on a huge variety of shapes and functions, providing its owner with a powerful hammer at one time or a delicate pair of forceps at another. The universal utility of the hand is even more enhanced by the ability to amplify the function of the hand by using tools. To understand and appreciate how the human brain controls movements of the hand, it is important to investigate both the healthy motor behavior and dysfunction during everyday manipulative tasks. This book provides a contemporary summary of the physiology and pathophysiology of the manipulative and exploratory functions of the human hand. With contributions from scientists and clinical researchers of biomechanics, kinesiology, neurophysiology, psychology, physical medicine and rehabilitation, it covers the development of healthy human grasping over the lifespan, the wide spectrum of disability in the pathological state and links basic motor research with modern brain sciences.

DENNIS NOWAK is a Neurologist and Neuroscientist based at the Hospital for Neurosurgery and Neurology, Kipfenberg, and the Department of Neurology, University of Cologne, Germany.

JOACHIM HERMSDÖRFER is an Engineer and Neuroscientist based at the Clinical Neuropsychology Research Group, Hospital München-Bogenhausen, in Munich, Germany.

Together, their research centers around physiological and pathophysiological sensorimotor processes, particulary during hand use.

SENSORIMOTOR CONTROL OF GRASPING: PHYSIOLOGY AND PATHOPHYSIOLOGY

Edited by

DENNIS A. NOWAK
Klinik Kipfenberg, Kipfenberg
Germany

JOACHIM HERMSDÖRFER
Hospital Munich-Bogenhausen,
Munich, Germany

CAMBRIDGE UNIVERSITY PRESS
Cambridge, New York, Melbourne, Madrid, Cape Town, Singapore, São Paulo, Delhi

Cambridge University Press
The Edinburgh Building, Cambridge CB2 8RU, UK

Published in the United States of America by Cambridge University Press, New York

www.cambridge.org
Information on this title: www.cambridge.org/9780521881579

First published 2009

Printed in the United Kingdom at the University Press, Cambridge

A catalog record for this publication is available from the British Library

ISBN 978-0-521-88157-9 hardback

Contents

The plates can be found between pages 112 and 113 and 368 and 369.

Contributors

Caterina Ansuini
Dipartimento di Psicologia Generale,
Università di Padova,
Via Venezia 8, 35131, Padova, Italy

B. Baur
EKN Clinical Neuropsychology Research
Group,
Hospital Munich-Bogenhausen,
Dachauer Str. 164, D-80992 Munich,
Germany

R. Martyn Bracewell
The Wolfson Centre for Clinical and
Cognitive Neuroscience,
University of Wales,
Bangor, UK

Thomas Brochier
Sobell Department of Motor Neuroscience
and Movement Disorders,
Institute of Neurology,
University College London,
London WC1N 3BG, UK
and
Institut de Neurosciences Cognitives de la
Méditerranée – INCM,
UMR 6193,
CNRS – Université de la Méditerranée,
Marseille, France

Umberto Castiello
Dipartimento di Psicologia Generale
Università di Padova
Via Venezia 8, 35131, Padova, Italy

Kelly J. Cole
Department of Integrative Physiology,
University of Iowa,
Iowa City, IA, USA

Daniel M. Corcos
Departments of Kinesiology and Nutrition,
Bioengineering, Physical Therapy,
University of Illinois at Chicago, Chicago, IL
and
Department of Neurological Sciences,
Rush Presbyterian St. Luke's Medical
Center, Chicago, IL, USA

Günther Deuschl
Department of Neurology
University Hospital Schleswig Holstein,
Campus Kiel,
Schittenhelmstr. 10,
24105 Kiel, Germany

H. Henrik Ehrsson
Department of Neuroscience,
Karolinska Institute,
Retzius väg 8, 177 77,
Stockholm, Sweden

Satoshi Endo
Behavioural Brain Sciences Centre,
School of Psychology,
University of Birmingham,
Birmingham, UK

Gereon R. Fink
Department of Neurology,
University of Cologne,
Kerpener Str. 62,
D-50924 Cologne, Germany

J. Randall Flanagan
Department of Psychology and Centre for
Neuroscience Studies,
Queen's University,
Kingston, ON, K7L 3N6, Canada

Tania S. Flink
Department of Kinesiology,
Arizona State University,
Tempe, AZ, USA

Hans Forssberg
Department of Women and Child Health,
Neuropediatric Research Unit,
Karolinska Institute and Stockholm Brain
Institute,
Astrid Lindgrens Barnsjukhus Q2:O7,
17176 Stockholm, Sweden

Kathleen M. Friel
Department of Neuroscience,
Columbia University,
New York, NY, USA

Andrew M. Gordon
Department of Biobehavioral Sciences,
Box 199,
Teachers College, Columbia University,
525 West 120th Street,
New York, NY 10027, USA

Christian Grefkes
Max-Planck-Institute for Neurological
Research,
Section "Neuromodulation &
Neurorehabilitation",
Cologne, Germany

Priska Gysin
Department of Biobehavioral Sciences,
Teachers College,
Columbia University,
New York, NY 10027, USA

Mark Hallett
Human Motor Control Section, NINDS,
NIH,
Building 10, Room 5N226
Bethesda, MD 20892-1428, USA

Joachim Hermsdörfer
EKN Clinical Neuropsychology Research
Group,
Hospital Munich-Bogenhausen
Dachauer Str. 164,
D-80992 Munich, Germany

Marc Jeannerod
Institut des Sciences Cognitives,
67 Boulevard Pinel,
69675, Bron, France

Roland S. Johansson
Physiology Section,
Department of Integrative Medical
Biology,
Umeå University,
SE-901 87 Umeå, Sweden

Jamie A. Johnston
Department of Kinesiology,
Arizona State University,
Tempe 85287, AZ, USA

Terry R. Kaminski
Department of Biobehavioral Sciences,
Teachers College,
Columbia University,
New York, NY 10027, USA

Mitsuo Kawato
JST, ICORP, Computational Brain Project,
4-1-8 Honcho,
Kawaguchi, Saitama, Japan and
ATR, Computational Neuroscience
Laboratories,
2-2-2 Hikaridai, Seika-cho Soraku-gun,
Kyoto 619-0288, Japan

Giacomo Koch
Clinica Neurologica,
Dipartimento di Neuroscienze,
Università di Roma Tor Vergata,
Via Montpellier 1,
00133 Rome, Italy
and
Fondazione,
S. Lucia IRCCS,
Via Ardetaina 306,
00179 Rome, Italy

Johann P. Kuhtz-Buschbeck
Institute of Physiology,
Kiel University,
Olshausenstrasse 40,
D 24098 Kiel, Germany

Catherine E. Lang
Program in Physical Therapy,
Program in Occupational Therapy,
Department of Neurology,
Washington University,
St. Louis, MO, USA

Mark L. Latash
Department of Kinesiology,
The Pennsylvania State University,
University Park,
PA 16802, USA

Susan J. Lederman
Department of Psychology,
Queen's University,
Kingston,
ON, Canada

Roger N. Lemon
Sobell Department of Motor
Neuroscience and Movement
Disorders,
Institute of Neurology,
University College London,
London WC1N 3BG, UK

Mario Manto
FNRS Neurologie,
ULB Erasme,
Brussells, Belgium

Kyle Merritt
Department of Psychology and
Centre for Neuroscience Studies,
Queen's University,
Kingston, ON K7L 3N6, Canada

Dennis A. Nowak
Klinik Kipfenberg,
Kindinger Str. 13,
D-85110 Kipfenberg, Germany

Erhan Oztop
JST, ICORP, Computational
Brain Project, 4-1-8 Honcho,
Kawaguchi, Saitama, Japan and
ATR, Computational Neuroscience
Laboratories,
2-2-2 Hikaridai, Seika-cho Soraku-gun,
Kyoto 619-0288, Japan

Jan Raethjen
Department of Neurology,
University Hospital Schleswig Holstein,
Campus Kiel,
Schittenhelmstr. 10,
24105 Kiel, Germany

Ralf Reilmann
Department of Neurology,
Universitätsklinik Münster (UKM),
Westfälische Wilhelms-University,
Albert-Schweitzer-Str. 33,
48129 Münster, Germany

Sarah Pirio Richardson
Department of Neurology,
University of New Mexico HSC,
MSC 10 5620,
1 University of New Mexico,
Albuquerque, NM 87131, USA and
Human Motor Control Section, NINDS,
NIH,
Building 10, Room 5N226,
Bethesda, MD 20892-1428, USA

John C. Rothwell
Sobell Department,
Institute of Neurology,
Queen Square,
London WC1N 3BG, UK

Marco Santello
Department of Kinesiology,
Arizona State University,
Tempe, AZ 85287-0404, USA

Marc H. Schieber
Departments of Neurology,
Neurobiology & Anatomy, and Physical
Medicine & Rehabilitation,
University of Rochester,
Rochester, NY, USA

H. R. Siebner
Department of Neurology,
University of Schleswig-Holstein, Kiel
Campus,
Neurozentrum,
Schittenhelmstrasse 10,
24105 Kiel, Germany

Allan M. Smith
Département de Physiologie,
Université de Montréal,
C.P. 6128 Succursale Centre ville,
Montréal, QC, Canada H3C 3T8

Rachel L. Spinks
Sobell Department of Motor Neuroscience
and Movement Disorders,
Institute of Neurology,
University College London,
London WC1N 3BG, UK

Matthew B. Spraker
Department of Bioengineering,
University of Illinois at Chicago,
Chicago, IL, USA

George E. Stelmach
Department of Kinesiology,
Arizona State University,
Tempe, AZ, USA

Lars Timmermann
Department of Neurology,
University Hospital Cologne,
Kerpener Str. 62,
50924 Cologne, Germany

Maria A. Umilta
Sobell Department of Motor Neuroscience
and Movement Disorders,
Institute of Neurology,
University College London,
London WC1N 3BG, UK and
Università di Parma,
Parma, Italy

David E. Vaillancourt
Departments of Kinesiology and Nutrition,
Bioengineering and Neurology and
Rehabilitation,
University of Illinois at Chicago,
Chicago, IL, USA

Brigitte Vollmer
Department of Women and Child Health,
Neuropediatric Research Unit,
Karolinska Institute and
Stockholm Brain Institute,
Astrid Lindgrens Barnsjukhus Q2:O7,
17176 Stockholm, Sweden

Roland Wenzelburger
Dänischenhagener Str. 12B,
D-24161 Altenholz, Germany

Alan M. Wing
Behavioural Brain Sciences Centre,
School of Psychology, University of
Birmingham,
Edgbaston,
Birmingham B15 2TT, UK

Vladimir M. Zatsiorsky
Department of Kinesiology,
The Pennsylvania State University,
University Park,
PA 16802, USA

Kirsten E. Zeuner
Department of Neurology,
University of Schleswig-Holstein,
Kiel Campus,
Neurozentrum,
Schittenhelmstr. 10,
24105 Kiel, Germany

Preface

The numerous skeletal and muscular degrees of freedom of the hand provide the human with an enormous dexterity that has not yet been achieved by any other species on earth. The human hand can take on a huge variety of shapes and functions, providing its owner with a powerful hammer at one time or a delicate pair of forceps at another. The universal utility of the hand is even more enhanced by the ability to amplify the function of the hand by using tools. True opposition between the thumb and index finger is only observed in humans, the great apes and Old World monkeys. The human thumb is much longer, relative to the index finger, than the thumb of other primates and this allows humans to grasp and manipulate objects between the tips of the thumb and index finger. Humans have more individuated muscles and tendons with which to control the digits and have evolved extensive cortical systems for controlling the hand. In addition to its manipulative function the hand is a highly sensitive perceptive organ, orchestrated by myriads of tactile and somatosensory receptors, which enables humans to perceive the world within their reach. Taken together all these phylogenetic developments have provided humans with the ability to interact with each other, make love and war, and also to shape the world. To understand and appreciate how the human brain controls movements of the hand, it is important to investigate both the healthy motor behavior of the hand and its dysfunction during everyday manipulative tasks.

Over the past three decades exciting novel achievements have enhanced our knowledge of the physiology and pathophysiology of human grasping. When trying to summarize what we know today about the physiology of human grasping we have to look back at the origins of its research. There is no doubt that the modern era of research on the kinematics and kinetics of human grasping started in the early 1980s with the epoch-making studies from the groups around Marc Jeannerod in Bron, France, and Roland Johansson in Umeå, Sweden. These researchers provided us with the first detailed descriptions of the kinematics of human grasping and the dynamic control of isometric grip forces when handling objects in the environment. Inspired from these early works, Alan Wing, Randy Flanagan, Hans Forssberg, Kelly Cole and Andrew Gordon, among others, carried on in this "orphan" field of research over the next decade. Thereafter, several scientists have been walking in the footsteps of these first-hour researchers, including ourselves. Consequently, the methodology of kinematic and kinetic analysis of grasping movements has rapidly found its way into clinics and aided in discovering the characteristics of impaired grasping in a huge variety of neurological, psychiatric and orthopedic disorders. Today, motor laboratories all over the world have established kinematic and kinetic investigation of grasping both in clinical and research settings and knowledge is still growing given the increasing number of citations each year in the PubMed database (www.ncbi.nlm.nih.gov/sites/).

It was our intention to bring together first-hour and last-generation neuroscientists and clinical researchers in the field to compile a contemporary summary about what we know today about the physiology and pathophysiology of the manipulative and exploratory functions of the human hand. The book is separated into four major sections: methodology, physiology of grasping, pathophysiology of grasping and therapy of impaired sensorimotor control of the hand. It covers the development of healthy human grasping over the lifespan and the wide spectrum of disability in the pathological state, and links basic motor research with modern brain sciences. The book focuses on, but is not limited to, grasping. Several additional aspects of the physiology and pathophysiology of fine motor performance of the hand, such as writing, multi-digit coordination and bimanual motor performance, are also covered. The book addresses scientists and clinical researchers from the areas of biomechanics, kinesiology, neurophysiology, psychology, physical medicine and rehabilitation. We are glad to have succeeded in pooling knowledge from "dinosaurs" in the field as well as from young scientists and clinical researchers from all over the world. This allows the book to contain basic knowledge from kinematic and kinetic recordings of the early days, and novel aspects regarding central control processes and models derived from more recent advances in technology, such as neurophysiology and neuroimaging.

When it comes to acknowledgments, we have to admit that there are many people without whom we certainly would not have arrived at this stage along our way through the world of grasping research. So we wish to direct our apologies to all those who are not mentioned here, despite their valuable support that is much appreciated. This is in particular to our team members, to all the doctoral students and to our clinical teachers. We wish to thank our families for their patience and constant support over the years. We both wish to dedicate this book to Norbert Mai, who inspired us to focus on the pathological aspects of grasping many years ago. Norbert died too early and we will always remember him for his visionary genius. Finally, we would like to thank Alison Evans, Anna-Marie Lovett and Martin Griffiths from Cambridge University Press for their assistance and guidance in making this project a success.

Part I
Methodology

1

Analysis of grip forces during object manipulation

JOACHIM HERMSDÖRFER

Summary

With the invention of strain gauges, isometric finger forces such as those produced during grasping an object could be measured continuously, precisely and without major constraints to the grip. In the precision grip between thumb and index finger, elementary performance aspects such as maximum grip force, ability to maintain a constant force, fast force changes or tracking of a dynamic target have been studied. In 1984, Johansson and Westling presented their paradigm based on the measurement of grip and load forces during grasping and lifting of an object. Their studies inspired a great deal of scientific interest in this aspect of fine motor control examined in healthy subjects as well as in patients with peripheral or central nervous system diseases. Research in this field progressed by introducing other motor tasks with specific demands on the control system, such as the compensation of inertial forces during movements of grasped objects. In addition, methods improved by technical developments such as 6-degree-of-freedom force/torque sensors, autonomous measurement devices, or force matrices to measure pressure distributions at grasping surfaces. Thus, measurements of isometric grip forces during object manipulation became a widely used method in neurophysiological and clinical motor sciences.

Control of isometric grip forces

Historically, the typical way to measure the force generated by the fingers or the whole hand was via compression of springs (e.g. Du Mensil de Rochemont, 1926). In addition, objects with known weights were used to load the hand or the fingers with a defined force (Truschel, 1913). With the invention of strain gauges in 1938, direct, practically isometric measurements of applied force became possible. Strain gauges change their electrical resistance during very small deformations in the range of hundredths of a millimeter. By fixing strain gauges on to elastic carriers such as metal beams and by using a special electrical circuit (Wheatstone Bridge) with differential electrical amplification, a voltage or current signal is generated that gives an immediate and continuous read-out of the applied force. Strain gauges, either on metallic or on silicon bases, still constitute the standard technique for continuous force measurements. Other electronic methods are based on piezoelectric,

Sensorimotor Control of Grasping: Physiology and Pathophysiology, ed. Dennis A. Nowak and Joachim Hermsdörfer. Published by Cambridge University Press. © Cambridge University Press 2009.

capacitive or optical principles. Still, forces can be measured without electronics by pure mechanical or hydraulic devices.

In a series of articles, Ghez and Gordon argued that the study of isometric force pulses avoids many of the problems associated with study of movements (Ghez & Gordon, 1987; Gordon & Ghez, 1987a, b). Thus, joint movements induce complex shifts of internal and external forces, for example by changing lever arms and corresponding torques. In addition, stretch reflexes may arise during movements and interfere with the effects of the central command. Finally, the relationship between electromyographic (EMG) signals and movement is complex and not uniquely determined (Ghez & Gordon, 1987). The authors showed that isometric force pulses produced by the elbow were isochronous, the rise time being independent from the force amplitude, even when accurate responses were required (Gordon & Ghez, 1987a; see also Freund & Büdingen, 1978). Many other studies employed paradigms with isometric force production by the arm, hand or fingers to study the principles of motor programming (e.g. Ulrich *et al.*, 1995) or the psychophysics of force perception, discrimination or matching (Stevens & Mack, 1959; Phillips, 1986).

Grasping an object between the pads of the thumb and the index finger has been considered as the prototype grip used for precision handling. This precision grip can be contrasted with a power grip that provides maximal contact of the palmar surfaces of fingers and hand with the manipulated object enabling transmission of high forces (Napier, 1956; Cutkosky & Howe, 1990). Both grip types have been subjected to numerous measurements of the produced grip force.

Power grip is primarily studied if the aim is to obtain a measure of maximum strength. Various devices have been developed to measure maximum force in a whole hand power grip. The most widely used clinical device is probably the Jamar Hand Dynamometer that measures strength with a hydraulic method and ergonomically formed handles (Mathiowetz *et al.*, 1985; Harth & Vetter, 1994). Normative data are available (Harth & Vetter, 1994; Hanten *et al.*, 1999) and the value of strength measurements in clinical disciplines like orthopedics and neurology is beyond doubt (Fees, 1986; Bohannon, 1989; Boissy *et al.*, 1999). Nevertheless caution was expressed against an over-generalization of strength measures in the context of neurological diseases (Bohannon, 1989; Jones, 1989; Hermsdörfer & Mai, 1996).

Maximum grip force may also be measured with a precision grip (Harth & Vetter, 1994; MacDermid *et al.*, 2001); however, more typically the precision grip is studied with an aim to examine the type of forces (usually well below the level of the subject's maximum strength) which are used when manipulating everyday-life objects. Thus, the relationship between the activity of hand and finger muscles contributing to a precision grip between thumb and index finger and the resulting grip force was established. A refined study of the EMG activity of 15 finger muscles detected correlations with low grip forces in the range between 1 and 3 N in ten muscles (Maier & Hepp-Reymond, 1995). In particular, intrinsic finger muscles showed a high correlation, extrinsic finger flexor muscles a moderate correlation, and extrinsic extensor muscles no correlation with low precision grip forces (see also Chapter 5). The relationship between grip force in a precision grip and cerebral activity was investigated in single cell recordings in monkeys and in neuroimaging studies of human subjects. In general, these studies showed a particularly strong relationship between the produced force and neural activity in the primary motor cortex, but also in the supplementary motor area and the premotor cortex (Hepp-Reymond *et al.*, 1978; Ashe, 1997; Cadoret & Smith, 1997; Cramer *et al.*, 2002; see Chapters 4 and 7).

Apart from relationships between force and neurophysiological parameters, specific aspects of force control have been investigated in clinical studies. For example, a motor task during which grip force had to be increased and decreased as fast as possible was tested in different groups of brain-damaged patients. These fast force changes proved highly sensitive to differentiate patients from healthy subjects and exhibited characteristic differences between hemiparetic and cerebellar patients (Avarello *et al.*, 1988; Mai *et al.*, 1988; Hermsdörfer *et al.*, 2003).

Another important aspect of grip force control is the ability to maintain a constant force. This was tested by providing a visual feedback of the produced force on a monitor (e.g. a vertical bar with force-linear length) and indicating a constant target force (e.g. a horizontal line). To study the role of visual control, feedback could be withdrawn. For this task, developmental effects were demonstrated in young children (Blank *et al.*, 1999) and decline with age was shown in elderly subjects (Vaillancourt *et al.*, 2003). In addition, deficits of constant grip force production were observed in different neurological patients such as patients with Parkinson's disease (Vaillancourt *et al.*, 2001), cerebellar patients (Mai *et al.*, 1988), or patients with somatosensory deafferentation (Rothwell *et al.*, 1982; see Chapter 19).

In addition, the ability to adapt the grip force to changing levels in the target force was examined. Such tracking tasks classically tested movements rather than forces with the aim to reveal the input–output characteristics of the human sensorimotor system (e.g. Navas & Stark, 1968; Miall *et al.*, 1985). When used to study the control of grip forces, one cursor on a computer screen typically represents the level of the target force and another cursor shows the currently produced grip force. Comparable to constant force production, tracking studies using isometric grip force as a feedback signal revealed effects of development (Blank *et al.*, 2000) and aging (Lazarus & Haynes, 1997). A trapezoid target signal was used in a study of motor training in stroke patients (Kriz *et al.*, 1995; see Chapter 29). Patients successfully decreased overshoots und undershoots of grip force during repeated tracking of the slow target.

Figure 1.1 compares some of these aspects of grip force control in three selected patients. The patients exhibit clear differences in performance across the different tasks. While patient Pat1 shows normal ability to maintain a constant force and to track a moving target, force changes were slowed and of small amplitude, and strength was reduced to 40% of the non-paretic hand. Patient Pat2 exhibited a nearly opposite pattern. His grip force was highly irregular during constant force production and even more irregular during tracking; however he performed force changes at a relatively high frequency and maximum force was in the lower range of his age group. Finally, patient Pat3 presented with sporadic force oscillations

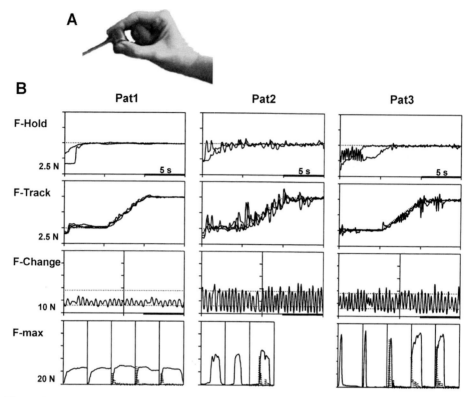

Figure 1.1. A. Force sensor grasped in a precision grip. B. Grip force of three patients with cerebral lesions (Pat1: infarction of medial cerebral artery; Pat2: hemorrhages in the brain stem; Pat3: head injury) in four tasks of isometric grip force control. F-Hold: maintenance of a low constant force; F-Track: tracking of a ramp-like increase of target force; F-Change: changes of grip force at maximum frequency; F-max: strength of precision grip. Only the impaired hand is shown. Modified from Hermsdörfer & Mai (1996).

during the precision tasks, a moderate slowing of force changes but high strength values (Hermsdörfer & Mai, 1996).

Grip forces during grasping and lifting of objects

In 1984 Johansson and Westling published their ingenious paradigm to study the control of grip force under natural conditions, namely the grasping and lifting of objects (Johansson & Westling, 1984; see Chapter 11). Figure 1.2 shows the device they developed for that purpose. The device is grasped with a precision grip between thumb and index finger at two disks. Appropriately mounted strain gauges measure the grip force and the load force. Note that at the beginning while the object is firmly on the ground load force is zero, it increases with a muscular activity that produces an upward lifting force, and equals the

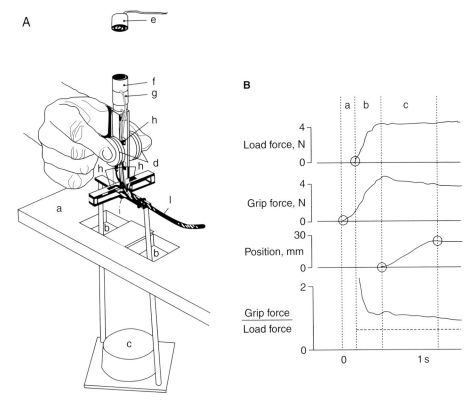

Figure 1.2. A. Device developed by Johansson and Westling to measure grip and load forces during grasping and lifting. The device is grasped with thumb and index finger at exchangeable disks (d), below which strain gauges are mounted to measure the grip forces of both fingers and the vertical load force (lifting force) (h). Additional sensors measure vertical acceleration (g) and position (lifting height) (f, e). The weight (c) can be exchanged. B. Load force, grip force, position and ratio between grip and load force during one typical grasping and lifting movement. The horizontal line in the grip force/load force diagram indicates the slip ratio (slip force related to load force). Time intervals a, b and c indicate the preload phase, the loading phase and the lifting (transitional) phase; from Johansson & Westling (1984).

weight of the device at the moment of lift-off. Then the load remains equal to the weight apart from possible inertial components due to acceleration. During stationary holding the grip forces of thumb and index finger are identical.

Most importantly, the grip force is not defined and can take any magnitude as long as it is higher than a certain minimum (slip force) to avoid slippage of the device (see Chapters 3 and 11). Thus the grip force could theoretically be established before any lifting force is produced and could be largely independent from the weight of the device. The study of grip force during lifting therefore reveals the response of the nervous system to the physical constraints defined by the task. Figure 1.2B demonstrates the most important characteristics of grip force control in this situation. At time zero the fingers contact the disks and grip force increases. After just a small increase of the grip force, the load force starts to increase as well. Both forces then increase until a change in the position signal indicates lift-off. The object is then lifted to the final height with a near-constant load (equalling weight) and a slight relaxation of grip force. The horizontal line in the grip force versus load force diagram indicates the minimum ratio necessary to prevent slipping of the object. It is obvious that grip force exerted when holding the object is only a bit higher than this minimum. The close temporal coupling between the grip and the load force and the precise scaling of the grip force to the load are characteristics of physiological grip-force control during grasping and lifting of objects. Thus, contrary to the theoretically possible control mode mentioned at the beginning of the paragraph, the system works highly economically and anticipates precisely the physical demands of the task (Johansson & Westling, 1984).

By exchanging the weight below the table (see Figure 1.2A) the weight of the whole device was varied experimentally. It was shown that grip forces are adjusted to the different weights (defining the load forces) with a nearly constant ratio between the grip force and the load force (Johansson & Westling, 1984; Westling & Johansson, 1984). Changing the surface material by exchanging the disks caused variations of the friction between fingers and object and resulted in different minimal grip forces required to prevent slippage of the object (Johansson & Westling, 1984; Westling & Johansson, 1984) (for the measurement of the coefficient of friction, see also Chapter 3). Grip forces were higher for more slippery surfaces with lower friction (e.g. silk) as opposed to materials with higher friction (e.g. sandpaper). Another relevant factor is the shape of the grasped surface. Experiments with changes of the inclination of the grip surface revealed increased grip force for more "Λ"-shaped test objects as opposed to lower grip force for "V"-shaped objects (Jenmalm & Johansson, 1997). In contrast, variation of curvature (convex versus concave with different radii) influenced the grip force only inconsistently (Jenmalm *et al.*, 1998).

In their experiments, Johansson and colleagues also investigated the effects of perturbations on the control of grasping and lifting. A clever and simple paradigm enabled the direct comparison between the effects of unpredictable and predictable perturbations (Johansson & Westling, 1988). The subject held the device stationary with eyes closed and a small mass was thrown from a defined height into a container fastened underneath the device. The mass was either released by the examiner at unpredictable time points, or it was released by the subject from the other hand. While in the unpredictable condition, grip force lagged the sudden load increase by latencies starting from 70 ms, anticipatory increases of the grip force were already obvious before impact when the mass was released by the subject (Turrell *et al.*, 1999; Delevoye-Turrell *et al.*, 2003; Nowak & Hermsdörfer, 2004). This shows how effectively sensory information is used in the control of grip force but also how effectively the control strategies are adapted to the demands of the task (Johansson, 1996; Flanagan & Johansson, 2002). A more thorough account of the control of grip forces in object grasping and lifting, with a particular emphasis on the sensory control mechanisms involved, is provided in Chapter 11.

Grasping–lifting paradigms, based on the methods developed by Johansson and colleagues, were used in numerous studies of human grasping and lifting such as studies on development (Forssberg *et al.*, 1991, 1992; Gordon *et al.*, 1992; see Chapter 17), on aging (Cole *et al.*, 1999; see Chapter 18), and on the effects of temporary sensory disturbances in healthy subjects (Johansson & Westling, 1984; Jenmalm & Johansson, 1997; Monzee *et al.*, 2003; see Chapter 19). The effects of nervous system damage on this task were investigated in various populations of patients with neurological diseases such as stroke (Hermsdörfer *et al.*, 2003; Wenzelburger *et al.*, 2005; Raghavan *et al.*, 2006; see Chapter 21), Parkinson's disease (Gordon *et al.*, 1997; Fellows *et al.*, 1998; Wenzelburger *et al.*, 2002; see Chapters 22 and 32), Huntington's disease (Gordon *et al.*, 2000; Schwarz *et al.*, 2001; see Chapter 23), cerebellar diseases (Müller & Dichgans, 1994; Fellows *et al.*, 2001; Nowak *et al.*, 2005; see Chapter 26), traumatic brain injury (see Chapter 24) or cerebral palsy (Eliasson *et al.*, 1992; Forssberg *et al.*, 1999; Gordon & Duff, 1999; Duque *et al.*, 2003; see Chapter 31).

One important methodological improvement for some research questions was achieved by 6-degrees-of-freedom force/torque sensors. These sensors measure the forces in the three spatial dimensions and the torques around the three spatial axes. This is technically achieved by a refined placing of the strain gauges on a specially designed metal frame. Cross-talk between the different channels is compensated by a controller device. Six-degrees-of-freedom sensors with an appropriate force range are available with relatively small dimensions and low weight (e.g. Nano 17, ATI Industrial Automation, 17×15 mm, 9 g). In addition to the measurement of the vertical load force and the orthogonal grip force (see the device in Figure 1.2A), these sensors would be capable of measuring the horizontal force which is tangential to the grip surfaces, the torque that would result from a sideward or forward–backward tilt of the whole device, and the torque that would result from an attempt to rotate the device around the vertical axis against a resistance (see also Chapter 3).

During the experimental condition depicted in Figure 1.2, the extra forces and torques will typically be small compared with the measured vertical load force and grip force. However, the execution may be less controlled when, for example, patients with movement disorders are tested. Torque can also be an experimental variable. For example, the torque during holding an object was varied by changing the horizontal distance of an extra weight connected to the grasping surface (Kinoshita *et al.*, 1997). The study showed that the torque was compensated by increased grip forces, and there was a linear relationship between both signals (see also Zatsiorsky *et al.*, 2003 and Chapter 3). In addition, measurements of torque allow calculation of the precision of grasping. If grip force is applied against a disk like the one depicted in Figure 1.2A and the point of force application is away from the center of the disks, torques arise. Measuring the torque and the grip force enables the calculation of the distance between finger contact and disk center (e.g. Monzee *et al.*, 2003; see also Chapter 3).

Grip forces during object movements

If we move a grasped object, inertial forces arise from the acceleration of the mass. These forces are in addition to the load resulting from the object's weight and have to be compensated by the applied grip force. Figure 1.3A shows schematically the resulting load from the vectorial summation of the load components due to weight and inertia during a discrete and a continuous vertical up-and-down movement with a grasped object. During the acceleration at the beginning of a vertical upward movement, weight and inertial loads are both downward-directed and add, while during deceleration near the end of the upward movement the inertia is upward-directed so that both loads subtract. During a vertical downward movement the sequence is reversed so that the movement starts with a low total load and ends with a high total load. As opposed to discrete movements with zero loads at the lower and upper stops, accelerations are maximal at the reversal points of continuous cyclic movements. In this condition the summation and subtraction of the load components due to weight and inertia result in maximum loads at the lower turning point and smaller loads at the upper turning point (see Figure 1.3A).

Flanagan, Wing and colleagues published the first detailed descriptions of force control during movements of a grasped object (Flanagan & Wing, 1993, 1995; Flanagan & Tresilian, 1994). Figure 1.3B shows recordings of the grip force during vertical movements that represent the typical findings in healthy subjects. In both conditions, the grip force varies in close synchrony with the load. This is mainly obvious by the timing of the grip force peaks which occur nearly simultaneously with the peaks in the load force. It has to be noted that grip force and load force are typically produced by different muscle synergies; synchronicity therefore is not a biomechanical effect, but results from a feedforward-control mode (Flanagan & Tresilian, 1994; Wolpert & Flanagan, 2001; Hermsdörfer et al., 2004). However, grip force does not slavishly follow every load change, as is obvious from the decrease of the load force to zero at the beginning of the downward acceleration that is associated with only a small dip in the grip force trace. In close correspondence to the findings during grasping and lifting, the studies of object movement with dynamic inertial loads proved the high economy and temporal precision of the anticipation of the grip force to physical loads.

In healthy subjects, the adaptation of movement-related grip-force control to environmental perturbations was studied with various paradigms such as artificial load properties (Flanagan & Wing, 1997; Flanagan et al., 2003), microgravity (Hermsdörfer et al., 2000; Augurelle et al., 2003), or Coriolis forces (Nowak et al., 2004b). The effect of changed somatosensory afferents was investigated in healthy subjects during anesthesia of the hand, during hand cooling, or during nerve compression (Nowak et al., 2001; Augurelle et al., 2003; Cole et al., 2003; Nowak & Hermsdörfer, 2003a; see Chapter 19). In patients with nervous system damage, various degrees of somatosensory disturbances were studied (Nowak & Hermsdörfer, 2003b; Nowak et al., 2004a; see Chapter 19). Comparable to the grasping and lifting task, the paradigm was used to study grip-force anticipation in different populations of patients with brain damage such as stroke (Hermsdörfer et al., 2003; Nowak et al., 2003; see Chapter 21), cerebellar diseases (Nowak et al., 2002; Rost et al., 2005; see Chapter 26), or Parkinson's disease (Nowak & Hermsdörfer, 2002).

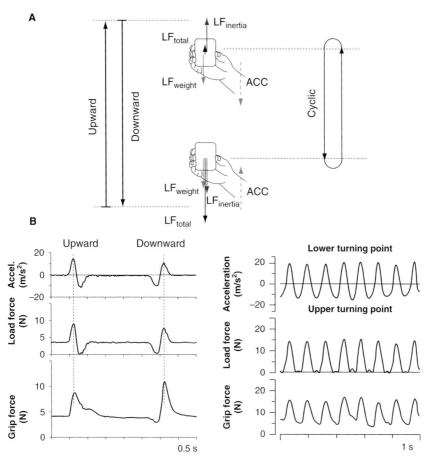

Figure 1.3. A. Illustration of the vectorial summation of the different load components during vertical movements of a grasped object. LF_{weight}: gravitational load (mass × gravity), ACC: vertical acceleration, $LF_{inertia}$: inertial load (mass × ACC), LF_{total}: total load (LF_{weight} + $LF_{inertia}$). Note that this illustration is restricted to vertical movements. When additional horizontal load components occur, particularly the component acting tangential to the grasping surface has to be considered in the calculation of the total load. B. Vertical acceleration of the object, total load force, and grip force during a single upward- and downward-directed movement (left side) and during continuous cyclic movements (right side) of a grasped object by a healthy subject.

Measurement of grip and load forces during object movements with standard equipment is usually hampered by a cable that connects the sensors with the electronics. In addition, for measurements outside the lab, in special environments, or for bedside clinical examinations, equipment is desirable that is easy to use, is quickly set up and contains only one or two units. For that purpose an autonomous instrumented manipulandum was developed (Philipp, 1999; Hermsdörfer *et al.*, 2003; Nowak & Hermsdörfer, 2005, 2006). The manipulandum has a cylindrical shape with a diameter of 9 cm, a width of 4 cm, and a mass of

372 g. It can easily be grasped at the two opposing surfaces. The housing contains one force transducer that measures grip force and three acceleration sensors that measure the acceleration in three spatial dimensions. Sensor properties were selected to fit the typical forces and accelerations produced during object manipulation (grip force 0–80 N, accuracy ± 0.1 N; accelerations ± 50 m/s², accuracy ± 0.2 m/s²). Autonomy of the device was reached by an in-built sensor supply, signal amplification, analog–digital conversion and data storage. The modularization of differential amplifiers and IC techniques allowed for a compact design with minimal volume. A microprocessor was appropriately programmed to control the process of data sampling and storage. Data are sampled at a frequency of 100 Hz and with a resolution of 12 bit. The flash memory enables the storage of up to 20 min of measurement. Small and light-weight rechargeable accumulators with a corresponding capacity are used. Care was taken to mount the accumulators in such a way inside the housing that the center of gravity was in the volumetric center. Thus no rotational torques occur when the object is grasped in the center of the grasping surfaces. Data acquisition is started and stopped manually by a micro-switch on the side of the housing. The autonomous grip object was used extensively in the various experiments by Nowak, Hermsdörfer and colleagues cited in the preceding paragraph.

Advances in micro-electronics enable an even higher integration of the electronic components. Future versions of the grip object will therefore have a lower weight, a smaller volume, and possess a higher data storage and accumulator capacity. In addition, a new version is realized with radio transmission so that as an alternative to internal storage, external storage and control by an adequately equipped PC or PDA is possible. Also the sensor equipment is extended by a second grip force sensor, enabling the separate assessment of both grasp surfaces, and a load sensor that directly measures self-generated or externally generated loads. Separate assessment of the grip force of the thumb and of the opposing fingers is helpful to measure temporal precision during early grasping or to control for object tilt. It must however be kept in mind that the measured signals are still inadequate to determine the complete biomechanics and dynamics of the system. The object movements should therefore be well controlled and non-measured force or torque contributions should be negligible.

Measurement of grip forces during writing

Grasping and lifting of an object or moving an object in the vertical direction are of course only two examples of an infinite number of fine motor tasks that necessitate precise control of the grip forces. The two tasks became so familiar for behavioral research because the load arising from the physical object characteristics and from the movement are well defined. In most natural object manipulations the biomechanical conditions are much more complicated when the manipulated object has an irregular shape, is moved in different directions, is tilted and rotated, or is not grasped at the center of gravity. In addition, while a substantial number of the above-summarized studies tested the precision grip between the thumb and the index finger, more fingers are used in most daily manipulations. Chapter 3 captures multi-finger grasp and presents a more profound approach to the biomechanics.

One limitation of the instrumented objects presented so far is that the placement of the sensors defines the point of force application and thus the necessary grasping location of the fingers. However, objects are grasped in different ways depending on factors such as the momentary weight (of a glass) or the task to perform with the object, or depending on individual anthropometrical characteristics. One solution to that problem is to measure forces at multiple points distributed at close range across the grasping surface. This can be technically achieved by sensor matrices that measure forces and pressure distributions in arrays of a certain size and with a certain spatial resolution. These sensor mats typically have the additional advantage of being thin (~1 mm) and flexible so that they can also be used in constricted environments and with curved surfaces.

We use force sensor matrices to measure the pen grip force during handwriting. Figure 1.4A shows the technical implementation. A force sensor matrix is wrapped around a writing pen (pliance system, novel, Munich). Eighty-eight force sensors are distributed across the whole length of the pen with a spatial resolution of $5 \times 10 \, mm^2$. The measurement principle is capacitive with the sensors embedded in elastomer material. The system is calibrated in the mounted configuration. In an area of $1 \, cm^2$, forces between 0.5 and 20 N can be measured. Hysteresis is better than 7%. The whole array can be sampled at frequencies up to 100 Hz with a frequency response better than 2 dB (0–100 Hz).

In Figure 1.4A a snapshot of a typical pressure distribution during writing a sentence is shown. The three fingers grasping the pen can be easily identified. In this grip, the thumb contributed more force and was located proximal to the index and the middle finger. In the background a small increase of pressure is obvious at the location where the rear part of the pen contacted the hand.

Figure 1.4B shows an application of the measurement of pen pressure distribution in a patient with writer's cramp. Script and pen grip force (integrated pressure) are displayed during writing a test sentence before and after a therapy intervention. It is obvious that the grip force was reduced substantially after the training while the characteristics of the script were largely unchanged. This large decrease of the grip force was achieved by a change of the finger configuration of the grip combined with a motor training. In this way, the characteristic disabling symptom in most cases of writer's cramp was effectively amelio-rated (see Baur *et al.*, 2006; see also Chapter 33).

In this last example, measurements of isometric grip force served as a method to quantify the effects of therapy at the level of the impairment providing highest ecological relevance. In addition, the capacity of grip force measurements to characterize physio-logical and pathophysiological performance was summarized in this chapter. As also emphasized in this chapter, grip force measurements can provide objective descriptions of a central aspect of fine motor performance against which other neurophysiological measures can be compared. As a consequence, measurements of isometric grip force became one of the most popular methods in neurophysiological and clinical motor science.

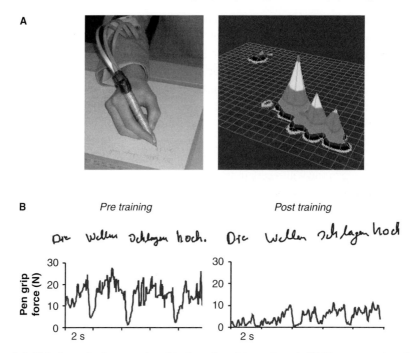

Figure 1.4. This figure is also reproduced in the color plate section. A. Writing pen equipped with a force sensor matrix to measure grip force during hand writing. Right panel: Pressure distribution over the pen surface at one time point during writing a sentence. The pen surface is displayed as a flat area. Peaks denote local maxima of grip force. B. Script and pen grip force of a patient with writer's cramp during writing a sentence (German "Die Wellen schlagen hoch", English: "The waves rise high") before and after a specific training (Schenk *et al.*, 2004; Baur *et al.*, 2006; see Chapter 33). Writing path and kinematics were assessed with a graphics tablet (Intous III, Wacom) and special software (CS, MedCom). Pen grip force was calculated at each time point from the integral of the pressure distribution.

References

Ashe, J. (1997). Force and the motor cortex. *Behav Brain Res*, **87**, 255–269.

Augurelle, A. S., Smith, A. M., Lejeune, T. & Thonnard, J. L. (2003). Importance of cutaneous feedback in maintaining a secure grip during manipulation of hand-held objects. *J Neurophysiol*, **89**, 665–671.

Avarello, M., Bolsinger, P. & Mai, N. (1988). Fast repetitive force changes in hemiparetic and cerebellar patients. *Eur Arch Psychiatry Neurolog Sci*, **237**, 135–138.

Baur, B., Schenk, T., Fürholzer, W. *et al.* (2006). Modified pen grip in the treatment of writer's cramp. *Hum Mov Sci*, **25**, 464–473.

Blank, R., Heizer, W. & Voss, H. v. (1999). Externally guided control of static grip forces by visual feedback – age and task effects in 3–6-year-old children and in adults. *Neurosci Lett*, **271**, 41–44.

Blank, R., Heizer, W. & Voss, H. v. (2000). Development of externally guided grip force modulation in man. *Neurosci Lett*, **286**, 187–190.

Bohannon, R. W. (1989). Is the measurement of muscle strength appropriate in patients with brain lesions? A special communication. *Phys Ther*, **69**, 225–230.

Boissy, P., Bourbonnais, D., Carlotti, M. M., Gravel, D. & Arsenault, B. A. (1999). Maximal grip force in chronic stroke subjects and its relationship to global upper extremity function. *Clin Rehabil*, **13**, 354–362.

Cadoret, G. & Smith, A. M. (1997). Comparison of the neuronal-activity in the SMA and the ventral cingulate cortex during prehension in the monkey. *J Neurophysiol*, **77**, 153–166.

Cole, K. J., Rotella, D. L. & Harper, J. G. (1999). Mechanisms for age-related changes of fingertip forces during precision gripping and lifting in adults. *J Neurosci*, **19**, 3238–3247.

Cole, K. J., Steyers, C. M. & Graybill, E. K. (2003). The effects of graded compression of the median nerve in the carpal canal on grip force. *Exp Brain Res*, **148**, 150–157.

Cramer, S. C., Weisskoff, R. M., Schaechter, J. D. *et al.* (2002). Motor cortex activation is related to force of squeezing. *Hum Brain Mapp*, **16**, 197–205.

Cutkosky, M. R. & Howe, R. D. (1990). Human grasp choice and robotic grasp analysis. In T. Iberall & S. T. Venkataraman (Eds.), *Dextrous Robot Hands* (pp. 5–31). New York, NY: Springer.

Delevoye-Turrell, Y. N., Li, F. X. & Wing, A. M. (2003). Efficiency of grip force adjustments for impulsive loading during imposed and actively produced collisions. *Quar J Exp Psychol A Hum Exp Psychol*, **56** A, 1113–1128.

Du Mensil de Rochemont, R. (1926). Über die dritte Komponente für die Wahrnehmung von Gliederbewegungen. *Z Biol*, **84**, 522ff.

Duque, J., Thonnard, J. L., Vandermeeren, Y. *et al.* (2003). Correlation between impaired dexterity and corticospinal tract dysgenesis in congenital hemiplegia. *Brain*, **126**, 732–747.

Eliasson, A. C., Gordon, A. M. & Forssberg, H. (1992). Impaired anticipatory control of isometric forecs during grasping by children with cerebral palsy. *Dev Med Child Neurol*, **34**, 216–225.

Fees, E. E. (1986). The need for reliability and validity in hand assessment instruments. *J Hand Surg*, **11A**, 621–622.

Fellows, S. J., Schwarz, M. & Noth, J. (1998). Precision grip and Parkinson's disease. *Brain*, **121**, 1771–1784.

Fellows, S. J., Ernst, J., Schwarz, M., Töpper, R. & Noth, J. (2001). Precision grip deficits in cerebellar disorders in man. *Clin Neurophysiol*, **112**, 1793–1802.

Flanagan, J. R. & Wing, A. M. (1993). Modulation of grip force with load force during point-to-point arm movements. *Exp Brain Res*, **95**, 131–143.

Flanagan, J. R. & Tresilian, J. (1994). Grip-load force coupling: a general control strategy for transporting objects. *J Exp Psychol Hum Perception Perform*, **20**, 944–957.

Flanagan, J. R. & Wing, A. M. (1995). The stability of precision grip forces during cyclic arm movements with a hand held load. *Exp Brain Res*, **105**, 455–464.

Flanagan, J. R. & Wing, A. M. (1997). The role of internal models in motion planning and control – evidence from grip force adjustments during movements of hand-held loads. *J Neurosci*, **17**, 1519–1528.

Flanagan, J. R. & Johansson, R. S. (2002). Hand movements. In *Encyclopedia of the Human Brain* (pp. 399–414). New York, NY: Elsevier Science.

Flanagan, J. R., Vetter, P., Johansson, R. S. & Wolpert, D. M. (2003). Prediction precedes control in motor learning. *Curr Biol*, **13**, 146–150.

Forssberg, H., Eliasson, A. C., Kinoshita, H., Johansson, R. S. & Westling, G. (1991). Development of human precision grip. I: Basic coordination of force. *Exp Brain Res*, **85**, 451–457.

Forssberg, H., Kinoshita, H., Eliasson, A. C. *et al.* (1992). Development of human precision grip. 2. Anticipatory control of isometric forces targeted for object's weight. *Exp Brain Res*, **90**, 393–398.

Forssberg, H., Eliasson, A.-C., Redon-Zouitenn, C., Mercuri, E. & Dubowitz, L. (1999). Impaired grip-lift synergy in children with unilateral brain lesions. *Brain*, **122**, 1157–1168.

Freund, H. J. & Büdingen, H. J. (1978). The relationship between speed and amplitude of the fastest voluntary contractions of human arm muscles. *Exp Brain Res*, **31**, 1–12.

Ghez, C. & Gordon, J. (1987). Trajectory control in targeted force impulses. I. Role of opposing muscles. *Exp Brain Res*, **67**, 225–240.

Gordon, A. M. & Duff, S. V. (1999). Fingertip forces during object manipulation in children with hemiplegic cerebral-palsy. I. Anticipatory scaling. *Dev Med Child Neurol*, **41**, 166–175.

Gordon, A. M., Forssberg, H., Johansson, R. S., Eliasson, A. C. & Westling, G. (1992). Development of human precision grip. 3. Integration of visual size cues during the programming of isometric forces. *Exp Brain Res*, **90**, 399–403.

Gordon, A. M., Ingvarsson, P. E. & Forssberg, H. (1997). Anticipatory control of manipulative forces in Parkinson's disease. *Exp Neurol*, **145**, 477–488.

Gordon, A. M., Quinn, L., Reilmann, R. & Marder, K. (2000). Coordination of prehensile forces during precision grip in Huntington's disease. *Exp Neurol*, **163**, 136–148.

Gordon, J. & Ghez, C. (1987a). Trajectory control in targeted force impulses. II. Pulse height control. *Exp Brain Res*, **67**, 241–252.

Gordon, J. & Ghez, C. (1987b). Trajectory control in targeted force impulses. III. Compensatory adjustments for initial errors. *Exp Brain Res*, **67**, 253–269.

Hanten, W. P., Chen, W. Y., Austin, A. A. *et al.* (1999). Maximum grip strength in normal subjects from 20 to 64 years of age. *J Hand Ther*, **12**, 193–200.

Harth, A. & Vetter, W. R. (1994). Grip and pinch strength among selected adult occupational groups. *Occup Ther Intern*, **1**, 13–28.

Hepp-Reymond, M. C., Wyss, U. R. & Anner, R. (1978). Neuronal coding of static force in the primate motor cortex. *J Physiol (Paris)*, **74**, 287–291.

Hermsdörfer, J. & Mai, N. (1996). Disturbed grip force control following cerebral lesions. *J Hand Ther*, **9**, 33–40.

Hermsdörfer, J., Marquardt, C., Philipp, J. *et al.* (2000). Moving weightless objects: grip force control during microgravity. *Exp Brain Res*, **132**, 52–64.

Hermsdörfer, J., Hagl, E., Nowak, D. A. & Marquardt, C. (2003). Grip force control during object manipulation in cerebral stroke. *Clin Neurophysiol*, **114**, 915–929.

Hermsdörfer, J., Hagl, E. & Nowak, D. A. (2004). Deficits of anticipatory grip force control after damage to peripheral and central sensorimotor systems. *Hum Mov Sci*, **23**, 643–662.

Jenmalm, P. & Johansson, R. S. (1997). Visual and somatosensory information about object shape control manipulative fingertip forces. *J Neurosci*, **17**, 4486–4499.

Jenmalm, P., Goodwin, A. W. & Johansson, R. S. (1998). Control of grasp stability when humans lift objects with different surface curvatures. *J Neurophysiol*, **79**, 1643–1652.

Johansson, R. S. (1996). Sensory control of dexterous manipulation in humans. In A. M. Wing, P. Haggard & J. R. Flanagan (Eds.), *Hand and Brain* (pp. 381–414). San Diego, CA: Academic Press.

Johansson, R. S. & Westling, G. (1984). Roles of glabrous skin receptors and sensorimotor memory control of precision grip when lifting rougher or more slippery objects. *Exp Brain Res*, **56**, 550–564.

Johansson, R. S. & Westling, G. (1988). Programmed and triggered actions to rapid load changes during precision grip. *Exp Brain Res*, **71**, 72–86.

Jones, L. A. (1989). The assessment of hand function: a critical review of techniques. *J Hand Surg*, **14**, 221–228.

Kinoshita, H., Backstrom, L., Flanagan, J. R. & Johansson, R. S. (1997). Tangential torque effects on the control of grip forces when holding objects with a precision grip. *J Neurophysiol*, **78**, 1619–1630.

Kriz, G., Hermsdörfer, J., Marquardt, C. & Mai, N. (1995). Feedback-based training of grip force control in patients with brain damage. *Arch Phys Med Rehab*, **76**, 653–659.

Lazarus, J. A. C. & Haynes, J. M. (1997). Isometric pinch force control and learning in older adults. *Exp Aging Res*, **23**, 179–199.

MacDermid, J. C., Evenhuis, W. & Louzon, M. (2001). Inter-instrument reliability of pinch strength scores. *J Hand Ther*, **14**, 36–42.

Mai, N., Bolsinger, P., Avarello, M., Diener, H.-C. & Dichgans, J. (1988). Control of isometric finger force in patients with cerebellar disease. *Brain*, **111**, 973–998.

Maier, M. A. & Hepp-Reymond, M. C. (1995). EMG activation patterns during force production in precision grip. I. Contribution of 15 finger muscles to isometric force. *Exp Brain Res*, **103**, 108–122.

Mathiowetz, V., Kashman, N., Volland, G. *et al.* (1985). Grip and pinch strength: normative data for adults. *Arch Phys Med Rehab*, **66**, 69–74.

Miall, R. C., Weir, D. J. & Stein, J. F. (1985). Visuomotor tracking with delayed visual feedback. *Neuroscience*, **16**, 511–520.

Monzee, J., Lamarre, Y. & Smith, A. M. (2003). The effects of digital anesthesia on force control using a precision grip. *J Neurophysiol*, **89**, 672–683.

Müller, F. & Dichgans, J. (1994). Dyscoordination of pinch and lift forces during grasp in patients with cerebellar lesions. *Exp Brain Res*, **101**, 485–492.

Napier, J. R. (1956). The prehensile movements of the human hand. *J Bone Joint Surg*, **38**, 902–913.

Navas, F. & Stark, L. (1968). Sampling or intermittency in hand control system dynamics. *Biophys J*, **8**, 252–302.

Nowak, D. A. & Hermsdörfer, J. (2002). Coordination of grip and load forces during vertical point-to-point movements with a grasped object in Parkinson's disease. *Behav Neurosci*, **116**, 837–850.

Nowak, D. A. & Hermsdörfer, J. (2003a). Digit cooling influences grasp efficiency during manipulative tasks. *Eur J Appl Physiol*, **89**, 127–133.

Nowak, D. A. & Hermsdörfer, J. (2003b). Selective deficits of grip force control during object manipulation in patients with reduced sensibility of the grasping digits. *Neurosci Res*, **47**, 65–72.

Nowak, D. A. & Hermsdörfer, J. (2004). Predictability influences finger force control when catching a free-falling object. *Exp Brain Res*, **154**, 411–416.

Nowak, D. A. & Hermsdörfer, J. (2005). Grip force behavior during object manipulation in neurological disorders: toward an objective evaluation of manual performance deficits. *Mov Disord*, **20**, 11–25.

Nowak, D. A. & Hermsdörfer, J. (2006). Objective evaluation of manual performance deficits in neurological movement disorders. *Brain Res Brain Res Rev*, **51**, 108–124.

Nowak, D. A., Hermsdörfer, J., Glasauer, S. *et al.* (2001). The effects of digital anaesthesia on predictive grip force adjustments during vertical movements of a grasped object. *Eur J Neurosci*, **14**, 756–762.

Nowak, D. A., Hermsdörfer, J., Marquardt, C. & Fuchs, H. H. (2002). Load force coupling during discrete vertical movements in patients with cerebellar atrophy. *Exp Brain Res*, **145**, 28–39.

Nowak, D. A., Hermsdörfer, J. & Topka, H. (2003). Deficits of predictive grip force control during object manipulation in acute stroke. *J Neurol*, **250**, 850–860.

Nowak, D. A., Glasauer, S. & Hermsdörfer, J. (2004a). How predictive is grip force control in the complete absence of somatosensory feedback? *Brain*, **127**, 182–192.

Nowak, D. A., Hermsdörfer, J., Schneider, E. & Glasauer, S. (2004b). Moving objects in a rotating environment: rapid prediction of Coriolis and centrifugal force perturbations. *Exp Brain Res*, **157**, 241–254.

Nowak, D. A., Hermsdörfer, J., Timmann, D., Rost, K. & Topka, H. (2005). Impaired generalization of weight-related information during grasping in cerebellar degeneration. *Neuropsychologia*, **43**, 20–27.

Philipp, J. (1999). *Ein Meßsystem zur Untersuchung der Feinmotorik beim Greifen und Bewegen von Gegenständen.* Dissertation, aus der Neurologischen Klinik der Ludwig-Maximilians-Universität München, erworben an der Medizinischen Fakultät der LMU München, München.

Phillips, C. G. (1986). *Movements of the Hand (Sherrington Lectures, vol. 17)*. Herndon, VA: Humanities Pr.

Raghavan, P., Krakauer, J. W. & Gordon, A. M. (2006). Impaired anticipatory control of fingertip forces in patients with a pure motor or sensorimotor lacunar syndrome. *Brain*, **129**, 1415–1425.

Rost, K. R., Nowak, D. A., Timman, D. T. & Hermsdörfer, J. (2005). Preserved and impaired aspects of predictive grip force control in cerebellar patients. *Clin Neurophysiol*, **116**, 1405–1414.

Rothwell, J. C., Traub, M. M., Day, B. L. *et al.* (1982). Manual motor performance in a deafferented man. *Brain*, **105**, 515–542.

Schenk, T., Baur, B., Steidle, B. & Marquardt, C. (2004). Does training improve writer's cramp? An evaluation of a behavioural treatment approach using kinematic analysis. *J Hand Ther*, **17**, 349–363.

Schwarz, M., Fellows, S. J., Schaffrath, C. & Noth, J. (2001). Deficits in sensorimotor control during precise hand movements in Huntington's disease. *Clin Neurophysiol*, **112**, 95–106.

Stevens, J. C. & Mack, J. D. (1959). Scales of apparent force. *J Exp Psychol*, **58**/5, 405–413.

Truschel, L. (1913). Experimentelle Untersuchung über Kraftempfindungen bei Federspannung und Gewichtshebungen. *Archiv Ges Psychol*, **28**, 183–273.

Turrell, Y. N., Li, F. X. & Wing, A. M. (1999). Grip force dynamics in the approach to a collision. *Exp Brain Res*, **128**, 86–91.

Ulrich, R., Wing, A. M. & Rinkenauer, G. (1995). Amplitude and duration scaling of brief isometric force pulses. *J Exp Psychol Hum Percept Perform*, **21**, 1457–1472.

Vaillancourt, D. E., Slifkin, A. B. & Newell, K. M. (2001). Visual control of isometric force in Parkinson's disease. *Neuropsychologia*, **39**, 1410–1418.

Vaillancourt, D. E., Larsson, L. & Newell, K. M. (2003). Effects of aging on force variability, single motor unit discharge patterns, and the structure of 10, 20, and 40 Hz EMG activity. *Neurobiol Aging*, **24**, 25–35.

Wenzelburger, R., Zhang, B. R., Pohle, S. *et al.* (2002). Force overflow and levodopa-induced dyskinesias in Parkinson's disease. *Brain*, **125**, 871–879.

Wenzelburger, R., Kopper, F., Frenzel, A. *et al.* (2005). Hand coordination following capsular stroke. *Brain*, **128**, 64–74.

Westling, G. & Johansson, R. S. (1984). Factors influencing the force control during precision grip. *Exp Brain Res*, **53**, 277–284.

Wolpert, D. M. & Flanagan, J. R. (2001). Motor prediction. *Curr Biol*, **11**, 729–732.

Zatsiorsky, V. M., Gao, F. & Latash, M. L. (2003). Prehension synergies: effects of object geometry and prescribed torques. *Exp Brain Res*, **148**, 77–87.

2

Kinematic assessment of grasping

UMBERTO CASTIELLO AND CATERINA ANSUINI

Summary

Research on grasping kinematics has proved to be particularly insightful in revealing important aspects of the motor control and selection processes underlying the control of hand action. The aim of this chapter is to provide a synthetic overview of the main kinematic techniques which have been utilized for describing and quantifying grasping movements. The first part of the chapter includes a brief description of the basic kinematic principles. The second part focuses on the main techniques used to perform kinematic analysis of grasping movements. Specifically, it describes video, optoelectronic, and sensor bending techniques. The third part is concerned with a brief review of studies which have characterized the kinematics of grasping not only in healthy adults but also in developmental, neuropsychological and comparative research. We conclude the chapter by discussing recent research issues and technical approaches which have recently started to emerge.

Introduction

Interest in the kinematic analysis of grasping was largely stimulated by the work of Marc Jeannerod who identified how specific kinematic landmarks modulate with respect to object properties to allow for both successful hand positioning and object grasping (Jeannerod, 1981, 1984). Since these early observations, this movement has been well characterized in those with no neurological damage. For subjects with central nervous system damage kinematic assessment allows a complete description of dysfunction and consequently assists with diagnosis and design of appropriate therapeutic regimes. Similarly, kinematic studies on how the human child develops the ability to grasp not only show the learning of a goal-directed skill but ultimately assist in the identification of abnormalities of motor control at each developmental stage. Finally, the investigation of grasping kinematics across species allows comparisons which illustrate the unique features of human movement.

In this interdisciplinary perspective, the present chapter will provide an overview of the principles underlying kinematic analysis, the types of kinematic analysis systems which have been used to investigate grasping movements and a brief review of the most significant findings obtained in various research domains.

Sensorimotor Control of Grasping: Physiology and Pathophysiology, ed. Dennis A. Nowak and Joachim Hermsdörfer. Published by Cambridge University Press. © Cambridge University Press 2009.

What is kinematics?

Kinematics considers human movement in terms of position and displacement (angular and linear) of body segments, center of gravity, acceleration and velocities of the whole body or segments of the body such as an upper limb or the trunk. A kinematic assessment will provide information on the relationship of parts of the body to each other. This is useful in measuring complex movements and it has provided the basis for understanding functional activities including hand grasping. Kinematic assessment uses anatomical terminology and a spatial reference system. For instance, the position of a hand joint would be described by a set of Y, X, Z coordinates that might represent the vertical, medial-lateral and anterior-posterior components or directions, respectively (Winter, 1991).

Kinematic analysis systems

Detected variables (sensors)

All measurement procedures ought not to alter the normal behavior of the system under investigation (Kelvin rule). Following such a rule, it makes sense for motion analysis techniques to be based on a photographic approach, specifically stereophotogrammetry, to allow both body sides and complete measurements.

Kinematic variables are sampled during time. The Shannon–Nyquist theorem states that an analog waveform signal, band limited, may be uniquely reconstructed, without error, from samples taken at equal time intervals. The sampling rate must be equal to, or greater than, twice the highest frequency component of the analog signal.

The frequency content of a movement such as grasping is below 10 Hz (Allard *et al.*, 1995). This is mainly due to the fact that the human body is not made of rigid links and the soft tissues act as a low-pass filter. Therefore the Shannon–Nyquist theorem gives a 20-Hz lower limit for data sampling. Knowledge regarding the frequency of a specific movement is a key point when planning data acquisition. This is because setting an inappropriate (e.g. too low) sampling frequency brings erroneous results (aliasing phenomena). In turn this suggests that high sampling frequencies may not be necessary when the field of application is clearly defined. High frequencies (e.g. more than 100 Hz) are generally needed only for sports applications.

It is important to define some parameters linked to the position measurement in order to distinguish the performance between different systems (Ehara *et al.*, 1997). They have been defined in different ways mainly because of technology changes. The following is an updated list of parameters and their definition: (i) *Resolution*, the minimum detectable movement of a marker; (ii) *Precision*, the standard deviation of random error of n samples; (iii) *Accuracy*, the standard deviation of the systematic error from n points equally distributed over the whole measuring volume. All the values obtained from both the resolution and the precision parameters have to be checked throughout that volume.

Video-based systems

A vast amount of qualitative information can be obtained from video recording. Human movement as a total pattern can be observed and re-observed. The relationship of all body parts to each other as well as the quality of the movement – whether it is fast or slow, uncoordinated or smooth – can be seen.

Quantitative data can be obtained by digitizing the video image and subjecting the data to computer processing and analysis. *Digitization* is the process whereby the image or parts of the image are converted to digital form so that the data can be manipulated by a computer. In order to be able to digitize film, it is helpful to place skin markers over major anatomical landmarks prior to filming. The process of digitizing can be undertaken either manually or by use of a computer software. This process involves viewing each frame (or field) of the videotape and identifying and storing the coordinates for each of the skin markers. This operation has to be performed for each frame of the film. The data thus obtained can be called upon when calculations are required. Manual digitizing is reliable and accurate, and human error is relatively small especially with experienced digitizers. Automatic digitization is also accurate and reliable, though it is necessary to undertake manual checks to ensure that the computer does not confuse two different markers when they cross in space. Direct measurements of an image on videotape taken from the video screen are subject to considerable error and should not be used as a method of quantifying human movement.

Computer-aided analysis of videotape can give a wide range of information and most systems now allow the analysis of movement in more than one plane. The computer-analysis software makes it possible to plot body coordinates (center of gravity etc.). Knowledge of the position of the center of gravity is important when considering the efficiency of movement. For example, smooth displacements of the center of gravity tend to indicate a more efficient movement than those where the center of gravity is subjected to extensive vertical displacement. The computer can also generate stick diagrams which are valuable as an initial qualitative analysis of the sequence of movement.

Optoelectronic techniques

Optoelectronic devices require markers to be placed on the body. The coordinates of these markers are tracked throughout the movement and calculations can then be made. Unlike other techniques such as video, these systems do not give a visual image of the subject, but simply a frame-by-frame representation of the position of each marker. From this, it is possible to produce computer-generated stick figures or graphs of the position of a joint showing motion range plotted against time. This gives good quantitative information but does not address the issue of quality of movement as there is no visual representation of the actual subject. Optoelectronic systems can be subdivided into two main classes: that using active markers and that using passive markers. Figure 2.1 represents an example of an optoelectronic system.

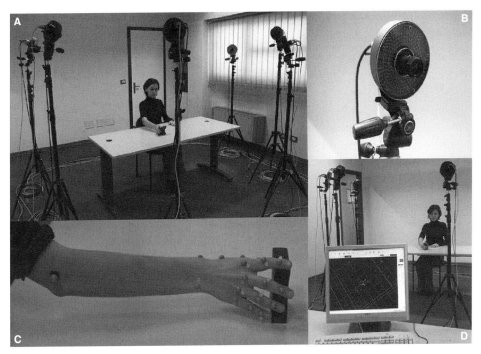

Figure 2.1. A. An example of camera positioning and calibrated space for the recording of a reach-to-grasp movement with an optoelectronic system (E-smart, BTS, Milan, Italy). B. Photograph representing an infrared camera in detail. C. An example of markers positioned on the anatomical landmarks classically considered for the kinematic analysis of a reach-to-grasp movement (i.e. fingers, wrist and elbow). D. Computer interface for the visualization of the reflective markers during data acquisition.

Systems using passive markers

Systems using passive markers rely on reflective markers (retro-reflective material on a plastic sphere) placed on the subject's skin. Some form of light, often infra-red, is transmitted towards the subject and the rays are reflected back off the markers to a series of "cameras" that record the marker position. A sufficient number of "cameras" needs to be placed around the subject so that each marker is visible to a minimum of two. Sampling frequency may vary from 50–250 Hz which enables the system to track the change in position of the markers and produces a reasonable record of the gross pattern of movement. The markers have no identity, leaving the system vulnerable if cross-over of markers occurs.

The accuracy of the system relies on human input to ensure that the computer accurately identifies which is for instance the wrist marker and which is the marker on the thumb, or it is dependent on a good quality computer software that is able to correctly process the incoming data. In order to avoid errors in the data processing, tracking procedures have to be used to identify and track the marker trajectories. For the most advanced systems, these procedures are able to track the markers by starting from a model defined by the user and identifying the representative points almost in real time. The use of different filters (i.e. with different wavelengths) for the camera and different marker color could potentially simplify the tracking algorithm.

Systems using active markers

The more expensive systems use active skin markers. Markers have their own small power pack that enables them to actively transmit infra-red rays to a receiving system of several "cameras". Since each marker has its own transmitting signal, the receiver picks up not only the position and displacement of the marker but can identify which marker it has picked up. This gives the advantage of differentiating between markers and removes the potential source of error that can occur when two markers cross over each other. In other words, no post-collection marker identification is needed, as time sequencing between marker illumination and detector reception uniquely identifies each light emitting diode (LED). Each marker is activated at a slightly different (in the order of microseconds) instant in time.

Potential errors in marker-based systems

Different kinds of errors can affect the measurement of marker-based systems (Cappozzo *et al.*, 1996): (i) stereophotogrammetric errors (the manual definition of all markers), coming from various sources linked to the specific hardware and software equipment used (their value is in the range of 0.5–2 mm); (ii) skin movement artefacts, owing to relative movement occurring between the skin and the underlying bone during grasping (their value is in the range of 10–25 mm for the most critical representative points) (Fuller *et al.*, 1997); (iii) marker repositioning, owing to the difficulty of repositioning the markers on the same representative relative point (their value is in the range of 5–15 mm).

One of the basic problems working with marker-based systems is to obtain the three-dimensional (3D) coordinates in an absolute reference system. Some kind of mapping is needed to allow for a transformation from camera coordinates to 3D space. The identification of the parameters to reconstruct such model is called *resection* while its use to compute 3D coordinates is called *intersection*.

For both the active and passive systems it is necessary to identify a reference point before measurement takes place. This enables the computer to calculate the absolute and relative positions of the markers in 3D. If only one marker is placed on an anatomical landmark of the body, the system can record the displacement of that landmark in 3D, giving the absolute and relative position. The application of two markers enables the system to calculate the distance between them relative to time. With three or more markers (a configuration classically used for reach-to-grasp movements), the angles at joints can be measured. Calculations of velocity and acceleration of limb segments can be performed on data from one or more markers.

DataGloves – bend-sensing technology

DataGloves are fully instrumented gloves that provide up to 22 high-accuracy joint-angle measurements. They use proprietary resistive bend-sensing technology to accurately transform hand and finger motion (as extracted from three sensors per finger, four sensors per abduction angle and one sensor for the palm) into real-time digital joint-angle data. Each sensor is extremely thin and flexible being virtually undetectable in the lightweight elastic glove. When there is a need to measure forearm position and orientation, as happens for many applications, the researchers usually rely on magnetic tracking sensors which can be interfaced with the glove.

A brief review of kinematic studies

Normal adult subjects

With the use of high speed film, Jeannerod (1981, 1984) made a series of observations which have led to a surge in research on human reach-to-grasp in various domains spanning from development to neuropsychology (for review see Castiello, 2005; see Chapter 10). He found that maximum grip aperture en route to a given target object increased with increasing object size (Figure 2.2A). Here maximum grip aperture was defined as the maximum distance reached by the index finger and thumb.

The majority of subsequent studies carried out with optoelectronic techniques continued on this two-digits analysis, looking at the relationship between maximum grip aperture and object size, but much of it dedicated to the investigation of other objects' properties, including fragility (Savelsbergh *et al.*, 1996), size of the contact surface (Bootsma *et al.*, 1994), texture (Weir *et al.*, 1991a) and weight (Gordon *et al.*, 1991; Weir *et al.*, 1991b). As recently reviewed (Smeets & Brenner, 1999), all of these factors influence the kinematics of grasping. Object weight constrains the positioning of the fingers: a larger grip aperture (i.e. distance between thumb and index finger) is necessary for heavier than for lighter objects. Grasping slippery objects requires a larger grip aperture earlier in the movement compared to grasping rough-surfaced objects (see also Chapter 10).

To date, the above-mentioned studies have given little attention to differences in the shape assumed by individual fingers when performing grasping movements to objects. The question of how the entire hand is shaped during the reach-to-object geometry has been chiefly addressed by using sensor-bending techniques (Santello & Soechting, 1998). Subjects were asked to reach, grasp and lift objects whose contours were varied to elicit different hand configurations at contact. The shape of the hand was defined as the pattern of angular excursions at the metacarpal and proximal interphalangeal joints of the thumb and fingers. It was found that the extent to which hand shape could predict object geometry increased in a monotonic fashion throughout the reach (Figure 2.2B).

By using similar techniques, Schettino *et al.* (2003) found that the hand pre-shaping is characterized by two epochs: an early phase during which grip selection is attained

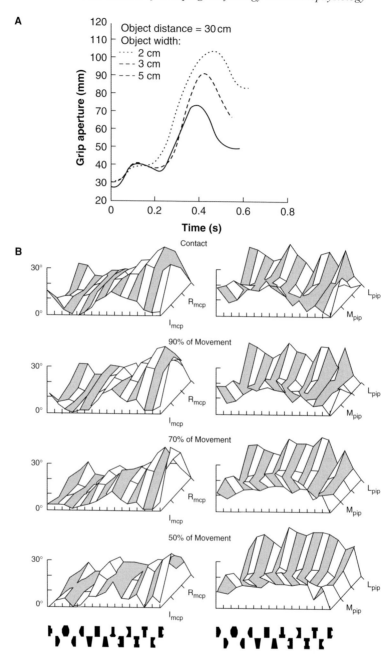

Figure 2.2. A. Grip aperture scaling against time for different sized objects. Note how maximum grip aperture increases with respect to object size and how the time at which maximum grip aperture is reached occurs earlier for smaller than for larger objects. B. The evolution of hand shape during reaching. Hand postures measured at different epochs during the movement (50, 70, 90 and 100% of movement time) are illustrated for each of the objects. Objects are arranged on the horizontal axis, with a progression from convex shapes (left) to concave ones (right). Oblique axis denotes metacarpal (mcp) (left) and proximal interphalangeal (pip) joints (right) for index (I), middle (M), ring (R) and little (L) fingers. Value 0° denotes the minimum value (most extended posture) for the 15 objects at each joint. Modified from Jakobson & Goodale (1991); Santello & Soechting (1998).

regardless of availability of visual feedback and a subsequent phase during which subjects may use visual feedback to finely modulate hand posture to an object's contours.

Development

In humans grasping movements are not present at birth. Their development occurs as a series of steps during ontogeny (see also Chapter 17). It is by 9 months of age that the hand begins to be shaped according to object size. Von Hofsten & Rönnqvist (1988) monitored the distance between the thumb and index finger in reaches performed by 5–6-month-old, 9-month-old, and 13-month-old infants by using optoelectronic techniques. They found that only the infants in the two older age groups did adjust the opening of the hand to the size of the target. However, the movement pattern for these groups was still very different from the adult pattern in which the maximum grip aperture during transport determined the point at which fingers started to close before contact with the target object (Figure 2.3A).

Regarding intermediate age levels, by using similar techniques, it has been shown that maximal finger aperture of 5-year-old children is larger than that of adults and in contrast to adults, grip aperture is scaled according to object size only in the absence of visual feedback (Zoia *et al.*, 2006). Furthermore, kinematic analysis of grasping in 12-year-old children (Kuhtz-Buschbeck *et al.*, 1998) revealed that these children were able to scale the grip aperture to various sizes of the target objects. This suggests that the development of prehensile skills lasts until the end of the first decade of life (see Chapter 17).

Neuropsychology

Striking evidence for a deficit in visually guided grasping has come from patients with optic ataxia. By using video techniques, Jeannerod (1986) found that the finger-grip aperture of patients with optic ataxia was abnormally large, and the usual correlation between maximum grip aperture and object size was missing. These findings were confirmed by applying optoelectronic techniques in a patient (V. K.) who showed an apparently normal early phase of hand opening during attempts to grasp an object, but an on-line control of grip aperture degenerated resulting in numerous secondary peaks in the grip aperture profile, rather than the single peak typical of a healthy subject (Jakobson *et al.*, 1991) (Figure 2.3B). The analysis of whole hand pre-shaping has also been applied to quantify the effect of neuro-logical disorders on processes of sensorimotor integration associated with modulation of hand shape to object geometry. In this respect, kinematic analysis of whole hand pre-shaping has been applied to quantify the effect of stroke and Parkinson's disease (PD) on processes of sensorimotor integration associated with modulation of hand shape to object geometry (Raghavan *et al.*, 2006; Schettino *et al.*, 2006). Results, obtained by using bend-sensing technology, indicate that both PD and stroke patients exhibit substantial impairments in hand pre-shaping. Specifically, these patients seem to heavily rely on visual feedback in order to shape the hand to an object's contours during a reach-to-grasp movement. In-depth

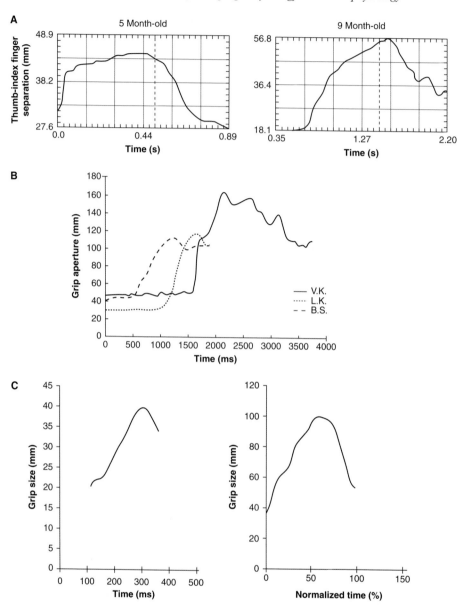

Figure 2.3. A. Two examples of reaches performed by 5- and 9-month-old infants towards either a 2.5 (left) or 3.5 cm (right) diameter object, respectively. Dashed lines indicate the time of contact. B. Grip aperture profiles from individual reaching trials executed by a patient with optic ataxia (V.K.) and by two age-matched control subjects (L.K. and B.S.). C. A comparison of grasping kinematics between a monkey and a human subject. The left and the right graphs represent grip size in a macaque monkey and a human subject, respectively. Modified from Von Hofsten & Rönnqvist (1988); Jakobson *et al.* (1991); Roy *et al.* (2000).

discussion regarding other grasping deficits in PD patients can be found elsewhere in this volume (see Chapter 22).

Comparative studies

To investigate the kinematic characteristics of grasping movements in non-human primates is important because researchers have often presumed an homology between the control mechanisms underlying grasping movement in humans and non-human primates.

By using optoelectronic techniques, Roy *et al.* (2000) found that, as in humans, macaques' maximum grip aperture was correlated with object size (Figure 2.3C). In another study, films of a rhesus monkey grasping small pieces of food were used to compare kinematics of grasping between macaques and humans. Results showed that monkeys exhibited an opening of the grip followed by closure before contact with the object as occurs in humans (Christel & Billard, 2002). Furthermore, the angular velocity and acceleration of the finger aperture were significantly higher in macaques than in humans, and macaques made smaller shoulder excursions than humans during the grasping movement.

A more recent study (Mason *et al.*, 2004) using optoelectronic techniques investigated the monkey's ability to conform hand configurations with respect to object shape. Results are consistent with that from human studies showing that the monkeys utilized a specific hand shape for each object rather than simply opening and closing the hand around the object.

Future directions

As this chapter indicates, kinematics has noticeably helped in making some progress for the characterization of grasping. Nonetheless, much remains unknown and the investigation of many important issues is still in its infancy. For instance, work in humans has shown that choosing a grip does not depend exclusively on the visual properties of the object, but it does depend heavily on the meaning attached to the object and what an individual intends to do with it. A series of recent studies is starting to show that intentional and contextual factors (e.g. end-goals and distractor objects, respectively) determine different patterns of hand shaping (Ansuini *et al.*, 2006, 2007). Furthermore, although enormous advances in our understanding of the links between the mind, the brain and behavior have been made in the last few decades, these have been largely based on studies in which people are considered as strictly isolated units. In this respect, kinematics is also proving to be a powerful method for the investigation of the mechanisms allowing skilful social interactions. A recent series of studies has revealed distinct kinematic patterns for actions performed in a social context involving two agents in a variety of tasks including cooperative and competing behaviors (Castiello, 2003; Georgiou *et al.*, 2007; Becchio *et al.*, 2007, 2008).

To undertake these and further theoretical issues, the best way forward seems to be represented by the combination of kinematics with different techniques. For instance, a series of recent studies has combined kinematics with transcranial magnetic stimulation

(Cattaneo *et al.*, 2005; Glover *et al.*, 2005; Tunik *et al.*, 2005; Davare *et al.*, 2006; Rice *et al.*, 2007; see also Chapter 6) and functional magnetic resonance imaging (Begliomini *et al.*, 2007; see also Chapter 7). Similarly, kinematics together with electrophysiological recording has provided important facts about the neural networks that mediate grasping (Gardner *et al.*, 2002; Raos *et al.*, 2006).

References

Allard, P., Stokes, I. A. F. & Blanchi, J. P. (Eds.) (1995). *Three-dimensional Analysis of Human Movement*. Human Kinetics, USA.

Ansuini, C., Santello, M., Massaccesi, S. & Castiello, U. (2006). Effects of end-goal on hand shaping. *J Neurophysiol*, **95**, 2456–2465.

Ansuini, C., Tognin V., Turella, L. & Castiello, U. (2007). Distractor objects affect fingers' angular distances but not fingers' shaping during grasping. *Exp Brain Res*, **178**, 194–205.

Becchio, C., Sartori, L., Bulgheroni, M. & Castiello, U. (2007). The case of Dr. Jekyll and Mr. Hyde: a kinematic study on social intention. *Conscious Cogn*, Epub ahead of print.

Becchio, C., Sartori, L., Bulgheroni, M. & Castiello, U. (2008). Both your intention and mine are reflected in the kinematics of my reach to grasp movement. *Cognition*, **106**, 894–912.

Begliomini, C., Wall, M. B., Smith, A. T. & Castiello, U. (2007). Differential cortical activity for precision and whole-hand visually guided grasping in humans. *Eur J Neurosci*, **25**, 1245–1252.

Bootsma, R. J., Marteniuk, R. G., MacKenzie, C. L. & Zaal, F. T. J. M. (1994). The speed-accuracy trade off in manual prehension: effects of movement amplitude, object size and object width on kinematic characteristics. *Exp Brain Res*, **98**, 535–541.

Cappozzo, A., Catani, F., Croce, U. D. & Leardini, A. (1996). Position and orientation in space of bones during movement: experimental artefacts. *Clin Biomech*, **2**, 90–100.

Castiello, U. (2003). Understanding other people's actions: intention and attention. *J Exp Psychol Hum Percept Perform*, **29**, 416–430.

Castiello, U. (2005). The neuroscience of grasping. *Nat Rev Neurosci*, **6**, 726–736.

Cattaneo, L., Voss, M., Brochier, T. *et al.* (2005). A cortico-cortical mechanism mediating object-driven grasp in humans. *Proc Nat Ac Sci*, **102**, 898–903.

Christel, M. I., & Billard, A. (2002). Comparison between macaques' and humans' kinematics of prehension: the role of morphological differences and control mechanisms. *Behav Brain Res*, **131**, 169–184.

Davare, M., Andres, M., Cosnard, G., Thonnard, J. L. & Olivier, E. (2006). Dissociating the role of ventral and dorsal premotor cortex in precision grasping. *J Neurosci*, **26**, 2260–2268.

Ehara, Y., Fujimoto, H., Miyazaki, S. *et al.* (1997). Comparison of the performance of 3D camera systems. *Gait Posture*, **5**, 251–255.

Fuller, J., Liu, L. J., Murphy, M. C. & Mann, R. W. (1997). A comparison of lower extremity skeletal kinematics measured with pin- and skin-mounted markers. *Human Movement Sci*, **16**, 219–242.

Gardner, E. P., Debowy, D. J., Ro, J. Y., Ghosh, S. & Babu, K. S. (2002). Sensory monitoring of prehension in the parietal lobe: a study using digital video. *Behav Brain Res*, **135**, 213–224.

Georgiou, I., Becchio, C., Glover, S. & Castiello, U. (2007). Different action patterns for cooperative and competitive behaviour. *Cognition*, **102**, 415–433.

Glover, S., Miall, R. C. & Rushworth, M. F. S. (2005). Parietal rTMS selectively disrupts the initiation of on-line adjustments to a perturbation of object size. *J Cogn Neurosci*, **17**, 124–136.

Gordon, A. M., Forssberg, H., Johansson, R. S. & Westling, G. (1991). The integration of haptically acquired size information in the programming of precision grip. *Exp Brain Res*, **83**, 483–488.

Jakobson, L. S. & Goodale, M. A. (1991). Factors affecting higher-order movement planning: a kinematic analysis of human prehension. *Exp Brain Res*, **86**, 199–208.

Jakobson, L. S., Archibald, Y., Carey, D. & Goodale, M. A. (1991). A kinematic analysis of reaching and grasping movements in a patient recovering from optic ataxia. *Neuropsychologia*, **29**, 803–809.

Jeannerod, M. (1981). Intersegmental coordination during reaching at natural visual objects. In J. Long & A. Baddeley (Eds.), *Attention and Performance IX* (pp. 153–168). Hillsdale, NJ: Lawrence Erlbaum.

Jeannerod, M. (1984). The timing of natural prehension movements. *J Mot Behav*, **16**, 235–254.

Jeannerod, M. (1986). Mechanisms of visuomotor coordination: a study in normal and brain-damaged subjects. *Neuropsychologia*, **24**, 41–78.

Kuhtz-Buschbeck, J. P., Stolze, H., Boczek-Funcke, A. & Illert, M. (1998). Development of prehension movements in children: a kinematic study. *Exp Brain Res*, **122**, 424–432.

Mason, C. R., Theverapperuma, L. S., Hendrix, C. M. & Ebner, T. J. (2004). Monkey hand postural synergies during reach-to-grasp in the absence of vision of the hand and object. *J Neurophysiol*, **91**, 2826–2837.

Raghavan, P., Santello, M., Krakauer, J. W. & Gordon, A. M. (2006). Shaping the hand to object contours after stroke. The control of fingertip position during whole hand grasping. *Soc Neurosci Abst*, **655**, 14.

Raos, V., Umiltà, M. A., Murata, A., Fogassi, L. & Gallese, V. (2006). Functional properties of grasping-related neurons in the ventral premotor area F5 of the macaque monkey. *J Neurophysiol*, **95**, 709–729.

Rice, N. J., Valyear, K. F., Goodale, M., Milner, A. D. & Culham, J. C. (2007). Orientation sensitivity to graspable objects: an fMRI adaptation study. *Neuroimage*, **36** (Suppl. 2), T87–T93.

Roy, A. C., Paulignan, Y., Farnè, A., Jouffrais, C. & Boussaoud, D. (2000). Hand kinematics during reaching and grasping in the macaque monkey. *Behav Brain Res*, **117**, 75–82.

Santello, M. & Soechting J. F. (1998). Gradual molding of the hand to object contours. *J Neurophysiol*, **79**, 1307–1320.

Savelsbergh, G. J. P., Steenbergen, B. & Van der Kamp, J. (1996). The role of fragility information in the guidance of the precision grip. *Hum Mov Sci*, **15**, 115–127.

Schettino, L. F., Adamovich, S. V. & Poizner, H. (2003). Effects of object shape and visual feedback on hand configuration during grasping. *Exp Brain Res*, **151**, 158–166.

Schettino, L. F., Adamovich, S. V., Hening, W. *et al.* (2006). Hand preshaping in Parkinson's disease: effects of visual feedback and medication state. *Exp Brain Res*, **168**, 186–202.

Smeets J. B. & Brenner, E. (1999). A new view on grasping. *Mot Control*, **3**, 237–271.

Tunik, E., Frey, S. H. & Grafton, S. T. (2005). Virtual lesions of the anterior intraparietal area disrupt goal-dependent on-line adjustments of grasp. *Nat Neurosci*, **8**, 505–511.

Von Hofsten, C. & Rönnqvist, L. (1988). Preparation for grasping an object: a developmental study. *J Exp Psychol Hum Percept Perform*, **14**, 610–621.

Weir, P. L., MacKenzie, C. L., Marteniuk, R. G. & Cargoe, S. L. (1991a). Is object texture a constraint on human prehension? Kinematic evidence. *J Mot Behav*, **23**, 205–210.

Weir, P. L., MacKenzie, C. L., Marteniuk, R. G., Cargoe, S. L. & Fraser, M. B. (1991b). The effects of object weight on the kinematics of prehension. *J Mot Behav*, **23**, 192–204.

Winter, D. A. (1991). *Biomechanics and Motor Control of Human Movement*, 2nd edn. Toronto: John Wiley & Sons.

Zoia, S., Pezzetta, E., Blason, L. *et al.* (2006). A comparison of the reach-to-grasp movement between children and adults: a kinematic study. *Dev Neuropsychol*, **30**, 719–738.

3

Digit forces in multi-digit grasps

VLADIMIR M. ZATSIORSKY AND MARK L. LATASH

Summary

This chapter briefly reviews four issues: (1) Single-digit contacts, (2) multi-digit grasps, (3) constraints on digit forces and (4) prehension synergies. In the section on the single-digit contacts the main models of contact (the point model, hard-finger contact and soft-finger contact) as well as slip prevention are considered. A subchapter on multi-digit grasps is a tutorial-like introduction to grasp mechanics, in particular to the concept of grasp matrices. The following issues are addressed: (a) vertically oriented prismatic grasps, (b) grasp matrices, (c) non-vertical prismatic grasps, (d) arbitrary grasps, (e) virtual finger and (f) internal forces. The constraints on digit forces are classified as mechanical and biological. Among the biological constraints *force deficit* and *finger enslaving* are addressed. *Inter-finger connection matrices* and their use for reconstruction of the neural command are discussed. *Prehension synergies*, experimental methods of their research, and the *principle of superposition* are described.

To manipulate hand-held objects, people exert forces on them. Performers grasp objects in different ways. A simplest classification of the grasps (or grips) includes two varieties: a *power grip* when the object is in contact with the palm of the hand (e.g. holding a tennis racket) and a *precision grip* when only the digit tips are in contact with the object. The present chapter deals with the precision grip.

Single-digit contacts

This section concentrates on two issues: modeling the digit contacts and slip prevention.

Modeling the digit contacts

Contact is a collection of adjacent points where two bodies touch each other. When a digit touches an object, the contact occurs over a certain area. Under the force the digit tip deforms and its contact area changes. The distance from the contact area to the distal phalangeal bone decreases. The point of force application may displace and the finger tip may roll over the contact surface. When analyzing digit forces, depending on the

Sensorimotor Control of Grasping: Physiology and Pathophysiology, ed. Dennis A. Nowak and Joachim Hermsdörfer. Published by Cambridge University Press. © Cambridge University Press 2009.

research task and available equipment, a researcher may choose one of several options (Mason & Salisbury, 1985):

(a) Neglect the described changes and assume that the digit tip force is exerted at a point that does not change its location during performance. The point of force application is usually assumed to coincide with the center of a force sensor. Such a model is called the point model, or the *point contact model*. In the point model with friction, the digit can exert a force but not a torque (moment of force) on the sensor. If the point model is selected, the researcher can use one-, two- or three-component force sensors.

(b) *Hard-finger contact*: the contact takes place over an area but the finger tip deformation and rolling is neglected. The point of force application does not change but the digit can exert a moment of force in the plane of contact, i.e. about an axis of rotation perpendicular to the contact surface. The moment production is possible because the tangential forces acting at different regions of the contact area may act in opposite directions and hence form a force couple, e.g. two equal and opposite forces. Moments generated by force couples are so-called *free moments*, they do not depend on the place of application and do not change under parallel displacement (see Zatsiorsky, 2002; Chapter 1).

(c) *Soft-finger contact*: the contact takes place over a certain area, the finger deforms and the point of force application can change during performance. It is assumed that the digits do not stick to the object and hence they can only push but not pull the object. Hence, the normal force exerted on the sensor (the force component that is perpendicular to the sensor surface) is uni-directional. As a result, the digits cannot exert force couples (free moments) on the sensor in the planes other than the plane of contact. An attempt to generate such a moment will result in a digit-tip rolling on the contact surface. With a soft-finger model the digit–object interaction is characterized by six variables: three orthogonal force components (the normal force component is uni-directional and the two tangential force components are bi-directional), free moment in the plane of contact, and two coordinates of the point of force application on the sensor. To obtain these data six-component force sensors are necessary; the sensors yield three orthogonal force component and three moment component values. The moments are however reported with respect to the sensor center (not with respect to the point of force application that can be different from the center). If f_Z is a normal force (along the Z-axis that is perpendicular to the sensor surface) and m_X is the moment about an axis X that goes through the sensor center in the plane of contact (Figure 3.1), the coordinate of force application along axis Y can be found as $y = m_X/f_Z$. The formula is valid because moment m_X is entirely due to force f_Z: the digit does not exert a free moment about an axis in the plane of contact and the tangential digit forces also do not exert moments about the axes in this plane.

The selection of the contact model (point contact, hard-finger contact or soft-finger contact) depends on the research task: whether the moments about axis Z and the coordinates of the point of digit force application are important. The displacement of the point of digit force application can be quite large: up to 12 mm for the thumb and 5–6 mm for the fingers (Zatsiorsky *et al.*, 2003a).

Slip prevention

In practical situations, both the tangential forces and friction act at the digit contacts. If tangential force is too large (or the friction is too low) the object will slip. In the majority of

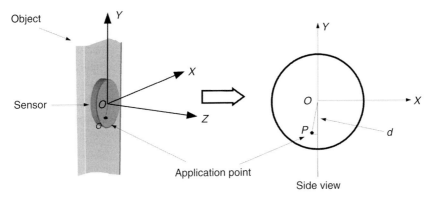

Figure 3.1. A sensor-fixed system of coordinates and the point of force application. *d* is the shortest distance from the force application point to the sensor center.

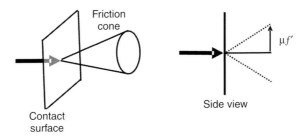

Figure 3.2. Geometric representation of the Coulomb friction model.

studies it is assumed that the skin obeys Coulomb's law. According to the law the force of friction is directly proportional to the load and does not depend on the area of contact or the skidding velocity. When the materials follow Coulomb's law, the coefficient of friction can be determined as the tangential force/normal force ratio. The coefficient itself is constant; its magnitude does not depend on the normal force. To which extent the skin obeys the law is not clear. Some researchers (Comaish & Bottoms, 1971; Bobjer *et al.*, 1993) reported that skin obeys Coulomb's law over a limited range of loads only, while others (Savescu *et al.*, 2007) found that the law is not valid for some materials, for example sand paper. Assuming that the Coulomb law is valid (such an assumption greatly simplifies the analysis), the admissible range of the tangential force magnitude that is below the slipping threshold is determined by $|f^t| \leq \mu f^n$, where $\mu > 0$ is the static coefficient of friction, f^t is the tangential force and f^n is the normal force. In three dimensions, this relation can be represented geometrically as a *friction cone* (Figure 3.2). To avoid slipping, the contact forces must lie in the friction cone (*FC*).

To quantify the risk of slipping two measures are suggested (Johansson & Westling, 1984; Westling & Johansson, 1984): (a) the tangential force/normal force ratio, to avoid slipping the ratio should be smaller than the coefficient of friction; and (b) the *safety margin* which can be estimated as a ratio (Burstedt *et al.*, 1999): *Safety margin* = $(f^n{}_i - |f^t{}_i|/\mu)/f^n{}_i$, where i is an arbitrary digit, f^n is the digit normal force, and f^t is the digit tangential force. For the vertical grasping, the reported safety margin values are between 0.3 and 0.5 (Burstedt *et al.*, 1999) or even as high as 0.6–0.7 (Pataky *et al.*, 2004b). Hence, when people manipulate objects they use a "better safe than sorry" strategy.

Performers adjust digit forces both to the load (tangential forces) and friction. When friction is low the grasping force (the normal force) increases (Johansson & Westling, 1984). When friction beneath each digit is different, i.e. some surfaces are more slippery than others, the performers adjust not only the normal forces but also the tangential forces to the local friction conditions: the more slippery the surface the larger the normal force and the smaller the tangential force (Aoki *et al.*, 2006; Niu *et al.*, 2007). The coordination of individual digit forces in such tasks is explained by the interaction of the local responses to friction and synergic responses necessary to maintain the equilibrium.

For non-vertical grasping, the safety margin concept becomes ambiguous; it does not capture the essence of the slip prevention. For the purposes of illustration, consider holding an object horizontally. In such a case, some digits support the load while others press downward. The digits of the first and the second group exert different normal forces while the tangential force is zero. Formally, the safety margin can be computed but it is not informative. For the non-vertical grasping, the "generalized safety margin" has been suggested (Pataky *et al.*, 2004c). It reduces to the safety margin for vertical grasps.

Multi-digit grasps: a gentle introduction to grasp mechanics

Commonly in science, complex multi-faceted issues are initially studied using simple examples. The simplest grasping tasks involve (a) two digits only, e.g. the thumb and the index finger, (b) parallel contact surfaces and (c) vertically oriented objects that are either at rest or are moved in a vertical direction. In such grasps (prismatic pinch grasps), the normal forces are exerted horizontally while the load force is directed vertically and hence is manifested as the shear or tangential force acting on the force sensors. The prismatic pinch grasps have traditionally been the object of the most intensive research (Johansson & Westling, 1984; Cole & Abbs, 1988; Jones & Hunter, 1992; Cole & Johansson, 1993; Cole & Beck, 1994; Johansson & Cole, 1994; Flanagan & Wing, 1995; Hermsdorfer *et al.*, 1999; Gordon *et al.*, 2000; Deutsch & Newell, 2002; Nowak & Hermsdorfer, 2002, 2003a, 2003b, 2006; Augurelle *et al.*, 2003; Monzee *et al.*, 2003; Blennerhassett *et al.*, 2006). An advantage of studying these tasks is the mechanical independence of the normal and tangential forces, in the sense that any change of the normal force is due to motor control and is not a mechanical artefact. In other grasps, for instance in the non-vertical grasps, the normal forces can be due to both gravity and motor control.

During recent years, however, research interests have shifted to include more complicated cases: multi-finger grasps, objects of diverse shape and mechanical properties that are differently oriented with respect to gravity and being moved in various directions (Kinoshita *et al.*, 1995; Nakazawa *et al.*, 1996; Flanagan *et al.*, 1999; Santello & Soechting, 2000; Baud-Bovy & Soechting, 2001, 2002; Latash *et al.*, 2003a; Rearick

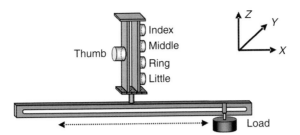

Figure 3.3. Experimental "inverted-T" handle/beam apparatus. Force components in *X*-direction are called the *normal* forces; force components in *Y* and *Z* directions are *shear* or *tangential forces*. Five 6-component force sensors are used to register individual digit forces and moments. The thumb opposes the four fingers. The T-shaped attachment allows varying the load and the external torque independently. The handle width, i.e. the horizontal projected distance between the surfaces of the thumb and finger sensors, is 2*r* (not shown in the figure).

et al., 2003; Pataky *et al.*, 2004a; Smith & Soechting, 2005; Dumon *et al.*, 2006; Pylatiuk *et al.*, 2006; Shim *et al.*, 2007). Studying such prehension tasks requires more complex approaches that may involve more complex mathematics. The goal of the present section is to introduce these approaches in a tutorial-like fashion. We start from relatively simple vertically oriented prismatic multi-finger grasps and planar cases.

Vertically oriented prismatic grasp

Consider maintaining in the air at rest an "inverted-T" handle, as shown in Figure 3.3. The contact surfaces are parallel, the handle is oriented vertically, and an external moment acts in the *plane of the grasp* (a vertical plane that contains all the points of digit contact with the object).

To maintain the handle at equilibrium three equations have to be satisfied:

$$0 = F_{th}^n - \left(F_i^n + F_m^n + F_r^n + F_l^n \right) \tag{3.1}$$

$$0 = \left(F_{th}^t + F_i^t + F_m^t + F_r^t + F_l^t \right) - L \tag{3.2}$$

$$0 = M - \left(\underbrace{F_{th}^n d_{th} + F_i^n d_i + F_m^n d_m + F_r^n d_r + F_l^n d_l}_{\text{Moment of the normal forces} \equiv M^n} + \underbrace{F_{th}^t r_{th} + F_i^t r_i + F_m^t r_m + F_r^t r_r + F_l^t r_l}_{\text{Moment of the tangential forces} \equiv M^t} \right) \tag{3.3}$$

where the subscripts *th, i, m, r* and *l* refer to the thumb, index, middle, ring and little finger, respectively; the superscripts *n* and *t* stand for the normal and tangential force components, respectively; *L* is load (weight of the object), and coefficients *d* and *r* stand for the moment arms of the normal and tangential force with respect to a pre-selected center, respectively. The center is commonly selected along the vertical midline (hence the moment arms of all the tangential forces are the same; they equal one half of the handle width) at the level of the center of the thumb force sensor. The moment arms of the digit normal forces are different for each digit. They equal the sum of (a) the vertical distance between the sensor center and the thumb center, and (b) the vertical displacement of the point of digit force application from the sensor center. It is assumed that the points of digit force application do not change during the performance. The equations reflect the balance of the forces in a horizontal direction – Equation (3.1); the balance of the forces in the vertical direction – Equation (3.2); and the balance of the moments of force – Equation (3.3), respectively. The normal forces should also satisfy the no-slip requirements, i.e. the digit forces should be inside the friction cones.

Grasp matrix

Equations (3.1–3.3) can be written as one matrix-vector equation. For arbitrary conditions, i.e. when relaxing the constraints of statics, the equation is:

$$
\begin{bmatrix} F_X \\ F_Y \\ M_{tot} \end{bmatrix} = \begin{bmatrix} 1 & -1 & -1 & -1 & -1 & 0 & 0 & 0 & 0 & 0 \\ 0 & 0 & 0 & 0 & 0 & 1 & 1 & 1 & 1 & 1 \\ d_{th} & d_i & d_m & d_r & d_l & -r & r & r & r & r \end{bmatrix} \begin{bmatrix} F^n_{th} \\ F^n_i \\ F^n_m \\ F^n_r \\ F^n_l \\ F^t_{th} \\ F^t_i \\ F^t_m \\ F^t_r \\ F^t_l \end{bmatrix} \tag{3.4a}
$$

or

$$
\mathbf{F}_{object} = [\mathbf{G}]\mathbf{F}_{digits} \tag{3.4b}
$$

where $\mathbf{F}_{object}$ is a (3×1) vector of the resultant force and moment acting on the object, $[\mathbf{G}]$ is a (3×10) *grasp matrix* (Mason & Salisbury, 1985), and $\mathbf{F}_{digits}$ is a (10×1) vector of the digit forces. The elements of the first two rows of $[\mathbf{G}]$ are the coefficients at the digit force values. Because in the position shown in Figure 3.3 the normal and tangential digit forces are along the X and Y axes of the global system of coordinates, the coefficients are either zeroes or equal plus or minus one. When the normal and tangential digit forces are not parallel to the X and Y axes, the coefficients equal the direction cosines, i.e. the projections of the unit vectors along the normal and tangential directions onto the X and Y axes. The elements of the last row in matrix $[\mathbf{G}]$ are the moment arms of the digit forces about axis Z through the origin of the system of coordinates. $[\mathbf{G}]$ is also known as the *matrix of moment arms*.

Equation (3.4) is a linear equation that allows for using the common methods of linear algebra. The equation is based, however, on a simplifying assumption that the elements of the grasp matrix are constant, i.e. the points of digit force applications do not displace during

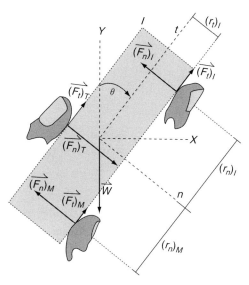

Figure 3.4. Non-vertical three-digit grasping. Coordinate systems and notation used. The local normal (n) and tangential (t) axes are coincident with the global horizontal (X) and vertical (Y), respectively, for orientation $\theta = 0$. The normal and tangential forces (F_n and F_t, respectively) are shown for the thumb and the index and middle fingers (T, I and M, respectively). Moment arms (r) about the origin are shown for F_n and F_t separately, but not for all digit forces. The origin of the n-t reference system is at the center of mass of the hand-held object. Reprinted by permission from Pataky *et al.* (2004b).

the period of observation. If they migrate, the elements of [**G**] are not constant any more and the equations become non-linear: variable values of digit forces are multiplied by the variable values of moment arms. This obstacle can be avoided if a (10×1) $\mathbf{F}_{digits}$ vector is expanded to a (15×1) vector where the added elements are the moments exerted by the individual digits with respect to the corresponding sensor centers. Matrix [**G**] in this case is 3×15.

Non-vertical prismatic grasps

In the non-vertical grasps, the gravity force is resisted not only by the tangential forces but also by a normal force exerted on the object. Equation (3.4) can be adjusted to such a task; the equation can also include the non-slip requirements. As an example consider a grasp shown in Figure 3.4 (Pataky *et al.*, 2004b). For simplicity, the example includes only three digits but the equations are written for an arbitrary number of digits N ($2 \leq N \leq 5$).

For the planar equilibrium of a grasped object with weight W three equilibrium equations have to be satisfied:

$$\sum_{i=1}^{N}(F_n)_i = W_n \qquad (3.5)$$

$$\sum_{i=1}^{N}(F_t)_i = W_t \qquad (3.6)$$

$$\sum_{i=1}^{N}(M_{nt})_i = M_{ext} \qquad (3.7)$$

where M_{nt} represents the moments exerted in the n–t plane, M_{ext} is the resultant moment necessary to maintain equilibrium, and W_n and W_t are the projections of the object's weight on the n and t axes, respectively:

$$W_n = sin|\theta| \qquad (3.8)$$
$$W_t = cos\theta \qquad (3.9)$$

Assuming hard finger–object contacts (constant points of force applications, no contact couples) and Coulomb friction, the normal forces $(F_n)_i$ are also subject to the following no-slip constraint:

$$(F_n)_i \geq (\mu_i)^{-1}\left|(F_t)_i\right| \qquad (3.10)$$

where μ_i is the coefficient of static friction between the ith finger and the grasped object. Equations (3.5–3.7) and (3.10) may be expressed in matrix form as:

$$[\mathbf{G}]\mathbf{F} = \mathbf{L} \qquad (3.11)$$
$$[\mathbf{A}]\mathbf{F} \leq \mathbf{S} \qquad (3.12)$$

where Equation (3.11) represents the equilibrium constraints (3.5–3.7) and Equation (3.12) represents the no-slip constraints (3.10). The vector $\mathbf{F}$ is the ($2N{\times}1$) vector of digit forces. The vector $\mathbf{S}$ is the ($2N{\times}1$) zero vector that constitutes the no-slip inequality constraints and $\mathbf{L}$ is the ($3{\times}1$) load vector that constitutes the right-hand side of the equilibrium constraints (Equations 3.5–3.7):

$$\mathbf{L} = [\mathbf{W_n}, \mathbf{W_t}, \mathbf{M_{ext}}]^{\mathbf{T}}$$

The grasp matrix $[\mathbf{G}]$ is ($3{\times}2N$) and the matrix $[\mathbf{A}]$ a ($2N{\times}2N$) square matrix. For the example in Figure 3.4, the terms of Equations 3.11 and 3.12 expand to:

$$\mathbf{F} = [(F_n)_T, (F_n)_I, (F_n)_M, (F_t)_T, (F_t)_I, (F_t)_M]^T$$
$$\mathbf{S} = [0, 0, 0, 0, 0, 0]^T$$

$$\mathbf{L} = [W_n, W_t, M_{ext}]^T$$

$$[\mathbf{G}] = \begin{bmatrix} 1 & -1 & -1 & 0 & 0 & 0 \\ 0 & 0 & 0 & 1 & 1 & 1 \\ (r_n)_T & (r_n)_I & (r_n)_M & (r_t)_T & (r_t)_I & (r_t)_M \end{bmatrix}$$

$$[\mathbf{A}] = \begin{bmatrix} -T & 0 & 0 & 1 & 0 & 0 \\ 0 & -I & 0 & 0 & 1 & 0 \\ 0 & 0 & -M & 0 & 0 & 1 \\ -1 & 0 & 0 & 0 & 0 & 0 \\ 0 & -1 & 0 & 0 & 0 & 0 \\ 0 & 0 & -1 & 0 & 0 & 0 \end{bmatrix}$$

where $(r_n)_i$ and $(r_t)_i$ are the moment arms of the normal and tangential forces, respectively. For compactness, moments are computed about the origin of the n–t coordinate system illustrated in Figure 3.4 (i.e. such that W does not produce a moment). The moment arms $(r_n)_i$ and $(r_t)_i$ can be expressed as $(1 \times N)$ vectors $\mathbf{r_n}$ and $\mathbf{r_t}$.

Arbitrary grasps

For a general 3D case, matrix [**G**] is 6×30. Its rows represent the three resultant force and three resultant moment components exerted on the object, respectively. Its columns represent digit force and moment components; six force and moment components exerted at a digit tip × five digits. The local digit moment components about the axes in the plane of contact can be replaced by the two coordinates of the point of force application.

Virtual finger

The forces of the four fingers can be reduced to a resultant force and a moment of force. This is equivalent to replacing a set of fingers with a *virtual finger*, VF (Arbib *et al.*, 1985; Iberall, 1987; Baud-Bovy & Soechting, 2001). A VF generates the same mechanical effect (the same *wrench*, see Zatsiorsky, 2002; Chapter 1) as a set of actual fingers. In various research, either individual digit forces are analyzed (the IF level) or the thumb and VF forces only are addressed (the VF level). There are substantial differences between the VF and IF forces: (a) the force directions are as a rule dissimilar (for a review see Zatsiorsky & Latash, 2007). The IF forces can be exerted in disparate directions such that only their resultant (i.e. VF) force is in the desired direction (Gao *et al.*, 2005a, b). (b) VF and IF forces adjust differently to modified task conditions (Zatsiorsky *et al.*, 2002a, b). (c) IF forces are much more variable than VF forces (Shim *et al.*, 2005). The desired performance at the VF level is achieved by a synergetic covariation among individual finger forces at the IF level.

Returning to Figure 3.1 and Equations (3.1)–(3.3), the requirements for keeping a vertically oriented object at equilibrium can be now summarized as follows. (1) Normal thumb and VF forces should be: (1a) equal in magnitude, (1b) large enough to prevent slipping, and (1c) generate a moment that matches the moment of the tangential forces. The moment control is achieved by varying the VF force magnitude and/or location of the point of the VF force application with respect to the thumb force application point (moment arm of the VF force). (2) Tangential thumb and VF forces: (2a) their sum should equal the weight of the hand-held object, and (2b) their difference – which is proportional to the moment of tangential forces – should match the moment of the normal forces such that the sum of the above moments is equal and opposite to the external moment acting on the object.

Internal forces

An *internal force* is a set of digit forces that does not perturb the object equilibrium (Mason & Salisbury, 1985; Murray *et al.*, 1994). The elements of an internal force vector cancel each other and, hence, do not contribute to the resultant force and moment acting on the object (the *manipulation force*). The best known example of the internal forces is the *grasp for*ce, two equal and opposite normal forces exerted by the thumb and VF against each other. Another possible example is equal and opposite tangential forces exerted by any two fingers. The resultant of such forces would be zero. Note that an internal force is not a single force; it is a set of forces and moments that, when acting together, generate a zero resultant force and a zero resultant moment. The dimensionality of such a vector equals the total number of the digit forces and moments, i.e. 30 in five-digit grasps (some of the elements of an internal force vector can be zero).

The internal forces lie in the so-called *null space* of the grasp matrix $[\mathbf{G}]$ (for a simple introduction to the concept of the null spaces see Zatsiorsky, 2002, Section 2.4). This is another way of saying that the elements of the internal force vectors cancel each other and do not affect the manipulation force. For the readers who prefer mathematical notation it can be said that the vector of internal forces $\mathbf{F}_i$ lies in the null space of $[\mathbf{G}]$; the null space of an *m* by *n* matrix $[\mathbf{G}]$ is a set of all vectors $\mathbf{F}$ in R^n such that $[\mathbf{G}]\mathbf{F} = \mathbf{0}$. (The symbol R^n designates an *n*-dimensional space of real numbers.) Any linear combination of the internal force vectors is also an internal force. In particular, multiplication of an internal force vector by a constant multiple does not affect the force cancelling and the sums of two or several internal force vectors are also in the null space. Hence, the internal force vectors are innumerable and looking for all of them does not make sense. However, it is possible to find independent vectors of a prescribed length, e.g. of unit length, spanning the null space of $[\mathbf{G}]$, the *orthonormal basis vectors*. Analysis of the grasp matrix $[\mathbf{G}]$ provides a convenient tool for discovering the orthonormal basis vectors. Because the rank of a 6×30 matrix is at most 6, the dimensionality of the null space of the grasp matrix (its nullity) is at least 24. A vector-by-vector analysis of all of them is a daunting task. Because of that, the internal force analysis is done only at the VF level. At this level, for the prismatic grasps, mathematical analysis yields the three internal forces (Gao *et al.*, 2005b): (a) the grasp force, (b) the internal moment, i.e. the *moment of normal force–moment of tangential force* combination, and (c) the twisting moments about the axes normal to the surfaces of the contacts. The latter combination cannot, however, be realized in single-hand grasping: people cannot twist the thumb and the finger(s) in opposite directions (in two-handed grasping this option can be realized).

In multi-digit grasping, the resultant force vector (manipulation force) and the vector of the internal force are mathematically independent (Kerr & Roth, 1986; Yoshikawa & Nagai, 1991). The mathematical independence of the above forces allows for their independent (decoupled) control. Such a decoupled control is realized in robotic manipulators (e.g. Zuo & Qian, 2000). The decoupled control saves computational resources. People, however, do not use this option; they modulate the internal forces with the manipulation force (Gao *et al.*, 2005c). When performers move a vertically oriented object in the vertical direction they vary the grip force in parallel with the load force (Johansson & Westling, 1984; Flanagan & Wing, 1993, 1995; Flanagan *et al.*, 1993; Flanagan & Tresilian, 1994; Kinoshita *et al.*, 1996; Nakazawa *et al.*, 1996; Gysin *et al.*, 2003; Gao *et al.*, 2005b, 2006; Zatsiorsky *et al.*, 2005b; see also Flanagan & Johansson, 2002 for a review). When manipulating an object the performers commonly change the moments of the normal and tangential force in opposite directions such that the resultant moment and the object orientation do not change (Gao *et al.*, 2005b). When a vertically oriented object is being moved horizontally the maximal grasping force is observed at the instances of minimal acceleration and maximal velocity; this is valid both for the three-digit grasps from above (Smith & Soechting, 2005) and prismatic grasps (Gao *et al.*, 2005b).

Constraints on digit forces

Digit forces during object manipulation are subject to mechanical and biological constraints. The mechanical constraints, such as the non-slipping constraint or the equilibrium constraints, are represented in equations (3.11) and (3.12). The biological constraints are not included in these equations. Besides a self-evident constraint that digit forces in grasping cannot exceed the force exerted during maximal voluntary contraction, there is a constraint imposed by *finger interdependence* (reviewed in Schieber & Santello, 2004). The finger interdependence is manifested as (a) *force deficit* and (b) *finger enslaving*.

The term *force deficit* refers to the fact that peak force generated by a finger in a multi-finger maximal voluntary contraction (MVC) task is smaller than its peak force in the single-finger MVC task (Ohtsuki, 1981; Zatsiorsky *et al.*, 1998). The deficit increases with the number of explicitly involved (master) fingers (Li *et al.*, 1998a). As follows from several studies (reviewed in Danion *et al.*, 2003), the deficit is similar across tasks with the same number of explicitly involved fingers. The deficit can be parameterized as the ratio of the actual total force in N-finger pressing tasks ($1 \leq N \leq 4$) to the predicted force computed as the sum of forces exerted in single-finger tasks. For two-finger tasks (IM, IR, IL, MR, ML and RL) the ratio ranged from 0.61 to 0.64, for three-finger tasks (IMR, IML, IRL and MRL) it ranged from 0.43 to 0.45, and for the four-finger task (IMRL) it was 0.38. Obviously, for one-finger tasks (I, M, R and L) the ratio was equal to 1. The dependence of the above ratio on N, the number of fingers explicitly involved in the task, is described by an empirical equation: *Force deficit ratio* $= 1/N^{0.712}$ (Danion *et al.*, 2003).

Enslaving refers to the fact that the fingers that are not required to produce any force by instruction involuntarily move or generate force (Kilbreath & Gandevia, 1994; Li *et al.*, 1998b; Zatsiorsky *et al.*, 2000; Kilbreath *et al.*, 2002). To demonstrate enslaving, turn your palm up and wiggle one of the fingers. You will see that other fingers also move. Finger enslaving is described by *inter-finger connection matrices* that relate hypothetic central commands to individual fingers with actual finger forces via a matrix equation (Zatsiorsky *et al.*, 1998; Li *et al.*, 2002; Danion *et al.*, 2003; Gao *et al.*, 2003; Latash *et al.*, 2003b):

$$\mathbf{F} = [\mathbf{W}]\mathbf{c} \tag{3.13}$$

where $\mathbf{F}$ is a (4×1) vector of the normal finger forces, $[\mathbf{W}]$ is a (4×4) inter-finger connection matrix whose elements depend on the number of fingers involved in the task, and $\mathbf{c}$ is a (4×1) vector of the *central* (*neural*) *commands*. The elements of vector $\mathbf{c}$ equal 1.0 if the finger is intended to produce maximal force (maximal voluntary activation) or 0.0 if the finger is not intended to produce force (no voluntary activation). Similar matrices were also computed for the finger interaction during maximal radial and ulnar deviation efforts (Pataky *et al.*, 2007).

For computing matrix $[\mathbf{W}]$ the maximal finger forces exerted in different tasks are recorded: the subjects are instructed to press on force sensors as hard as possible with either one, two, three or four fingers, using all possible finger combinations. Then, the inter-finger connection matrices are computed by artificial neural networks (Zatsiorsky *et al.*, 1998; Li *et al.*, 2002; Gao *et al.*, 2003, 2004; Latash *et al.*, 2003b) or they are estimated by simple algebraic equations (Danion *et al.*, 2003). The elements of the matrix are the weight coefficients. When multiplied by commands c_i they represent: (a) elements on the main diagonal: the forces exerted by finger i in response to the command sent to this finger (*direct*, or *master, forces*); (b) elements in rows: force of finger i due to the commands sent to all the fingers (sum of the direct and enslaved forces); and (c) elements in the columns: forces exerted by all four fingers due to a command sent to one of the fingers (a *mode*). The following equation (Danion *et al.*, 2003) may serve as an example:

$$\begin{pmatrix} F_I \\ F_M \\ F_R \\ F_L \end{pmatrix} = \frac{1}{N^{0.712}} \begin{bmatrix} 49.1 & 14.0 & 9.4 & 7.7 \\ 10.5 & 38.0 & 16.5 & 6.7 \\ 5.5 & 12.9 & 29.9 & 15.0 \\ 2.7 & 4.1 & 10.6 & 24.8 \end{bmatrix} \begin{pmatrix} c_I \\ c_M \\ c_R \\ c_L \end{pmatrix}$$

The off-diagonal elements in the matrix represent the finger enslaving. An empirical coefficient $1/N^{0.712}$ corrects for the force deficit (N is number of fingers in the task). This empirical equation has been determined for young healthy male subjects.

Force generated by a finger in prehension arises from the command sent to this finger ("direct" finger force) as well as from the commands sent to other fingers (enslaved force). The direct finger forces can be computed as the product $w_{ii}c_i$ ($i = 1, 2, 3, 4$) where w_{ii} is a diagonal element of the weight matrix, and c_i is a command intensity to this finger. To perform these computations commands c_i to individual fingers should be known; in other words, the command vector $\mathbf{c}$ should be *reconstructed*. If matrix $[\mathbf{W}]$ is known and actual finger forces in a prehension task are recorded, the vector of neural commands $\mathbf{c}$ is reconstructed by inverting equation (3.13): $\mathbf{c} = [\mathbf{W}]^{-1}\mathbf{F}$ (Zatsiorsky *et al.*, 2002b). Note that matrix $[\mathbf{W}]$ is 4×4 and hence is invertible. When the vector $\mathbf{c}$ is reconstructed, forces generated by individual fingers can be decomposed into components that are due to (a) direct commands to the targeted fingers, and (b) the enslaving effects, i.e. the commands sent to other fingers.

Prehension synergies

Prehension synergies have been defined as conjoint changes in finger forces and moments during multi-finger gripping tasks (Santello & Soechting, 2000; Rearick & Santello, 2002; Rearick *et al.*, 2003; Baud-Bovy & Soechting, 2002; Zatsiorsky *et al.*, 2003b; Zatsiorsky & Latash, 2004; Pataky *et al.*, 2004a, b; Shim *et al.*, 2003, 2005). Some of these adjustments are dictated by mechanical and biological constraints discussed above, whereas others are results of choice by the CNS.

To study prehension synergies researchers have employed several experimental techniques. They: (a) inflicted external perturbations (Cole & Abbs, 1987, 1988; Gao *et al.*, 2005b); (b) inflicted self-perturbations, in particular varied the number of grasping fingers during prehension (Budgeon, 2007); (c) recorded correlations among output variables in single trials of long duration (Santello & Soechting, 2000; Vaillancourt *et al.*, 2002); (d) varied the task parameters, in particular the object geometry, friction, resisted torque and/or load (Zatsiorsky *et al.*, 2002a, b, 2003; Aoki *et al.*, 2006, 2007; Niu *et al.*, 2007) and (e) studied trial-to-trial variability (Shim *et al.*, 2003, 2005; Shim & Park, 2007).

A general conclusion is that during prehension the individual digit force variations are always inter-related; i.e. they manifest the prehension synergies. For instance, people do not perform the same prehension task in one and the same way; in each trial, the digit forces are different. However, a change in one elemental variable is compensated by an adjustment in another variable(s) such that the total output stays put (Figure 3.5). As it was already mentioned, some of the adjustments are mechanically necessitated while others are not. For instance to maintain the object equilibrium, the sum of the VF and thumb tangential forces should be constant and equal to the load force (Equation 3.2). As seen in Figure 3.5, the VF and thumb forces are on the line that represents a constant total tangential force. However, the precise location of the VF–thumb values along the line is not specified by the task mechanics and can vary among the trials (the "sausages" along the line). The different location of the force–force values along the line correspond to different moments of the tangential forces. These moments should be matched by moments of normal forces such that the sum of the moments counterbalances the external torque acting on the handle (Equation 3.3). In summary, some of the variables, e.g. sum of the tangential forces, are prescribed by the task mechanics whilst others, e.g. relative contribution of the moments of the normal and tangential forces into the total moment, are a matter of choice made by the central controller.

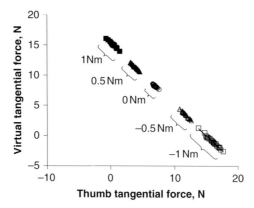

Figure 3.5. Tangential forces of the thumb versus tangential forces of the virtual finger across 25 trials. A representative subject. The data were collected with a setup shown in Figure 3.3. The subjects performed 25 trials at each of the external torques -1.0 Nm, -0.5 Nm, 0 Nm, 0.5 Nm and 1.0 Nm while the total load was always 14.9 N. In individual trials the forces were different. However, they were along the same line such that their sum was always constant. All the coefficients of correlation were -1.00. Because the moment of the tangential forces is proportional to the difference between the tangential forces, different location of the forces along the line is indicative of different moments of the tangential forces. The trial-to-trial variations of the moment of tangential forces were compensated by the matching changes of the moments of normal forces (not shown in the figure). Adapted by permission from J. K. Shim, M. L. Latash, and V. M. Zatsiorsky (2003). Prehension synergies: tria-to-trial variability and hierarchical organization of stable performance. *Exp Brain Res* **152**, 173–184.

The prehension synergy can be viewed as a hierarchy of synergies with at least two sub-synergies visibly manifested. The sub-synergies correspond to the two internal forces discussed previously, the grip force and the internal moment. Such an understanding of prehension control agrees well with the so-called *principle of superposition*, suggested in robotics (Arimoto *et al.*, 2001). According to the principle, some skilled actions can be decomposed into several elemental actions that can be controlled independently by several controllers. In particular, dexterous manipulation of an object by two soft-tip robot fingers can be realized by a linear superposition of two commands, one command for the stable grasping and the second one for regulating the object orientation. In robotics, such a decoupled control decreases the computation time. When applied to human performers and multi-finger grasps, the principle claims that forces and moments during prehension are defined by two independent commands: "Grasp the object stronger/weaker to prevent slipping" and "Maintain the rotational equilibrium of the object." The effects of the two commands are summed up. The applicability of the principle to multi-finger prehension in humans has been confirmed in several studies (Shim *et al.*, 2003, 2005; Zatsiorsky & Latash, 2004; Zatsiorsky *et al.*, 2004; Gao *et al.*, 2006; Latash & Zatsiorsky, 2006; Shim & Park, 2007).

Multi-finger prehension is an example of a mechanically redundant task: the same resultant forces on the object can be exerted by different digit forces. Because the forces can be recorded, the prehension is an advantageous object for studying the motor redundancy problem, the Bernstein problem (Bernstein, 1967). It was hypothesized that the central controller uses similar strategies to confront motor redundancy in various tasks, e.g. comparable strategies are used to control redundant muscles acting at a joint and redundant fingers acting on the object (Li *et al.*, 1998a). If this supposition is correct, studying multi-finger prehension may help in understanding how the redundant motor systems are controlled.

References

Aoki, T., Niu, X., Latash, M. L. & Zatsiorsky, V. M. (2006). Effects of friction at the digit-object interface on the digit forces in multi-finger prehension. *Exp Brain Res*, **172**, 425–438.

Aoki, T., Latash, M. L. & Zatsiorsky, V. M. (2007). Adjustments to different local friction in multi-finger prehension. *J Motor Behav*, **39**, 276–290.

Arbib, M. A., Iberall, T. & Lyons, D. (1985). Coordinated control programs for movements of the hand. In A. W. Goodwin & I. Darian-Smith (Eds.), *Hand Function and the Neocortex* (pp. 111–129). Berlin: Springer Verlag.

Arimoto, S., Tahara, K., Yamaguchi, M., Nguyen, P. T. A. & Han, H. Y. (2001). Principles of superposition for controlling pinch motions by means of robot fingers with soft tips. *Robotica*, **19**, 21–28.

Augurelle, A. S., Smith, A. M., Lejeune, T. & Thonnard, J. L. (2003). Importance of cutaneous feedback in maintaining a secure grip during manipulation of hand-held objects. *J Neurophysiol*, **89**, 665–671.

Baud-Bovy, G. & Soechting, J. F. (2001). Two virtual fingers in the control of the tripod grasp. *J Neurophysiol*, **86**, 604–615.

Baud-Bovy, G. & Soechting, J. F. (2002). Factors influencing variability in load forces in a tripod grasp. *Exp Brain Res*, **143**, 57–66.

Bernstein, N. A. (1967). *The Co-ordination and Regulation of Movements*. Oxford: Pergamon Press.

Blennerhassett, J. M., Carey, L. M. & Matyas, T. A. (2006). Grip force regulation during pinch grip lifts under somatosensory guidance: comparison between people with stroke and healthy controls. *Arch Phys Med Rehabil*, **87**, 418–429.

Bobjer, O., Johansson, S. E. & Piguet, S. (1993). Friction between hand and handle. Effects of oil and lard on textured and non-textured surfaces; perception of discomfort. *Appl Ergon* **24**, 190–202.

Budgeon, M. K. (2007). *Prehension synergies during finger manipulation*. Unpublished Masters Thesis. The Pennsylvania State University.

Burstedt, M. K., Flanagan, J. R. & Johansson, R. S. (1999). Control of grasp stability in humans under different frictional conditions during multidigit manipulation. *J Neurophysiol*, **82**, 2393–2405.

Cole, K. J. & Abbs, J. H. (1987). Kinematic and electromyographic responses to perturbation of a rapid grasp. *J Neurophysiol*, **57**, 1498–1510.

Cole, K. J. & Abbs, J. H. (1988). Grip force adjustments evoked by load force perturbations of a grasped object. *J Neurophysiol*, **60**, 1513–1522.

Cole, K. J. & Beck, C. L. (1994). The stability of precision grip force in older adults. *J Motor Behav* **26**, 171–177.

Cole, K. J. & Johansson, R. S. (1993). Friction at the digit-object interface scales the sensorimotor transformation for grip responses to pulling loads. *Exp Brain Res*, **95**, 523–532.

Comaish, S. & Bottoms, E. (1971). The skin and friction: deviations from Amonton's laws, and the effects of hydration and lubrication. *Br J Dermatol*, **84**, 37–43.

Danion, F., Schoner, G., Latash, M. L. *et al.* (2003). A mode hypothesis for finger interaction during multi-finger force-production tasks. *Biol Cybern*, **88**, 91–98.

Deutsch, K. M. & Newell, K. M. (2002). Children's coordination of force output in a pinch grip task. *Dev Psychobiol*, **41**, 253–264.

Dumont, C. E., Popovic, M. R., Keller, T. & Sheikh, R. (2006). Dynamic force-sharing in multi-digit task. *Clin Biomech (Bristol, Avon)*, **21**, 138–146.

Flanagan, J. R. & Johansson, R. S. (2002). Hand movements. In V. S. Ramshandran (Ed.), *Encyclopaedia of the Human Brain* (pp. 399–414). San Diego, CA: Academic Press.

Flanagan, J. R. & Tresilian, J. (1994). Grip-load force coupling: a general control strategy for transporting objects. *J Exp Psychol Hum Percept Perform*, **20**, 944–957.

Flanagan, J. R. & Wing, A. M. (1993). Modulation of grip force with load force during point-to-point arm movements. *Exp Brain Res*, **95**, 131–143.

Flanagan, J. R. & Wing, A. M. (1995). The stability of precision grip forces during cyclic arm movements with a hand-held load. *Exp Brain Res*, **105**, 455–464.

Flanagan, J. R., Tresilian, J. & Wing, A. M. (1993). Coupling of grip force and load force during arm movements with grasped objects. *Neurosci Lett*, **152**, 53–56.

Flanagan, J. R., Burstedt, M. K. & Johansson, R. S. (1999). Control of fingertip forces in multidigit manipulation. *J Neurophysiol*, **81**, 1706–1717.

Gao, F., Li, S., Li, Z. M., Latash, M. L. & Zatsiorsky, V. M. (2003). Matrix analyses of interaction among fingers in static force production tasks. *Biol Cybern*, **89**, 407–414.

Gao, F., Latash, M. L. & Zatsiorsky, V. M. (2004). Neural network modeling supports a theory on the hierarchical control of prehension. *Neural Comput Appl*, **13**, 352–359.

Gao, F., Latash, M. L. & Zatsiorsky, V. M. (2005a). Control of finger force direction in the flexion-extension plane. *Exp Brain Res*, **161**, 307–315.

Gao, F., Latash, M. L. & Zatsiorsky, V. M. (2005b). Internal forces during object manipulation. *Exp Brain Res*, **165**, 69–83.

Gao, F., Latash, M. L. & Zatsiorsky, V. M. (2005c). In contrast to robots, in humans internal and manipulation forces are coupled (ThP01-18). In *Proceedings of 2005 9th IEEE International Conference on Rehabilitation Robotics* (pp. 404–407). Chicago, IL, USA.

Gao, F., Latash, M. L. & Zatsiorsky, V. M. (2006). Maintaining rotational equilibrium during object manipulation: linear behavior of a highly non-linear system. *Exp Brain Res*, **169**, 519–531.

Gordon, A. M., Quinn, L., Reilmann, R. & Marder, K. (2000). Coordination of prehensile forces during precision grip in Huntington's disease. *Exp Neurol*, **163**, 136–148.

Gysin, P., Kaminski, T. R. & Gordon, A. M. (2003). Coordination of fingertip forces in object transport during locomotion. *Exp Brain Res*, **149**, 371–379.

Hermsdorfer, J., Marquardt, C., Philipp, J. *et al.* (1999). Grip forces exerted against stationary held objects during gravity changes. *Exp Brain Res*, **126**, 205–214.

Iberall, T. (1987). The nature of human prehension: Three dexterous hands in one. In *Proceedings of 1987 IEEE International Conference on Robotics and Automation* (pp. 396–401). Raleigh, NC, USA.

Johansson, R. S. & Cole, K. J. (1994). Grasp stability during manipulative actions. *Can J Physiol Pharmacol*, **72**, 511–524.

Johansson, R. S. & Westling, G. (1984). Roles of glabrous skin receptors and sensorimotor memory in automatic control of precision grip when lifting rougher or more slippery objects. *Exp Brain Res*, **56**, 550–564.

Jones, L. A. & Hunter, I. W. (1992). Changes in pinch force with bidirectional load forces. *J Motor Behav*, **24**, 157–164.

Kerr, J. R. & Roth, B. (1986). Analysis of multifingered hands. *J Robotics Res*, **4**, 3–17.

Kilbreath, S. L. & Gandevia, S. C. (1994). Limited independent flexion of the thumb and fingers in human subjects. *J Physiol*, **479**, 487–497.

Kilbreath, S. L., Gorman, R. B., Raymond, J. & Gandevia, S. C. (2002). Distribution of the forces produced by motor unit activity in the human flexor digitorum profundus. *J Physiol*, **543**, 289–296.

Kinoshita, H., Kawai, S. & Ikuta, K. (1995). Contributions and co-ordination of individual fingers in multiple finger prehension. *Ergonomics*, **38** (6), 1212–1230.

Kinoshita, H., Kawai, S., Ikuta, K. & Teraoka, T. (1996). Individual finger forces acting on a grasped object during shaking actions. *Ergonomics*, **39** (2), 243–256.

Latash, M. L. & Zatsiorsky, V. M. (2006). Principle of superposition in human prehension. In S. Kawamura & M. Swinin (Eds.), *Advances in Robot Control: From Everyday Physics to Human-Like Movements* (pp. 249–261). New York, NY: Springer.

Latash, M. L., Danion, F., Scholz, J. F., Zatsiorsky, V. M. & Schoner, G. (2003a). Approaches to analysis of handwriting as a task of coordinating a redundant motor system. *Hum Mov Sci*, **22**, 153–171.

Latash, M. L., Gao, F. & Zatsiorsky, V. M. (2003b). Similarities and differences in finger interaction across typical and atypical populations. *J Appl Biomech*, **19**, 264–270.

Li, Z. M., Latash, M. L. & Zatsiorsky, V. M. (1998a). Force sharing among fingers as a model of the redundancy problem. *Exp Brain Res*, **119**, 276–286.

Li, Z. M., Latash, M. L., Newell, K. M. & Zatsiorsky, V. M. (1998b). Motor redundancy during maximal voluntary contraction in four-finger tasks. *Exp Brain Res*, **122**, 71–77.

Li, Z. M., Zatsiorsky, V. M., Latash, M. L. & Bose, N. K. (2002). Anatomically and experimentally based neural networks modeling force coordination in static multi-finger tasks. *Neurocomputing*, **47**, 259–275.

Mason, M. T. & Salisbury, J. K. (1985). *Robot Hands and the Mechanics of Manipulation.* Cambridge, MA: MIT Press.

Monzee, J., Lamarre, Y. & Smith, A. M. (2003). The effects of digital anesthesia on force control using a precision grip. *J Neurophysiol*, **89**, 672–683.

Murray, R. M., Li, Z. & Sastry, S. S. (1994). *A Mathematical Introduction to Robotic Manipulation.* Boca Raton, FL: CRC Press.

Nakazawa, N., Uekita, Y., Inooka, H. & Ikeura, R. (1996). Experimental study on human grasping force. In *IEEE International Workshop on Robot and Human Communication* (pp. 280–285). Tsukuba, Japan.

Niu, X., Latash, M. L. & Zatsiorsky, V. M. (2007). Prehension synergies in the grasps with complex friction patterns: local vs. synergic effects and the template control. *J Neurophysiol*, **98**, 16–28.

Nowak, D. A. & Hermsdorfer, J. (2002). Coordination of grip and load forces during vertical point-to-point movements with a grasped object in Parkinson's disease. *Behav Neurosci*, **116**, 837–850.

Nowak, D. A. & Hermsdorfer, J. (2003a). Digit cooling influences grasp efficiency during manipulative tasks. *Eur J Appl Physiol*, **89**, 127–133.

Nowak, D. A. & Hermsdorfer, J. (2003b). Sensorimotor memory and grip force control: does grip force anticipate a self-produced weight change when drinking with a straw from a cup? *Eur J Neurosci*, **18**, 2883–2892.

Nowak, D. A. & Hermsdorfer, J. (2006). Predictive and reactive control of grasping forces: on the role of the basal ganglia and sensory feedback. *Exp Brain Res*, **173**, 650–660.

Ohtsuki, T. (1981). Inhibition of individual fingers during grip strength exertion. *Ergonomics*, **24**, 21–36.

Pataky, T. C., Latash, M. L. & Zatsiorsky, V. M. (2004a). Tangential load sharing among fingers during prehension. *Ergonomics*, **47**, 876–889.

Pataky, T. C., Latash, M. L. & Zatsiorsky, V. M. (2004b). Prehension synergies during nonvertical grasping, I: experimental observations. *Biol Cybern*, **91**, 148–158.

Pataky, T. C., Latash, M. L. & Zatsiorsky, V. M. (2004c). Prehension synergies during nonvertical grasping, II: Modeling and optimization. *Biol Cybern*, **91**, 231–242.

Pataky, T. C., Latash, M. L. & Zatsiorsky, V. M. (2007). Finger interaction during maximal radial and ulnar deviation efforts: experimental data and linear neural network modeling. *Exp Brain Res*, **179**, 301–312.

Pylatiuk, C., Kargov, A., Schulz, S. & Doderlein, L. (2006). Distribution of grip force in three different functional prehension patterns. *J Med Eng Technol*, **30**, 176–182.

Rearick, M. P. & Santello, M. (2002). Force synergies for multifingered grasping: effect of predictability in object center of mass and handedness. *Exp Brain Res*, **144**, 38–49.

Rearick, M. P., Casares, A. & Santello, M. (2003). Task-dependent modulation of multi-digit force coordination patterns. *J Neurophysiol*, **89**, 1317–1326.

Santello, M. & Soechting, J. F. (2000). Force synergies for multifingered grasping. *Exp Brain Res*, **133**, 457–467.

Savescu, A., Latash, M. L. & Zatsiorsky, V. M. (2007). A method to determine friction at the finger tips. *J Appl Biomech*, **24**, 43–50.

Schieber, M. H. & Santello, M. (2004). Hand function: peripheral and central constraints on performance. *J Appl Physiol*, **96**, 2293–2300.

Shim, J. K. & Park, J. (2007). Prehension synergies: principle of superposition and hierarchical organization in circular object prehension. *Exp Brain Res*, **180**, 541–556.

Shim, J. K., Latash, M. L. & Zatsiorsky, V. M. (2003). Prehension synergies: trial-to-trial variability and hierarchical organization of stable performance. *Exp Brain Res*, **152**, 173–184.

Shim, J. K., Latash, M. L. & Zatsiorsky, V. M. (2005). Prehension synergies in three dimensions. *J Neurophysiol*, **93**, 766–776.

Shim, J. K., Huang, J., Hooke, A. W., Latash, M. L. & Zatsiorsky, V. M. (2007). Multi-digit maximum voluntary torque production on a circular object. *Ergonomics*, **50**, 660–675.

Smith, M. A. & Soechting, J. F. (2005). Modulation of grasping forces during object transport. *J Neurophysiol*, **93**, 137–145.

Vaillancourt, D. E., Slifkin, A. B. & Newell, K. M. (2002). Inter-digit individuation and force variability in the precision grip of young, elderly, and Parkinson's disease participants. *Motor Control*, **6**, 113–128.

Westling, G. & Johansson, R. S. (1984). Factors influencing the force control during precision grip. *Exp Brain Res*, **53**, 277–284.

Yoshikawa, T. & Nagai, K. (1991). Manipulating and grasping forces in manipulation by multifingered robot hands. *IEEE Trans Robotics Autom*, **7**, 67–77.

Zatsiorsky, V. M. (2002). *Kinetics of Human Motion*. Champaign, IL: Human Kinetics.

Zatsiorsky, V. M. & Latash, M. L. (2004). Prehension synergies. *Exerc Sport Sci Rev*, **32**, 75–80.

Zatsiorsky, V. M. & Latash, M. L. (2007). Multi-finger prehension: an overview. *J Motor Behav*, **40**, 446–475.

Zatsiorsky, V. M., Li, Z. M. & Latash, M. L. (1998). Coordinated force production in multi-finger tasks: finger interaction and neural network modeling. *Biol Cybern*, **79** (2), 139–150.

Zatsiorsky, V. M., Li, Z. M. & Latash, M. L. (2000). Enslaving effects in multi-finger force production. *Exp Brain Res*, **131**, 187–195.

Zatsiorsky, V. M., Gregory, R. W. & Latash, M. L. (2002a). Force and torque production in static multifinger prehension: biomechanics and control. I. Biomechanics. *Biol Cybern*, **87**, 50–57.

Zatsiorsky, V. M., Gregory, R. W. & Latash, M. L. (2002b). Force and torque production in static multifinger prehension: biomechanics and control. II. Control. *Biol Cybern*, **87**, 40–49.

Zatsiorsky, V. M., Gao, F. & Latash, M. L. (2003a). Finger force vectors in multi-finger prehension. *J Biomech*, **36**, 1745–1749.

Zatsiorsky, V. M., Gao, F. & Latash, M. L. (2003b). Prehension synergies: effects of object geometry and prescribed torques. *Exp Brain Res*, **148**, 77–87.

Zatsiorsky, V. M., Latash, M. L., Gao, F. & Shim, J. K. (2004). The principle of superposition in human prehension. *Robotica*, **22**, 231–234.

Zatsiorsky, V. M., Gao, F. & Latash, M. L. (2005a). Motor control goes beyond physics: differential effects of gravity and inertia on finger forces during manipulation of hand-held objects. *Exp Brain Res*, **162**, 300–308.

Zatsiorsky, V. M., Gao, F. & Latash, M. L. (2005b). Prehension stability: experiments with expanding and contracting handle. *J Neurophysiol*, **95**, 2513–2529.

Zuo, B. R. & Qian, W. H. (2000). A general dynamic force distribution algorithm for multifingered grasping. *IEEE Trans Syst Man Cybernetics Part B – Cybernetics*, **30**, 185–192.

4

Recordings from the motor cortex during skilled grasping

THOMAS BROCHIER AND ROGER N. LEMON

Summary

This chapter offers an overview of the most recent techniques for recording of cortical activity in the awake, behaving monkey. We review the different types of signals that can be extracted from extracellular cortical recordings made with microelectrodes. We also discuss how these signals can be related to dexterous hand movements. This leads us to consider the functional organization of the motor cortex for the control of the distal muscles during grasp.

Introduction

The unique ability of human and non-human primates to interact with their environment is dependent upon the skilled use of the hands for grasping and manipulation of objects. The grasping of objects requires continuous interaction between the sensory processing of the object's physical properties and the motor mechanisms controlling the shape of the hand and the positioning of the hand and digits upon the object. Over the past 30 years, intracortical extracellular recording techniques in the awake monkey have been an essential tool to investigate the organization of the cortical circuits involved in the control of grasp. It has been shown that multiple areas in the parietal and frontal lobes contribute to the transformation from sensory inputs to motor outputs for efficient grasp. This cortical network influences the spinal circuitry that controls the distal hand and digit muscles. Part of this corticospinal control is mediated by direct cortico-motoneuronal (CM) projections from the primary motor cortex (M1) onto motoneurons innervating hand muscles. In this chapter, we will describe the different technologies available for the recording of cortical activity in the monkey performing hand movements. We will present the various types of signals that can be extracted from extracellular cortical recordings and how they relate to the monkey's motor behavior.

Methods for recording of cortical activity in the awake monkey

Intracortical recording is a powerful tool to investigate the functional properties of the cortical motor network. The basic principle is to record extracellularly the electrical activity

Sensorimotor Control of Grasping: Physiology and Pathophysiology, ed. Dennis A. Nowak and Joachim Hermsdörfer Published by Cambridge University Press. © Cambridge University Press 2009.

generated by the neurons in the vicinity of the tip of a metal microelectrode advanced into the brain of the awake animal. By using microelectrodes with a very small tip size (less than 10 μm) and a high impedance (> 1 MΩ at 1 kHz), the spiking activity of a single neuron can be isolated from other neighboring activity. Alternatively, electrodes with a larger exposed tip and lower impedance (< 0.5 MΩ) are more appropriate for the recording of local field potentials (LFPs) which represent the net dendritic synaptic activity of a large number of neurons in the vicinity of the electrode tip (Mitzdorf, 1985). A number of different single- and multiple-electrode techniques are now available.

The single-electrode recording technique (Lemon, 1984) has been widely used to analyze the activity of the motor areas in the monkey performing different types of grasping task. With this technique, signals from single neurons are collected sequentially in different recording sessions over a period of weeks or months. A single electrode is advanced carefully through the exposed dura mater above a specific cortical target. The electrode is connected to a hydraulic or similar type of microdrive which is then used to move the electrode in small steps through the cortical tissue. The cortical signals that are recorded are analyzed in relation with the monkey's behavior to infer the functional role of physiologically or anatomically defined areas of the brain. However, these single-electrode techniques are unsuitable to analyze in real time the complex interactions taking place between groups of neurons in a given area or between different areas of the motor network.

To make it possible to study these neuronal interactions, multi-electrode recording drives have been designed to control independently a number of electrodes so that the activity of small populations of neurons can be recorded simultaneously from the cortex of the awake animal. Several systems with rather different mechanical and electronic characteristics are commercially available for recording with up to 16 electrodes (e.g. Thomas Recording® Eckhorn system, Alpha-Omega Engineering® MT, Plexon® NAN). In a typical recording session, the electrode tips are brought to the surface of the exposed dura mater in a protective guide tube and the electrodes are then lowered one by one to penetrate the dura and reach their final position in the cortex. Scarring of the exposed dura can make transural penetration with fine microelectrodes difficult, but this can be overcome by treating the dura with anti-mitotic compounds (Spinks *et al.*, 2003). For the optimal yield of experimental data from these systems, the signal from each electrode needs to be carefully monitored for the stable recording of single neuron activity or LFPs during execution of the behavioral tasks. Since continuous quality control of signals from many different electrodes cannot be performed by a single experimenter, many investigators save the analog signals recorded (rather than on-line discriminated events) in order to allow accurate off-line discrimination and analysis. In our view, a small team of experimenters is preferable to ensure that the yield of data from each recording session far exceeds the yield that would be obtained using standard single electrode techniques (see Baker *et al.*, 1999a). These constraints set a limit on the maximum number of electrodes that can be successfully used using the multi-electrode drives. Another constraint is that the electrodes have to be removed from the cortex at the end of every recording session. This prevents the recording of the same population of neurons over long periods of time which would be required for the analysis of the cortical phenomena taking place on a longer time scale such as learning or cortical plasticity. Against this must be set two further advantages of this approach. First, it allows the investigator to search for particular types or classes of neurons, which may be identified by their synaptic or antidromic responses (see Lemon, 1984; Baker *et al.*, 1999a). Second, the degree of damage to the cortex is probably much less than when the microelectrodes are permanently implanted (see below).

An alternative approach is to use chronically implanted electrodes for the recording of stable populations of neurons over days or weeks. Different types of implants have been designed that can be used for different purposes. Multi-contact silicon probes can be used for multi-site recording in the deep layers of the cortex or in the deep part of the cortical sulci (e.g. NeuroNexus®). Small size arrays of up to 100 electrodes are also available for high-density recordings in limited cortical areas (Plexon Plextrodes®, Cyberkinetics® micro-electrode arrays). To implant the chronic electrodes or arrays in the cortex usually requires significant surgical expertise. This involves incision of the dura mater, penetration of the electrodes in the cortex with the lowest possible compression of the tissue and wiring of the ensemble towards a connector attached to the skull. Within the first week of implantation, more than 50% of the chronic electrodes can record from at least one isolated neuron (Nicolelis *et al.*, 2003). However, due to cell death and gliosis around the implant, the quality of the recording drops gradually over longer periods of time. Future development will aim to obtain a more stable quality of recording by using a chronic implant with independently moveable electrodes (Jackson & Fetz, 2007), which importantly allow for searching for different cell types (see above). Also, new developments hold the promise for the possible recording of cortical activity in the freely moving animal using wire-free telemetry technology to transmit the data from the chronic implant to a remote receiver (Jackson *et al.*, 2006). These techniques are already being used for the development of cortical neuroprosthetic devices in paralyzed human patients (Trucculo *et al.*, 2008). Exciting new developments may use "organic electronic" devices to record long-term activity (Berggren *et al.*, 2007).

Useful signals for decoding cortical activity during grasp

In the primate, CM neurons of M1 play an essential role in the coordination of muscle activity during precision grip (Muir & Lemon, 1983). However the level of activity of a given muscle is not directly represented in the activity of CM neurons (Bennett & Lemon, 1994). This observation suggests that a more complex relationship takes place between different levels of movement representation within M1 and the control of hand muscles for grasping (Brochier & Umiltà, 2007). The spiking activity of single neurons (SUA) and other signals such as multi-neuron spiking activity (MUA) and local field potentials (LFPs) recorded from the same cortical electrodes have been related to the performance of hand movements (Baker *et al.*, 1999b; Asher *et al.*, 2007; Stark & Abeles, 2007). In addition, it has been shown that the information content of spiking activity depends not only on the discharge frequency of the neurons (rate coding), but also on the temporal coordination among groups of neurons (temporal coding) or on the synchronization between SUA and LFPs (Riehle *et al.*, 1997, 2000; Spinks *et al.*, 2008). The following section briefly describes how the modulations of SUA (rate and temporal coding), MUA and LFPs have been used to investigate the functional properties of the motor network for the control of grasp.

The analysis of SUA provided some essential insight into the cortical control of complex hand movements. In a pioneering study, Smith *et al.* (1975) analyzed the activity of primary motor cortex (M1) neurons for the control of different levels of force in an isometric precision-grip task. They observed that within the finger representation of M1, neurons modulate their activity in relation with increasing or decreasing force levels in close temporal relationship with the activity of hand and finger muscles. Subsequent studies (Buys *et al.*, 1986; Lemon *et al.*, 1986; Maier *et al.*, 1993; Bennett & Lemon, 1994) have detailed the special contribution of the CM neurons of M1 to the control of distal muscles for the production of force. One important finding is that the activity in CM cells discloses a variable relationship with the activity of its target muscles during distinct phases of the grasp. These observations suggest that the production of relatively independent finger movements originates from a distributed motor command that is differently weighted for distinct motoneuron pools. This descending command would be modulated through complex interactions between the corticospinal and spinal circuits (Bennett & Lemon, 1994). Other studies have compared the contribution of distinct cortical areas to the control of force during precision grip (Wannier *et al.*, 1991; Hepp-Reymond *et al.*, 1994; Cadoret & Smith, 1997; Salimi *et al.*, 1999; Boudreau *et al.*, 2001). Neurons coding for force have been observed in the primary somatosensory and motor cortex as well as in all the secondary motor areas (cingulate motor cortex, supplementary motor area, premotor cortex) but with specific functional and temporal relationships to the task for each of these areas. It is therefore suggested that in addition to the direct control exerted by M1 over the spinal circuitry, a complex network of interconnected areas is involved in the control of force during grasp. These effects are probably mediated both by the corticospinal projections from each of these areas, as well as through their reciprocal connections with M1 and its corticospinal output.

Spiking activity of single neurons has also been extensively used to study the control of hand pre-shaping during grasp which is another important aspect of hand movements (Murata *et al.*, 1997; Raos *et al.*, 2004, 2006; Gardner *et al.*, 2007a, b; Umiltà *et al.*, 2007). There is clear selectivity in the discharge of both motor and premotor cortex neurons for particular types of grasp, and this is recognizable both at the level of single neurons (compare the firing pattern of the two neurons during the three different grasps shown in Figure 4.1D) and at the population level (Umiltà *et al.*, 2007). Again, it has been observed that a large network of cortical areas including the anterior part of the parietal cortex (area AIP) and the ventral part of the premotor cortex (area F5) contribute to the transformation of the visual information about an object into a descending command controlling the coordinated activation of hand and finger muscles for hand shaping (Brochier & Umiltà, 2007).

In contrast to the important number of studies looking at SUA, few attempts have been made to analyze the modulation of LFPs during grasping movements. Murthy & Fetz (1996)

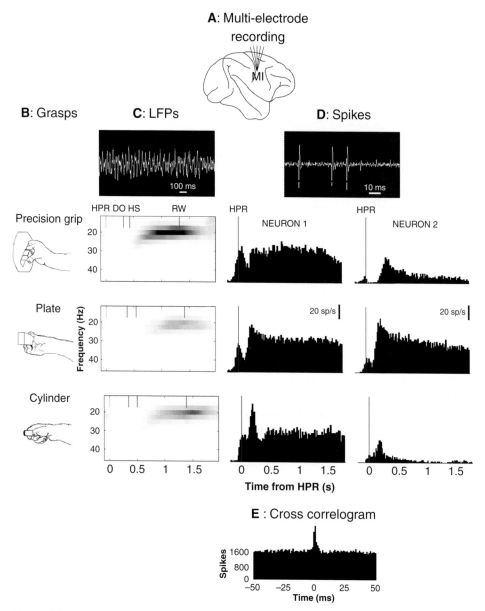

Figure 4.1. Intracortical recordings from the primary motor cortex (M1). A: Recordings are obtained from independently controlled electrodes positioned in the deep layers of the MI hand area in the monkey performing a reach-to-grasp task. This involved releasing a starting home pad to reach out, grasp and pull an object into a position window and hold it for 1 second to get a food reward. B: Typical hand postures used by the monkey for grasping of three differently shaped objects (precision grip, plate and cylinder). C: LFP data. The top panel shows a sample of raw data filtered for LFP recordings (10–250 Hz). Bottom panels show the spectrograms averaged from 50 trials of each of the three different grasps aligned on the movement onset (HPR). DO, HS and RW indicate the mean time of the object displacement onset, the

observed that episodes of LFP oscillations around 25 Hz occur when a monkey is retrieving raisins from a Klüver Board in a free-moving paradigm. Baker *et al.* (1997) confirmed that when the monkey is performing a precision grip task, oscillations in the beta frequency range (15–35 Hz) show consistent task-dependent modulations and are coherent with oscillations in the EMG recorded from distal muscles. These oscillations are mainly present during the hold period of the task when the monkey is maintaining a constant level of force against a lever or during steady grasp of an object (Figure 4.1B and 4.1C). These findings, along with the observation that stimulation of the pyramidal tract evokes a phase-locked resetting in the cortical LFP (Jackson *et al.*, 2002), provide striking evidence of a functional link between beta range oscillations and corticospinal function. Asher *et al.* (2007) reported that LFP recorded from different subdivisions of the parietal cortex show task-dependent modulations in a reach-to-grasp task but across a wider range of frequencies. Interestingly in all the frequency bands, a large proportion of the recorded LFPs were tuned for a given type of grasp (Asher *et al.*, 2007). Similar tuning of the LFPs during grasping of differently shaped objects has been observed from recordings in the primary motor cortex and area F5 (Spinks *et al.*, 2005; see Figure 4.1C). These large-scale oscillations in the motor network would be required to maintain the motor set appropriate for each category of grasp used by the monkey. Since LFP signals are easy to record and show more stability than SUA, they represent an interesting source of signals that can be used for the decoding of brain activity during grasp. However, LFPs recorded from neighboring electrodes are often correlated and cannot be used as non-redundant sources of information about ongoing brain activity. Stark & Abeles (2007) suggest that as an alternative to LFPs, an estimate of multi-unit activity (MUA) from the motor cortex offers a less redundant signal. Multi-unit activity reflects the superimposed activity of a small population of neurons around the tip of the electrode following a 300–6000 Hz band-pass filtering of the raw data signal. When compared with SUA or LFPs, MUA are very stable over time and provide a reliable source of information for the decoding of cortical activity during reach and grasp movements (Stark & Abeles, 2007).

Another important aspect is the role of synchronous firing between single neurons for the processing of information in the motor systems. It has been proposed that the flow of information in a neuronal network is facilitated by the combined activation of neuron assemblies that will produce synchronous EPSPs on common post-synaptic neurons (Vaadia *et al.*, 1995). Such synchronous activity has been observed between pairs of neurons

---

Figure 4.1. (cont.)
start of the hold period and the completion of a successful trial, respectively. A peak in the power (mostly in the beta range of frequency 15–30 Hz) occurs during the hold period of the task. For this particular cortical site, the LFP was maximal for the precision grip. D: SUA data. The top panel shows a sample of raw data filtered for SUA recordings (1–10 kHz). Bottom panels show the task-related activity in two simultaneously recorded neurons recorded from two different electrodes in M1. The peri-event time histograms are aligned on HPR (60 trials per grasp for each unit; bin size 20 ms). E: Cross-correlation histogram of cell activity for the two neurons in D (bin size, 1 ms). A large peak in the histogram is observed at lag 0, indicating some synchronization in the spiking activity of the two neurons.

of the motor cortex in the monkey performing a precision grip task (Baker *et al.*, 1999b, 2001) or during grasping of differently shaped objects (see Figure 4.1E). The changes in the level of synchronous firing during different periods of the task demonstrate the functional relevance of synchrony for the control of hand movements by the primary motor cortex. One hypothesis is that synchronous firing of cortical neurons would be an efficient way of driving the activity of the motoneurons during the static part of an action such as the hold period of a precision grip (Kilner *et al.*, 2002). In line with this assumption, it has been reported that M1 CM neurons with closely related muscle fields exhibit greater levels of spike synchronization than CM neurons with non-overlapping muscle fields (McKiernan *et al.*, 2000; Jackson *et al.*, 2003). To further investigate the functional relevance of synchronization among assemblies of MI neurons sharing the same muscle targets, future research will be needed to compare the modulation of synchrony during grasping movements towards differently shaped objects.

Conclusions

The recording of extracellular activity in the awake monkey is an essential tool to investigate the functional organization of the motor network for the control of hand and finger movements. Nowadays, a large range of recording techniques is available to the experimenter to collect cortical neuronal activity during movement preparation and execution. This includes single- and multi-electrode recording techniques using independently controlled electrodes or chronically implanted wires and arrays. Different signals such as SUA, MUA and LFPs with different information contents can be extracted from the cortical recordings. These signals are providing new insights into the the complexity of the central mechanisms involved in the control of grasping and manipulation of objects.

References

Asher, I., Stark, E., Abeles, M. & Prut, Y. (2007). Comparison of direction and object selectivity of local field potentials and single units in macaque posterior parietal cortex during prehension. *J Neurophysiol*, **97**, 3684–3695.

Baker, S. N., Olivier, E. & Lemon, R. N. (1997). Coherent oscillations in monkey motor cortex and hand muscle EMG show task-dependent modulation. *J Physiol*, **501**, 225–241.

Baker, S. N., Philbin, N., Spinks, R. *et al.* (1999a). Multiple single unit recording in the cortex of monkeys using independently moveable microelectrodes. *J Neurosci Methods*, **94**, 5–17.

Baker, S. N., Kilner, J. M., Pinches, E. M. & Lemon, R. N. (1999b). The role of synchrony and oscillation in the motor output. *Exp Brain Res*, **128**, 109–117.

Baker, S. N., Spinks, R., Jackson, A. & Lemon, R. N. (2001). Synchronization in monkey motor cortex during a precision grip task. I. Task-dependent modulation in single-unit synchrony. *J Neurophysiol*, **85**, 869–85.

Bennett, K. M. & Lemon, R. N. (1994). The influence of single monkey cortico-motoneuronal cells at different levels of activity in target muscles. *J Physiol*, **477**, 291–307.

Berggren, M., Nilsson, D. & Robinson, N. D. (2007). Organic materials for printed electronics. *Nat Mater*, **6**, 3–5.

Boudreau, M. J., Brochier, T., Paré, M. & Smith, A. M. (2001). Activity in ventral and dorsal premotor cortex in response to predictable force-pulse perturbations in a precision grip task. *J Neurophysiol*, **86**, 1067–78.

Brochier, T. & Umiltà, M. A. (2007). Cortical control of grasp in non-human primates. *Curr Opin Neurobiol*, **17**, 637–643.

Buys, E. J., Lemon, R. N., Mantel, G. W. & Muir, R. B. (1986). Selective facilitation of different hand muscles by single corticospinal neurones in the conscious monkey. *J Physiol*, **381**, 529–49.

Cadoret, G. & Smith, A. M. (1997). Comparison of the neuronal activity in the SMA and the ventral cingulate cortex during prehension in the monkey. *J Neurophysiol*, **77**, 153–66.

Gardner, E. P., Ro, J. Y., Babu, K. S. & Ghosh, S. (2007a). Neurophysiology of prehension. II. Response diversity in primary somatosensory (S-I) and motor (M-I) cortices. *J Neurophysiol*, **97**, 1656–70.

Gardner, E. P., Babu, K. S., Reitzen, S. D. *et al.* (2007b). Neurophysiology of prehension. I. Posterior parietal cortex and object-oriented hand behaviors. *J Neurophysiol*, **97**, 387–406.

Hepp-Reymond, M. C., Hüsler, E. J., Maier, M. A. & Ql, H. X. (1994). Force-related neuronal activity in two regions of the primate ventral premotor cortex. *Can J Physiol Pharmacol*, **72**, 571–579.

Jackson, A. & Fetz, E. E. (2007). Compact movable microwire array for long-term chronic unit recording in cerebral cortex of primates. *J Neurophysiol*, **98**, 3109–3118.

Jackson, A., Spinks, R. L., Freeman, T. C. B., Wolpert, D. M. & Lemon, R. N. (2002). Rhythm generation in monkey motor cortex explored using pyramidal tract stimulation. *J Physiol*, **541**, 685–699.

Jackson, A., Gee, V. J., Baker, S. N. & Lemon, R. N. (2003). Synchrony between neurons with similar muscle fields in monkey motor cortex. *Neuron*, **38**, 115–125.

Jackson, A., Moritz, C. T., Mavoori, J., Lucas, T. H. & Fetz, E. E. (2006). The Neurochip BCI: towards a neural prosthesis for upper limb function. *IEEE Trans Neural Syst Rehabil Eng*, **14**, 187–190.

Kilner, J. M., Alonso-Alonso, M., Fisher, R. & Lemon, R. N. (2002). Modulation of synchrony between single motor units during precision grip tasks in humans. *J Physiol*, **541**, 937–948.

Lemon, R. N. (1984). Methods for neuronal recording in conscious animals. In *IBRO Handbook Series: Methods in Neurosciences*. Chichester, UK: John Wiley.

Lemon, R. N., Mantel, G. W. & Muir, R. B. (1986). Corticospinal facilitation of hand muscles during voluntary movement in the conscious monkey. *J Physiol*, **381**, 497–527.

Maier, M. A., Bennett, K. M., Hepp-Reymond, M. C. & Lemon, R. N. (1993). Contribution of the monkey corticomotoneuronal system to the control of force in precision grip. *J Neurophysiol*, **69**, 772–785.

McKiernan, B. J., Marcario, J. K., Karrer, J. H. & Cheney, P. D. (2000). Correlations between corticomotoneuronal (CM) cell postspike effects and cell-target muscle covariation. *J Neurophysiol*, **83**, 99–115.

Mitzdorf, U. (1985). Current source-density method and application in cat cerebral cortex: investigation of evoked potentials and EEG phenomena. *Physiol Rev*, **65**, 37–100.

Muir, R. B. & Lemon, R. N. (1983). Corticospinal neurons with a special role in precision grip. *Brain Res*, **261**, 312–316.

Murata, A., Fadiga, L., Fogassi, L. *et al.* (1997). Object representation in the ventral premotor cortex (area F5) of the monkey. *J Neurophysiol*, **78**, 2226–2230.

Murthy, V. N. & Fetz, E. E. (1996). Oscillatory activity in sensorimotor cortex of awake monkeys: synchronization of local field potentials and relation to behavior. *J Neurophysiol*, **76**, 3949–3967.

Nicolelis, M. A., Dimitrov, D., Carmena, J. M. *et al.* (2003). Chronic, multisite, multielectrode recordings in macaque monkeys. *Proc Natl Acad Sci USA*, **100**, 11041–11046.

Raos, V., Umiltá, M. A., Gallese, V. & Fogassi, L. (2004). Functional properties of grasping-related neurons in the dorsal premotor area F2 of the macaque monkey. *J Neurophysiol*, **92**, 1990–2002.

Raos, V., Umiltá, M. A., Murata, A., Fogassi, L. & Gallese, V. (2006). Functional properties of grasping-related neurons in the ventral premotor area F5 of the macaque monkey. *J Neurophysiol*, **95**, 709–729.

Riehle, A., Grün, S., Diesmann, M. & Aertsen, A. (1997). Spike synchronization and rate modulation differentially involved in motor cortical function. *Science*, **278**, 1950–1953.

Riehle, A., Grammont, F., Diesmann, M. & Grün, S. (2000). Dynamical changes and temporal precision of synchronized spiking activity in monkey motor cortex during movement preparation. *J Physiol Paris*, **94**, 569–582.

Salimi, I., Brochier, T. & Smith, A. M. (1999). Neuronal activity in somatosensory cortex of monkeys using a precision grip. I. Receptive fields and discharge patterns. *J Neurophysiol*, **81**, 825–834.

Smith, A. M., Hepp-Reymond, M. C. & Wyss, U. R. (1975). Relation of activity in precentral cortical neurons to force and rate of force change during isometric contractions of finger muscles. *Exp Brain Res*, **23**, 315–332.

Spinks, R. L., Baker, S. N., Jackson, A., Khaw, P. T. & Lemon, R. N. (2003). The problem of dural scarring in recording from awake behaving monkeys: a solution using 5-flurouracil. *J Neurophysiol*, **90**, 1324–1332.

Spinks, R. L., Kraskov, A., Brochier, T., Umilta, M. A. & Lemon, R. N. (2008). Selectivity for grasp in local field potential and single neuron activity recorded simultaneously from M1 and F5 in the awake macaque monkey. *J. Neurosci*, **28**, 10961–10971.

Stark, E. & Abeles, M. (2007). Predicting movement from multiunit activity. *J Neurosci*, **27**, 8387–8394.

Truccolo, W., Friehs, G. M., Donoghue, J. P. & Hochberg, L. R. (2008). Primary motor cortex tuning to intended movement kinematics in humans with tetraplegia. *J Neurosci*, **28**, 1163–1178.

Umiltà, M. A., Brochier, T., Spinks, R. L. & Lemon, R. N. (2007). Simultaneous recording of macaque premotor and primary motor cortex neuronal populations reveals different functional contributions to visuomotor grasp. *J Neurophysiol*, **98**, 488–501.

Vaadia, E., Haalman, I., Abeles, M. *et al.* (1995). Dynamics of neuronal interactions in monkey cortex in relation to behavioural events. *Nature*, **373**, 515–518.

Wannier, T. M., Maier, M. A. & Hepp-Reymond, M. C. (1991). Contrasting properties of monkey somatosensory and motor cortex neurons activated during the control of force in precision grip. *J Neurophysiol*, **65**, 572–589.

5

Recording of electromyogram activity in the monkey during skilled grasping

THOMAS BROCHIER, RACHEL L. SPINKS, MARIA A. UMILTA
AND ROGER N. LEMON

Summary

This chapter provides a brief presentation of the available techniques for electromyogram (EMG) recordings in the awake monkey using chronically implanted electrodes. We illustrate how this technique can be used for the analysis of the monkey's motor behavior during dexterous grasp. We also investigate how the grasp specificity of EMG activity can be related to the activity of a population of pyramidal tract neurons (PTNs) recorded from the hand area of the primary motor cortex (M1).

Introduction

The ability to grasp and manipulate objects of various sizes and shapes is essential for a large range of human activities. The debilitating loss of skilled hand movements following stroke, spinal injury and many other pathological disorders results in a marked loss of autonomy for the affected patient. The characteristic structure of the human hand provides this organ with a unique combination of motor and sensory capacities that underpin the control of manual dexterity. The anatomy of the hand includes some 27 different bones, and some 39 different muscles located either in the forearm (extrinsic muscles) or in the hand itself (intrinsic muscles; Tubiana, 1981). Special features of bony structures in the hand contribute directly to dexterity, and are important for rotation of the human thumb during precision grip (Tallis, 2004). The muscular control of the multi-articulate hand presents some demanding biomechanical solutions. Thus, the hand and digit segments (i.e. the phalanges) are controlled by pairs of antagonist muscles acting at each joint and also by a combination of muscles whose tendons cross multiple joints. The coordinated activation of this muscular complex supports the flexibility of hand movements (see Schieber, 1995). From a sensory viewpoint, the high concentration of sensory nerve endings in the muscles, joints and skin of the human hand provides a rich source of feedback for the control of movement.

The increased dexterity of the hand in human and non-human primates has also been linked with pronounced changes of the neural control system operating at cortical and sub-cortical levels (Kuypers, 1978; Porter & Lemon, 1993). These changes include the development of a specific cortical network for the control of distal movements (Jeannerod et al.,

Sensorimotor Control of Grasping: Physiology and Pathophysiology, ed. Dennis A. Nowak and Joachim Hermsdörfer. Published by Cambridge University Press. © Cambridge University Press 2009.

1995). A number of structures in this cortical network project to the spinal cord through the corticospinal tract (CST; Dum & Strick, 1991, 2002). In the most dextrous primate species, a monosynaptic fast-conducting projection from motor cortex to spinal motoneurons is present (Nakajima *et al.*, 2000; Lemon & Griffiths, 2005). In the macaque monkey, this cortico-motoneuronal (CM) component is relatively small in terms of numbers of axons, but is regarded as an essential component for the skilled control of distal movements (Bennett & Lemon, 1994, 1996). This species is therefore one of the best available models to understand how the motor system is organized for the control of muscle activity during hand movement in humans. Of course, EMG activity represents only one aspect of the biomechanics of hand function, and needs to be complemented by studies of hand and digit kinematics and dynamics.

The techniques used for EMG recordings in the awake monkey are guided by two main requirements. To fully capture the mechanisms involved in the coordination of distal muscle activity, recordings should be obtained simultaneously from as many muscles as possible. In addition, it is desirable to obtain stable EMG recordings over long periods of time (weeks or months). This latter point is critical if one seeks to establish the relationship between the activity of distal muscles and pattern of discharge among a population of neurons at the cortical level. In this chapter, we describe a study of EMG activity recorded with the method originally described by Miller *et al.* (1993). We show how a representative sample of 10–12 muscles can accurately describe the diversity of hand movements performed by the macaque monkey in a reach-to-grasp task. We also report how these EMG signals can be analyzed in relation with the activity of a population of neurons recorded from the primary motor cortex.

Methods for chronic EMG recordings in the awake monkey

Several techniques are available for chronic EMG recordings from relatively large numbers of muscles over long periods of time. Special care has always been taken to minimize the surgical stress for the animal and to avoid infection-related rejections of the implant. We obtained EMG data from up to 12 forelimb muscles implanted following the methods described by Miller *et al.* (1993). All procedures were carried out in accordance with the UK Home Office regulations. All surgeries were performed under deep general anesthesia, induced with 10 mg/kg ketamine (im), and maintained with 2–2.5% isoflurane in 50:50 O_2–N_2O.

In this approach an EMG electrode assembly was prepared presurgically and carefully sterilized before implantation. This assembly consisted of a set of small silastic patch electrodes (1×1 cm) attached by way of 25–50 cm long multistrand, stainless steel wires, insulated with Teflon and soldered to a 25-way female miniature D-connector supported in a Dacron patch (see Miller *et al.*, 1993). The electrode contact is an exposed 2–3 mm section of the wire firmly supported onto the silastic patch. In the deeply anesthetized monkey, the patch electrodes were implanted on to the exposed surfaces of 12 shoulder, arm, hand and digit muscles. Each electrode was subcutaneously routed from a skin incision in the back of the animal to a small incision above its target muscle. Electrical stimulation was used to test for correct placement of the patch before it was sutured with 4–0 silk or PDS sutures onto the surface of the muscle. Wounds were closed with steel sutures (which were removed after the first week or so). The final stage of the surgery involved suturing the Dacron patch to the back of the monkey at the mid-scapula level, which prevented the monkey from interfering with the implant, but gave easy access for connection to external EMG amplifiers during recording sessions (Figure 5.1). This was done through a male 25-way connector and lightweight connecting cable 2 m in length.

The implanted muscles were: first dorsal interosseous (1DI), abductor pollicis longus (AbPL), thenar (Th; this probably combines recordings from the short abductor, flexor and adductor of the thumb), abductor digiti minimi (AbDM), palmaris longus (PL), flexor digitorum sublimis (FDS), flexor digitorum profundis (FDP), extensor digitorum communis (EDC), extensor digitorum 4,5 (ED45), flexor carpi ulnaris (FCU), brachioradialis (BR) and anterior deltoid (AD). Using this technique, we obtained stable EMG recordings from 10–12 muscles for at least 6 months. Infection of the back connector was prevented by regular inspection of the Dacron patch and by local antibiotic treatment when necessary. The correct location of the patch electrodes was confirmed by both stimulation through the back connector and observing the evoked movement, and by post-mortem dissection. Using the same technique, Morrow & Miller (2003) reported stable recordings of EMG activity from 23 proximal and distal upper limb muscles.

Closely related techniques are routinely used in other laboratories for chronic EMG implants. In some studies, the EMG electrodes are made from pairs of isolated steel wires with the bare tip transcutaneously inserted into various arm, wrist and hand muscles (Belhaj-Saïf *et al.*, 1998; McKiernan *et al.*, 1998; Jackson *et al.*, 2006b, 2007). The leads were fixed to the skin with a drop of cyano-acrylate glue (Jackson *et al.*, 2007) or with elastomeric tape (Belhaj-Saïf *et al.*, 1998; McKiernan *et al.*, 1998). The animal had to wear loose-fitting, long-sleeved jackets for protection of the wires and the connectors located either next to the muscle or on the back. Such techniques removed the need for skin incisions over the target muscles but resulted in more unstable and shorter periods of recording. Park *et al.* (2000) refined these techniques by tunneling the wires subcutaneously towards modular connectors on the arm or towards a larger skull connector. This yielded stable recording of EMG activity from as many as 24 muscles over periods of 6 and 10 months (Park *et al.*, 2000).

Although not yet ready for practical application, the future of this approach is undoubtedly to use wire-free telemetry systems to transmit the EMG data from the implanted electrodes to a nearby receiver (see Jackson *et al.*, 2006a) which will obviate the need for any connectors or devices externalized through a skin incision, to enable recording from the implant. This should greatly increase the length of time that the implant remains viable and considerably improve the animal's toleration of the implant, through reduced tissue trauma and infection.

Few studies have examined the coordination of intrinsic and extrinsic hand muscles for the control of grasp in humans. Weiss & Flanders (2004) have analyzed the activity of hand and forearm muscles in human subjects producing 52 specific hand shapes. Electromyogram activity was recorded using adhesive bipolar electrodes attached to the skin. Using these surface electrodes, only a limited number of muscles can be recorded simultaneously (up to 7

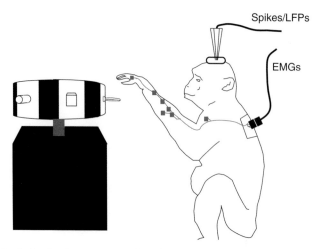

Figure 5.1. Methods for simultaneous recording of electromyograms (EMGs) and cortical spikes and LFPs during skilled grasping in the macaque monkey. Electromyogram electrodes chronically implanted in arm and hand muscles are routed subcutaneously to a back connector at the mid-scapula level. During recording sessions, EMGs are connected to external amplifiers and recorded along with cortical activity while the monkey performed the task. This task involves reaching and grasping a set of differently shaped objects mounted on separated sectors of a rotating carousel.

in Weiss & Flanders, 2004). Also due to the skin conductance, part of the signal on each channel is expected to originate from muscles in the vicinity of the target muscle. The stability and the selectivity of the recordings are greatly improved by using bipolar intramuscular needle electrodes directly inserted in the belly of the muscle (Bawa & Lemon, 1993). However this later technique is best suited to analyze the activity of extrinsic hand muscles with the forearm in fixed position and it has not been used for EMG recordings from human subjects of multiple muscles in an active reach-to-grasp task. To date, EMG recording in the awake monkey using chronic implanted electrodes is the best available model to study the pattern of activation of arm, hand and finger muscles during grasp.

Data collection and analysis

We analyzed the muscle activation patterns during skilled hand movements in a monkey (M39, 5 kg) trained in a reach-to-grasp task towards differently shaped objects. A trial started when the monkey pressed two home-pads located at the waist with the left and right hand, respectively. One of six objects mounted in independent sectors of a motor driven carousel was presented to the monkey (Figure 5.1). Following a visual presentation period of 1–1.5 s, the monkey had to reach out, grasp and pull the object with the left hand and hold it for 1 s within a position window to get a fruit reward (Brochier *et al.*, 2004; Umilta *et al.*, 2007). After object release, the carousel was rotated for presentation of a new object before the next trial was initiated. The set of objects was selected to induce the animal to use a variety of distinct hand postures for grasping (Brochier *et al.*, 2004). Electromyograms and cortical signals (spikes/LFPs recorded with a multiple electrode system) were recorded simultaneously while the monkey was performing the task (Umilta *et al.*, 2007).

Electromyogram

The signals from the chronic EMG implant were recorded differentially, amplified 2000×, highpass filtered at 30 Hz (Neurolog EMG amplifier, NL824, Digitimer Ltd, UK) and sampled at 5 kHz using an A/D interface (PCI-6071E, National Instruments). Since some pairs of electrodes were implanted on closely adjacent muscles (for instance EDC and ED45), the level of physical cross-talk between the recorded signals had to be carefully controlled. This cross-talk results from volume conduction in the biological tissue and a high level of cross-talk makes EMG recordings from nearby muscles partly redundant. To estimate the level of cross-talk between pairs of muscles, we used cross-correlation between unrectified EMG signals over periods of 100 s of recording during which the monkey used a variety of hand postures to perform the task. Redundant signals from two nearby electrode pairs would produce a cross-correlation in which the r value is close to 1 at zero-lag. On the other hand, independent signals would return an r value close to 0. The r value was less than 0.2 for all the muscle pairs recorded in M39, suggesting that despite the proximity of some pairs of electrodes, the EMG signals reflected the activity of independent muscles.

Cortical neurons

While recording multi-channel EMG activity, we also used a Thomas Recording multi-electrode drive to record from a small population of neurons located in the hand region of the contralateral primary motor cortex (M1). Two stimulating electrodes were implanted in the pyramidal tract for antidromic identification of the cortical neurons whose axons project towards the spinal cord (pyramidal tract neurons, PTNs). The timing of key behavioral events in the trained tasks was recorded together with spike, LFP and EMG data: this included the timing of home-pad release (HPR), the onset of object displacement (DO) and an analog record of object displacement. Single units were discriminated offline by principal component analysis on spike waveforms and cluster cutting (Eggermont, 1990).

Electromyogram-based identification of object-specific grasp in the awake monkey

The kinematic of hand movements for grasping differently shaped objects have been analyzed in human and non-human primates (Santello *et al.*, 1998; Roy *et al.*, 2000; Mason *et al.*, 2001, 2004). It has been suggested that to simplify the problem of hand movement control, a reduced number of postural synergies is used for grasping a large set of objects. These observations raise the issue of the degree of task-specificity of the arm, hand and digit muscles during grasping of differently shaped objects. We addressed this question by comparing the patterns

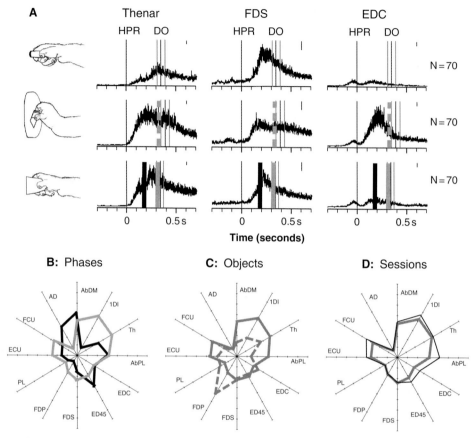

Figure 5.2. Comparison of the modulation of EMG activity during grasp of differently shaped objects. A. Rectified EMG activity averaged across trials for three muscles and the three grasps illustrated on the left (small cylinder, precision grip and large cone). HPR: release of right home pad; DO: object displacement onset (gray vertical lines around DO indicate the 95% confidence interval for this event). Calibration bar: 20 μV. N indicates the number of trials in the averages shown. B–D, Polar plots showing the patterns of EMG activity for 12 simultaneously recorded muscles. Each point indicates the level of EMG activity normalized to the unit length for a given muscle in a given epoch. B. Comparison of the EMG patterns in two different epochs for grasp of the large cone, halfway between HPR and DO (epoch E5, thick black line in A and B) and just before DO (epoch E10, thick grey line in A and B). C. Comparison of the EMG patterns for two different grasps in epoch E10; large cone (thick gray line in A and C) and precision grip (dotted gray line in A and C). D. Comparison of the EMG patterns in two independent recording sessions (session 1, thick gray line in A and D; session 2, black line in D) for grasp of the large cone in epoch E10. See text for abbreviations of muscle names.

of EMG activitation for 12 hand and arm muscles at different time points during grasp. Figure 5.2A illustrates the averaged modulation of EMG activity for three simultaneously recorded muscles (thenar, FDS and EDC) and for grasping of three different objects that required three different types of grip (small cylinder, precision grip and large cone). For each of the three muscles, the form and amplitude of their activity was distinct for the three grasps. Also, for a given object, the grasp-related EMG activity was different for the three muscles. For instance, FDS was more active than thenar or EDC during grasping of the small cylinder, with a later onset of activity after movement initiation (home-pad release, HPR).

To compare the different grasps on the basis of activity in the 12 recorded EMG channels, each object was represented as an N dimensional muscle vector (NDMV), with each dimension describing the averaged and normalized level of activity of a given muscle. For each muscle, the normalization was carried out by dividing the averaged EMG signal for each grasp by the peak of EMG activity measured across all grasps. This normalization process used the same amplitude scale (0–1) and allowed comparison between levels of EMG activity across muscles (Brochier *et al.*, 2004). The NDMV are represented on polar plots in Figure 5.2B–D. In these plots, each point on a branch represents the normalized level of activity of a given muscle and for a given time window (epoch) during the grasp. The epochs were defined using a normalized time scale in which the duration of each epoch corresponded to $1/10^{th}$ of the total time interval between HPR and DO (typically 300–500 ms depending on the object grasped). Figure 5.2B shows superimposed NDMVs computed in two different epochs during grasping of the large cone; half way between HPR and DO (epoch E5) and just before DO (epoch E10). The lack of similarity between the two NDMVs suggests that the recorded muscles were involved quite differently in the successive phases of the reach to grasp task. Figure 5.2C compares the NDMVs for two different grasps (precision grip and large cone) at the same point in time, just before DO. Whereas a strong activation of FDP accompanied precision grip, more activation of the intrinsic muscles (AbDM, 1DI and Th) was observed for grasp of the large cone. These data demonstrate that the selected objects evoked a specific combination of EMG activity in the different muscles. In the last polar plot (Figure 5.2D), the NDMVs computed from data recorded in two separate experimental sessions for the same grasp and at the same point in time (just before DO) are superimposed. This plot shows that the EMG patterns were highly consistent and characterized the stereotyped and overtrained hand shapes and movements used by the monkey to grasp the differently shaped objects.

Object/grasp specificity of EMG and cortical activity

In order to assess the degree of grasp differentiation in the EMGs, we calculated the Euclidian distance (*D*) between every possible pair of grasp-related NDMVs within each epoch (Brochier *et al.*, 2004). For NDMVs V_i and V_j, *D* was computed as follows:

$$D_{ij}^2 = \left(V_i - V_j\right)'\left(V_i - V_j\right) = \sum_{k=1}^{N} \left(V_{ik} - V_{jk}\right)^2$$

where $(V_i - V_j)'$ is the transpose of $(V_i - V_j)$. The Euclidian distance varies in accordance with the degree of similarity between pairs of grasps, and is represented by the black and white matrix in Figure 5.3A for epoch E5 (mid-way between HPR and DO). To maximize the

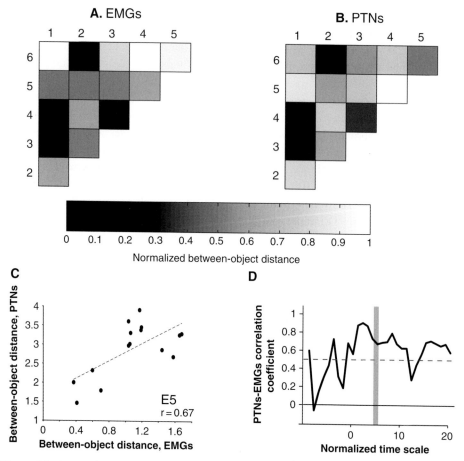

Figure 5.3. Grasp differentiation at the cortical and EMG levels. A–B. Between-object normalized Euclidian distance for every possible pair of objects computed from the activity of 12 forelimb muscles (A) or from the activity of 65 M1-PTNs recorded in independent sessions (B) in epoch E5 (vertical gray bar in D). C. Correlation of the between-object distance at the EMG and cortical levels in epoch E5. D. Modulation of the EMG-PTN distance correlation along time. The time scale is normalized (epoch E0, HPR; epoch E10, DO, see text for details). The horizontal dotted line indicates the level of significance for the positive correlations ($P < 0.05$).

contrast between pairs of grasps, the Euclidian distance has been normalized between 0 (shortest distance among all possible pairs of object, in black) and 1 (longest distance, in white). The EMG patterns for grasp 1 and 4 (insertion of the fingers to grasp a ring and a cube, respectively) showed the greatest degree of similarity (shortest distance in black) but are highly differentiated from the EMG patterns for grasp 6 (large disk, longest distance in white). To investigate the role of MI in controlling the differentiation of the EMG patterns, we compared the between-grasp activity at both EMG and cortical levels. In a separate study we reported that different types of grasps are represented by a different level of activity within a large population of M1 neurons (Umilta *et al*., 2007). Here, we question if this grasp differentiation in cortical activity reflects the encoding of muscle patterns for grasping. The matrix in Figure 5.3B represents the Euclidian distance between every possible pair of grasps calculated from the level of activity in 65 PTNs in epoch E5. This matrix shares common features with the EMG matrix in Figure 5.3A. Again, grasp 1 and 4 (insertion grasps) are the most similar but are both differentiated from grasp 6 (large disk). To compare the level of object differentiation at the EMG and PTN levels in epoch E5, we plotted the between-object distance for the EMGs and PTNs for all possible pairs of grasps. The correlation was significant ($P < 0.05$) suggesting a causal relationship between the level of grasp differentiation in M1 and the EMGs. The variation of the coefficient of correlation across all the grasp-related epochs is presented in Figure 5.3D. A peak of correlation is observed early after HPR while the monkey is shaping the hand to grasp the object. A high level of correlation is maintained during the late reaching phase and object holding, suggesting a tight coupling between EMG and PTN activity during hand movements.

Conclusion

Chronic EMG implants in arm and hand muscles provide a powerful tool for the study of the monkey's motor behavior during dexterous movement. Recent refinement of this technique allows the recording of a stable signal from a large number of muscles over a long period of time. This type of information is essential for the study of the central mechanisms controlling the distal muscles for grasping and manipulation of objects.

Acknowledgments

This study was supported by the Wellcome Trust, MRC and EU project QLRT-1999–00448; COSPIM. We thank Helen Lewis, Samantha Shepherd, Victor Baller, Chris Seers, Robin Richards, David MacManus, Ed Bye and Jim Turton for their technical assistance.

References

Bawa, P. & Lemon, R. N. (1993). Recruitment of motor units in response to transcranial magnetic stimulation in man. *J Physiol*, **471**, 445–464.
Belhaj-Saïf, A., Karrer, J. H. & Cheney, P. D. (1998). Distribution and characteristics of poststimulus effects in proximal and distal forelimb muscles from red nucleus in the monkey. *J Neurophysiol*, **79**, 1777–1789.
Bennett, K. M. & Lemon, R. N. (1994). The influence of single monkey cortico-motoneuronal cells at different levels of activity in target muscles. *J Physiol*, **477**, 291–307.
Bennett, K. M. & Lemon, R. N. (1996). Corticomotoneuronal contribution to the fractionation of muscle activity during precision grip in the monkey. *J Neurophysiol*, **75**, 1826–1842.

Brochier, T., Spinks, R. L., Umilta, M. A. & Lemon, R. N. (2004). Patterns of muscle activity underlying object-specific grasp in the macaque monkey. *J Neurophysiol*, **92**, 1770–1782.

Dum, R. P. & Strick, P. L. (1991). The origin of corticospinal projections from the premotor areas in the frontal lobe. *J Neurosci*, **11**, 667–89.

Dum, R. P. & Strick, P. L. (2002). Motor areas in the frontal lobe of the primate. *Physiol Behav*, **77**, 677–682.

Eggermont, J. J. (1990). Neural responses in primary auditory cortex mimic psychophysical, across-frequency-channel, gap-detection thresholds. *J Neurophysiol*, **84**, 1453–1463.

Kuypers, H. G. (1978). The motor system and the capacity to execute highly fractionated distal extremity movements. *Electroencephalogr Clin Neurophysiol Suppl*, **34**, 429–431.

Jackson, A., Moritz, C. T., Mavoori, J., Lucas, T. H. & Fetz, E. E. (2006a). The Neurochip BCI: towards a neural prosthesis for upper limb function. *IEEE Trans Neural Syst Rehabil Eng*, **14**, 187–190.

Jackson, A., Mavoori, J. & Fetz, E. E. (2006b). Long-term motor cortex plasticity induced by an electronic neural implant. *Nature*, **444**, 56–60.

Jackson, A., Mavoori, J. & Fetz, E. E. (2007). Correlations between the same motor cortex cells and arm muscles during a trained task, free behavior, and natural sleep in the macaque monkey. *J Neurophysiol*, **97**, 360–374.

Jeannerod, M., Arbib, M. A., Rizzolatti, G. & Sakata, H. (1995). Grasping objects: the cortical mechanisms of visuomotor transformation. *Trends Neurosci*, **18**, 314–320.

Lemon, R. N. & Griffiths, J. (2005). Comparing the function of the corticospinal system in different species: organizational differences for motor specialization? *Muscle Nerve*, **32**, 261–279.

Mason, C. R., Gomez, J. E. & Ebner, T. J. (2001). Hand synergies during reach-to-grasp. *J Neurophysiol*, **86**, 2896–2910.

Mason, C. R., Theverapperuma, L. S., Hendrix, C. M. & Ebner, T. J. (2004). Monkey hand postural synergies during reach-to-grasp in the absence of vision of the hand and object. *J Neurophysiol*, **91**, 2826–2837.

McKiernan, B. J., Marcario, J. K., Karrer, J. H. & Cheney, P. D. (1998). Corticomotoneuronal postspike effects in shoulder, elbow, wrist, digit, and intrinsic hand muscles during a reach and prehension task. *J Neurophysiol*, **80**, 1961–1980.

Miller, L. E., van Kan, P. L., Sinkjaer, T. *et al.* (1993). Correlation of primate red nucleus discharge with muscle activity during free-form arm movements. *J Physiol*, **469**, 213–243.

Morrow, M. M. & Miller, L. E. (2003). Prediction of muscle activity by populations of sequentially recorded primary motor cortex neurons. *J Neurophysiol*, **89**, 2279–2288.

Nakajima, K., Maier, M. A., Kirkwood, P. A. & Lemon, R. N. (2000). Striking differences in transmission of corticospinal excitation to upper limb motoneurons in two primate species. *J Neurophysiol*, **84**, 698–709.

Park, M. C., Belhaj-Saif, A. & Cheney, P. D. (2000). Chronic recording of EMG activity from large numbers of forelimb muscles in awake macaque monkeys. *J Neurosci Methods*, **96**, 153–60.

Porter, R. & Lemon, R. N. (1993). *Corticospinal Function and Voluntary Movement*. Oxford: Clarendon Press.

Roy, A. C., Paulignan, Y., Farne, A., Jouffrais, C. & Boussaoud, D. (2000). Hand kinematics during reaching and grasping in the macaque monkey. *Behav Brain Res*, **117**, 75–82.

Santello, M., Flanders, M. & Soechting, J. F. (1998). Postural hand synergies for tool use. *J Neurosci*, **18**, 10105–10115.

Schieber, M. H. (1995). Muscular production of individuated finger movements: the roles of extrinsic finger muscles. *J Neurosci*, **15**, 284–297.

Tallis, R. (2004). *The Hand. A Philosophical Enquiry into Human Being*. Edinburgh: Edinburgh University Press.

Tubiana, R. (1981). In R. Tubiana (Ed.), *The Hand*. Philadelphia, PA: Saunders.

Umilta, M. A., Brochier, T. G., Spinks, R. L. & Lemon, R. N. (2007). Simultaneous recording of macaque premotor and primary motor cortex neuronal populations reveals different functional contributions to visuomotor grasp. *J Neurophysiol*, **98**, 488–501.

Weiss, E. J. & Flanders, M. (2004). Muscular and postural synergies of the human hand. *J Neurophysiol*, **92**, 523–535.

6

Transcranial magnetic stimulation investigations of reaching and grasping movements

GIACOMO KOCH AND JOHN C. ROTHWELL

Summary

Transcranial magnetic stimulation (TMS) has emerged as a suitable technique to investigate the network of cortical areas involved in human grasp/reach movements. Applied over the primary motor cortex (M1), TMS reveals the pattern of activation of different muscles during complex reaching-to-grasp tasks. Repetitive TMS (rTMS) used to induce "virtual lesions" of other cortical areas has allowed investigation of other cortical structures such as the ventral premotor cortex (PMv), dorsal premotor cortex (PMd) and the anterior intra-parietal sulcus (aIPS). Each of these appears to contribute to specific aspects of reaching, grasping and lifting objects. Finally, twin-coil TMS studies can illustrate the time course of operation of parallel intracortical circuits that mediate functional connectivity between the PMd, PMv, the posterior parietal cortex and the primary motor cortices.

Introduction

The ease with which we can make reach-to-grasp movements conceals a good deal of the underlying complexity of the task. Thus, the target of the reach must be located in space; a decision must be made about the most appropriate type and orientation of grasp according to the weight and shape of the object; and the timing of the reaching movement of the arm must be synchronized with the opening of the hand so that the object can be grasped as effectively and quickly as possible (for a review see Castiello, 2005; see also Chapters 2 and 10).

Studies in primates with localized lesions suggest that in addition to the primary motor cortex, two key cortical areas seem to be involved in transforming visual information about the position and characteristics of the target into the sequence of motor commands needed to perform the task: area F5 and the anterior intraparietal cortex (AIP). Area F5 forms the rostral part of the monkey ventral premotor cortex (human homolog: PMv); the AIP is situated in the rostral part of the posterior bank of the intraparietal sulcus and is directly connected to area F5 (Jeannerod *et al.*, 1995). A similar bilateral network of frontoparietal areas, including the posterior parietal cortex (PPC), the ventral (PMv) and the dorsal (PMd) premotor cortex (Binkofski *et al.*, 1998, 1999; Ehrsson *et al.*, 2000, 2001; Kuhtz-Buschbeck *et al.*, 2001) have also been shown to be activated in functional imaging studies of grasping

Sensorimotor Control of Grasping: Physiology and Pathophysiology, ed. Dennis A. Nowak and Joachim Hermsdörfer. Published by Cambridge University Press. © Cambridge University Press 2009.

in healthy humans. The precise homolog of the macaque AIP in humans is still debated. Several studies suggest that it lies within the anterior intraparietal sulcus (aIPS). Patients with circumscribed lesions to the aIPS show marked deficits in hand preshaping during visually guided reach-to-grasp movements, whereas reaching remains relatively intact (Binkofski *et al.*, 1998) and several functional neuroimaging studies indicate that focal activation within the aIPS of the healthy brain occurs in association with visually guided grasping (Ehrsson *et al.*, 2000; Culham *et al.*, 2003).

However, neither lesion studies nor functional imaging approaches provide information about the time course of task-related activity in a cortical area. Recently, transcranial magnetic stimulation has emerged as a suitable technique to investigate the time course over which these areas are involved in grasp/reach movements in humans. This is because a single TMS pulse interferes with ongoing activity at the stimulated site for 50–150 ms, and hence can be viewed as producing a short-lasting and reversible functional lesion of that area (Walsh & Cowey, 2000). The duration of the effect can be extended by giving several stimuli at intervals of 100 ms or so. Changes in behavior induced by this procedure can therefore reveal information about the role and time of operation of the area of cortex that underwent TMS in a given task. In addition to this method, paired pulse TMS may be used to test how stimulation of one area affects the excitability of the motor cortex. For example, a TMS pulse to PMd, PMv or AIP changes motor cortical excitability several milliseconds later, indicating that there is a functional connection between these areas. Testing how the effectiveness of this connection changes during a task gives information about how and when the connection is active.

Transcranial magnetic stimulation studies of the primary motor cortex

In the original studies by Lemon *et al.* (1995, 1996) single-pulse TMS was used to investigate corticospinal influences during a task in which human subjects had to reach out to grasp and lift an object. Transcranial magnetic stimulation applied to the hand area of the motor cortex was delivered during different phases of the task. Different muscles showed varied patterns of modulation across successive phases of the task: throughout the reach there was a strong excitatory corticospinal drive to brachioradialis and anterior deltoid, which contribute to hand transport, and to the extrinsic hand muscles, which orientate the hand and fingertips. In contrast, the intrinsic hand muscles appeared to receive their strongest cortical input as the digits close around and first touch the object. These results showed a significant and heterogeneous variation in cortical activity during a complex task requiring reaching, grasping and lifting objects.

The lifting of such objects requires fine motor control; too much force can damage the object or result in an excessive lifting movement, and too little force can cause the object to slip away. Throughout life, we build internal representations for the weight of different objects (Wolpert & Flanagan, 2001). In cases when the weight is lighter than expected, somatosensory information related to lift-off will generate corrective forces to stabilize the object (Johansson & Westling, 1988). In cases when the weight is heavier than expected, the absence of an expected lift-off will generate corrective forces to overcome gravity on the object (Johansson & Westling, 1988).

Chouinard *et al.* (2005) demonstrated that low-frequency rTMS applied over the primary motor and dorsal premotor cortices influenced differentially the anticipatory scaling of forces. When applied over the primary motor cortex, repetitive stimulation disrupted the scaling of forces based on information acquired during a previous lift, suggesting that the primary motor cortex not only plays a role in planning different phases of reaching and grasping, but is also engaged in forming memory traces associated with a recent experience. Experiments by Nowak *et al.* (2005), showing that 20 s of rTMS applied over the hand area of the dominant M1 increased the grip-force output when next lifting a familiar object, were consistent with this idea. Conversely when applied over the PMd, repetitive stimulation disrupted the scaling of forces based on arbitrary color cues, indicating a role of this area in translating arbitrary sensory cues into motor programs.

Finally, a recent study by Schabrun *et al.* (2008) investigated the role of the primary motor (M1) and sensory (S1) cortices during a grip-lift task using inhibitory transcranial magnetic theta-burst stimulation (TBS). Disruption of sensory cortex function by TBS increased the delay between initial contact with the object and the initiation of the lift (i.e. pre-load duration), suggesting that disrupting sensory cortex delays the integration of sensory signals arising from contact with the object into the motor plan. In contrast, disruption to the motor cortex produced a clear deficit in normal anticipatory control of grip-force scaling. These findings provide further evidence of the specific contribution of M1 to anticipatory force scaling. In addition, they provide evidence for the contribution of the sensory cortex to object manipulation, suggesting that sensory information is not necessary for optimal functioning of anticipatory control but plays a key role in triggering subsequent phases of the motor plan.

"Virtual lesion" studies of aIPS

The first TMS studies of this area were performed by Desmurget *et al.* (1999), who examined the effect of single-pulse TMS to a region just caudal to the aIPS on the ability to correct reach-to-point movements in response to abrupt changes in target location. Under normal conditions, the eye and the hand usually move towards the target together, with the eye arriving first because of its much lower inertia. If the target is moved during the saccade, subjects may be unaware of the change due to saccadic suppression yet their hand paths are corrected by a smooth adjustment of trajectory so that the target is reached correctly. Remarkably, path adjustments were disrupted in four of five subjects when TMS was applied over the PPC.

To define the role of aIPS in the grasp component of a visually guided reach-to-grasp movement, Tunik *et al.* (2005) asked subjects to reach out and grasp an object that could be rotated after the arm had started to move. On these perturbed trials, subject had either to adjust the amplitude of their grip or its orientation in order to grasp the object correctly. Single-pulse TMS to aIPS produced a delay of, on average, 88 ms in time of peak aperture and 146 ms increase in movement time (Tunik *et al.*, 2005). The TMS induced "grasp-related" deficits if adjustment of grip size was the goal and "forearm-related" deficits were produced if adjustment of forearm orientation was the goal. Notably, aperture deficits were only apparent after magnetic stimulation over aIPS within 65 ms of object perturbation, indicating that aIPS is causally involved in rapid, goal-dependent updating of reach-to-grasp actions.

In a second study from the same group, Rice *et al.* (2006) examined whether aIPS was involved only in correcting for perturbed tasks or whether it had a more general role in executing the reach-to-grasp movement. They also delivered TMS to the anterior (aIPS), middle (mIPS) and caudal (cIPS) IPS in order to obtain information about the spatial specificity of the stimulus sites. Subjects were required to reach and grasp an object in a stable context (no-perturbation task) or under conditions in which an adaptive response was required (perturbation task). In the no-perturbation task, it was shown that TMS to aIPS (but not mIPS or cIPS) disrupted grasping when applied during the execution phase of the movement. In the perturbation task, it was shown that TMS to aIPS (but not mIPS or cIPS) disrupted grasping when applied during the correction phase of the movement (Rice *et al.*, 2006). Namely, TMS decreased time of maximum grip aperture and increased velocity of grip aperture in unperturbed trials, while it increased distance of maximum grip aperture (defined as the three-dimensional distance between the index and thumb) and velocity in perturbed trials. They suggested that the results were compatible with the idea that aIPS computes a continuous on-line comparison of the intended and actual movement and feeds this information forwards to motor areas controlling task execution (see Figure 6.1).

Similar results were obtained by Glover *et al.* (2005) when rTMS was applied over the left intraparietal sulcus as participants reached to grasp a small or large illuminated cylinder with their right hand. On some trials, the illumination could suddenly switch from the small to large cylinder, or vice-versa. Small–large switches were associated with relatively early grip aperture adjustments, whereas large–small switches were associated with relatively late grip aperture adjustments. When rTMS at 10 Hz was applied early in the movement, it disrupted on-line adjustments to small–large target switches, but not to large–small switches. Conversely, when rTMS was applied late in the movement, it disrupted adjustments to large–small target switches but not to small–large switches. The timing of the disruption by rTMS appeared linked to the initiation of the adjustment. It was concluded that the left parietal lobe plays a critical role in initiating an on-line adjustment to a change in target size, but not in executing that adjustment.

The most recent study by Davare *et al.* (2007) examined how two to five pulses of TMS affected the performance of a grip-lift task in which subjects had to reach out and grasp then lift an instrumented object. During movement preparation, TMS over aIPS had distinct consequences on precision grasping of either hand depending on its time of occurrence: TMS applied 270–220 ms before the fingers contacted the manipulandum increased the variability and the mean location of the points where thumb and index finger contacted the target (hand shaping). In contrast, if TMS was applied later, 170–120 ms before contact time, then

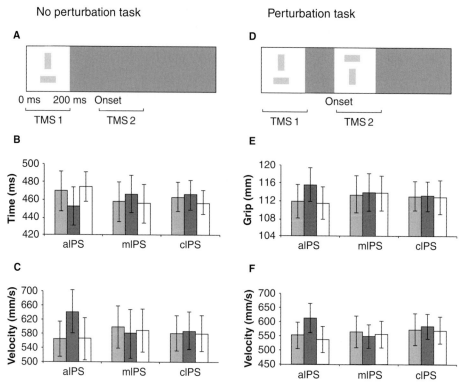

Figure 6.1. Effect of transcranial magnetic stimulation applied over different cortical areas on grasp control during planning and execution of unperturbed (A–C) and perturbed (D–F) trial types. Subjects had to reach out and grasp with forefinger and thumb a target that was presented either vertically or horizontally, and viewed through liquid-crystal shutter glasses. The latter involved a wider grip than the former. The upper panel shows the time course of events. In the unperturbed task (A), subjects could see the object for 200 ms (white box), but thereafter the glasses were opaque. They had to reach out to grasp the object without vision. The TMS was applied either during the viewing phase, or after the beginning of the arm movement (onset). In the perturbed task (D), subjects were shown the object as before and had to reach out to grasp it. However, after onset of movement, the object could be seen again for 200 ms and could have rotated to the opposite orientation, necessitating on-line adjustment of grip aperture. The TMS was applied either during the second viewing phase or just after the glasses had become opaque for the second time. The bars below show the time to reach maximum grip aperture (B), maximum grip aperture (E) and the peak velocity of grip aperture (C, F). White bars are control trials with no TMS; light gray bars are trials when TMS was applied during the first (unperturbed trials A–C) or second (perturbed trials D–F) viewing phases; dark gray bars are when TMS was applied during movement execution. Results adapted from Rice *et al.* (2006).

it caused subjects to grip the manipulandum with more force than usual (grip-force scaling). The lateralization of these two processes in aIPS was also strikingly different: whereas a bilateral lesion of aIPS was necessary to impair hand shaping, only a unilateral lesion of the left but not right aIPS altered grip-force scaling in either hand. The study shows that, during movement preparation, aIPS is responsible for processing two distinct, temporally dissociated, precision-grasping parameters, regardless of the hand in use. This indicates that the contribution of aIPS to hand movements is "effector-independent," a finding that may explain the invariance of grasping movements performed with either hand.

"Virtual lesion" studies of premotor involvement in reach-to-grasp

The experiments described above show that the aIPS is involved in executing and correcting a grasping action. The experiments described in this section show how TMS has been used to test the involvement of premotor areas of cortex in similar tasks.

Davare *et al.* (2006) asked subjects to grasp a manipulandum and to lift it at a natural speed with the right hand to a height of ~ 20 cm after an auditory "Go" signal. In a first series of experiments, a short train of five TMS pulses (10 Hz, 500 ms) were delivered at the same time as the Go signal over the the left/right PMv or the left/right PMd. TMS over PMv (but not PMd) increased the time taken by subjects to move their hand to the manipulandum as well as the interval between the contact times of the index finger and thumb on the object. TMS over left PMv, but not right PMv or PMd, also increased the variability and increased the mean distance between the contact points of thumb and index on the object, similarly to the effect of TMS over aIPS. The conclusion was that PMv was involved in executing the grip phase of the movement. In contrast, TMS over PMd affected the lifting portion of the task by increasing the time between onset of grip force and onset of lift. In further experiments, a pair of TMS pulses (5 ms interval) was applied to disrupt function at specific times in the course of the movement. TMS over left PMv produced effects only when it was delivered early in the movement (50 ms or 100 ms after the Go signal, corresponding, respectively, to 230 and 180 ms before finger contact with the manipulandum). On the contrary, left PMd was found to intervene later (150 and 200 ms after the Go signal, 160–110 ms before subjects started to lift the manipulandum). Therefore this study reinforces the view that the premotor areas, and PMv in particular, play a key role in visuomotor transformations required to generate grasping movements. However, the availability of visual feedback and the fact that TMS was only delivered after the Go signal in that study made it impossible to dissociate planning from execution.

Cattaneo *et al.* (2005) confirmed the involvement of PMv in grasping movements using a rather more indirect technique. When two threshold TMS pulses are applied to the motor cortex with an interstimulus interval of about 1.2 or 2.5 ms, the resulting MEP in hand muscles is greatly facilitated in comparison with the size of the response to each stimulus alone. The effect is thought to reflect generation of I-wave inputs to corticospinal neurons set up by the first pulse (Ziemann & Rothwell, 2000). These occur at intervals of about 1.2 ms so that if a second pulse is applied at a multiple of this, facilitation occurs. Cattaneo *et al.* (2005) measured the amount of facilitation in different hand muscles at an interstimulus interval of 2.5 ms at different times after presentation of an object to be grasped by the subject. Before onset of any EMG activity in the grasp, Cattaneo *et al.* (2005) saw increases in facilitation specifically in muscles that would be active in the forthcoming grasp, depending on the type of object that was presented. They argued that this modulation was likely to reflect inputs from PMv to M1 that had been activated by the sight of the object. This would then produce particular patterns of excitability in output zones of the corticospinal tract that could be used in the forthcoming task. In a recent study by the same group these changes in intracortical excitability were found to occur only if the object was specified by current visual input at the moment of grasp initiation. In contrast, when subjects had to remember the target object, even for only 1 s, a different neural network for motor preparation seemed to be used (Prabhu *et al.*, 2007), suggesting that these circuits contribute to action selection only when immediate sensory information specifies which action to make.

Paired-pulse studies of PMd involvement in reach-to-grasp

While the studies above reveal the contribution of PMv, PMd and aIPS in generating grasping movements and in on-line corrections after perturbations, they do not provide information about the functional connectivity of these non-primary motor areas with the primary motor cortex that would explain how their activity may modulate the spatial pattern of output from primary motor areas preceding execution of a movement. A different TMS approach has recently been proposed to address this crucial issue.

Previous combined TMS/PET, TMS/fMRI and paired TMS investigations have shown that TMS not only changes neural activity at the site of stimulation, but also affects interconnected cortical and subcortical areas (Ferbert *et al.*, 1992; Ugawa *et al.*, 1995; Paus *et al.*, 1997; Strafella *et al.*, 2003; Lee *et al.*, 2003; Bestmann *et al.*, 2004, 2005). Paired-pulse transcranial magnetic stimulation (pp-TMS) provides a unique opportunity in humans to probe inputs to the primary motor cortex from other areas of the motor system. A conditioning stimulus (CS) is first used to activate putative pathways to the motor cortex from the site of stimulation, while a second, test stimulus (TS), delivered over the primary motor cortex a few milliseconds later, probes any changes in excitability that are produced by the input. Depending on the intensity of the conditioning stimulus and the interstimulus interval (ISI) both facilitation and inhibition may be detected in the primary motor cortex (M1), ipsilateral or contralateral to the site of conditioning. Previous studies have been conducted with the CS delivered over the contralateral M1 (Ferbert *et al.*, 1992), the cerebellum (Ugawa *et al.* 1995), and the premotor cortex (Civardi *et al.*, 2001; Mochizuki *et al.*, 2004; Baumer *et al.*, 2006; Koch *et al.*, 2007a) confirming the existence of the pathways from these areas to M1 in humans. For instance, Mochizuki *et al.* (2004) found that a conditioning TMS pulse over the right PMd at 90 or 110% of the resting motor threshold (RMT) reduced the amplitude of motor-evoked potentials (MEPs) in hand muscles elicited by a second TMS pulse to the contralateral M1. The effect was seen best if the interstimulus interval was 8–10 ms. The opposite effect, facilitation of contralateral MEPs, was found by Bäumer *et al.* (2006) when they applied left PMd conditioning stimuli of lower intensity (80% active motor threshold, AMT) at ISI = 8 ms.

The advantage of probing these pathways with TMS methods is that the response to a TMS conditioning pulse depends on the excitability of the pathway at the time the stimulus is applied. Thus changes in the effectiveness of the conditioning pulse give an indication of how the excitability of the connection changes over time. In a recent study, Koch *et al.* (2006) used these methods to test both the inhibitory and facilitatory connections between the PMd and contralateral M1 during a behavioral task requiring selection of action. They hypothesized that if they were physiologically relevant they would show temporally specific changes in their excitability at times when PMd contributes to task performance. In fact, previous studies showed that the PMd in the left hemisphere is dominant for selection of actions (Schluter *et al.*, 2001; Rushworth *et al.*, 2003), since the right PMd is active only for movements made by the left hand whereas the left PMd is active for movements of either hand (Schluter *et al.*, 2001). Furthermore, TMS of the right PMd only disrupts the selection of left-hand movements whereas TMS of the left PMd disrupts the selection of movements that will be made by either hand (Schluter *et al.*, 1998; Johansen-Berg *et al.*, 2002). In both cases, TMS had the largest effect on performance if it was applied in the earlier part of the task in contrast with the disruptive effect of TMS over the M1 which is maximal late in the reaction period. Thus activity in the PMd seems crucial during early decisional processes involved in selection of movements.

In the study by Koch *et al.* (2006) subjects were required to contract the right or left first dorsal interosseus (FDI) muscle as quickly as possible, performing a rapid isometric squeeze of the block of the left or the right hand as soon they heard a cue sound (auditory choice RT task). The facilitatory and inhibitory PMd–M1 interactions were tested 50, 75, 100, 125, 150 and 200 ms after the cue sound.

Koch *et al.* (2006) demonstrated that the interhemispheric interactions between left PMd and right M1 changed their excitability during the reaction period of the task. Facilitatory connections were activated 75 ms after a tone that indicated subjects should move the left hand whereas inhibitory connections were more excitable 100 ms after a tone indicating a movement of the right hand (while the left hand remained stationary). These connections were modulated only for muscles that might be involved in the upcoming movement; no effects were observed in non-involved muscles. In contrast, results obtained from right PMd conditioning left M1 were slightly different, since the interhemispheric interactions between right PMd and left M1 at rest were mainly inhibitory and a similar profile of transcallosal inhibition as for left PMd–right M1 interactions was observed during the reaction period of the task, but no facilitation was evident at any time (see Figure 6.2). The authors suggested that the contribution of PMd to control of movements of the ipsilateral hand is due at least in part to activity in these pathways: the left but not right PMd may facilitate movements of the ipsilateral hand that are about to be made; and that PMd of both hemispheres may also suppress movements that have been prepared but are not used.

Recently, interactions at rest have been discovered also between the PPC and the ipsilateral motor cortex (Koch *et al.*, 2007b). At rest, clear facilitation of the ipsilateral M1 may be induced by cIPS TMS with specific parameters of stimulation, while there seems to be an inhibitory pathway originating from aIPS (Koch *et al.*, 2007b). Following this line of research, we have tested how this connectivity depends on current motor plans, applying TMS not only at rest, but also shortly after a left-arm reach towards a leftward or rightward location had been auditorily cued (yet prior to reach initiation) (Koch *et al.*, 2008). The same

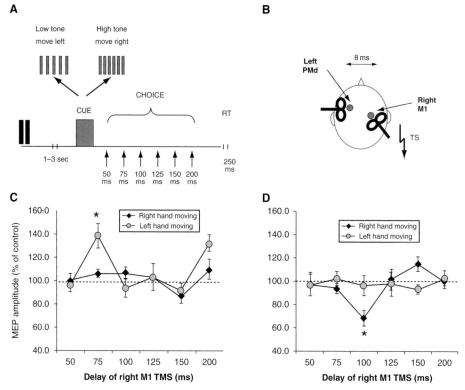

Figure 6.2. A. Schematic representation of the choice reaction time task. Subjects were required to activate right or left FDI muscle as quickly as possible, performing a rapid isometric contraction of either hand as soon they hear a cue sound. The TMS was delivered over right M1 at different delays (50, 75, 100, 125, 150, 200 ms) after the cue sound (test stimulus, TS). The intensity of TS was adjusted to evoke an MEP of approximately 1 mV peak-to-peak in the relaxed left FDI. In half of the trials M1 TMS was preceded by a conditioning stimulus (CS) delivered 8 ms before over the contralateral PMd (B). For Left PMd–right M1 connectivity, facilitatory connections (C) were evident 75 ms after a tone that indicates subjects should move the left hand whereas inhibitory connections (D) manifested 100 ms after a tone indicating a movement of the right hand. Methods and results adapted from Koch *et al.* (2006).

paired-pulse TMS during the planning phase of our reach task revealed facilitatory influences only when planning a leftward rather than rightward reach, at two specific time intervals (50 ms and 125 ms) after the auditory cue. The earlier reach-direction-specific facilitatory influence from PPC on M1 was replicated with subjects blindfolded. If instead non-blindfolded subjects had to plan directional saccades, rather than reach movements, the effect disappeared. In this study reach planning did not modulate paired-pulse TMS effects for dorsal premotor cortex (PMd) on ipsilateral M1. Thus the results showed that functional connectivity between human parietal and motor cortex is enhanced for the early stages of planning a reach in the contralateral direction.

A variety of TMS methods are therefore now available to study the time course of involvement as well as the functional connectivity of areas active during reach to grasp. They illustrate the time course of operation of parallel intracortical circuits and cortico-cortical connections between the PMd, PMv, PPC and the primary motor cortices.

References

Baumer, T., Bock, F., Koch, G. *et al.* (2006). Magnetic stimulation of human premotor or motor cortex produces interhemispheric facilitation through distinct pathways. *J Physiol (Lond)*, **572**, 857–868.

Bestmann, S., Baudewig, J., Siebner, H. R., Rothwell, J. C. & Frahm, J. (2004). Functional MRI of the immediate impact of transcranial magnetic stimulation on cortical and subcortical motor circuits. *Eur J Neurosci*, **19**, 1950–1962.

Bestmann, S., Baudewig, J., Siebner, H. R., Rothwell, J. C. & Frahm, J. (2005). BOLD MRI responses to repetitive TMS over human dorsal premotor cortex. *Neuroimage*, **28**, 22–29.

Binkofski, F., Dohle, C., Posse, S. *et al.* (1998). Human anterior intraparietal area subserves prehension. *Neurology*, **50**, 1253–1259.

Binkofski, F., Buccino, G., Posse, S. *et al.* (1999). A fronto-parietal circuit for object manipulation in man: evidence from an fMRI-study. *Eur J Neurosci*, **11**, 3276–3286.

Castiello, U. (2005). The neuroscience of grasping. *Nat Rev Neurosci*, **6**, 726–736.

Cattaneo, L., Voss, M., Brochier, T. *et al.* (2005). A cortico-cortical mechanism mediating object-driven grasp in humans. *Proc Natl Acad Sci USA*, **102**, 898–903.

Chen, R. (2004). Interactions between inhibitory and excitatory circuits in the human motor cortex. *Exp Brain Res*, **154**, 1–10.

Chen, R., Yung, D. & Li, J. Y. (2003). Organization of ipsilateral excitatory and inhibitory pathways in the human motor cortex. *J Neurophysiol*, **89**, 1256–1264.

Civardi, C., Cantello, R., Asselman, P. & Rothwell, J. C. (2001). Transcranial magnetic stimulation can be used to test connections to primary motor areas from frontal and medial cortex in humans. *Neuroimage*, **14**, 1444–1453.

Chouinard, P. A., Leonard, G. & Paus, T. (2005). Role of the primary motor and dorsal premotor cortices in the anticipation of forces during object lifting. *J Neurosci*, **25**, 2277–2284.

Culham, J. C. *et al.* (2003). Visually guided grasping produces fMRI activation in dorsal but not ventral stream brain areas. *Exp Brain Res*, **153**, 180–189.

Davare, M., Andres, M., Cosnard, G., Thonnard, J. L. & Olivier, E. (2006). Dissociating the role of ventral and dorsal premotor cortex in precision grasping. *J Neurosci*, **26**, 2260–2268.

Davare, M., Andres, M., Clerget, E., Thonnard, J. L. & Olivier, E. (2007). Temporal dissociation between hand shaping and grip force scaling in the anterior intraparietal area. *J Neurosci*, **27**, 3974–3980.

Desmurget, M., Epstein, C. M., Turner, R. S. *et al.* (1999). Role of the posterior parietal cortex in updating reaching movements to a visual target. *Nat Neurosci*, **2**, 563–567.

Ehrsson, H. H., Fagergren, A., Jonsson, T. *et al.* (2000). Cortical activity in precision- versus power-grip tasks: an fMRI study. *J Neurophysiol*, **83**, 528–553.

Ehrsson, H. H., Fagergren, E. & Forssberg, H. (2001). Differential fronto-parietal activation depending on force used in a precision grip task: an fMRI study. *J Neurophysiol*, **85**, 2613–2623.

Ferbert, A., Priori, A., Rothwell, J. C. *et al.* (1992). Interhemispheric inhibition of the human motor cortex. *J Physiol (Lond)*, **453**, 525–546.

Glover, S., Miall, R. C. & Rushworth, M. F. S. (2005). Parietal rTMS selectively disrupts the initiation of on-line adjustments to a perturbation of object size. *J Cogn Neurosci*, **17**, 124–136.

Jeannerod, M., Arbib, A., Rizzolatti, G. & Sakata, H. (1995). Grasping objects: the cortical mechanisms of visuomotor transformation. *Trends Neurosci*, **18**, 314–320.

Johansen-Berg, H., Rushworth, M. F., Bogdanovic, M. D. *et al.* (2002). The role of ipsilateral premotor cortex in hand movement after stroke. *Proc Natl Acad Sci USA*, **99**, 14518–14523.

Johansson, R. S. & Westling, G. (1988). Coordinated isometric muscle commands adequately and erroneously programmed for the weight during lifting task with precision grip. *Exp Brain Res*, **71**, 59–71.

Kuhtz-Buschbeck, J. P., Ehrsson, H. H. & Forssberg, H. (2001). Human brain activity in the control of fine static precision grip forces: an fMRI study. *Eur J Neurosci*, **14**, 382–390.

Koch, G., Franca, M., Fernandez Del Olmo, M. *et al.* (2006). Time course of functional connectivity between dorsal premotor and contralateral motor cortex during movement selection. *J Neurosci*, **26**, 7452–7459.

Koch, G., Franca, M., Mochizuki, H. *et al.* (2007a). Interactions between pairs of transcranial magnetic stimuli over the human left dorsal premotor cortex differ from those seen in primary motor cortex. *J Physiol*, **578**, 551–562.

Koch, G., Fernandez Del Olmo, M., Cheeran, B. *et al.* (2007b). Focal stimulation of the posterior parietal cortex increases the excitability of the ipsilateral motor cortex. *J Neurosci*, **27**, 6815–6822.

Koch, G., Fernandez Del Olmo, M., Cheeran, B. *et al.* (2008). Functional interplay between posterior parietal and ipsilateral motor cortex revealed by twin-coil transcranial magnetic stimulation during reach planning toward contralateral space. *J Neurosci*, **28**, 5944–5953.

Lee, L., Siebner, H. R., Rowe, J. B. *et al.* (2003). Acute remapping within the motor system induced by low-frequency repetitive transcranial magnetic stimulation. *J Neurosci*, **23**, 5308–5318.

Lemon, R. N., Johansson, R. S. & Westling, G. (1995). Corticospinal control during reach, grasp, and precision lift in man. *J Neurosci*, **15**, 6145–6156.

Lemon, R. N., Johansson, R. S. & Westling, G. (1996). Modulation of corticospinal influence over hand muscles during gripping tasks in man and monkey. *Can J Physiol Pharmacol*, **74**, 547–558.

Mochizuki, H., Huang, Y. Z. & Rothwell, J. C. (2004). Interhemispheric interaction between human dorsal premotor and contralateral primary motor cortex. *J Physiol*, **561**, 331–338.

Nowak, D. A., Voss, M., Huang, Y. Z., Wolpert, D. M. & Rothwell, J. C. (2005). High-frequency repetitive transcranial magnetic stimulation over the hand area of the primary motor cortex disturbs predictive grip force scaling. *Eur J Neurosci*, **22**, 2392–2396.

Paus, T., Jech, R., Thompson, C. J. *et al.* (1997). Transcranial magnetic stimulation during positron emission tomography: a new method for studying connectivity of the human cerebral cortex. *J Neurosci*, **17**, 3178–3184.

Prabhu, G., Voss, M., Brochier, T. *et al.* (2007). Excitability of human motor cortex inputs prior to grasp. *J Physiol*, **581**, 189–201.

Rice, N. J., Tunik, E. & Grafton, S. T. (2006). The anterior intraparietal sulcus mediates grasp execution, independent of requirement to update: new insights from transcranial magnetic stimulation. *J Neurosci*, **26**, 8176–8182.

Rushworth, M. F., Johansen-Berg, H., Gobel, S. M. & Devlin, J. T. (2003). The left parietal and premotor cortices: motor attention and selection. *Neuroimage*, **20**, S89–S100.

Schabrun, S. M., Ridding, M. C. & Miles, T. S. (2008). Role of the primary motor and sensory cortex in precision grasping: a transcranial magnetic stimulation study. *Eur J Neurosi*, **27**, 750–756.

Schluter, N. D., Rushworth, M. F., Passingham, R. E. & Mills, K. R. (1998). Temporary interference in human lateral premotor cortex suggests dominance for the selection of movements. A study using transcranial magnetic stimulation. *Brain*, **121**, 785–799.

Schluter, N. D., Krams, M., Rushworth, M. F. & Passingham, R. E. (2001). Cerebral dominance for action in the human brain: the selection of actions. *Neuropsychologia*, **39**, 105–113.

Strafella, A. P., Paus, T., Fraraccio, M. & Dagher, A. (2003). Striatal dopamine release induced by repetitive transcranial magnetic stimulation of the human motor cortex. *Brain*, **126**, 2609–2615.

Tunik, E., Frey, S. H. & Grafton, S. T. (2005). Virtual lesions of the anterior intraparietal area disrupt goal-dependent on-line adjustments of grasp. *Nat Neurosci*, **8**, 505–511.

Ugawa, Y., Uesaka, Y., Terao, Y., Hanajima, R. & Kanazawa, I. (1995). Magnetic stimulation over the cerebellum in humans. *Ann Neurol*, **37**, 703–713.

Walsh, V. & Cowey, A. (2000). Transcranial magnetic stimulation and cognitive neuroscience. *Nat Rev Neurosci*, **1**, 73–79.

Wolpert, D. M. & Flanagan, J. R. (2001). Motor prediction. *Curr Biol*, **11**, R729–R732.

Ziemann, U. & Rothwell, J. C. (2000). I-waves in motor cortex. *J Clin Neurophysiol*, **17**, 397–405.

7

Neuroimaging of grasping

H. HENRIK EHRSSON

Summary

The last 10 years have seen major advances in the functional magnetic resonance imaging (fMRI) of the brain's role in grasping. A number of technical problems related to artefacts produced by arm movements and the registration of movements and fingertip forces have been solved. Reproducible activation of key areas involved in grasping, such as the ventral premotor cortex and the anterior part of the intraparietal sulcus, has been reported. More than that, fMRI seems to be capable of detecting biologically relevant activity in all the cortical and subcortical structures involved in the control of reaching, grasping and manipulation. Importantly, imaging has also been able to identify how activity in these areas supports key sensorimotor control mechanisms used in human dexterous manipulation. In particular, the anticipatory and reactive control of grip forces during object manipulation has been associated with specific neuronal responses in motor, parietal and cerebellar areas. Particularly interesting new lines of research include the use of effective connectivity analyses to characterize the neural interactions between the nodes in the frontoparietal circuits, and the combination of computational neuroscience approaches and functional imaging.

Functional magnetic resonance imaging is one of the most important techniques available to cognitive neuroscientists. It is a non-invasive, relatively inexpensive, whole-brain imaging modality that can be used to investigate the brain basis of perception, action and cognition with an anatomical resolution of 2–3 mm. In this chapter we will describe the contribution of this method to the understanding of human grasping and object manipulation. We will focus on two paradigms that have been widely used in the field: visually guided reach-to-grasp and precision grip manipulation of objects (between index finger and thumb). Imaging grasping represents a particular challenge to neuroimagers in terms of paradigm design, analysis and the elimination of artefacts, therefore we will also describe approaches that have been developed to overcome these obstacles.

Early neuroimaging studies

The first neuroimaging study of what we loosely refer to as "grasping" was a paper by Roland & Larsen (1976). These authors used a 254-channel dynamic gamma camera to

Sensorimotor Control of Grasping: Physiology and Pathophysiology, ed. Dennis A. Nowak and Joachim Hermsdörfer. Published by Cambridge University Press. © Cambridge University Press 2009.

measure clearance from the hemisphere of xenon 133 (133Xe) injected into the carotid artery while participants explored the tactile shapes of objects with their fingers. The authors reported increased blood flow in the primary sensorimotor cortex and the premotor cortex during this tactile exploration. This was the first human brain-imaging study where subjects manually interacted with an object.

Despite the early beginning, however, it was not until 1996 that imaging was used directly to map human brain activity during grasping (Grafton *et al.*, 1996a; Matsumura *et al.*, 1996). These authors engaged in this research using positron emission tomography (PET), which is a technique that measures changes in regional cerebral blood flow (rCBF) by the introduction into the brain of substances tagged with radionuclides that emit positrons, such as O_{15}-water or O_{15}-butanol. Both studies employed similar paradigms where subjects either had to reach and point towards an object (in the Matsumura study the pointing finger touched the object), or reach and grasp it with the right fingers. By subtracting the reach-to-point condition from the reach-to-grasp one, Matsumura *et al.* noted increased blood flow in the bilateral posterior parietal cortex and the bilateral dorsal premotor cortex (Matsumura *et al.*, 1996), and Grafton detected significant grasp-related responses in the left lateral parietal operculum (SII region) (Grafton *et al.*, 1996a). In addition, Grafton and colleagues described how both tasks activated areas in the contralateral primary sensorimotor cortex, cerebellum, dorsal premotor cortex, supplementary motor area, superior posterior parietal cortex and occipital areas relative to the baseline, when the participants simply looked at the object.

In these early PET studies of grasping it was noted that several key areas were not activated (Grafton *et al.*, 1996b). Most notably, the ventral premotor cortex (PMv) and the anterior part of the intraparietal sulcus were not significantly active. It had previously been suggested that these areas were key nodes for visually guided reach-to-grasp in non-human primates (Jeannerod *et al.*, 1995). Neurons in both of these areas code the size, shape and orientation of objects and the specific type of grips needed to grasp them (Rizzolatti *et al.*, 1988; Taira *et al.*, 1990; Sakata *et al.*, 1995; Murata *et al.*, 1997, 2000) (see Chapter 4). Partly as a consequence of these negative findings, ongoing discussions took place in the literature about the tentative location of the human PMv (see Rizzolatti *et al.*, 1996; Fink *et al.*, 1997). As it happens, the negative results were probably attributable to the limited number of scans that can be collected from a single subject with PET as only about 10–12 scans can be registered under the safety restrictions regarding the total dose of radioactive tracer that can be administered to a volunteer. In this respect, fMRI has the advantage that, in theory, there is no limit to the number of scans that can be acquired. Indeed, in a typical fMRI experiment, several hundred, and sometimes over a thousand, scans are collected in a single individual. This substantially reduces the standard errors of the means for the estimated condition-specific effects, thereby increasing the sensitivity of the statistical analysis. As will be described in the text below, several recent fMRI studies have shown reproducible and robust activity in both PMv and the anterior intraparietal sulcus during grasping.

Human dexterity relies heavily on the capacity to manipulate small objects between the tips of the index fingers and thumbs, the so called "precision grip" (Napier, 1956) (see

Chapters 1, 12 and 13). Precision grips require independent finger movements and fine control of the direction and magnitude of the fingertip forces (Johansson & Cole, 1992; Flanagan *et al.*, 1999). The differences in sensitivity between PET and fMRI were probably also an important factor in the two first imaging studies of precision grips (Ehrsson *et al.*, 2000; Kinoshita *et al.*, 2000). Kinoshita and colleagues used PET to register increases in rCBF in the left primary motor cortex, supplementary motor area, right inferior parietal cortex and cerebellum when subjects repeatedly lifted and released an object with a precision grip between index finger and thumb, relative to the resting baseline. With the exception of the inferior parietal activation, these are motor structures that are always active during voluntary motor tasks (Roland & Zilles, 1996), irrespective of the particular task used. Ehrsson and colleagues used fMRI to scan a precision grip task where subjects were required to apply fine isometric grip forces to a test object between right index finger and thumb (Ehrsson *et al.*, 2000). They compared this to a power grip task where the participants compressed a plastic cylinder between all digits in a palmar opposition grasp. Both tasks activated the core executive motor areas (the contralateral left primary motor cortex, the bilateral SMA and right anterior lateral cerebellum), but the precision grip task produced significantly stronger activation in the right PMv (with statistical trend in the left PMv), the right anterior part of the intraparietal sulcus, and the bilateral inferior parietal cortex (supramarginal gyrus). In another early fMRI study, Binkofski and colleagues (Binkofski *et al.*, 1999) scanned people during a tactile shape exploration paradigm (similar to the one used by Roland and Larsen described above), and found bilateral activation of the PMv, anterior IPS and supramarginal gyrus. A recent fMRI study (Schmitz *et al.*, 2005) employing a precision-grip-lift paradigm similar to the one used by Kinoshita and colleagues found activation in the left ventral premotor cortex (areas 44 and 6), and right supramarginal gyrus when lifting objects of different weights. In summary, fMRI seems to be a powerful imaging modality to detect biologically relevant signals in the premotor and posterior parietal cortices during grasping.

Functional magnetic resonance imaging of grasping

The physiological basis of the BOLD signal

So what is the physiological basis of the fMRI signal? fMRI detects and localizes neural activity by examining changes in regional blood flow in the brain. In an active region of the brain, the supply of oxygenated blood is greater than its consumption, leading to a higher than normal ratio of oxygenated to deoxygenated blood. Because these two forms of hemoglobin have different magnetic properties, they produce different magnetic signals. This is the blood-oxygenation level dependent signal (BOLD) that serves as an index of neural activity (Kwong *et al.*, 1992; Ogawa *et al.*, 1992).

Recent years have seen substantial advances in our understanding of the cellular basis of the BOLD signal (for a recent review, see Logothetis & Wandell, 2004). Briefly, there is a close relationship between the BOLD signal and the overall level of synaptic activity in cortical populations of neurons, as measured with local field potential (Logothetis *et al.*, 2001; Logothetis, 2003). Thus the BOLD signal is strongly related to the presynaptic activity of interneurons and afferent inputs to the neuronal population in question. There is also a fairly good relation between the spike rate of neurons and the BOLD signal, at least in the cortex (Logothetis *et al.*, 2001; Rasch *et al.*, 2008). Because most cortical neurons are excitatory, the BOLD signal can generally be assumed to reflect increased information processing, although cerebellar circuits (Lauritzen & Gold, 2003) and certain cortical perceptual suppression mechanisms (Wilke *et al.*, 2006) may prove to be exceptions in this case. It is reassuring that an increasing number of fMRI studies in non-human primates have found good correspondence between BOLD activations and neurophysiological data (Orban *et al.*, 2004; Tsao *et al.*, 2006; Durand *et al.*, 2007).

Technical challenges

The fMRI of grasping presents a number of technical challenges. First, arm movements induce distortions of the magnetic field, which, in turn, lead to artefacts in the images. Second, arm movements can cause postural adjustments involving neck muscles and small head movements that disturb the images. Third, in a standard MRI experiment, the person lies in a supine position with their head inside the head-coil of the scanner. Thus their view of the object in front of them is obscured and can only be made available with the help of mirrors. The introduction of mirrors could change the pattern of brain activity because of the additional neural processing related to the spatial transformation of the visual information. Finally, in studies of grasping and object manipulation, it is often essential to measure the forces applied between the fingertips and the contact surfaces of the object. However, the strong magnetic fields in the scanner room and radio-frequency pulses emitted by the scanner can cause interference in electronic measurement devices. Conversely, the electronic devices tend to introduce radio-frequency noise that can reduce image quality and introduce artefacts in the data. The combination of these technical problems made the initial progress of fMRI studies of grasping slower than in other areas of cognitive neuroimaging.

The problem of how to implement visually guided reach-to-grasp in the scanner environment turned out to have two key solutions. The first of these was to ensure participants had a direct view of the manipulated object by tilting their head inside the head-coil at an angle of about 25–30 degrees (Culham *et al.*, 2003) (as shown in Figure 7.1A). Although participants in this position no longer have their head in the center of the magnetic field inside the head-coil, thereby losing some signal quality, the images are still good enough to enable relevant active areas to be detected. This approach, however, can be uncomfortable for some participants in the longest experiments because of neck muscle fatigue from having their head tilted, and from effort related to maintaining an oblique gaze. A further development to address these issues was recently reported by Ivan Toni's team in Holland (Grol *et al.*, 2007). They tilted the head coil rather than the head inside the coil (see Figure 7.1B). This set-up is more comfortable for the participant, both in terms of the neck posture and gaze, but requires a phased-array coil as receiver and body coil as transmitter. Further, tilting the head-coil is

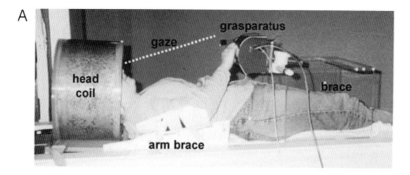

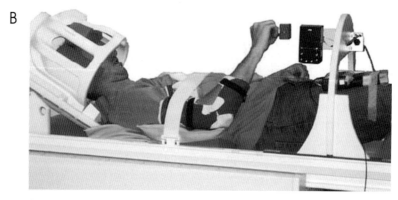

Figure 7.1. This figure is also reproduced in the color plate section. Set-ups used to scan reach-to-grasp. A. In Culham *et al.* (2003) study the participants have their heads tilted inside the head-coil enabling them to see the object directly. B. A further development of this approach, as used in the Toni lab (Grol *et al.*, 2007), is to tilt the head-coil. In many scanners this provides a clearer view of the object and is more comfortable for the participant.

associated with signal loss in proportion to the tilt from the horizontal place (the B0-direction), but in combination with high field imaging and phase-array head-coils, standard in the latest generation of scanners, signal quality is usually good anyway.

The second main problem in the reach-to-grasp experiments was from the artefacts induced by movements of the arm. These included the distortions of the magnetic field and head movements correlated with the upper limb movement mentioned above. The elimination of these artefacts turned out to have a quite elegant solution involving the very time course of the BOLD signal itself. The BOLD signal is delayed 6 s with respect to the neuronal activity, whereas the movement-induced artefacts are instantaneous (both field distortions and head movement artefacts). Thus in the fMRI data the artefacts occur *before* the BOLD signal related to the task allowing the task-relevant activations to be characterized. Technically speaking, by employing a (sparse) event-related design where the BOLD signal evoked by individual reaching movements is modeled, one can effectively temporally decouple the task-relevant activations from the artefacts (Culham *et al.*, 2003). This approach is also useful in experiments of reactive grip responses to unexpected perturbations as these are particularly likely to induce movements of the head (Ehrsson *et al.*, 2007). It should be pointed out that movements of the head that are smaller than one voxel (about 3 mm) and not correlated with the paradigm are very well corrected for in the postprocessing of the images (e.g. SPM or Brain Voyager).

As mentioned above, it is often important to measure the forces applied at the contact surfaces of the digits and the object. The fingertip-forces provide important information about the sensorimotor control processes involved in human dexterous manipulation (see below). In the MRI environment this can be achieved using non-magnetic fiber-optic transducers to measure the forces employed on the manipulated object (see Figure 7.2A) (Schmitz *et al.*, 2005). As the participant grasps the instrumented test object between index finger and thumb (in a precision grip), the grip forces (applied normal to the surface) and load forces (applied tangentially) are recorded. The position of the object and the hand can also be registered and the weight of the object lifted can be altered. By using this type of set-up, a substantial amount of information has been gathered about the neural basis of fingertip force control during object lifting and manipulation (Ehrsson, 2001; Ehrsson *et al.*, 2001, 2003; Kuhtz-Buschbeck *et al.*, 2001; Schmitz *et al.*, 2005; Jenmalm *et al.*, 2006).

Contribution of fMRI to the neuroscience of grasping

It is beyond the scope of this chapter to provide a full account of the cortical physiology of grasping and manipulation (for a recent review, see Castiello, 2005; Castiello & Begliomini, 2008). Instead we will briefly summarize two areas of research where imaging experiments have been particularly informative. The intention here is to give the reader an idea of the sort of questions imaging is able to address and the type of answers it provides.

Visually guided reach-to-grasp and frontoparietal circuits

The identification of the frontoparietal areas that are active when people reach out to grasp objects guided by vision has been the goal of many imaging studies (the visually guided reach-to-grasp paradigm) (Grafton *et al.*, 1996b; Matsumura *et al.*, 1996; Binkofski *et al.*, 1998; Culham *et al.*, 2003; Culham & Valyear, 2006; Begliomini *et al.*, 2007a, 2007b; Cavina-Pratesi *et al.*, 2007). These have revealed several active foci within the intraparietal sulcus (Figure 7.3). The responses in the most anterior part of the intraparietal sulcus correspond well with the anterior intraparietal area (AIP) as defined in the brains of non-human primates (Binkofski *et al.*, 1998; Culham *et al.*, 2003; Culham & Valyear, 2006; Begliomini *et al.*, 2007a, 2007b; Cavina-Pratesi *et al.*, 2007). More posterior foci within the intraparietal sulcus may correspond to other regions involved in reaching in macaques, such as the "parietal reach region," MIP or LIP, although great uncertainty still remains about these relationships (Culham & Valyear, 2006). Robust responses during grasping are also seen in the human ventral premotor cortex in the frontal lobes (inferior area 6 and area 44).

A

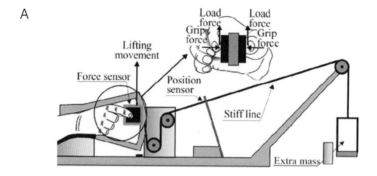

B

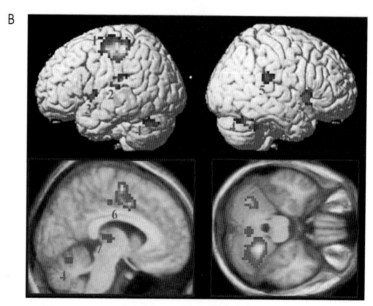

Figure 7.2. This figure is also reproduced in the color plate section. A. An instrumented apparatus to measure grip forces and load forces applied to a test object held between the index finger and thumb (precision grip) (used in Schmitz *et al.*, 2005; Jenmalm *et al.*, 2006). The position of the object can also be registered and the weight of the object manipulated. B. The brain activation during repeated lifting of the test object with a precision grip with the right hand (without visual feedback). Activations in the left primary sensorimotor cortex (1), the parietal operculum (SII region) (2), the bilateral ventral premotor cortex (the inferior precentral sulcus and inferior frontal gyrus pars opercularis) (3), the anterior lateral cerebellum (4), the right supramarginal gyrus in the inferior parietal lobe (5), the left supplementary motor area and the cingulate motor area on the medial wall of the frontal lobes (6), and the left ventral thalamus (7). From Schmitz *et al.* (2005).

Human imaging experiments of grasping and the lifting or squeezing of objects with precision grips often report multiple foci in the PMv, in the inferior part of the precentral gyrus (area 6), in the inferior part of the precentral sulcus (the border zone between area 44 and area 6) and in the inferior frontal gyrus pars opercularis (area 44) (Ehrsson *et al.*, 2000,

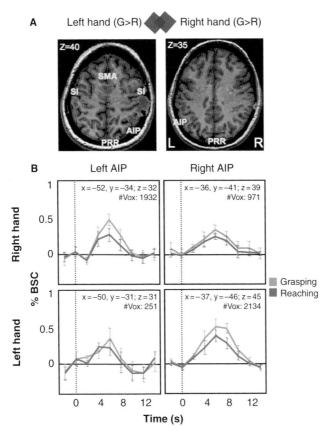

Figure 7.3. This figure is also reproduced in the color plate section. A. Functional magnetic resonance imaging activity in the contralateral human anterior intraparietal area (hAIP; encircled in orange) during visually guided reach-to-grasp using the right hand (blue) or the left hand (pink). The activation maps are produced by contrasting a reach-to-grasp condition (G) and a reach-only condition (R). B. The BOLD signal changes (% BSC on the y-axis) from single reach to grasp events are plotted (green) and compared to the response during a reaching without grasping control condition (red) over time (x-axis) (from (Culham *et al.*, 2006). The coordinates in standard space for the plotted hAIP clusters are indicated. (SMA, supplementary motor area; S1, primary somatosensory cortex; PCu, precuneus).

2001, 2003; Ehrsson, 2001). It is still unclear how these foci relate to the rostral (area F5) and caudal (area F4) sections of PMv in the monkey brain. Rizzolatti and colleagues have proposed that area 44 is the functionally equivalent area of monkey F5, and inferior area 6 the equivalent of F4 (Rizzolatti *et al.*, 2002).

Imaging is not only trying to confirm what was already known from experiments in non-human primates. New experiments have begun characterizing the differences in response patterns between the premotor and anterior intraparietal areas in greater detail (Begliomini *et al.*, 2007a, 2007b), taking advantage of the obvious fact that neuroimaging experiments

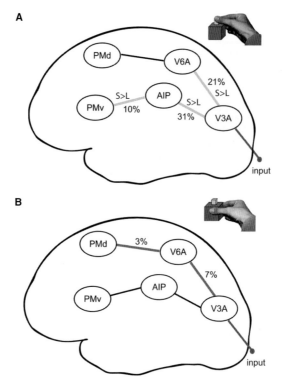

Figure 7.4. This figure is also reproduced in the color plate section. Functional magnetic resonance imaging can be used to investigate changes in effective connectivity between frontoparietal areas during grasping. Dynamic causal modeling was used to characterize the coupling of the time-courses of activity in key areas involved in visually guided reach-to-grasp (from Grol *et al.*, 2007). The effective connectivity between the anterior intraparietal area and the ventral premotor cortex was particularly strong (green lines) during the grasping of small objects using a precision grip, as compared with grasping a larger object which tended to increase coupling between a tentative area V6A (the posterior branch of the dorsal end of the parieto-occipital sulcus) and the dorsal premotor cortex (red lines).

record neural activity in many areas simultaneously. Another new development in imaging is to use fMRI to characterize changes in the effective connectivity between the frontal and parietal areas during grasping (Grol *et al.*, 2007). Grol and colleagues recently described how the inter-area coupling between the anterior intraparietal cortex and ventral premotor cortex was particularly strong when grasping small objects using a precision grip, consistent with the hypothesis that transmission of information processing in this circuit is important for grasping (see Figure 7.4). This technique can thus be used to investigate how frontal and parietal areas communicate when producing grasping behavior. In conclusion, imaging experiments using the visually guided reach-to-grasp paradigm have reported findings that are consistent with the state of knowledge in the animal literature and experiments of this kind are increasingly beginning to address new questions about the function of the frontoparietal circuits.

Sensorimotor control mechanisms for dexterous manipulation

A second fruitful line of research has taken as its starting point knowledge obtained from behavioral experiments in humans. This tradition has relied heavily on versions of the precision-grip-lift paradigm that were introduced by Roland Johansson and Göran Westling to investigate the control of dexterous manipulation (Johansson & Westling, 1984; Johansson & Cole, 1992) (see Chapters 1 and 11). The imaging experiments follow-ing this tradition have sought to identify the neuronal correlates of key sensorimotor control mechanisms used in human dexterous manipulation. For example, we know that anticipa-tory control is a fundamental policy employed by the central nervous system when lifting objects (Johansson, 1996; Flanagan *et al.*, 2006) (see Chapter 12). When reaching out to lift a familiar object, the finger-tip forces are programmed in advance to ensure that they are appropriate for the anticipated weight of the object. We also know that somatosensory feedback during different phases of the action is critical for the detection of mismatches between anticipated and afferent feedback, and for the triggering of pre-programmed corrective force responses to stabilize the grip if excessive or inadequate forces are applied (Johansson & Cole, 1992; Flanagan *et al.*, 2006; see Chapter 11).

Jenmalm and colleagues used fMRI to identify the neural correlates of mismatches between the anticipated and the actual weight of objects during a precision-grip lift task, and two different types of pre-programmed corrective responses associated with such mismatch events (Jenmalm *et al.*, 2006). The authors used an event-related design that allowed them to characterize the BOLD responses associated with individual lift-and-release actions. The results revealed different activity patterns in the cerebellum, motor cortex and inferior parietal cortex, three areas that have been hypothesized to be important for anticipatory motor control (see Figure 7.5). The inferior parietal cortex seemed to signal the discrepancy between the predicted and actual feedback, regardless of whether the object was heavier or lighter than expected. The cerebellar and motor responses seemed to be related to the specific type of pre-programmed corrective responses used to inhibit excessive force (cerebellum) and to prevent slipping by increasing the force (motor cortex), respectively.

The anticipatory control policy is also expressed in the coupling of grip and load forces during the manipulation of stable objects, and other fMRI experiments have been designed to localize neuronal activity reflecting this force coordination by introducing various control conditions where grip or load forces are applied in isolation. Activity related to the conditions where coordinated grip and load forces are required has been reported in the parietal cortex (Ehrsson *et al.*, 2003) and posterior lateral cerebellum (Kawato *et al.*, 2003; Boecker *et al.*, 2005), two regions that are anatomically connected (Clower *et al.*, 2001, 2005), possibly forming parieto-cerebellar circuits for anticipatory coupling of grip and load forces.

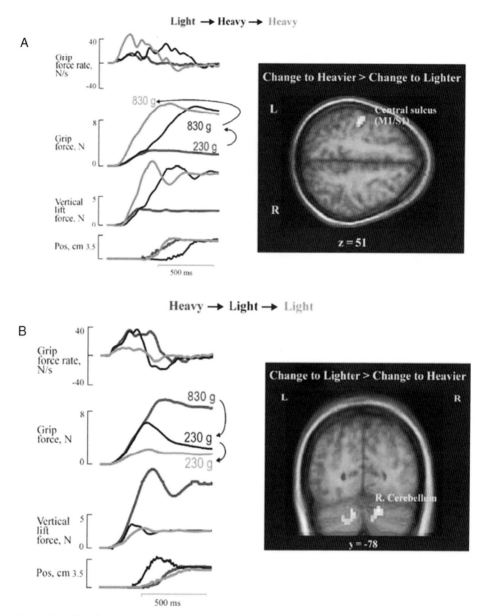

Figure 7.5. This figure is also reproduced in the color plate section. Event-related fMRI study of neuronal correlates of unexpected changes in object weight during lifting. Participants repeatedly lifted a test object with their right index finger (precision grip) without visual feedback. In some trials, the weight of the object was unpredictably changed between a light weight (230 g) and a heavy one (830 g). The characteristic corrective force patterns associated with these trials are shown to the left. Mismatches between anticipated and actual weight were associated with activation of the right supramarginal gyrus in the inferior parietal lobe, irrespective of whether the object was lighter or heavier than expected (not shown). A. When the object was heavier than expected, there was an additional specific response in the primary motor cortex, presumably reflecting the "pulsatile" increase in the grip force needed to achieve lift-off and the target force (see black grip and lift force curves). B. Trials where the object was lighter than expected, and excessive forces had to be reduced (see black grip force curves), were associated with specific activation of the bilateral lateral cerebellar hemispheres (B, right panel). From Jenmalm *et al.* (2006).

Another important control strategy is reactive control of fingertip forces, which is used to stabilize unexpected perturbations, or when handling unpredictable "active" objects such as when holding a dog on a leash. This type of control relies heavily on somatosensory feedback from the fingertips (Johansson & Cole, 1992). Imaging experiments have begun to characterize the neural responses associated with reactive control and have detected responses in the primary motor cortex and anterior lateral cerebellar hemispheres associated with this (Ehrsson *et al.*, 2006).

Finally, the precision-grip-lift paradigm has provided an important model for computational movement neuroscience, which aims to explain the computational principles that underlie skilled movements (Kawato & Wolpert, 1998; Wolpert & Ghahramani, 2000) (see Chapter 9). Increasingly, the theoretical models provided in this area of research have served as a framework by neuroimagers for developing neuroanatomical hypotheses (Kawato *et al.*, 2003; Boecker *et al.*, 2005; Schmitz *et al.*, 2005; Bursztyn *et al.*, 2006; Jenmalm *et al.*, 2006). However, these hypotheses are still at the stage of looking at differences between conditions, such as "localizing the internal model" or "localizing the central grip-load force coupling," which is fine in itself; however, the real potential in this direction lies in the capacity of computational models to derive formal hypotheses about BOLD responses, i.e. concerning responses that do not simply involve the differences between two or more conditions or a simple linear relationship between movement parameters and neural activity. Such approaches have turned out to be very successful in other areas of cognitive neuro-imaging (O'Doherty *et al.*, 2003) and could offer a powerful approach for testing hypotheses concerning the computational principles used by the brain circuits to control grasping and skilled movement more generally.

References

Begliomini, C., Wall, M. B., Smith, A. T. & Castiello, U. (2007a). Differential cortical activity for precision and whole-hand visually guided grasping in humans. *Eur J Neurosci*, **25**, 1245–1252.

Begliomini, C., Caria, A., Grodd, W. & Castiello, U. (2007b). Comparing natural and constrained movements: new insights into the visuomotor control of grasping. *PLoS ONE*, **2**, e1108.

Binkofski, F., Dohle, C., Posse, S. *et al.* (1998). Human anterior intraparietal area subserves prehension: a combined lesion and functional MRI activation study. *Neurology*, **50**, 1253–1259.

Binkofski, F., Buccino, G., Posse, S. *et al.* (1999). A fronto-parietal circuit for object manipulation in man: evidence from an fMRI-study. *Eur J Neurosci*, **11**, 3276–3286.

Boecker, H., Lee, A., Muhlau, M. *et al.* (2005). Force level independent representations of predictive grip force-load force coupling: a PET activation study. *Neuroimage*, **25**, 243–252.

Bursztyn, L. L., Ganesh, G., Imamizu, H., Kawato, M. & Flanagan, J. R. (2006). Neural correlates of internal-model loading. *Curr Biol*, **16**, 2440–2445.

Castiello, U. (2005). The neuroscience of grasping. *Nat Rev Neurosci*, **6**, 726–736.

Castiello, U. & Begliomini, C. (2008). The cortical control of visually guided grasping. *Neuroscientist*, **14**, 157–170.

Cavina-Pratesi, C., Goodale, M. A. & Culham, J. C. (2007). FMRI reveals a dissociation between grasping and perceiving the size of real 3D objects. *PLoS ONE*, **2**, e424.

Clower, D. M., West, R. A., Lynch, J. C. & Strick, P. L. (2001). The inferior parietal lobule is the target of output from the superior colliculus, hippocampus, and cerebellum. *J Neurosci*, **21**, 6283–6291.

Clower, D. M., Dum, R. P. & Strick, P. L. (2005). Basal ganglia and cerebellar inputs to 'AIP'. *Cereb Cortex*, **15**, 913–920.

Culham, J. C. & Valyear, K. F. (2006). Human parietal cortex in action. *Curr Opin Neurobiol*, **16**, 205–212.

Culham, J. C., Danckert, S. L., DeSouza, J. F. *et al.* (2003). Visually guided grasping produces fMRI activation in dorsal but not ventral stream brain areas. *Exp Brain Res*, **153**, 180–189.

Culham, J. C., Cavina-Pratesi, C. & Singhal, A. (2006). The role of parietal cortex in visuomotor control: what have we learned from neuroimaging? *Neuropsychologia*, **44**, 2668–2684.

Durand, J. B., Nelissen, K., Joly, O. *et al.* (2007). Anterior regions of monkey parietal cortex process visual 3D shape. *Neuron*, **55**, 493–505.

Ehrsson, H. H. (2001). Neural correlates of skilled movement: functional mapping of the human brain with fMRI and PET. PhD thesis at Department of Woman and Child Health. Stockholm: Karolinska University Press.

Ehrsson, H. H., Fagergren, A., Jonsson, T. *et al.* (2000). Cortical activity in precision- versus power-grip tasks: an fMRI study. *J Neurophysiol*, **83**, 528–536.

Ehrsson, H. H., Fagergren, E. & Forssberg, H. (2001). Differential fronto-parietal activation depending on force used in a precision grip task: an fMRI study. *J Neurophysiol*, **85**, 2613–2623.

Ehrsson, H. H., Fagergren, A., Johansson, R. S. & Forssberg, H. (2003). Evidence for the involvement of the posterior parietal cortex in coordination of fingertip forces for grasp stability in manipulation. *J Neurophysiol*, **90**, 2978–2986.

Ehrsson, H. H., Fagergren, A., Ehrsson, G. O. & Forssberg, H. (2007). Holding an object: neural activity associated with fingertip force adjustments to external perturbations. *J Neurophysiol*, **97**, 1342–1352.

Fink, G. R., Frackowiak, R. S., Pietrzyk, U. & Passingham, R. E. (1997). Multiple nonprimary motor areas in the human cortex. *J Neurophysiol*, **77**, 2164–2174.

Flanagan, J. R., Burstedt, M. K. & Johansson, R. S. (1999). Control of fingertip forces in multidigit manipulation. *J Neurophysiol*, **81**, 1706–1717.

Flanagan, J. R., Bowman, M. C. & Johansson, R. S. (2006). Control strategies in object manipulation tasks. *Curr Opin Neurobiol*, **16**, 650–659.

Grafton, S. T., Arbib, M. A., Fadiga, L. & Rizzolatti, G. (1996a). Localization of grasp representations in humans by positron emission tomography. 2. Observation compared with imagination. *Exp Brain Res*, **112**, 103–111.

Grafton, S. T., Fagg, A. H., Woods, R. P. & Arbib, M. A. (1996b). Functional anatomy of pointing and grasping in humans. *Cereb Cortex*, **6**, 226–237.

Grol, M. J., Majdandzic, J., Stephan, K. E. *et al.* (2007). Parieto-frontal connectivity during visually guided grasping. *J Neurosci*, **27**, 11877–11887.

Jeannerod, M., Arbib, M. A., Rizzolatti, G. & Sakata, H. (1995). Grasping objects: the cortical mechanisms of visuomotor transformation. *Trends Neurosci*, **18**, 314–320.

Jenmalm, P., Schmitz, C., Forssberg, H. & Ehrsson, H. H. (2006). Lighter or heavier than predicted: neural correlates of corrective mechanisms during erroneously programmed lifts. *J Neurosci*, **26**, 9015–9021.

Johansson, R. S. (1996). Sensory control of dexterous manipulation. In A. M. Wing, P. Haggard & J. R. Flanagan (Eds.), *Hand and Brain: The Neurophysiology and Psychology of Hand Movements* (pp. 381–412). San Diego, CA: Academic Press.

Johansson, R. S. & Westling, G. (1984). Roles of glabrous skin receptors and sensorimotor memory in automatic control of precision grip when lifting rougher or more slippery objects. *Exp Brain Res*, **56**, 550–564.

Johansson, R. S. & Cole, K. J. (1992). Sensory-motor coordination during grasping and manipulative actions. *Curr Opin Neurobiol*, **2**, 815–823.

Kawato, M. & Wolpert, D. (1998). Internal models for motor control. *Novartis Found Symp*, **218**, 291–304; discussion 304–297.

Kawato, M., Kuroda, T., Imamizu, H. *et al.* (2003). Internal forward models in the cerebellum: fMRI study on grip force and load force coupling. *Prog Brain Res*, **142**, 171–188.

Kinoshita, H., Oku, N., Hashikawa, K. & Nishimura, T. (2000). Functional brain areas used for the lifting of objects using a precision grip: a PET study. *Brain Res*, **857**, 119–130.

Kuhtz-Buschbeck, J. P., Ehrsson, H. H. & Forssberg, H. (2001). Human brain activity in the control of fine static precision grip forces: an fMRI study. *Eur J Neurosci*, **14**, 382–390.

Kwong, K. K., Belliveau, J. W., Chesler, D. A. *et al.* (1992). Dynamic magnetic resonance imaging of human brain activity during primary sensory stimulation. *Proc Natl Acad Sci USA*, **89**, 5675–5679.

Lauritzen, M. & Gold, L. (2003). Brain function and neurophysiological correlates of signals used in functional neuroimaging. *J Neurosci*, **23**, 3972–3980.

Logothetis, N. K. (2003). The underpinnings of the BOLD functional magnetic resonance imaging signal. *J Neurosci*, **23**, 3963–3971.

Logothetis, N. K. & Wandell, B. A. (2004). Interpreting the BOLD signal. *Annu Rev Physiol*, **66**, 735–769.

Logothetis, N. K., Pauls, J., Augath, M., Trinath, T. & Oeltermann, A. (2001). Neurophysiological investigation of the basis of the fMRI signal. *Nature*, **412**, 150–157.

Matsumura, M., Kawashima, R., Naito, E. *et al.* (1996). Changes in rCBF during grasping in humans examined by PET. *Neuroreport*, **7**, 749–752.

Murata, A., Fadiga, L., Fogassi, L. *et al.* (1997). Object representation in the ventral premotor cortex (area F5) of the monkey. *J Neurophysiol*, **78**, 2226–2230.

Murata, A., Gallese, V., Luppino, G., Kaseda, M. & Sakata, H. (2000). Selectivity for the shape, size, and orientation of objects for grasping in neurons of monkey parietal area AIP. *J Neurophysiol*, **83**, 2580–2601.

Napier, J. R. J. (1956). The prehensile movements of the human hand. *J Bone Joint Surg*, **38B**, 902–913.

O'Doherty, J. P., Dayan, P., Friston, K., Critchley, H. & Dolan, R. J. (2003). Temporal difference models and reward-related learning in the human brain. *Neuron*, **38**, 329–337.

Ogawa, S., Tank, D. W., Menon, R. *et al.* (1992). Intrinsic signal changes accompanying sensory stimulation: functional brain mapping with magnetic resonance imaging. *Proc Natl Acad Sci USA*, **89**, 5951–5955.

Orban, G. A., Van Essen, D. & Vanduffel, W. (2004). Comparative mapping of higher visual areas in monkeys and humans. *Trends Cogn Sci*, **8**, 315–324.

Rasch, M. J., Gretton, A., Murayama, Y., Maass, W. & Logothetis, N. K. (2008). Inferring spike trains from local field potentials. *J Neurophysiol*, **99**, 1461–1476.

Rizzolatti, G., Camarda, R., Fogassi, L. *et al.* (1988). Functional organization of inferior area 6 in the macaque monkey. II. Area F5 and the control of distal movements. *Exp Brain Res*, **71**, 491–507.

Rizzolatti, G., Fadiga, L., Matelli, M. *et al.* (1996). Localization of grasp representations in humans by PET: 1. Observation versus execution. *Exp Brain Res*, **111**, 246–252.

Rizzolatti, G., Fogassi, L. & Gallese, V. (2002). Motor and cognitive functions of the ventral premotor cortex. *Curr Opin Neurobiol*, **12**, 149–154.

Roland, E. & Larsen, B. (1976). Focal increase of cerebral blood flow during stereognostic testing in man. *Arch Neurol*, **33**, 551–558.

Roland, P. E. & Zilles, K. (1996). Functions and structures of the motor cortices in humans. *Curr Opin Neurobiol*, **6**, 773–781.

Sakata, H., Taira, M., Murata, A. & Mine, S. (1995). Neural mechanisms of visual guidance of hand action in the parietal cortex of the monkey. *Cereb Cortex*, **5**, 429–438.

Schmitz, C., Jenmalm, P., Ehrsson, H. H. & Forssberg, H. (2005). Brain activity during predictable and unpredictable weight changes when lifting objects. *J Neurophysiol*, **93**, 1498–1509.

Taira, M., Mine, S., Georgopoulos, A. P., Murata, A. & Sakata, H. (1990). Parietal cortex neurons of the monkey related to the visual guidance of hand movement. *Exp Brain Res*, **83**, 29–36.

Tsao, D. Y., Freiwald, W. A., Tootell, R. B. & Livingstone, M. S. (2006). A cortical region consisting entirely of face-selective cells. *Science*, **311**, 670–674.

Wilke, M., Logothetis, N. K. & Leopold, D. A. (2006). Local field potential reflects perceptual suppression in monkey visual cortex. *Proc Natl Acad Sci USA*, **103**, 17507–17512.

Wolpert, D. M. & Ghahramani, Z. (2000). Computational principles of movement neuroscience. *Nat Neurosci*, **3** Suppl., 1212–1217.

8

Functional magnetic resonance imaging studies of the basal ganglia and precision grip

MATTHEW B. SPRAKER, DANIEL M. CORCOS AND
DAVID E. VAILLANCOURT

Summary

Grasping behavior has been well studied in both human and non-human primates. Studies have revealed a classic grasping circuit that involves several regions, such as the motor, prefrontal and parietal cortices. However, the functional contribution of the basal ganglia to grasping control is often overlooked. This is surprising because many basal ganglia disorders (e.g. Parkinson's disease) have been experimentally associated with deficits in grasping control. Recent work in our laboratory used fMRI to demonstrate that the caudate, putamen, internal and external segments of the globus pallidus (GPi and GPe, respectively), and subthalamic nucleus (STN) participate in circuits that independently regulate the selection and scaling of parameters important for grasping. These findings provide new evidence that grasping must be considered as a behavior that is processed in both cortical and subcortical structures.

Introduction

Prehension remains one of the most important functions of primate motor systems. The remarkable adaptability and effortlessness with which primates can reach for and grasp objects of variable size, shape and mass has had unequivocal evolutionary importance. Nevertheless, it is widely accepted that even simple reach-to-grasp movements pose considerable challenges for the primate sensorimotor system (Johnson-Frey, 2003). During the prehension of a given object, individuals must use visual (i.e. object distance, direction) and somatosensory information (i.e. joint angle) to transport the hand to the object location via a precisely aimed reaching movement. An important study which characterized prehension kinematically in humans used video to demonstrate features of the entire reach-to-grasp movement (Jeannerod, 1984) (see Chapters 2 and 10). Since then, studies of prehension have shown that a number of object properties significantly affect reaching and grasping kinematics, such as fragility, size, texture and weight (Smeets & Brenner, 1999). In addition, it has been hypothesized that the individual's intended activity for which the object will be used can also affect the kinematics of prehension (Castiello, 2005). Finally, appropriate

Sensorimotor Control of Grasping: Physiology and Pathophysiology, ed. Dennis A. Nowak and Joachim Hermsdörfer. Published by Cambridge University Press. © Cambridge University Press 2009.

levels of force must be applied as the hand makes contact with the object and lifts the object during manipulation (Flanagan *et al.*, 2006) (see Chapters 11 and 12).

Grasping poses a formidable challenge for the CNS because these kinetic variables must be rapidly adjusted to grasp unique items many times a day. This becomes quite apparent from clinical observations of patients with movement disorders such as Parkinson's disease (PD), in which both reach and grasp motor deficits have been identified (Castiello *et al.*, 1993; Alberts *et al.*, 1998, 2000; Rearick *et al.*, 2002; Bertram *et al.*, 2005; Nowak & Hermsdorfer, 2006; Rand *et al.*, 2006; see also Chapters 22, 23 and 32). Also, when PD patients grasp a pen during writing, micrographia has been identified as an important functional deficit (Contreras-Vidal *et al.*, 1995). Moreover, patients with other movement disorders including Huntington's disease (Gordon *et al.*, 2000) and dystonia (Nowak *et al.*, 2005) overestimate grip force greater than the task-appropriate level. Since the pathology of these movement disorders is related to dysfunction of the basal ganglia (DeLong, 1990; Abbruzzese & Berardelli, 2003), this raises questions about the role of the basal ganglia in precision grip force.

Here we discuss three recent studies from our laboratory that segregate the function of individual basal ganglia nuclei by examining their differential role in selecting and regulating the production of precision grip force. First, we focus on work that demonstrates that the caudate and anterior putamen of the basal ganglia select the amplitude of grip force pulses whereas the posterior putamen, GPi and STN regulate the production of force pulses (Vaillancourt *et al.*, 2007). Second, we focus on studies that demonstrate that segregated circuits in the basal ganglia independently scale the duration, rate of change and amplitude of grip-force output (Vaillancourt *et al.*, 2004; Spraker *et al.*, 2007). Finally, we discuss these findings in the context of the classic cortical grasping circuit to demonstrate that grasping is controlled by a network of several cortical and subcortical structures.

Selection of grip force output and the basal ganglia

The work of Vaillancourt and colleagues (2007) used a block-design functional magnetic resonance imaging (fMRI) paradigm to examine whether specific nuclei of the basal ganglia are involved in producing the same series of force pulses or selecting different levels of force, or if they are involved in both. The experimental setup and grasping configuration is shown in Figure 8.1A and B. We compared the blood oxygenation level dependent (BOLD) signal during three experimental tasks: (1) hold, (2) similar pulse and (3) select different pulse. Also, each task was completed under internally and externally guided conditions where force amplitude was either scaled by the subject internally or based on a visual target, respectively.

As shown in Figure 8.2A, the hold condition required subjects to generate steady-state grip force at 15% of their maximum voluntary contraction (MVC) for 30 s. The pulse condition required subjects to generate a force pulse to 15% MVC for 2 s, rest for 1 s, and repeat 10 times in the 30 s block. The BOLD signal during the pulse condition was compared with the hold condition to determine the signal changes associated with producing

A

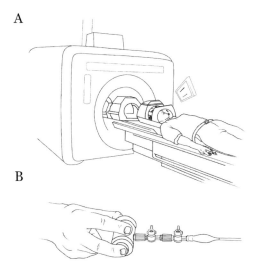

B

Figure 8.1. Depicts the experimental setup used in studies discussed in this chapter. A. Subjects lay supine in the scanner and receive visual feedback from a projection screen situated overhead. B. Depicts the hand configuration used to produce force against the pinch grip device. The subject's thumb and middle digit contacted the device at all times during the experiments. Reprinted from Vaillancourt *et al.* (2004) *Neuroimage*, **23**, 175, with permission from Elsevier.

force pulses relative to a sustained precision grip. This did not require subjects to select amongst different force amplitudes because they could only produce force to one predetermined level. The select condition required subjects to generate a similar series of force pulses as in the pulse condition, except the force pulses varied in amplitude within the block. The subjects were trained to produce force pulses with amplitudes that varied between 5% and 40% MVC with the average across the 10 pulses per block equal to 15% MVC. The BOLD signal during the select condition was compared with that during the pulse condition to determine the changes in BOLD response associated with actively selecting which force level to produce. In addition, the internally guided condition required subjects to internally select which force level to produce while the externally guided condition did not. Thus, the internally guided condition had both variable amplitude pulse and selection components while the externally guided condition had only variable amplitude pulse components to the task. We trained subjects so that the average force output and the duration of force were constant across these two conditions.

Figure 8.2B and C show the summary of findings from this study for the basal ganglia. We found that individual regions of the basal ganglia could be segregated into three functional groups based on the BOLD response related to the three tasks. First, the caudate had increased activation during the free selection of force amplitude but not during the production of force pulses with the same force amplitude. Second, the STN, GPi and the posterior putamen increased in activation during the production of force pulses with the same force amplitude, but maintained a constant level of activation during the internal selection of

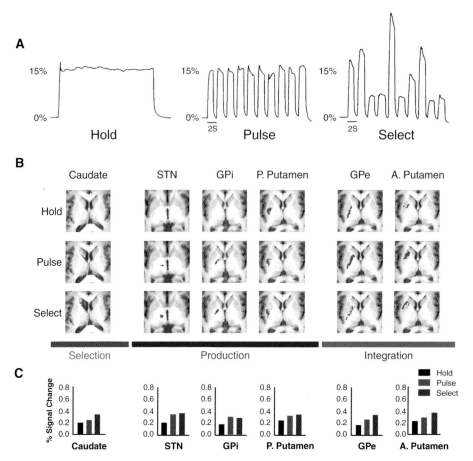

Figure 8.2. This figure is also reproduced in the color plate section. A. Force output time series for hold, similar pulse and the select different pulse conditions. The data were collected from one subject during the internally guided condition (externally guided condition not shown). During the hold task (left), the subject maintained force output at 15% of his or her MVC for 30 s. During the similar pulse task (middle), the subject produced 10 force pulses (2 s) at the level of 15% MVC during the 30 s block. During the select different pulse condition (right), the subject self-selected different force amplitudes for each of 10 pulses (2 s) where the mean force output for all 10 pulses was approximately 15% MVC. B. Depicts a summary of basal ganglia areas with significant BOLD activation patterns across the three tasks. These activation patterns allowed for the segregation of these areas into three functional groups: selection (caudate), production (STN, GPi, posterior putamen), and integration (GPe, anterior putamen). C. Depicts group mean percent BOLD signal change for the hold (black), pulse (red), and select (blue) conditions. Reprinted from Vaillancourt *et al.* (2007) *Neuroimage*, **36**, 793, with permission from Elsevier.

different amplitude force pulses. These regions are caudal to the caudate nucleus, so caudal areas of the basal ganglia process basic motor tasks like the production of force pulses while rostral areas process tasks with more of a cognitive element (i.e. selecting amplitude). This rostro-caudal gradient has been previously observed in the basal ganglia and cortex using various paradigms (Kawashima *et al.*, 1994; Deiber *et al.*, 1996; Krams *et al.*, 1998; Thoenissen *et al.*, 2002; Gerardin *et al.*, 2004; Lehericy *et al.*, 2006). For instance, Gerardin and colleagues (2004) showed that movement selection, preparation and execution were associated with activation in the caudate, anterior putamen and posterior putamen, respectively. Our work extends the concept of cognitive and basic functions processed in a rostro-caudal gradient throughout the basal ganglia to a function involved in grasping.

We also observed a third pattern of activation in which the lateral portion of the globus pallidus (GPe) and the anterior putamen increased during the production of force pulses with the same amplitude and further increased when subjects freely selected the amplitude of force (Figure 8.2B, C). This observation suggests that these nuclei integrate the two behaviors and play a role in both force pulse production and force amplitude selection. This provides support in humans for the concept that GPe is well suited for processing a multitude of movement parameters (Brotchie *et al.*, 1991). GPe may integrate afferent neural signals from the striatum and STN to then regulate GPi and substantia nigra reticulata which are the output nuclei of the basal ganglia (Alexander *et al.*, 1986).

In summary, the caudate nucleus regulates selecting grip force amplitude while the posterior putamen, GPi and STN regulate the production of force against the object being grasped. Also, the anterior putamen and GPe process signals that integrate circuits under-lying force amplitude selection and grip force production. A similar rostro-caudal gradient associated with cognitive and basic motor functions has been observed in both the cortex and basal ganglia while using a variety of behavioral tasks. Therefore, we propose that rostral cortical and basal ganglia areas are involved in high-order grasping functions (i.e. determining object mass to select force amplitude) while caudal areas are involved in basic grasping functions (i.e. scaling the selected force amplitude).

Role of the basal ganglia in scaling the duration, rate and amplitude of force

We have also investigated how individual nuclei of the basal ganglia differentially control both the duration of force and the rate of change of force (Vaillancourt *et al.*, 2004). In this study, we experimentally varied the duration of force and the rate of change of force while holding the amplitude of force constant. This fMRI experiment was performed using the same experimental setup presented in Figure 8.1A and B. We used a block design with four different conditions: (1) fast pulse, (2) fast hold, (3) medium hold and (4) slow hold (Figure 8.3A). All of the conditions required subjects to produce grip force at 25% MVC, and the subjects received online visual feedback of the target and cursor on the screen. The subjects were required to produce force by modulating the rate of change of force and duration of force consistent with the specific condition as shown in Figure 8.3A.

During the fast pulse condition, subjects were trained to produce a force pulse (~500 ms) as quickly and accurately as possible to the target force amplitude between brief periods of rest (2.5 s). The fast hold condition also required subjects to produce grip force to the target as quickly and accurately as possible. However, once the force amplitude reached the target level, the subjects were required to maintain the target force amplitude for the duration of the

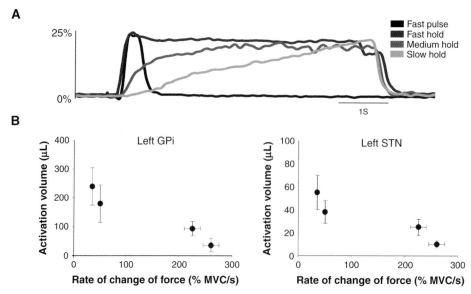

Figure 8.3. A. Depicts the four force contractions produced in the four different task conditions with the right hand. The fast pulse condition (black) was a force pulse (< 500 ms). During the fast hold condition (dark gray), subjects produced force quickly to the target level (25% MVC) and maintained force for 3 s. During the medium hold condition (gray), subjects produced force more slowly than the fast hold condition. Subjects produced force the most slowly during the slow hold condition (light gray). B. Shows the group mean volume of activation in the GPi and STN plotted against the corresponding group mean peak rate of change of force (%/s) associated with each condition. Reprinted from Vaillancourt *et al.* (2004) *Neuroimage*, **23**, 175, with permission from Elsevier.

trial period (3 s). The BOLD signal during the fast pulse condition was compared with that during the fast hold condition to determine which brain regions are involved in regulating the duration of a force pulse. The medium hold condition required subjects to implicitly control the rate of change of force to reach the target more slowly compared with the fast hold condition. The slow hold condition required subjects to control the rate of change of force so that they reached the target grip force amplitude as close as possible to the end of the trial period (~3 s).

   The basal ganglia nuclei were separated into two functional groups based on the BOLD response observed during the four different conditions, which each subject completed with his or her right hand. First, activation in the bilateral putamen and GPe increased from the fast pulse to the fast hold condition, but reached a plateau across the fast hold, medium hold and slow hold conditions. This indicates that the putamen and GPe may participate in regulating the duration of grip force production. In support of this hypothesis, we found that r^2 values from the linear regression between BOLD activation volume and duration were high for putamen and GPe. Second, the left STN and bilateral GPi increased in activation volume across the fast hold, medium hold and slow hold conditions but did not change significantly between the fast pulse and fast hold conditions. These findings indicate that GPi and STN may regulate the rate of change of force during pinch grip independently of the duration of pinch grip force. Figure 8.3B shows this scaling relation and indicates that a small portion of the GPi and STN is activated during the fast, ballistic grip contractions. In contrast, contractions with slow grip force production led to an increased metabolic demand in the GPi and STN. In order to explain these findings, we suggest that grip force contractions involve feedback mechanisms that input into the STN from the motor cortex. Afferent projections to the STN include motor and premotor cortex, GPe, thalamus and brainstem inputs (Hamani *et al.*, 2004). In particular, the convergence of mesial and lateral premotor input along with primary cortical innervations provides direct access from the motor cortex to the STN (Nambu *et al.*, 1996, 2000). A smooth ramping contraction requires that the subsequent instantaneous output force amplitude be scaled with respect to the current instantaneous force amplitude. Indeed, the hyper-direct cortico-subthalamo-pallidal pathway (Nambu *et al.*, 2000) would be appropriate for real-time feedback processing. Since we found minimal volume of activation in the GPi and STN during fast, ballistic grip force contractions, we also propose that fast force production is "preprogrammed" and requires little feedback from the motor cortex. This conclusion is consistent with early hypotheses that the basal ganglia act as a ramp generator by adjusting the velocity of smooth continuous voluntary movement (Kornhuber, 1971; DeLong & Strick, 1974). This hypothesis is strengthened by clinical observations of patients with basal ganglia disorders (i.e. Parkinson's disease). Our hypothesis suggests that ballistic movements require a small volume of STN and GPi activation, so abnormal and excessive signaling from the basal ganglia to the motor cortex would slow down movement. Indeed, patients with Parkinson's disease have slow movements related to excessive inhibition from the GPi (Albin *et al.*, 1989; DeLong, 1990). Thus, feedback mechanisms that are preferentially activated in the STN and GPi during slow movement in healthy individuals may be interfering with the ballistic movements that are intended by the patients.

It is noteworthy that several studies have found either no relation (Mink & Thach, 1987; Hamada *et al.*, 1990; Brotchie *et al.*, 1991) or a positive relation (Rao *et al.*, 1993; Sadato *et al.*, 1997; Turner *et al.*, 1998, 2003; Berns *et al.*, 1999; Onla-Or & Winstein, 2001; Taniwaki *et al.*, 2003) between basal ganglia activation and movement velocity or frequency. For instance, a PET study that investigated movement at three different frequencies (0.1, 0.4 and 0.7 Hz) found that left putamen and GP had a positive correlation between regional cerebral blood flow (rCBF) and movement velocity (Turner *et al.*, 1998). Other work by Turner and colleagues (2003) also revealed a positive correlation between rCBF and movement extent in the bilateral putamen and ipsilateral GP. Similarly, Taniwaki and colleagues (2003) used fMRI to show that the ipsilateral putamen increased in percent signal change with increasing movement frequency.

There are two main differences between the observations in these studies (references above) and the activation patterns observed in the work from our laboratory (Vaillancourt *et al.*, 2004). First, these studies suggest that activation in both putamen and globus pallidus is related to movement frequency (references cited above). Our work demonstrates that GPi and STN scale in activation with rate of change of force (Vaillancourt *et al.*, 2004). Second, these studies suggest a positive relation between basal ganglia activation and movement velocity or frequency. This is in contrast to our work (Vaillancourt *et al.*, 2004), since we found a negative relation between GPi and STN activation and rate of change of force. An important factor that may account for the different findings is that the activation pattern of the muscle varies for different movement frequencies. For instance, the muscle force required to move the index finger at fast rates (i.e. 3 Hz) is much greater than the muscle force required to move the index finger at slow rates (i.e. 0.1 Hz). In addition, the duration over which the muscle exerts a significant degree of force is shorter during a fast movement compared with a slow movement (Gottlieb *et al.*, 1989; Todorov, 2000). Therefore, as movement frequency changes, the amplitude and duration of muscle force must also change. A recent fMRI study in our laboratory demonstrates a positive relation between activation in bilateral GPi and STN and pinch grip force amplitude (Spraker *et al.*, 2007). In addition, the work of Vaillancourt and colleagues (2004) demonstrates a positive relation between activation in bilateral putamen and GPe with duration of force. Thus, we propose that the observed positive relation between activation in the basal ganglia and movement frequency (references cited above) is due to the combined effects of basal ganglia circuits that independently control force amplitude (GPi and STN) and duration (putamen and GPe).

In summary, this section has described different circuits in the basal ganglia nuclei that regulate different aspects of grasping force output. First, the bilateral putamen and GPe regulate the duration of grip-force production. Second, the GPi and STN regulate the amplitude of grip-force output during grasping. Finally, the GPi and STN also regulate the rate of change of force, but become activated primarily during slow grip-force contractions.

Summary and conclusions

The classic grasping circuit identified through single cell recordings in monkeys includes regions such as the anterior intraparietal sulcus (AIP), premotor cortex (F5) and primary motor cortex (F1) (Castiello, 2005) (see Chapters 4, 6 and 13). Anterior intraparietal neurons seem to represent the entire action sequence involved in grasping, whereas F5 neurons may code the particular segment of the action (Rizzolatti *et al.*, 1998; Murata *et al.*, 2000). Also, lesions to F1 produce deficits in the control of individual fingers, which disrupt normal grasping behaviors (Lawrence & Kuypers, 1968a, 1968b; Lawrence & Hopkins, 1976). Elegant work in humans has also shown the importance of AIP and the premotor cortex in grasping behaviors (Binkofski *et al.*, 1998) (see Chapter 6). The main sub-cortical structure that has been associated with grasping is the intermediate zone of the cerebellum where most of the cells recorded in this region were associated with reaching to grasp an object rather than moving the limb while holding a handle (Gibson *et al.*, 1996). The current chapter has described recent work that examined how the selection and regulation of grip force output is processed in distinct nuclei of the basal ganglia. This work extends our understanding of the grasping circuit to involve the basal ganglia (see also Chapter 22). Moreover, the work described in this chapter suggests that distinct circuits involved in regulating parameters of grasping are represented in the basal ganglia. The rostral nuclei in the basal ganglia regulate the selection of force amplitude levels. The caudal nuclei regulate the rate of change of force and duration of force, and scale the amplitude of grip-force output. Finally, the intermediary nuclei, such as GPe, integrate grasping-related functions from the rostral and caudal basal ganglia structures. Therefore, grasping cannot be considered a behavior that is processed distinctly in the cortex. Instead grasping is a behavior that requires network activity from multiple cortical and subcortical structures. As shown in other chapters in this book, when cortical and subcortical dysfunction occurs in humans this leads to characteristic changes in grasping behavior (see Chapters 21 to 27).

References

Abbruzzese, G. & Berardelli, A. (2003). Sensorimotor integration in movement disorders. *Mov Disord*, **18**, 231–240.

Alberts, J. L., Tresilian, J. R. & Stelmach, G. E. (1998). The co-ordination and phasing of a bilateral prehension task. The influence of Parkinson's disease. *Brain*, **121**, 725–742.

Alberts, J. L., Saling, M., Adler, C. H. & Stelmach, G. E. (2000). Disruptions in the reach-to-grasp actions of Parkinson's patients. *Exp Brain Res*, **134**, 353–362.

Albin, R. L., Young, A. B. & Penney, J. B. (1989). The functional anatomy of basal ganglia disorders. *Trends Neurosci*, **12**, 366–375.

Alexander, G. E., Delong, M. R. & Strick, P. L.(1986). Parallel organization of functionally segregated circuits linking basal ganglia and cortex. *Annu Rev Neurosci*, **9**, 357–381.

Berns, G. S., Song, A. W. & Mao, H. (1999). Continuous functional magnetic resonance imaging reveals dynamic nonlinearities of "dose-response" curves for finger opposition. *J Neurosci*, **19**, RC17.

Bertram, C. P., Lemay, M. & Stelmach, G. E. (2005). The effect of Parkinson's disease on the control of multi-segmental coordination. *Brain Cogn*, **57**, 16–20.

Binkofski, F., Dohle, C., Posse, S. *et al*. (1998). Human anterior intraparietal area subserves prehension: a combined lesion and functional MRI activation study. *Neurology*, **50**, 1253–1259.

Brotchie, P., Iansek, R. & Horne, M. K. (1991). Motor function of the monkey globus pallidus. 2. Cognitive aspects of movement and phasic neuronal activity. *Brain*, **114**, 1685–1702.

Castiello, U. (2005). The neuroscience of grasping. *Nat Rev Neurosci*, **6**, 726–736.

Castiello, U., Stelmach, G. E. & Lieberman, A. N. (1993). Temporal dissociation of the prehension pattern in Parkinson's disease. *Neuropsychologia*, **31**, 395–402.

Contreras-Vidal, J. L., Teulings, H. L. & Stelmach, G. E. (1995). Micrographia in Parkinson's disease. *Neuroreport*, **6**, 2089–2092.

Deiber, M. P., Ibanez, V., Sadato, N. & Hallett, M. (1996). Cerebral structures participating in motor preparation in humans: a positron emission tomography study. *J Neurophysiol*, **75**, 233–247.

DeLong, M. R. (1990). Primate models of movement disorders of basal ganglia origin. *Trends Neurosci*, **13**, 281–285.

DeLong, M. R. & Strick, P. L. (1974). Relation of basal ganglia, cerebellum, and motor cortex units to ramp and ballistic limb movements. *Brain Res*, **71**, 327–335.

Flanagan, J. R., Bowman, M. C. & Johansson, R. S. (2006). Control strategies in object manipulation tasks. *Curr Opin Neurobiol*, **16**, 650–659.

Gerardin, E., Pochon, J. B., Poline, J. B. *et al.* (2004). Distinct striatal regions support movement selection, preparation and execution. *Neuroreport*, **15**, 2327–2331.

Gibson, A. R., Horn, K. M., Stein, J. F. & Van Kan, P. L. (1996). Activity of interpositus neurons during a visually guided reach. *Can J Physiol Pharmacol*, **74**, 499–512.

Gordon, A. M., Quinn, L., Reilmann, R. & Marder, K. (2000). Coordination of prehensile forces during precision grip in Huntington's disease. *Exp Neurol*, **163**, 136–148.

Gottlieb, G. L., Corcos, D. M. & Agarwal, G. C. (1989). Strategies for the control of voluntary movements with one mechanical degree of freedom. *Behavioral and Brain Sciences*, 189–250.

Hamada, I., Delong, M. R. & Mano, N. (1990). Activity of identified wrist-related pallidal neurons during step and ramp wrist movements in the monkey. *J Neurophysiol*, **64**, 1892–1906.

Hamani, C., Saint-Cyr, J. A., Fraser, J., Kaplitt, M. & Lozano, A. M. (2004). The subthalamic nucleus in the context of movement disorders. *Brain*, **127**, 4–20.

Jeannerod, M. (1984). The timing of natural prehension movements. *J Mot Behav*, **16**, 235–254.

Johnson-Frey, S. H. (2003). What's so special about human tool use? *Neuron*, **39**, 201–204.

Kawashima, R., Roland, P. E. & O'Sullivan, B. T. (1994). Fields in human motor areas involved in preparation for reaching, actual reaching, and visuomotor learning: a positron emission tomography study. *J Neurosci*, **14**, 3462–3474.

Kornhuber, H. H. (1971). Motor functions of cerebellum and basal ganglia: the cerebellocortical saccadic (ballistic) clock, the cerebellonuclear hold regulator, and the basal ganglia ramp (voluntary speed smooth movement) generator. *Kybernetik*, **8**, 157–162.

Krams, M., Rushworth, M. F., Deiber, M. P., Frackowiak, R. S. & Passingham, R. E. (1998). The preparation, execution and suppression of copied movements in the human brain. *Exp Brain Res*, **120**, 386–398.

Lawrence, D. G. & Hopkins, D. A. (1976). The development of motor control in the rhesus monkey: evidence concerning the role of corticomotoneuronal connections. *Brain*, **99**, 235–254.

Lawrence, D. G. & Kuypers, H. G. (1968a). The functional organization of the motor system in the monkey. I. The effects of bilateral pyramidal lesions. *Brain*, **91**, 1–14.

Lawrence, D. G. & Kuypers, H. G. (1968b). The functional organization of the motor system in the monkey. II. The effects of lesions of the descending brain-stem pathways. *Brain*, **91**, 15–36.

Lehericy, S., Bardinet, E., Tremblay, L. *et al.* (2006). Motor control in basal ganglia circuits using fMRI and brain atlas approaches. *Cereb Cortex*, **16**, 149–161.

Mink, J. W. & Thach, W. T. (1987). Preferential relation of pallidal neurons to ballistic movements. *Brain Res*, **417**, 393–398.

Murata, A., Gallese, V., Luppino, G., Kaseda, M. & Sakata, H. (2000). Selectivity for the shape, size, and orientation of objects for grasping in neurons of monkey parietal area AIP. *J Neurophysiol*, **83**, 2580–2601.

Nambu, A., Takada, M., Inase, M. & Tokuno, H. (1996). Dual somatotopical representations in the primate subthalamic nucleus: evidence for ordered but reversed body-map transformations from the primary motor cortex and the supplementary motor area. *J Neurosci*, **16**, 2671–2683.

Nambu, A., Tokuno, H., Hamada, I. *et al.* (2000). Excitatory cortical inputs to pallidal neurons via the subthalamic nucleus in the monkey. *J Neurophysiol*, **84**, 289–300.

Nowak, D. A. & Hermsdorfer, J. (2006). Predictive and reactive control of grasping forces: on the role of the basal ganglia and sensory feedback. *Exp Brain Res*, **173**, 650–660.

Nowak, D. A., Rosenkranz, K., Topka, H. & Rothwell, J. (2005). Disturbances of grip force behaviour in focal hand dystonia: evidence for a generalised impairment of sensory-motor integration? *J Neurol Neurosurg Psychiatry*, **76**, 953–959.

Onla-Or, S. & Winstein, C. J. (2001). Function of the 'direct' and 'indirect' pathways of the basal ganglia motor loop: evidence from reciprocal aiming movements in Parkinson's disease. *Brain Res Cogn Brain Res*, **10**, 329–332.

Rand, M. K., Smiley-Oyen, A. L., Shimansky, Y. P., Bloedel, J. R. & Stelmach, G. E. (2006). Control of aperture closure during reach-to-grasp movements in Parkinson's disease. *Exp Brain Res*, **168**, 131–142.

Rao, S. M., Binder, J. R., Bandettini, P. A. *et al.* (1993). Functional magnetic resonance imaging of complex human movements. *Neurology*, **43**, 2311–2318.

Rearick, M. P., Stelmach, G. E., Leis, B. & Santello, M. (2002). Coordination and control of forces during multifingered grasping in Parkinson's disease. *Exp Neurol*, **177**, 428–442.

Rizzolatti, G., Luppino, G. & Matelli, M. (1998). The organization of the cortical motor system: new concepts. *Electroencephalogr Clin Neurophysiol*, **106**, 283–296.

Sadato, N., Ibanez, V., Campbell, G. *et al.* (1997). Frequency-dependent changes of regional cerebral blood flow during finger movements: functional MRI compared to PET. *J Cereb Blood Flow Metab*, **17**, 670–679.

Smeets, J. B. & Brenner, E. (1999). A new view on grasping. *Motor Control*, **3**, 237–271.

Spraker, M. B., Yu, H., Corcos, D. M. & Vaillancourt, D. E. (2007). Role of individual basal ganglia nuclei in force amplitude generation. *J Neurophysiol*, **98**, 821–834.

Taniwaki, T., Okayama, A., Yoshiura, T. *et al.* (2003). Reappraisal of the motor role of basal ganglia: a functional magnetic resonance image study. *J Neurosci*, **23**, 3432–3438.

Thoenissen, D., Zilles, K. & Toni, I. (2002). Differential involvement of parietal and precentral regions in movement preparation and motor intention. *J Neurosci*, **22**, 9024–9034.

Todorov, E. (2000). Direct cortical control of muscle activation in voluntary arm movements: a model. *Nat Neurosci*, **3**, 391–398.

Turner, R. S., Grafton, S. T., Votaw, J. R., Delong, M. R. & Hoffman, J. M. (1998). Motor subcircuits mediating the control of movement velocity: a PET study. *J Neurophysiol*, **80**, 2162–2176.

Turner, R. S., Desmurget, M., Grethe, J., Crutcher, M. D. & Grafton, S. T. (2003). Motor subcircuits mediating the control of movement extent and speed. *J Neurophysiol*, **90**, 3958–3966.

Vaillancourt, D. E., Mayka, M. A., Thulborn, K. R. & Corcos, D. M. (2004). Subthalamic nucleus and internal globus pallidus scale with the rate of change of force production in humans. *Neuroimage*, **23**, 175–186.

Vaillancourt, D. E., Yu, H., Mayka, M. A. & Corcos, D. M. (2007). Role of the basal ganglia and frontal cortex in selecting and producing internally guided force pulses. *Neuroimage*, **36**, 793–803.

9

Models for the control of grasping

ERHAN OZTOP AND MITSUO KAWATO

Summary

This chapter underlines the multifaceted nature of reach and grasp behavior by reviewing several computational models that focus on selected features of reach-to-grasp movements. An abstract meta-model is proposed that subsumes previous modeling efforts, and points towards the need to develop computational models that embrace all the facets of reaching and grasping behavior.

Introduction

Hand transport and hand (pre)shaping are basic components of primate grasping. The different views on their dependence and coordination lead to different explanations of human control of grasping. One can view these two components as being controlled independently but coordinated so as to achieve a secure grasp. The alternative view is that the hand and the arm are taken as a single limb and controlled using a single control mechanism. Needless to say, this distinction is not very sharp; but it becomes a choice to be made by a control engineer when it is necessary to actually implement a grasp controller. The experimental findings so far point towards the view that human grasping involves independent but coordinated control of the arm and the hand (see Jeannerod *et al.*, 1998) (see also Chapter 10). However, reports against this view do exist as it has been suggested that human grasping is a generalized reaching movement that involves movement of digits so as to bring the fingers to their targets on the object surface (Smeets & Brenner, 1999, 2001). Although theoretically both control mechanisms are viable, from a computational viewpoint, the former is more likely. Learning and/or optimization of a single gigantic controller is very difficult; dealing with smaller and simpler controllers (i.e. for the hand and the arm) and coordinating them according to the task requirements seems more plausible (Kawato & Samejima, 2007).

Monkey neurophysiology and human brain-imaging studies help us delineate the brain regions that are involved in grasp planning and execution (see Chapters 4, 6, 7 and 8); however, it is far from known *how* exactly these regions work together to sustain a grasping mechanism that exhibits the range of properties observed in adult or infant reach and grasp.

Sensorimotor Control of Grasping: Physiology and Pathophysiology, ed. Dennis A. Nowak and Joachim Hermsdörfer. Published by Cambridge University Press. © Cambridge University Press 2009.

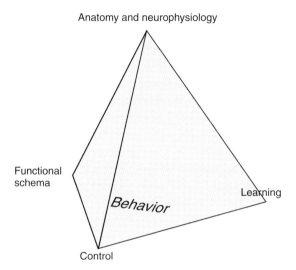

Anatomy and neurophysiology

Functional
schema

Learning

Behavior

Control

Figure 9.1. The conceptual tetrahedron illustrating the multitude of ways of approaching primate reach-to-grasp.

The computational studies often focus on a selected subset of the properties of primate grasping. Among these, many consider an adult grasping system that is modeled with a low-to-moderate level of biological realism so as to match the experimental, mostly behavioral, data. Some other studies attempt to explain how the skill of grasping can be attained via learning, often employing techniques from machine learning. It is unfortunate that, to date, there are no computational models that can explain the full extent of reach and grasp movements in terms of development, neural mechanisms and behavioral markers in a single framework. So, here we propose a meta-model that considers those previous studies as points inside (or on) the conceptual tetrahedron (see Figure 9.1) whose corners are identified by (1) anatomy and neurophysiology, (2) schemas representing brain function, (3) learning (infant to adulthood transition of grasping) and (4) control (the optimization and control principles of reach-to-grasp).

In what follows, we review existing modeling efforts that can be considered to be representatives for the corners of the conceptual tetrahedron shown in Figure 9.1. First, we briefly review the cortical areas involved in grasp control (anatomy and neurophysiology) and introduce the FARS model that addresses the neural ingredients of reach-to-grasp in terms of functional schemas (Fagg & Arbib, 1998). Then, we move to the models that attempt to explain the learning aspect of grasping. We present a developmentally oriented model that learns finger configurations for stable grasping (Oztop *et al.*, 2004), and a model that synthesizes human-like grasp using human motion capture data (Uno *et al.*, 1995; Iberall & Fagg, 1996). For the control aspect of grasping, we first present the Hoff–Arbib model that explains the coordination of the timing of transport and preshape (Hoff & Arbib, 1993), and an internal model explanation of the load-force–grip-force coupling (Kawato, 1999).

Neurophysiological considerations and the FARS model

Neurophysiological data indicate that the parietal cortex is involved in visuomotor aspects of manual manipulative movements (Wise *et al.*, 1997). In particular, the anterior intraparietal area (AIP) of macaque monkeys discharges in response to viewing and/or grasping of 3D objects representing object features relevant for grasping (Sakata *et al.*, 1995, 1998) (see also Chapters 6, 7 and 13). Area AIP has strong recurrent connections with the rostral part of the ventral premotor cortex (area F5) in the macaque (Luppino *et al.*, 1999). The ventral premotor cortex is involved in grasp planning and execution (Rizzolatti *et al.*, 1990), and projects to motoneurons that control finger muscles (Dum & Strick, 1991). The activity of neurons in the primary motor cortex (area F1) when contrasted to the premotor activity suggests that the primary motor cortex may be more involved in dynamic aspects of movement, executing "instructions" sent by the premotor cortex. Thus, it is generally accepted that the anterior intraparietal area–ventral premotor cortex–primary motor cortex network (AIP–F5–F1 circuit in short) is responsible for grasp planning and execution (Gallese *et al.*, 1994; Jeannerod *et al.*, 1995; Fagg & Arbib, 1998; Fogassi *et al.*, 2001). Cerebral cortex coordinates the execution of the AIP–F5–F1 circuit with the transport of the hand, which is mediated via a similar parietal-to-motor pathway. For representation of the space for action the areas in and around the intraparietal sulcus play a key role. Ventral, medial and lateral intraparietal areas represent the space in different coordinate frames (Colby & Duhamel, 1996; Duhamel *et al.*, 1998; Colby & Goldberg, 1999). Although MIP appears to be the main area responsible for the representation of a reach target, we hold that LIP and VIP should be involved in representing targets as well (e.g. it is likely that the goal of a slapping action triggered by a mosquito bite is registered in area VIP). Therefore we take the liberty of collectively using VIP/MIP/LIP (ventral/medial/lateral intraparietal areas) as the regions responsible for reach target representation. This representation is used by the caudal part of the ventral premotor cortex (area F4) for planning and executing the transport phase of the grasping action. In short we have a (VIP/MIP/LIP)–F4–F1 pathway for reaching and AIP–F5–F1 for grasping (see Figure 9.2A) in macaque monkeys. Strong evidence suggests that the human brain has a similar organization for reaching and grasping, with homologous areas (Culham *et al.*, 2006; Culham & Valyear, 2006). The human homolog[1] of macaque AIP appears to be the area located at the junction of the anterior part of the intraparietal sulcus and the inferior postcentral sulcus (see Culham *et al.*, 2006 and citations therein). The other macaque intraparietal areas also have their homologs in or around the human intraparietal sulcus although these are not as well established as AIP (Culham *et al.*, 2006). There is strong evidence that the human area 44 is the homolog of the monkey F5 (the rostral part of the ventral premotor cortex) with similar motor and cognitive functions (Rizzolatti *et al.*, 2002). The homology of monkey area F4 is not well established yet, however a likely candidate is the ventral part of area 6 neighboring area 44.

Based on the properties of the summarized parietofrontal reach and grasp areas, Fagg & Arbib (1998) proposed a schema model of grasp planning and execution (FARS,

A

B

Pre training   Post training

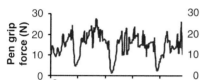

Figure 1.4

A

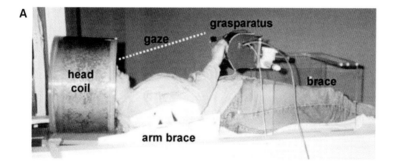

B

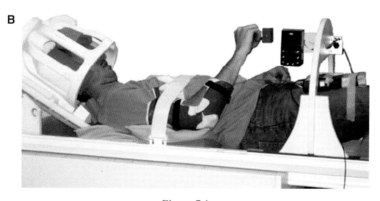

Figure 7.1

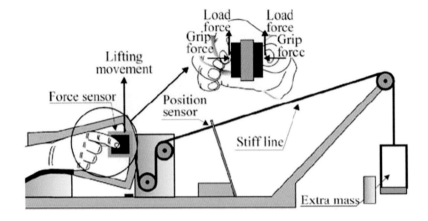

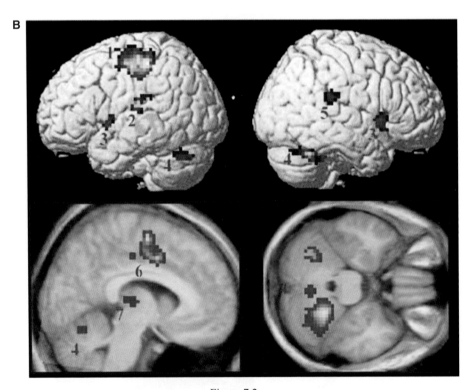

Figure 7.2

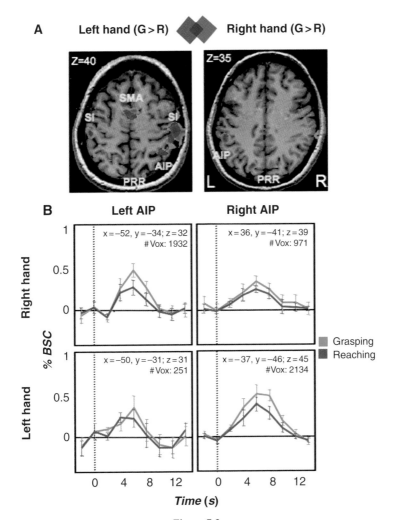

Figure 7.3

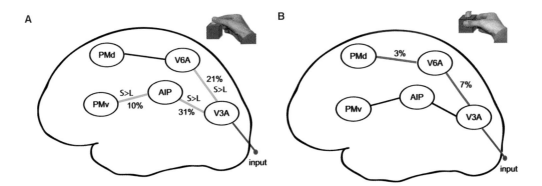

Figure 7.4

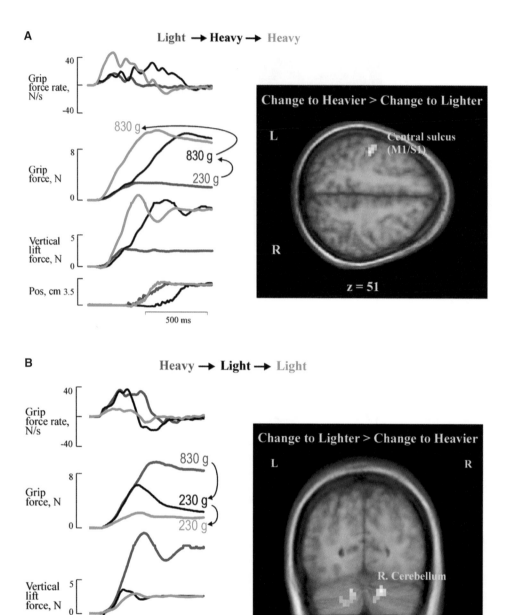

Figure 7.5

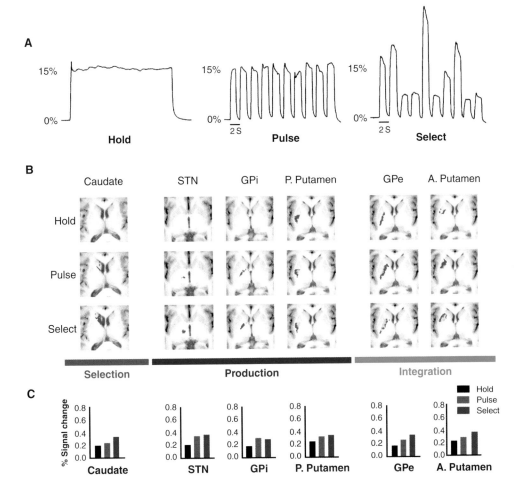

Figure 8.2

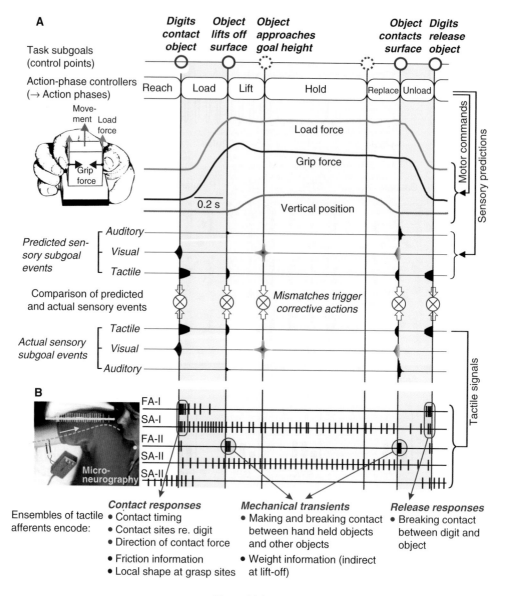

Figure 11.1

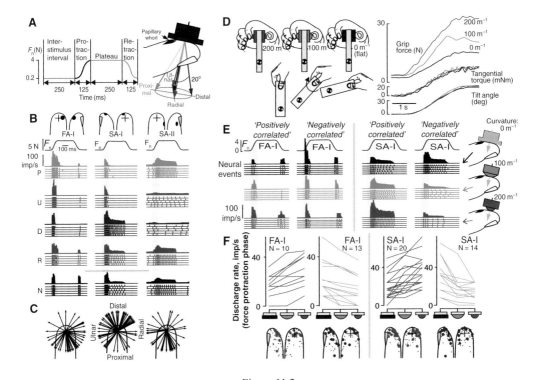

Figure 11.2

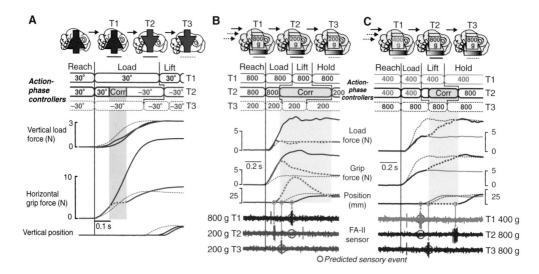

Figure 11.3

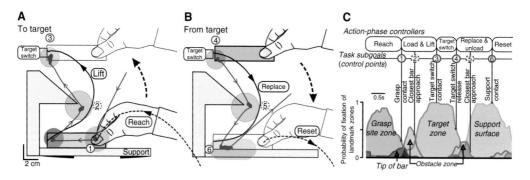

Figure 11.4

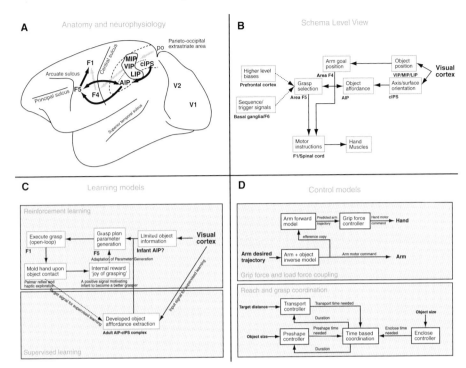

Figure 9.2. A. The cortical organization of reach and grasp pathways are shown. The darker arrows indicate the grasp-related visuomotor transformation pathways, whereas the lighter arrows indicate the projections mediating the transport component. B. The schema-level organization of grasping according to the FARS model (Fagg & Arbib, 1998). C. Upper panel: According to the ILGM model (Oztop *et al.*, 2004) infant grasp development is mediated by the internal reward generated when an object is grasped. Lower panel: How the result of ILGM learning can bootstrap visuomotor development (Oztop *et al.*, 2006). D. Upper panel: How the central nervous system adjusts the grip force according to the predicted load force using inverse and forward models (Kawato, 1999). Lower panel: How the 'time-to-completion' signal can be used for coordinating transport and hand controllers (Hoff & Arbib, 1993)

Fagg–Arbib–Sakata–Rizzolatti model). According to the model (see Figure 9.2B), AIP converts the information relayed by the dorsal and ventral visual stream into a set of representations, called the "object affordances." These affordances then are forwarded to area F5, which selects the suitable affordance given task constraints. Then the selected affordance is reported back to AIP, virtually establishing a memory. The execution then unfolds with a sequencing mechanism that monitors phases of the grasp action (such as "maximum aperture reached" and "object contact"). The sequencing is postulated to be implemented by basal ganglia and the presupplementary motor area (area F6) via inhibition and disinhibition mechanisms. Figure 9.2B illustrates the FARS model's main functional units (schemas) and their interrelation.

Learning: grasp development in infants

On the one hand, there are many experimental studies on the attributes of adult and infant reaching; on the other, there are several models of reaching to grasp (Arbib & Hoff, 1994) and robotic grasp synthesis techniques (Shimoga, 1996). However, there are few computational models that combine the empirical data on infant motor development relating to cortical structures. In this direction, Oztop *et al.* (2004) has proposed the Infant learning to grasp model (ILGM) that attempts to give an account of infant grasp learning that could be mapped onto cortical structures and capture infant motor development data.

Infants exhibit a crude ability to reach at birth (von Hofsten, 1982) (see also Chapter 17), that transforms into better controlled reaching and simple grasping by 4–5 months, and adult-like reaching and grasping by 9 months, achieving precision grasping by 12–18 months (von Hofsten, 1984). Infants 12 weeks of age make hand contact with glowing and sound-making objects under lighted and dark conditions with similar frequency, and the onset of successful grasping under these two conditions takes place at approximately the same age of 15–16 weeks (Clifton *et al.*, 1993). This indicates that reaches of the neonate elicited by vision of an object (counterintuitively) can be executed without vision of the hand. Between 9 and 13 months of age, reaches become better matched to the target object, with earlier hand orientation and anticipatory grasp closure relative to object orientation and size (Lockman *et al.*, 1984; von Hofsten & Ronnqvist, 1988; Newell *et al.*, 1993). These data suggest that infants initially use an open-loop control (i.e. ballistic) strategy that transports the hand to the target object followed by a haptic grasping phase (molding of the hand to match the object shape). Perhaps, later in development, the initial open-loop grasping provides the training stimuli for development of a visual feedback grasping circuit that is a prerequisite for delicate manipulation (see Oztop *et al.*, 2006).

In the ILGM, a grasp plan is defined with the triplet (p, r, v) generated by the computational modules of Hand position (p), Wrist rotation (r) and Virtual fingers (v). The triplet (p, r, v) represents a minimal set of kinematic parameters specifying basic grasp actions. The movement execution mechanism first transports the hand to the location p while the wrist is rotated according to r. After this phase, the hand is transported towards the object center. So, p in effect determines the "approach direction" of the grasping movement. On contact with the object, the fingers specified with v enclose the object simulating the palmar reflex (when simulating infants of 4–6 months old, v engages all the fingers). If this enclosure results in a stable grasp, a positive reward stimulus is relayed back to the computational modules that generated p, r, v parameters. The connection weights among the modules are updated to encourage such output next time a similar object is presented. A failure to contact with the object or an unstable grasp produces a negative reward signal that causes changes in the connection weights to discourage such p, r, v output. Thus, during learning, the ILGM discovers which orientations and approach directions are appropriate for a particular object.

The ILGM not only captures some interesting features of infant grasp development, but also maps the modules of the model to the brain areas that might be involved in the associated functions.[2] The ILGM is a systems-level model based on the broad organization of the primate visuomotor network, where visual features are extracted by the parietal cortex, and used by the premotor cortex to generate high-level motor signals that drive the lower motor centers for movement generation (see Figure 9.2C, upper panel). The feedback arising from the object contact is used to modify grasp generation mechanisms within the premotor cortex. The model consists of four modules: the Object Information/Input module, the Grasp Learning/Generation module, the Movement Execution module, and the Grasp Evaluation module. The Grasp Learning/Generation module contains the aforementioned computational layers that generate the grasp plan (*Virtual finger*, *Hand position* and *Wrist rotation*). These layers are motivated by the *Preshape*, *Approach vector* and *Orient* grasping schemas proposed by Iberall & Arbib (1990). The Object Information module is postulated to be located in the parietal cortex (probably AIP; Sakata *et al.*, 2005), with the function of extracting object affordances. The affordances are relayed to the Grasp Learning module that is postulated to be located in the ventral premotor cortex that is known to be involved in grasp programming (Jeannerod *et al.*, 1995; Luppino *et al.*, 1999). The grasp plan output by the Grasp Learning module is then used by the Movement Execution module for actual grasp execution. The primary motor cortex and spinal motor circuitry is postulated to undertake the actual execution task. The Movement Execution module also implements the palmar reflex upon object contact. The sensory stimuli generated by the execution of the grasp plan are integrated and evaluated by the Movement Evaluation module that is postulated to be located in the primary somatosensory cortex. Output of the somatosensory cortex, the reinforcement signal, is used to adapt the parietal–premotor circuit's internal parameters. Grasp evaluation is implemented in terms of grasp stability: a grasp attempt that misses the target or yields an unstable object enclosure produces a negative reinforcement signal.

The ILGM simulations showed that a limited set of behaviors (reaching and grasp reflex) coupled with simple haptic feedback (holding of object) was enough for a goal-directed trial and error learning mechanism to yield an interesting set of grasping behaviors. The experiments with the model showed that power grasping would be the dominant mode of grip in the early stages of learning, and as learning progresses several different type of grips would be added to the infant's grasp repertoire (Oztop *et al.*, 2004). In addition, the simulations showed that the task constraints and the environment shape the infant grasp repertoire: when the ILGM was simulated in a situation where only small objects on top of a table were presented, precision pinch became the dominant mode of grasping, as a tiny object over a hard surface could not be picked up by power grasping (Oztop *et al.*, 2004).

Learning: hand configurations suitable for the target object

Grasping an object in one's hand requires at least two forces to be applied to the object. Iberall *et al.* (1986) used the term opposition to describe basic forms of force application patterns. (1) Pad opposition occurs when an object is held between a set of fingers and the thumb, as in holding a peanut with the index finger and the thumb. (2) Palm opposition occurs when an object is held with fingers opposing the palm, as in holding a large hammer. (3) Side opposition occurs when the thumb's volar surface opposes radial sides of the fingers, as in holding a key. This classification, in effect, transforms a complex, high degree-of-freedom problem into a lower dimensional problem of (1) determining which opposition(s) are to be used for a given object, and (2) implementing the selected opposition(s), in terms of how many fingers should be involved and what forces be produced. The set of fingers, thumb or the palm surface that is involved in providing oppositional forces are called the virtual fingers (Iberall *et al.*, 1986; see also Chapter 3 and Baud-Bovy & Soechting, 2001). Within this framework grasping proceeds as follows: (1) object properties are perceived, (2) object is located in space, (3) opposition to be used is determined, (4) the virtual fingers are set up (i.e. which fingers to involve are decided), (5) hand aperture is determined, (6) the grasp is executed (preshape and enclose). Experimental evidence suggests that the choice of fingers to use depends on many factors including object properties, the manipulation required after the grip, environmental constraints and the anatomy of the forearm. Iberall used a feedforward neural network to determine which opposition to use for a given set of task requirements (object properties + force and precision requirements), and which fingers to contribute in the selected opposition (see Iberall & Fagg, 1996 and citations therein). After training the neural network could produce the opposition necessary for the given object properties and task requirements as input.

A conceptually similar, but more advanced approach to train an artificial neural network for producing the "right" finger configuration given an object was proposed by Uno *et al.* (1995). The neural network proposed was a five-layer information compression network that was trained to reproduce the given input at the output (i.e. an identity mapping was to be learned). In this kind of network, when the number of units in the middle layer is chosen less than the input, the middle layer acts as a bottle neck so that the network effectively performs a non-linear principal component analysis.[3] The input and output consisted of visual (object information) and motor information (joint angles). The input and output training set was prepared by recording a set of successful grasping actions of a human actor. After the network was trained it formed a compressed, and thus a multimodal representation of the grasping actions that "encoded" compatible object properties and finger configurations. The grasp generation was performed by an optimization process where for a given visual input (x) a complementary motor code (y) was searched so that the network retained its identity mapping property (that is (x, y) would be mapped to (x, y)) and a specified optimization criterion was satisfied, which could tilt the bias towards certain grasp types (i.e. precision pinch vs. power grasp). The simulations showed that the network can not only produce the taught grasps but also generalize to objects with different dimensions.

The concept of multimodal representation appears to be a promising and a logical target for modeling as it is supported by monkey electrophysiology. The neurons in the monkey anterior intraparietal area (AIP) are involved in grasp planning and have multimodal responses, encoding a mixture of object features (i.e. object affordances) and executed grasp properties (Sakata *et al.*, 1995, 1998; Murata *et al.*, 1996, 2000). A classic view related to synaptic plasticity and learning in the brain is Hebbian learning ("when an axon of cell A is near enough to excite a cell B and repeatedly or persistently takes part in firing it, some growth process or metabolic change takes place in one or both cells such that A's efficiency, as one of the cells firing B, is increased."). A similar mechanism may lead to emergence of multimodal representation in the cerebral cortex (Keysers & Perrett, 2004; Oztop *et al.*, 2005a; Chaminade *et al.*, 2008).

Uno *et al.*'s model requires a set of successful grasping examples to become functional. Therefore, from a developmental perspective, we can say that it models the stage where infants grasp objects with a rudimentary grasping circuit (4–6 months of age). So, the Infant learning to grasp model (Oztop *et al.*, 2004) presented above and Uno *et al.*'s model are complementary in that ILGM learning provides the successful examples that Uno *et al.*'s model requires to form multimodal representations and to function as a more elaborate grasp-planning circuit. In the same vein, Oztop *et al.* (2006) also proposed a neural network model of AIP neurons, where a combination of self-organizing map and a three-layer feedforward network "learned" from the performance of the rudimentary grasping ability provided by ILGM.

Control: coordination of reach and grasp

One of the characteristics of reach-to-grasp movement is that during the execution of the movement when the target location is suddenly changed not only the transport phase of the movement but also the kinematics and timing of the preshape are altered (Paulignan *et al.*, 1991a, 1991b; Roy *et al.*, 2006; see also Chapters 2 and 10). Likewise, when the target object size is suddenly changed not only the finger kinematics but also the transport phase of the movement is affected (Paulignan *et al.*, 1991a, 1991b; Roy *et al.*, 2006). The Hoff–Arbib model attempts, and to a large extent succeeds in explaining the temporal relation of hand transport kinematics with the finger aperture kinematics observed during reach-to-grasp movements. Hoff & Arbib (1993) have postulated the existence of a higher-level schema that coordinates the controllers ("schemas") for reach and grasp. The overall control of the movement is achieved by a modular decomposition of (1) transport, (2) preshape and (3) enclose controllers as shown in the lower panel of Figure 9.2D. The higher-level schema receives "time-to-completion" information from the reach and grasp schemas. The schema that needs a longer time to complete is allowed its time, whereas the others are slowed down. Although the controllers were built upon the minimum-jerk model, the key is the coordination of the controllers rather than the particular choice for the actual implementation of each controller. Perhaps the model would predict human grasping behavior better when one of the more recent models e.g. "minimum variance" (Harris & Wolpert, 1998) or the "TOPS" (Miyamoto *et al.*, 2004) model was used instead of the minimum-jerk model. The Hoff–Arbib model accounts for the smooth corrections in response to sudden position and object size alterations observed in human reach-to-grasp movements. Moreover, the Hoff–Arbib model, although originally developed for reach and grasp, seems to account for the temporal invariance property observed in bimanual actions that require complex hierarchical and temporal coordination (Weiss & Jeannerod, 1998; Weiss *et al.*, 2000).

Control: internal models and load force–grip force coupling

Internal models, when used in the context of the central nervous system, refer to neural mechanisms that mimic input–output relationships of the limbs and external objects. A forward internal model predicts the sensory consequences of a given motor command (i.e. corollary discharge), thus providing a mechanism to overcome the delay (that is undesirable for control) involved in sensing the actual sensory outcome using sensory receptors. Forward models also can be used in mental simulation (Oztop *et al.*, 2005b) of actions as well as attenuating the sensory stimulus generated by one's own movement (Blakemore *et al.*, 1998). An inverse model starts from a desired sensory state and outputs the motor command that will achieve the desired state. Experimental evidence from human neuro-imaging and monkey neurophysiology indicate that the cerebellum is involved in acquiring inverse models through motor learning (Flanagan & Wing, 1997; Imamizu *et al.*, 2000; Kawato *et al.*, 2003; Bursztyn *et al.*, 2006; Kawato, 2008; see also Chapter 26). The so-called "grip force–load force coupling" (Johansson, 1996) that is observed when the hand is moved voluntarily while an object is being held by finger(s) opposing the thumb demon-strates that the central nervous system employs forward models in sensory motor control. Since the movement of the hand induces varying load forces on the held object, the central nervous system has to adjust the grip force such that the object does not slip and an unnecessarily large force is avoided. So the grip force modulation has similar temporal waveform as the load force and is usually associated with a phase advance indicating that the grip-force modulation is anticipatory. This is so because the sensory delay in relaying the change in load force to the central nervous system is of the range 50–100 ms.

Kawato (1999) proposed a control model that explains the "grip force–load force coupling." Imaging studies verified that the components of the model can be located in the cerebellum and cerebral cortex; in particular it was shown that right and superior cerebellum may be the locus of the forward model that allows an anticipatory grip force modulation to occur (Kawato *et al.*, 2003). In addition, a PET imaging study by Boecker *et al.* (2005) indicated the existence of modular representations for predictive force coupling in the cerebellum, which are applicable to different environmental contexts (see also Nowak *et al.*, 2007 and Chapter 26). According to Kawato's model, there are at least three key computational elements (see Figure 9.2D, upper panel): the Arm controller, the Grip controller, and the Forward model. The Arm controller controls the arm, hand and the object held. This is usually considered an inverse model that produces feedforward motor commands given a desired arm trajectory.[4] The Grip controller produces hand motor commands to keep the object firm in the hand by computing the grip force necessary given the arm trajectory: first the load force is derived using the arm trajectory,[5] and then the grip force is easily estimated from the load force using the friction coefficient (this depends on the object/finger contact, here it is assumed to be known) and a safety margin scale factor. The Forward model uses the efference copy of the command sent to the arm muscles and predicts the arm state in the future. This future state is relayed to the Grip controller which calculates the required grip force *at the right time*, that is, before the object actually experiences the load force that the motion of the arm will cause. In this way, the central nervous system guarantees the stability of the object even when the hand movements are very fast, where a feedback grip force controller would fail to react fast enough, and thus lose the grip on the object. This conceptual model will work fine for a single object as long as the forward model and the inverse models embedded in the controllers can be learned. In fact, there are various computational architectures that can satisfy these learning requirements and handle multiple objects. One particularly well suited to this context is the Modular selection and identification for control (MOSAIC) model (Wolpert & Kawato, 1998; Wolpert *et al.*, 2003). MOSAIC is a modular adaptive controller that can learn the dynamics of the controlled limb and multiple objects with different dynamic properties (Haruno *et al.*, 2001). Therefore it removes the multiple object limitation, and can be used to implement the model outlined above (Kawato, 1999).

Discussion

In primates, including humans, the brain areas that contribute to planning and execution of reach and grasp movements are relatively well known. Yet, the question of "how" still remains to be answered. This is the point where computational modeling comes into play. If a proposed working mechanism (conceptual model) cannot be spelled out in computational terms and implemented on a computer, then it is very likely that the proposal is wrong, or missing critical components. (Of course, this argument is rather philosophical; one may be spiritual or believe that the way the brain computes cannot be emulated with a Turing-machine.)

Grasping and reaching being a complex behavior, it is not trivial to model in its entirety. Adult behavior gives us clues about the intrinsic optimizations carried on by the cerebral cortex in planning grasping and reaching. Detailed experimentation with the growing infant tells us about the phases of neural and motor development. First, the random-looking reaches of the newborn become directed towards visual and auditory stimuli by 2 months. The occasional contact of the hand with the objects in the environment triggers the palmar reflex. Infants use these "coincidental" grasps to learn how to orient or preshape their hands for a small set of objects by 6 months of age. These early grasps become more adult-like grasps by the age of 9 months; the transport phase now reflects the effects of the target object: during the transport the hand is preshaped and oriented according to the target size and orientation.

But then, how is the intrinsic optimization that appears to be central to adult reaching and grasping acquired together with infants' motor development? Are development and tuning for optimal behavior mediated by different cortical mechanisms that follow different time courses? Motor development stabilizes after childhood; but obviously we retain our ability to learn new skills such as using new tools and inventing new grasping skills. So, how does the adult motor learning ability compare with the infant motor learning? One speculation is that the immature motor apparatus (for the hand and the arm) – the hardware – of an infant imposes restrictions on the "motor learning and optimization mechanisms" – the software – that undertakes the tasks of (1) learning to grasp, (2) learning to act optimally, (3) staying adaptive for new task requirements (i.e. what is "optimal" is not fixed, but can be redefined by context). The corticospinal system is the main neural substrate for independent finger control (Triggs *et al.*, 1998). In infant primates, the development of corticospinal projections terminating in the ventral horn on motor neurons innervating hand muscles is essential for independent finger control (Bortoff & Strick, 1993), and lesion of the corticospinal tract prevents the development of independent finger movement (Lawrence & Hopkins, 1976). The myelination of the corticospinal tract and the enlargement of the diameter of the corticospinal axons (which both contribute to the conduction velocity) are main components of the cortical motor maturation. Therefore, the myelination and the enlargement of the diameter of the corticospinal axons might be a mechanism by which the "motor learning and optimization mechanism" are forced to follow a temporarily staged learning regime as seen in the developing infant. In fact, myelination of the corticospinal tract is far from complete by the end of the second postnatal year, and the increase in the conduction velocity of the corticospinal tract goes well beyond childhood (see Lemon *et al.*, 1997 and citations therein).

Although considerable knowledge has been gathered on individual properties of human grasping (i.e. at corners of the conceptual tetrahedron), we do not have a single and concise picture of human grasping. We believe that it is only when we can combine (1) the development of grasping, (2) the intrinsic optimality principles and (3) the adaptation capability of human reach and grasp within a neurophysiologically plausible computational model, that it will be possible to make predictions that may shed light on the impairments of reach and grasp, and mediate the development of smart prosthetics and brain–machine interfaces.

Notes

1 We use homology in a somewhat loose sense; in the text it generally indicates functional equivalence rather than the strict definition adopted in evolutionary biology.
2 With the assumption that monkey and human grasp development follow similar paths, the model is specified in terms of macaque monkey nomenclature. In the macaque monkey, a specialized circuit in the parietal area AIP extracts object affordances relevant to grasping (Sakata *et al.*, 1998; Murata *et al.*, 1996, 2000) and relays this information to the premotor cortex where contextual and intention-related bias signals are also integrated for grasp selection/execution (see Fagg & Arbib, 1998). It is very likely that a similar circuit exists in the human (see Jeannerod *et al.*, 1995) and is adapted during infancy for subsequent acquisition of adult grasp skills.
3 Here it is assumed that the network activation functions are non-linear (e.g. sigmoidal activation function).
4 The net motor command arriving at the muscles is a summation of the feedforward command and the signals from the feedback loops at the cortical and subcortical structures. However for simplicity we talk about a single inverse dynamic controller for the arm.
5 Note that it is also possible to split the Grip force controller into two modules, the first one computing the load force based on the arm trajectory, and the second one deriving a grip force based on the output of the first module, the load force (see Kawato *et al.*, 2003).

References

Arbib, M. A. & Hoff, B. (1994). Trends in neural modeling for reach to grasp. In K. M. B. Bennett & U. Castiello (Eds.), *Insights Into The Reach to Grasp Movement* (pp. 311–344). Amsterdam: North-Holland.

Baud-Bovy, G. & Soechting, J. F. (2001). Two virtual fingers in the control of the tripod grasp. *J Neurophysiol*, **86**, 604–615.

Blakemore, S. J., Wolpert, D. M. & Frith, C. D. (1998). Central cancellation of self-produced tickle sensation. *Nat Neurosci*, **1**(7), 635–640.

Boecker, H., Lee, A., Muhlau, M. *et al.* (2005). Force level independent representations of predictive grip force-load force coupling: a PET activation study. *Neuroimage*, **25**, 243–252.

Bortoff, G. A. & Strick, P. L. (1993). Corticospinal terminations in two New-World primates – further evidence that corticomotoneuronal connections provide part of the neural substrate for manual dexterity. *J Neurosci*, **13**, 5105–5118.

Bursztyn, L. L., Ganesh, G., Imamizu, H., Kawato, M. & Flanagan, J. R. (2006). Neural correlates of internal-model loading. *Curr Biol*, **16**, 2440–2445.

Chaminade, T., Oztop, E., Cheng, G. & Kawato, M. (2008). From self-observation to imitation: visuomotor association on a robotic hand. *Brain Res Bull*, **75**, 775–784.

Clifton, R. K., Muir, D. W., Ashmead, D. H. & Clarkson, M. G. (1993). Is visually guided reaching in early infancy a myth? *Child Development*, **64**(4), 1099–1110.

Colby, C. L. & Duhamel, J. R. (1996). Spatial representations for action in parietal cortex. *Cogn Brain Res*, **5**, 105–115.

Colby, C. L. & Goldberg, M. E. (1999). Space and attention in parietal cortex. *Ann Rev Neurosci*, **22**, 319–349.

Culham, J. C. & Valyear, K. F. (2006). Human parietal cortex in action. *Curr Opin Neurobiol*, **16**, 205–212.

Culham, J. C., Cavina-Pratesi, C. & Singhal, A. (2006). The role of parietal cortex in visuomotor control: what have we learned from neuroimaging? *Neuropsychologia*, **44**, 2668–2684.

Duhamel, J. R., Colby, C. L. & Goldberg, M. E. (1998). Ventral intraparietal area of the macaque: congruent visual and somatic response properties. *J Neurophysiol*, **79**, 126–136.

Dum, R. P. & Strick, P. L. (1991). The origin of corticospinal projections from the premotor areas in the frontal lobe. *J Neurosci*, **11**, 667–689.

Fagg, A. H. & Arbib, M. A. (1998). Modeling parietal-premotor interactions in primate control of grasping. *Neural Networks*, **11**, 1277–1303.

Flanagan, J. R. & Wing, A. M. (1997). The role of internal models in motion planning and control: evidence from grip force adjustments during movements of hand-held loads. *J Neurosci*, **17**, 1519–1528.

Fogassi, L., Gallese, V., Buccino, G. *et al.* (2001). Cortical mechanism for the visual guidance of hand grasping movements in the monkey – a reversible inactivation study. *Brain*, **124**, 571–586.

Gallese, V., Murata, A., Kaseda, M., Niki, N. & Sakata, H. (1994). Deficit of hand preshaping after muscimol injection in monkey parietal cortex. *Neuroreport*, **5**, 1525–1529.

Harris, C. M. & Wolpert, D. M. (1998). Signal-dependent noise determines motor planning. *Nature*, **394**, 780–784.

Haruno, M., Wolpert, D. M. & Kawato, M. (2001). MOSAIC model for sensorimotor learning and control. *Neural Computat*, **13**, 2201–2220.

Hoff, B. & Arbib, M. A. (1993). Models of trajectory formation and temporal interaction of reach and grasp. *J Motor Behav*, **25**, 175–192.

Iberall, T. & Arbib, M. A. (1990). Schemas for the control of hand movements: an essay on cortical localization. In M. A. G (Ed.), *Vision and Action: the Control of Grasping*. Norwood, NJ: Ablex.

Iberall, T. & Fagg, A. H. (1996). Neural network models for selecting hand shapes. In
A. M. Wing, P. Haggard & J. R. Flanagan (Eds)., *Hand and Brain: The
Neurophysiology and Psychology of Hand Movements* (pp. 243–264). New York, NY:
Academic Press.

Iberall, T., Bingham, G. & Arbib, M. (1986). Opposition space as a structuring concept for
the analysis of skilled hand movements. In H. Heuer & C. Fromm (Eds.), *Generation
and Modulation of Action Patterns* (pp. 158–173). Berlin: Springer-Verlag.

Imamizu, H., Miyauchi, S., Tamada, T. *et al.* (2000). Human cerebellar activity reflecting an
acquired internal model of a new tool. *Nature*, **403**, 192–195.

Jeannerod, M., Arbib, M. A., Rizzolatti, G. & Sakata, H. (1995). Grasping objects – the
cortical mechanisms of visuomotor transformation. *Trends Neurosci*, **18**, 314–320.

Jeannerod, M., Paulignan, Y. & Weiss, P. (1998). Grasping an object: one movement,
several components. *Novartis Found Symp*, **218**, 5–16; discussion 16–20.

Johansson, R. S. (1996). Sensory control of dexterous manipulation in humans. In
A. M. Wing, P. Haggard & J. R. Flanagan (Eds.), *Hand and Brain: The Neurophysiology
and Psychology of Hand Movements* (pp. 381–414). New York, NY: Academic Press.

Kawato, M. (1999). Internal models for motor control and trajectory planning. *Curr Opin
Neurobiol*, **9**, 718–727.

Kawato, M. (2008). Cerebellum: models. In L. R. Squire (Ed.), *Encyclopedia of
Neuroscience*. Amsterdam, the Netherlands: Elsevier Science.

Kawato, M. & Samejima, K. (2007). Efficient reinforcement learning: computational
theories, neuroscience and robotics. *Curr Opin Neurobiol*, **17**, 205–212.

Kawato, M., Kuroda, T., Imamizu, H. *et al.* (2003). Internal forward models in the
cerebellum: fMRI study on grip force and load force coupling. In C. Prablanc,
D. Pelisson & Y. Rossetti (Eds.), *Progress in Brain Research: Neural Control of Space
Coding and Action Production*, vol. 142 (pp. 171–188). Amsterdam, the Netherlands:
Elsevier Science.

Keysers, C. & Perrett, D. I. (2004). Demystifying social cognition: a Hebbian perspective.
Trends Cogn Sci, **8**, 501–507.

Lawrence, D. G. & Hopkins, D. A. (1976). The development of motor control in the rhesus
monkey: evidence concerning the role of corticomotoneuronal connections. *Brain*, **99**,
235–254.

Lemon, R. N., Armand, J., Olivier, E. & Edgley, S. A. (1997). Skilled action and the
development of the corticospinal tract in primates. In K. J. Connolly & H. Forssberg
(Eds.), *The Neurophysiology and Neuropsychology of Motor Development*
(pp. 162–176). Cambridge: Cambridge University Press.

Lockman, J., Ashmead, D. H. & Bushnell, E. W. (1984). The development of anticipatory
hand orientation during infancy. *J Exp Child Psychol*, **37**, 176–186.

Luppino, G., Murata, A., Govoni, P. & Matelli, M. (1999). Largely segregated parietofrontal
connections linking rostral intraparietal cortex (areas AIP and VIP) and the ventral
premotor cortex (areas F5 and F4). *Exp Brain Res*, **128**, 181–187.

Miyamoto, H., Nakano, E., Wolpert, D. & Kawato, M. (2004). TOPS (Task Optimization in
the Presence of Signal-dependent noise) model. *Systems and Computers in Japan*
(Translated from Denshi Tsushin Gakkai Ronbunshi, J85-D-II, 940–949), **35**, 48–58.

Murata, A., Gallese, V., Kaseda, M. & Sakata, H. (1996). Parietal neurons related to
memory-guided hand manipulation. *J Neurophysiol*, **75**, 2180–2186.

Murata, A., Gallese, V., Luppino, G., Kaseda, M. & Sakata, H. (2000). Selectivity for the
shape, size, and orientation of objects for grasping in neurons of monkey parietal area
AIP. *J Neurophysiol*, **83**, 2580–2601.

Newell, K. M., McDonald, P. V. & Baillargeon, R. (1993). Body scale and infant grip configurations. *Dev Psychobiol*, **26**, 195–205.

Nowak, D. A., Topka, H., Timmann, D., Boecker, H. & Hermsdorfer, J. (2007). The role of the cerebellum for predictive control of grasping. *Cerebellum*, **6**, 7–17.

Oztop, E., Bradley, N. S. & Arbib, M. A. (2004). Infant grasp learning: a computational model. *Exp Brain Res*, **158**, 480–503.

Oztop, E., Chaminade, T., Cheng, G. & Kawato, M. (2005a). Imitation Bootstrapping: Experiments on a Robotic Hand. *IEEE-RAS International Conference on Humanoid Robots*, Tsukuba, Japan.

Oztop, E., Wolpert, D. & Kawato, M. (2005b). Mental state inference using visual control parameters. *Brain Res Cogn Brain Res*, **22**, 129–151.

Oztop, E., Imamizu, H., Cheng, G. & Kawato, M. (2006). A computational model of anterior intraparietal (AIP) neurons. *Neurocomputing*, **69**, 1354–1361.

Paulignan, Y., MacKenzie, C., Marteniuk, R. & Jeannerod, M. (1991a). Selective perturbation of visual input during prehension movements. 1. The effects of changing object position. *Exp Brain Res*, **83**, 502–512.

Paulignan, Y., Jeannerod, M., MacKenzie, C. & Marteniuk, R. (1991b). Selective perturbation of visual input during prehension movements. 2. The effects of changing object size. *Exp Brain Res*, **87**, 407–420.

Rizzolatti, G., Gentilucci, M., Camarda, R. M. *et al.* (1990). Neurons related to reaching-grasping arm movements in the rostral part of area-6 (Area-6a-Beta). *Exp Brain Res*, **82**, 337–350.

Rizzolatti, G., Fogassi, L. & Gallese, V. (2002). Motor and cognitive functions of the ventral premotor cortex. *Curr Opin Neurobiol*, **12**, 149–154.

Roy, A., Paulignan, Y., Meunier, M. & Boussaoud, D. (2006). Prehension movements in the macaque monkey: effects of perturbation of object size and location. *Exp Brain Res*, **169**, 182–193.

Sakata, H., Taira, M., Murata, A. & Mine, S. (1995). Neural mechanisms of visual guidance of hand action in the parietal cortex of the monkey. *Cereb Cortex*, **5**, 429–438.

Sakata, H., Taira, M., Kusunoki, M. *et al.* (1998). Neural coding of 3D features of objects for hand action in the parietal cortex of the monkey. *Philos Trans R Soc Lond B Biol Sci*, **353**, 1363–1373.

Sakata, H., Tsutsui, K.-I. & Taira, M. (2005). Toward an understanding of the neural processing for 3D shape perception. *Neuropsychologia*, **43**, 151–161.

Shimoga, K. B. (1996). Robot grasp synthesis algorithms: a survey. *Int J Robotics Res*, **15**, 230–266.

Smeets, J. B. & Brenner, E. (1999). A new view on grasping. *Motor Control*, **3**, 237–271.

Smeets, J. B. & Brenner, E. (2001). Independent movements of the digits in grasping. *Exp Brain Res*, **139**, 92–100.

Triggs, W. J., Yathiraj, S., Young, M. S. & Rossi, F. (1998). Effects of task and task persistence on magnetic motor-evoked potentials. *J Contemp Neurol* (http://mitpress.mit.edu/e-journals/JCN/abstracts/003/cn3-2.html), **1998(2A)**.

Uno, Y., Fukumura, N., Suzuki, R. & Kawato, M. (1995). A computational model for recognizing objects and planning hand shapes in grasping movements. *Neural Networks*, **8**, 839–851.

von Hofsten, C. (1982). Eye-hand coordination in the newborn. *Dev Psychol*, **18**, 450–461.

von Hofsten, C. (1984). Developmental changes in the organization of prereaching movements. *Dev Psychol*, **20**, 378–388.

von Hofsten, C. & Ronnqvist, L. (1988). Preparation for grasping an object: a developmental study. *J Exp Psychol Hum Percept Perform*, **14**, 610–621.

Weiss, P. & Jeannerod, M. (1998). Getting a grasp on coordination. *News Physiol Sci*, **13**, 70–75.

Weiss, P. H., Jeannerod, M., Paulignan, Y. & Freund, H. J. (2000). Is the organisation of goal-directed action modality specific? A common temporal structure. *Neuropsychologia*, **38**, 1136–1147.

Wise, S. P., Boussaoud, D., Johnson, P. B. & Caminiti, R. (1997). Premotor and parietal cortex: corticocortical connectivity and combinatorial computations. *Ann Rev Neurosci*, **20**, 25–42.

Wolpert, D. M. & Kawato, M. (1998). Multiple paired forward and inverse models for motor control. *Neural Networks*, **11**, 1317–1329.

Wolpert, D. M., Doya, K. & Kawato, M. (2003). A unifying computational framework for motor control and social interaction. *Philos Trans R Soc Lond B Biol Sci*, **358**, 593–602.

Part II

The physiology of grasping

10

The study of hand movements during grasping. A historical perspective

MARC JEANNEROD

Summary

A satisfactory description of human hand movements during the action of grasping was not available until the early 1980s. Kinematic parameters extracted from the displacements of anatomical landmarks located on the hand were used to differentiate between a transport component carrying the hand at the target location and a grasp component shaping the finger according to the object shape and size. These parameters, including the maximum grip aperture (MGA) are now currently adopted for testing vision for action in normal subjects, including in children at different stages of their visuomotor development, and in a wide range of pathological disorders affecting goal-directed movements.

Introduction: hand grasping movements before 1980

The hand is both a sensory and a motor organ. On the sensory side, in Sherrington's terms, it is the fovea of the somesthetic system, to the same extent as the center of the retina is the fovea of the visual system. The hand explores the haptic world by touching, grasping and manipulating objects in the same way as eye movements explore the visual world by displacing the retina between fixation points. The sensory and motor functions of the hand are complementary with one another. The movements of the fingers contribute to the exploration and perception of object shape and texture during manipulation and conversely sensory cues arising from the skin receptors contribute to the control of hand movements. One of the best illustrations of this two-way collaboration (what Sherrington called "active touch;" see Phillips, 1986) is provided by the automatic control of grip force by the fast-adapting receptors of the glabrous skin of the finger pads during handling objects (Johansson & Westling, 1984; see Mountcastle, 2005).

The sensory and motor functions of the primate hand are the end result of evolutionary changes. In humans, bipedal stance, by liberating the hand from the constraints of locomotion, has coincided with the development of a powerful thumb equipped with three unique muscles lacking in African apes; one for flexing the thumb, another one for forcibly opposing it to the other fingers, and one for stabilizing it across its knuckle (Bradshaw, 1997). The powerful and opposable thumb, together with the high degree of independence

Sensorimotor Control of Grasping: Physiology and Pathophysiology, ed. Dennis A. Nowak and Joachim Hermsdörfer. Published by Cambridge University Press. © Cambridge University Press 2009.

of the movements of the other fingers relative to each other, enables an extensive number of prehensile and manipulative activities. Furthermore, the evolution of the hand has been paralleled by changes in the primate central nervous system, where the proportion of motor and sensory cortical areas devoted to the hand has massively expanded, as compared with carnivores and rodents. The primate brain is endowed with a number of unique features which testify to the high degree of specialization of hand movements in these species. First, the hand muscles' motoneurons are monosynaptically connected with the hand area of primary motor cortex. Second, a large amount of cortico-cortical fibers directly connect the hand premotor cortex with several posterior parietal areas, which accounts for the precise sensori-motor control of hand movements, both by the somatosensory and the visual modalities.

Despite its broadly acknowledged importance in primate behavior and cognition, many aspects of hand function have barely received attention until recently. Whereas extensive and comprehensive studies have been devoted to the neurophysiological aspects of hand motor control in anesthetized animal preparations since the time of Sherrington, little attention was given to the movements themselves. Obviously, the complexity of hand and finger movements (up to 25 degrees of freedom) must have been an obstacle for a complete description of these movements. The problem raised by these many degrees of freedom, however, was not only technical, it was also, and mainly, conceptual: the number of degrees of freedom could not be reduced and the proper studies could not be started before a functional description of the hand activities, and specifically of the hand prehensile move-ments, was given. The systematic description of hand postures by Napier represented a first step in this direction. Considering the prehensile movements of the hand, Napier (1956) made the remark that, "At first sight it would seem that the prehensile activities of the hand are so extensive and varied that a simple analysis would not be feasible." But, he continues, "A study of the normal hand suggests that there are, in fact, only two distinct patterns of movement in man and that these, either separately or in combination, provide the anatomical basis for all prehensile activities [...]." Stability of prehension, a prerequisite for further manipulative activity, can be achieved in one of two ways: "(1) The object may be held in a clamp formed by the partly flexed fingers and the palm, counter pressure being applied by the thumb lying more or less in the plane of the palm. This is referred to as the power grip. (2) The object may be pinched between the flexor aspects of the fingers and the opposing thumb. This is called the precision grip" (Napier, 1956, p. 903). This description accounts for most situations involving complex tool use, including for bimanual actions where one hand (usually the left hand in right handers) firmly holds an object with a power grip, whereas the other hand makes the fine manipulation with a tool held in a precision grip.

Napier's description, however, only considered the final static postures held by the hand in hand/object interactions, and did not address the dynamic aspect of how these postures were achieved. Crucially, the simple fact that one cannot grasp an object without reaching for it, and that the hand, in order to achieve its task, has to be transported to the location of the object or surface, was not taken into account (Figure 10.1). The idea that prehension in fact involves two coordinated and collaborative motor components, hand transportation and finger grip, became central to the study of hand function in the 1970s. The main incentive for

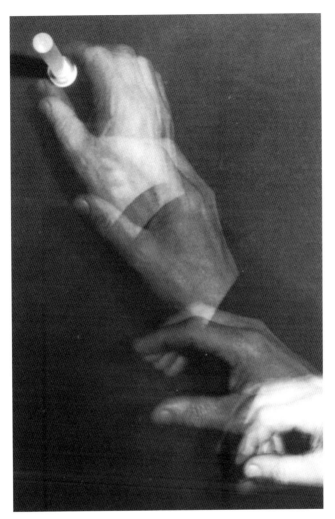

Figure 10.1. Chronophotograph of a prehension movement. Note opening and closure of the finger grip prior to contact with the object. Also note that the shaping of the hand during transport is independent from the orientation of the hand, which changes as it approaches the object.

these studies was given by a series of neuropsychological findings in monkeys. Hans Kuypers and his colleagues showed that split-brain monkeys which were forced to use an ipsilateral eye-hand control to retrieve food morsels from a Klüver board were only able to reach for the board, but could not shape their hand to grasp the food morsels. By contrast, when using a contralateral eye-hand control, they were able to perform accurate grasps (Brinkman & Kuypers, 1972; Haaxma & Kuypers, 1975). These findings extended to visually goal-directed movements the notion of a distinct functional organization of the motor control for the proximal and the distal segments of the upper limb (Lawrence & Kuypers, 1968): the distal hand and finger movements were exclusively controlled by the

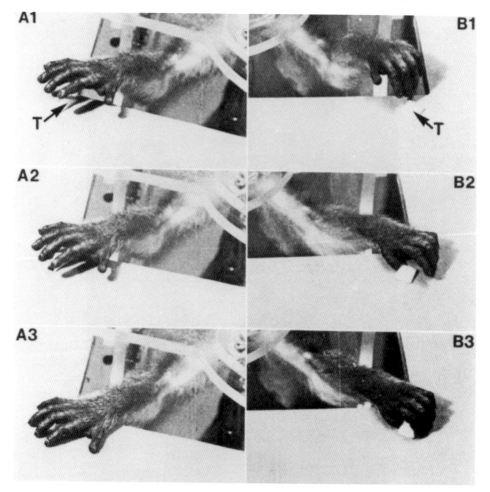

Figure 10.2. Effect of a left-sided lesion of parietal area 7 in a macaque monkey. Note difference between the ipsilateral (left) and the contralateral (right) hand during a prehension movement directed at a food target (T). The contalateral hand (A) fails to shape according to the target size. Data from Faugier-Grimaud *et al.* (1978).

contralateral primary motor cortex via the crossed corticospinal pathway, whereas the more proximal shoulder and elbow movements were controlled by both contralateral and ipsilateral motor cortices via indirect bilateral corticospinal projections. At about the same time, the visuomotor function of the parietal cortex was discovered by Vernon Mountcastle and his colleagues, who demonstrated the existence, in posterior parietal areas, of neurons specifically activated during visually directed reaching and manipulation (Mountcastle *et al.*, 1975). Finally, Simone Faugier-Grimaud and her colleagues showed that monkeys with a lesion of the posterior parietal areas containing these neurons were impaired in reaching and grasping visual objects with their contralateral hand (Figure 10.2) (Faugier-Grimaud *et al.*,

1978). Thus, in a few years, the stage was set for a comprehensive study of hand grasping movements and their neural substrate.

The Brandeis Meeting (1978)

The idea of splitting prehension into components, one for transporting the hand toward the object (the transportation, or reach, component), and another one for grasping and handling the object (the manipulation, or grasp, component) also reflected the renewal of the study of the visual system around 1968. Several authors working in amphibians, rodents or primates had proposed a division of labor between the sub-cortical visual system, which was assigned the role of localizing visual targets in space, and the cortical visual system specialized in identifying the visual stimuli (Ingle, 1967; Trevarthen, 1968). Most of the people interested in this line of research attended a meeting at Brandeis University in June 1978. At that meeting, I presented the preliminary results of our study with B. Biguer on prehension movements. In our paper, we proposed to divide prehension into "visuomotor channels," such that "Each channel would deal with a specific aspect of the visual stimulus and, when activated, would release a motor program adapted to the input pattern" (Jeannerod & Biguer, 1982, p. 399). One of these channels, the "object" channel, corresponded to the grasping component: its role was to encode the intrinsic properties of the object relevant to grasping and manipulating it (e.g. size, shape, orientation), and to shape the hand in anticipation of these properties. Another channel was the "space" channel, which corresponded to the reach component. Its role was to encode the spatial localization of the object with respect to the body, and to transport the hand accordingly (note that the terms "object" and "space" channels had already been used by Mishkin in a different context, see below). The arguments supporting the existence of visuomotor channels were drawn from two main sources. First, developmental studies in infants suggested an asynchronous maturation of two distinct behavioral components for reaching and grasping, respectively. Second, the effects of cortical lesions in adults suggested that the two channels could be dissociated from one another. This point, however, remained inconclusive in our paper: although we correctly assigned the two channels to mechanisms located within the parietal lobe, we had not at our disposal clinical material with lesions sufficiently small to affect one or the other channel separately (Jeannerod & Biguer, 1982). At any rate, the assignment of visuomotor mechanisms to parietal cortex rather than to subcortical structures was in agreement with the revised Two-visual systems model, which was presented by Ungerleider and Mishkin at that same meeting (Ungerleider & Mishkin, 1982). Mortimer Mishkin, on the basis of monkey experiments, had provided since 1972 anatomic and behavioral evidence for a segregation of visual functions into two parallel cortico-cortical pathways, the occipito-parietal, or dorsal pathway, and the occipito-temporal, or ventral pathway. The dorsal pathway was assigned by Ungerleider and Mishkin a function in the perception of spatial relationships between objects, while the ventral pathway was assigned the function of object recognition. Mishkin's work, which was going to have a deep influence on neuropsychological thinking (see Paillard & Beaubaton, 1974), was later completed by introducing visuomotor function as a predominant feature of the dorsal pathway (Milner & Goodale, 1995).

A method for reducing the number of degrees of freedom

The complete behavioral description of the two components was proposed at another meeting in Cambridge (UK) in 1980 (due to publication delays of the proceedings of the Brandeis meeting, this second study appeared first in the literature; Jeannerod, 1981). The data published in the 1981 and 1982 papers were obtained using a cinematographic method which is worth describing in some detail before describing the results. The main concern in this study was to reduce the number of degrees of freedom of the arm. To this aim, a standardized reach and grasp movement was adopted where the subject was moving his/her hand from a starting position near the body axis and reached for an object placed at different distances in the sagittal plane. Objects were wooden cylinders or spheres of different sizes, placed at the same horizontal level as the starting position. Subjects were instructed to reach, grasp and raise the object at a natural rate, using a precision grip opposing the thumb and the index finger (and additional fingers if required by the object size). This arrangement resulted in a curved trajectory of the hand which was raised from its starting position and then lowered down at the object location. In most cases, the starting hand posture and the object orientation were combined so as to minimize wrist rotation.

Sixteen-millimetre films were taken using a cine-camera running at 50 frames per second. The camera was placed so that the radial side of the arm appeared on the films. The films were subsequently analyzed manually frame by frame using a Lafayette projector to project the image of the hand onto a screen with a one-to-one magnification. The position of three anatomical landmarks was plotted directly on the screen across successive frames: one landmark representing the position of the wrist, another one at the tip of the index finger and the third one at the tip of the thumb (Figure 10.3). Movement time (MT) was measured as the number of frames between the first detectable wrist movement and the first detectable movement of the object. The wrist kinematics were calculated by plotting the distance between successive positions of the wrist along the movement trajectory, as a function of time: this provided a direct measure of changes in instantaneous velocity and acceleration of the wrist. In addition, for each frame, the distance between the tip of the thumb and the tip of the index finger was measured, which allowed reconstruction of the change in grip size as a function of time. Standard parameters were extracted from these measures. For the transportation component, these parameters were: Time to peak acceleration (TPA), Time to peak velocity (TPV), Time to peak deceleration (TPD). For the manipulation component: Maximum grip aperture (MGA). These same parameters are still in use nowadays for describing hand movements, but the lengthy frame by frame analysis has been replaced by automatic computation of the position of markers attached to the wrist and to the extremities of the thumb and index finger, using high resolution motion analyzers (see also Chapter 2).

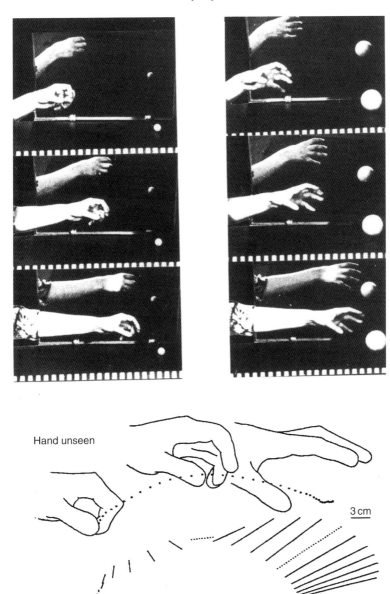

Figure 10.3. Upper part: Film records of human prehension movements. Note early shaping of the hand during transport. Also note different hand shapes corresponding to objects of different sizes. For each movement, three frames have been selected from a film taken at 50 frames/s.

Lower part: Frame by frame processing of one prehension movement directed at the large object shown on the upper part. Movement executed in the open-loop condition (subject sees the object, not her hand). The dots represent successive positions (every 20 ms) of the wrist marker during the transportation of the hand. Note the low velocity phase close to the object. The lines represent successive apertures of the finger grip (every 40 ms). Note progressive opening of finger grip, followed by closure during the low velocity phase. The three superimposed drawings represent hand postures redrawn from film frames during the same movement. Data from Jeannerod (1981).

It followed from these results that prehension movements could not be reduced to a description of the final posture of the hand on the object. Rather, a large amount of information became available in the anticipatory processes that led to this final posture. The fact that the hand posture progressively shapes during the reach (hand pre-shaping) appeared to be an essential feature of prehension.

The kinematic description of grasping

The study of the grasping component in itself revealed something of particular interest. Indeed, the main surprise in the 1981 study was the serendipitous finding of a systematic change in grip size as the hand approached the object. In the terms of the 1981 paper, "The first obvious feature of this [manipulation] component was that the fingers opened up to a maximum grip aperture, which was a function of the anticipated object's size, although it was always greater than that which the actual size of the object would require. Hence, after the maximum grip aperture had been reached, the fingers began to close in anticipation of contact with the object" (Jeannerod, 1981, p. 161). The existence of this biphasic profile of the finger movements (opening followed by closure) is a robust finding which represents a definite characteristic of prehension in humans and in primates in general (see below). Of particular interest is the tight relationship of MGA to the size of the object. In the 1981 study, where objects of only two different sizes were used as targets for the movements, the MGA was found to be larger for the larger object than for the smaller one, and to exceed the actual object size by about 20% (see Figure 10.3, upper part). Later studies where the object size was systematically varied showed a linear relationship between MGA and object size (e.g. Marteniuk *et al.*, 1990).

There are two points here. First, the relationship with object size testifies to the existence of an anticipatory computation, by the visual system, of the intrinsic properties of the object, like its shape or size. The fact that these properties are likely to be processed by the occipitoparietal–dorsal pathway suggests that this pathway, and not only the ventral pathway, is equipped for processing object shape, at least when it is relevant to visuomotor transformation. In other words, the same properties of an object are processed differently, according to whether this object is a target for grasping or a stimulus for perceptual identification and recognition. Various modalities of response, including verbal responses, can be used for studying the process of recognition, whereas MGA practically stands as the only objective index for the study of visuomotor transformation during grasping. For this reason, MGA eventually became a tool for studying visuomotor transformation in different circumstances, including in patients presenting impairments in visuomotor or motor functions.

The second point to be considered is that MGA, in addition to being scaled to object size prior to contact, exceeds object size by about 20%. Several explanations have been proposed for this phenomenon of exaggerated opening of the finger grip. A simple explanation refers to the biomechanics of the finger opposition during grasping: it is safer to apply the opposition forces by approaching the object perpendicularly than tangentially to its surface.

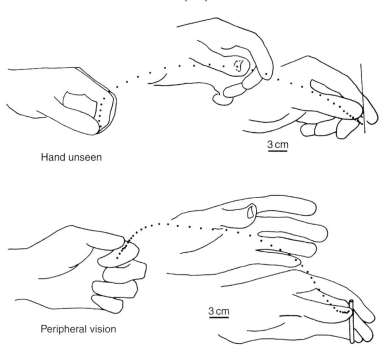

Figure 10.4. Effect of visual feedback conditions on prehension movements. Upper part: Film processing of one movement directed at a small object (an insect pin) in the open-loop condition (subject sees the object, not her hand). Note preservation of the full pattern of prehension, including terminal accuracy (from Jeannerod, 1981). Lower part: movement directed at a small object presented in the peripheral visual field (subject sees both the hand and the object). Note preservation of the pattern of the transportation component, contrasting with lack of hand shaping and awkward final grasp (unpublished data).

An alternative explanation is that the extra opening represents a safety margin taking into account the possible terminal errors affecting the transportation component. Should this explanation be correct, the safety margin should increase in conditions where terminal errors are more likely to occur. One such condition is suppression of visual feedback during the movement. As it was shown in another paper published in 1984, the final phase of prehension is under visual control. Suppressing visual feedback during the movement, although it does not affect the global pattern of prehension, affects the final phase of the movement, with the trajectory of the reach becoming more variable: in this case the observed increase in MGA may well correspond to a compensatory increase of the safety margin (Figure 10.4) (Jeannerod, 1984).

It is noteworthy that the same biphasic pattern (aperture of the finger grip followed by closure) of the grip formation is also observed during natural prehension movements in monkeys. The kinematic profiles of transportation and grasping obtained in macaques by Alice Roy and her coworkers are strikingly similar to those found in human prehension (Roy *et al.*, 2002) (see also Chapter 2). This similarity of human and simian grasping includes the fact that, also in monkeys, MGA is scaled to object size (see Figure 10.2, right). However, due to the brevity of their thumb, most non-human primates perform the precision grip with a side opposition rather than with a pad opposition (Napier, 1960; Christel, 1993; Spinozzi *et al.*, 2004). It is therefore likely that MGA represents a widespread feature of object acquisition, which reflects the encoding, by the visual system, of the physical parameters of the object, irrespective of the effector used to take the object. Indeed, a similar pattern has been observed for the gape aperture in birds picking up seeds with their beak (Klein *et al.*, 1985), or in humans opening their mouth while eating pieces of food of different sizes (Castiello, 1997).

The coordination and timing of the two components

The fact that the single action of grasping an object includes distinct components raises the question of how these components are coordinated with one another. In order to achieve a stable grasp, the relative timing of the wrist displacement and of the finger closure has to obey very strict constraints. Records clearly show that the wrist velocity falls to zero at the time of contact of the fingers with the object. If the hand displacement had not come to a stop at the time when the fingers closed, the object would be pushed forward along the direction of the reach; if the fingers had closed before contact, the object would be bumped and the grasp would fail; conversely, if fingers closed too late, the contact with the object would be made with the palm of the hand, resulting in an awkward palmar grasp. This is indeed what may happen in pathological conditions (e.g. Jeannerod *et al.*, 1994).

The relative timing of the two components was an early concern in the grasping studies. In the 1982 paper, we had noticed that, although the transportation component (as reflected by the activity of the biceps brachii) started first, the arm displacement was frequently observed to lag the first detectable finger movements, a difference which we attributed to the heavier load of the arm compared with the fingers (Figure 10.5) (Jeannerod & Biguer, 1982). At the other end of the movement, the transportation component was marked by a sharp deceleration, followed by a brief reacceleration until the velocity of the hand finally dropped down to zero at the time of contact of the fingers with the object. The onset of this low-velocity phase of the transportation component, at about 70% of movement time, was found to co-vary with the parameter MGA, i.e. the time at which the finger grip began to close (Jeannerod, 1981). Globally, as already mentioned, this temporal organization was little affected by the suppression of visual feedback during the movement (Jeannerod, 1984).

The presence of a correlation, independently from the presence of visual feedback, between onset of finger closure and slowing down of the arm transportation (Figure 10.6), thus suggested the existence of an additional mechanism for synchronizing and coordinating the two components. This was indeed one of the important features of the model designed by Michael Arbib to account for the separate and coordinated programming of the motor "schemas" involved in the action of grasping (Arbib, 1981). The hypothesis of a correlation mechanism implied that, if one of the components were perturbed during the execution of the movement by an unexpected event (e.g. a brisk change in the position or the size of the

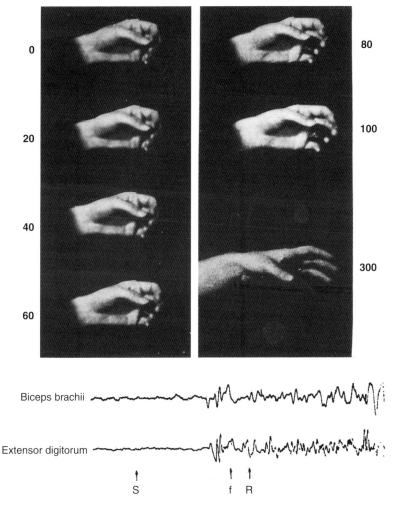

Figure 10.5. Upper part: single frames of a prehension movement taken from a film at 50 frames/s. The numbers next to each frame indicate the time position of the frame in movement time. The frame at 0 ms corresponds to the frame prior to the first detectable finger movement, clearly detectable on the next two frames (20 and 40 ms). In this movement, the first sign of hand transportation was detected on the frame at 100 ms, i.e. 80 ms after onset of finger movements.

Lower part: Electromyogram recordings of biceps brachii and extensor digitorum at the onset of the prehension movement shown in the upper part. Note synchronous onset of the contraction of the two muscles. Letters below the drawing represent the temporal position of events taken from the film: S: presentation of the target object; f: first detectable finger movement; R: onset of the reaching movement. Data from Jeannerod & Biguer (1982).

object for the transportation component, or a change in object size or orientation for the grasping component), this should also affect the other component. It was predicted, in the 1981 paper, that, if the spatial position of the object would be altered after the initiation of the movement, the dynamics of the grip should be modified to fit with those of the

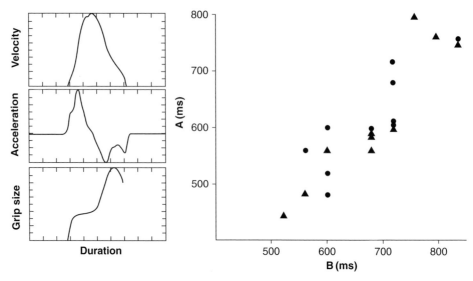

Figure 10.6. Time relation between the two components of prehension. Left, upper two diagrams: averaged velocity and acceleration profiles of 20 transportation components in one subject. Lower diagram: averaged change in grip aperture during the same 20 movements. Right: plot of the time occurrence of maximum grip size (B, in milliseconds) versus the onset of the low velocity phase of the transportation component (A, in milliseconds) for each of the same 20 movements. The correlation $A = f(B)$ is $r = .81$, $P < 0.001$. Data from Jeannerod (1984).

transportation (Jeannerod, 1981). This prediction was indeed fully verified by Yves Paulignan and his coworkers in their perturbation experiments beginning in 1990 (Paulignan *et al.*, 1991a, 1991b). Nowadays, the precise nature of this coordination mechanism still remains largely unknown.

Conclusion

The possibility of providing a satisfactory description of hand movements with only three landmarks and a few parameters extracted from these landmarks opened a new field of research. The psychophysical description of grasping enriched the study of visuomotor behavior which had been until then limited to a description based on pointing movements (see Jeannerod, 1988). The influence of experimental conditions (e.g. presence or absence of visual feedback, effect of movement speed) and the influence of the goal of the movement (e.g. effect of changing size, shape or orientation of the object) were systematically investigated in normal subjects, including in children at different stages of their visuomotor development. In pathological conditions, this description provided new insight into a wide range of disorders affecting goal-directed movements, such as central motor disorders (e.g. Parkinson's disease), visuomotor disorders (e.g. optic ataxia) or higher-order motor disorders like apraxia.

References

Arbib, M. A. (1981). Perceptual structures and distributed motor control. In V. B. Brooks (Ed.), *Handbook of Physiology, Section I: The Nervous System, Vol. 2: Motor Control* (pp. 1449–1480). Baltimore, MD: Williams and Wilkins.

Bradshaw, J. L. (1997). *Human Evolution. A Neuropsychological Perspective*. Hove, UK: Psychology Press.

Brinkman, J. & Kuypers, H. G. J. M. (1972). Split-brain monkeys: cerebral control of ipsilateral and contralateral arm, hand and finger movements. *Science*, **176**, 536–539.

Castiello, U. (1997). Arm and mouth coordination during the eating action in humans. A kinematic analysis. *Experimental Brain Research*, **115**, 552–556.

Christel, M. (1993). Grasping techniques and hand preferences in Hominoidea. In H. Preuschoft & D. J. Chivers (Eds.), *Hands of Primates* (pp. 91–108). New York, NY: Springer-Verlag.

Faugier-Grimaud, S., Frenois, C. & Stein, D. G. (1978). Effects of posterior parietal lesions on visually guided behavior in monkeys. *Neuropsychologia*, **16**, 151–168.

Haaxma, R. & Kuypers, H. G. J. M. (1975). Intrahemispheric cortical connections and visual guidance of hand and finger movements in the rhesus monkey. *Brain*, **98**, 239–260.

Ingle, D. J. (1967). Two visual mechanisms underlying the behavior of fish. *Psychologische Forschung*, **31**, 44–51.

Jeannerod, M. (1981). Intersegmental coordination during reaching at natural visual objects. In J. Long & A. Baddeley (Eds.), *Attention and Performance IX* (pp. 153–168). Hillsdale, NJ: Lawrence Erlbaum.

Jeannerod, M. (1984). The timing of natural prehension movements. *J Motor Behaviour*, **16**, 235–254.

Jeannerod, M. (1988). *The Neural and Behavioural Organization of Goal-directed Movements*. Oxford, UK: Oxford University Press.

Jeannerod, M. & Biguer, B. (1982). Visuomotor mechanisms in reaching within extrapersonal space. In D. Ingle, M. A. Goodale & R. Mansfield (Eds.), *Advances in the Analysis of Visual Behavior* (pp. 387–409). Boston, MA: MIT Press.

Jeannerod, M., Decety, J. & Michel, F. (1994). Impairment of grasping movements following a bilateral posterior parietal lesion. *Neuropsychologia*, **32**, 369–380.

Johansson, R. S. & Westling, G. (1984). Roles of glabrous skin receptors and sensorimotor memory in automatic control of precision grip when lifting rougher or more slippery objects. *Exp Brain Res*, **56**, 550–564.

Klein, B. G., Deich, J. D. & Zeigler, H. P. (1985). Grasping in the pigeon (*Columba livia*): final common path mechanism. *Behav Brain Res*, **18**, 201–213.

Lawrence, D. G. & Kuypers, H. G. J. M. (1968). The functional organization of the motor system in the monkey. I. The effects of bilateral pyramidal lesions. *Brain*, **91**, 1–14.

Marteniuk, R. G., Leavitt, J. L., MacKenzie, C. L., & Athenes, S. (1990). Functional relationships between grasp and transport components in a prehension task. *Hum Move Sci*, **9**, 149–176.

Milner, A. D. & Goodale, M. A. (1995). *The Visual Brain in Action*. Oxford, UK: Oxford University Press.

Mountcastle, V. B. (2005). *The Sensory Hand. Neural Mechanisms of Somatic Sensation*. Cambridge, MA: Harvard University Press.

Mountcastle, V. B., Lynch, J. C., Georgopoulos, A. Sakata, H. & Acuna, C. (1975). Posterior parietal association cortex of the monkey: command functions for operations within extra-personal space. *J Neurophysiol*, **38**, 871–908.

Napier, J. R. (1956). The prehensile movements of the human hand. *J Bone Joint Surg*, **38B**, 902–913.

Napier, J. R. (1960). Studies of the hands of living primates. *Proc Zoo Soc Lond*, **134**, 647–657.

Paillard, J. & Beaubaton, D. (1974). Problèmes posés par les contrôles moteurs ipsilatéraux après déconnexion hémisphérique chez le singe. In F. Michel & B. Schott (Eds.), *Les Syndromes de Disconnexion Calleuse chez l'Homme* (pp. 137–171). Lyon, France: Hôpital Neurologique.

Paulignan, Y., MacKenzie, C., Marteniuk, R. & Jeannerod, M. (1991a). Selective perturbation of visual input during prehension movements. I. The effects of changing object position. *Exp Brain Res*, **83**, 502–512.

Paulignan, Y., Jeannerod, M., MacKenzie, C. & Marteniuk, R. (1991b). Selective perturbation of visual input during prehension movements. 2. The effects of changing object size. *Exp Brain Res*, **87**, 407–420.

Phillips, C. G. (1986). *Movements of the Hand*. The Sherrington Lecture XVII. Liverpool, UK: Liverpool University Press.

Roy, A. C., Paulignan, Y., Meunier, M. & Boussaoud, D. (2002). Prehension movements in the macaque monkey. Effect of object size and location. *J Neurophysiol*, **88**, 1491–1499.

Spinozzi, G., Truppa, V. & Lagana, T. (2004). Grasping behaviour in tufted capuchin monkeys (*Cebus paella*). Grip types and manual laterality for picking up a small food item. *Am J Phys Anthropol*, **125**, 30–41.

Trevarthen, C. B. (1968). Two mechanisms of vision in primates. *Psych Forsch*, **31**, 299–337.

Ungerleider, L. & Mishkin, M. (1982). Two cortical visual systems. In D. J. Ingle, M. A. Goodale & R. J. W. Mansfield (Eds.), *Advances in the Analysis of Visual Behavior* (pp. 549–586). Cambridge, MA: MIT Press.

11

Sensory control of object manipulation

ROLAND S. JOHANSSON AND J. RANDALL FLANAGAN

Summary

Series of action phases characterize natural object manipulation tasks where each phase is responsible for satisfying a task subgoal. Subgoal attainment typically corresponds to distinct mechanical contact events, either involving the making or breaking of contact between the digits and an object or between a held object and another object. Subgoals are realized by the brain selecting and sequentially implementing suitable action-phase controllers that use sensory predictions and afferents signals in specific ways to tailor the motor output in anticipation of requirements imposed by objects' physical properties. This chapter discusses the use of tactile and visual sensory information in this context. It highlights the importance of sensory predictions, especially related to the discrete and distinct sensory events associated with contact events linked to subgoal completion, and considers how sensory signals influence and interact with such predictions in the control of manipulation tasks.

Sensory systems supporting object manipulation

In addition to multiple motor systems (arm, hand, posture), most natural object manipulation tasks engage multiple sensory systems. Vision provides critical information for control of task kinematics. In reaching, we use vision to locate objects in the environment and to identify contact sites for the digits that will be stable and advantageous for various actions we want to perform with the grasped object (Goodale *et al.*, 1994; Santello & Soechting, 1998; Cohen & Rosenbaum, 2004; Cuijpers *et al.*, 2004; Lukos *et al.*, 2007). For example, when we pick up a hammer to drive in a nail, we will likely use different grasp sites than when picking it up to give it to another person. When reaching, people naturally direct their gaze to visible targets and looking at the target enables optimal use of visual feedback of hand position to guide the hand (Paillard, 1996; Land *et al.*, 1999; Saunders & Knill, 2004). In addition, proprioceptive and/or motor signals related to gaze position can be used to guide the hand; even when the hand is not visible, directing gaze to the target improves reaching accuracy (Prablanc *et al.*, 1986, 2003). Once grasped, we often move the object to make or break contact with other objects or surfaces (e.g. in lift, transport and place tasks) or to

Sensorimotor Control of Grasping: Physiology and Pathophysiology, ed. Dennis A. Nowak and Joachim Hermsdörfer. Published by Cambridge University Press. © Cambridge University Press 2009.

contact and impose forces on other objects (e.g. when using tools such as a hammer, screwdriver or wrench). Studies of eye movements in object manipulation have shown that gaze fixations also play an important role in providing information for planning and control of motions with objects in hand (Ballard *et al.*, 1992; Land *et al.*, 1999; Johansson *et al.*, 2001). In addition to providing information for motion planning, visual cues about the identity, size and shape of an object can provide information about its mechanical properties that is useful for predicting the forces required for successful manipulation. For example, visual cues related to object weight and mass distribution can be used to predict magnitudes of required fingertip forces (Gordon *et al.*, 1991, 1993; Wing & Lederman, 1998; Salimi *et al.*, 2003) and visual cues about the shape of grasp surfaces can be used to predict stable fingertip force directions (Jenmalm & Johansson, 1997; Jenmalm *et al.*, 2000).

However, vision is of limited utility when objects are out of sight or partially occluded and for assessing contact sites for digits contacting the backside of objects. Furthermore, vision only provides indirect information about mechanical interactions between the hand and objects. That is, the use of vision in manipulation relies on learned associations (statistical correlations) between visual cues and their mechanical meaning. Such associations are grounded in movement–effect relationships evaluated through signals in sensors that transmit veridical information about mechanical interactions between the body and objects in the environment. The tactile modality directly provides information about mechanical interactions between our hands and objects and plays a pivotal role in the learning, planning and control of dexterous object manipulation tasks. Indeed, people with impaired digital sensibility have great difficulty even with routine tasks performed under optimal visual guidance. For example, they often drop objects, may easily crush fragile objects, and have tremendous difficulties with everyday activities such as buttoning a shirt or picking up a match (see Chapter 19). In humans, the density of tactile innervation is highest in body surface areas that typically contact objects, i.e. the palmar surfaces of the hands, the soles of the feet, and the tongue and lips. Furthermore, for the hand and the foot, the density is highest at the most distal segments. For the human hand, about 2000 tactile afferents innervate each fingertip whereas some 10 000 afferent neurons innervate all of the remaining glabrous skin areas of the digits and the palm (Johansson & Vallbo, 1979). Four different types of tactile afferents encode complementary aspects of the deformations of the soft tissues when the hands interact with objects (Johansson & Vallbo, 1983; Vallbo & Johansson, 1984) (for further details see Fig. 11.1B). Overall, these sensors have evolved for extracting – rapidly and with high spatiotemporal fidelity – features of dynamic mechanical events that occur on top of the low frequency and often large forces typically present when holding and moving hand-held objects (Johansson & Westling, 1987; Westling & Johansson, 1987; Macefield *et al.*, 1996). These dynamic tactile signals reliably encode various aspects of contact events around which most object manipulation tasks are organized.

Comparatively little is known about the contribution of proprioceptive and auditory signals in the control of manipulation tasks. Like vision, these modalities can only provide indirect information about mechanics. Proprioception signals related to muscle length, joint angle and muscle force do not directly code the contact state between the hands and objects, and the sensitivity of non-digital mechanoreceptive afferents (e.g. musculotendinous afferents) to fingertip events is very low in comparison to that of tactile sensors (cf. Macefield & Johansson, 1996 and Macefield *et al.*, 1996; see also Häger-Ross & Johansson, 1996).

Contact events and action goals in manipulation tasks

Natural object manipulation tasks usually involve a series of action phases that accomplish specific goals (or task subgoals) typically associated with mechanical contact events. For example, consider the task of lifting, holding and replacing a box on a tabletop. This task involves a series of action phases separated by contact events involving either the making or breaking of contact (Figure 11.1A) (Johansson & Westling, 1984; see also Chapter 1). Thus, the goal of the initial reach phase is marked by the digits contacting the box and the goal of the subsequent load phase (during which increasing vertical load forces and horizontal grip forces are applied under isometric conditions) is marked by the breaking of contact between the object in hand and the support surface. These and subsequent contact events give rise to discrete sensory signals from one or more sensory modalities. For example, when the box is replaced on the table, the contact between the box and surface gives rise to discrete tactile and auditory signals and, if the box is in the field of view, visual signals as well. That is, subgoal attainments are generally associated with distinct signals in one or more sensory modalities, each providing an afferent neural signature that encodes the timing as well as the characteristics of the mechanical interaction of the corresponding contact event.

A given object manipulation task can be represented as a sensory plan wherein a sequence of sensory goals is specified in one or more sensory modalities (Flanagan *et al.*, 2006). To achieve the sensory goals, the brain selects and executes a corresponding sequence of basic actions, or action-phase controllers (Figure 11.1A). To be accurate, when possible the controllers use knowledge of object properties, combined with information about the current state of the system (including the initial configuration of the motor apparatus and objects in the environment), to predictively adapt the motor output with reference to attainment of their sensory goals. For example, during the load phase of lifting, people normally scale the rate of change of force output to the predicted weight of the object.

In addition to generating motor commands, implemented action-phase controllers generate predictions about the sensory consequences of the motor output, including sensory signals associated with contact events. By comparing predicted and actual sensory signals, task progression is monitored (Figure 11.1A). Contact events, which denote completion of action goals, represent critical *sensorimotor control points* in this respect because they give rise to discrete and distinct sensory signals in one or more modalities. If a mismatch occurs in one or more modalities because of misleading initial state information or unpredicted external or internal events, the brain can launch corrective actions (or smart reflexes), the nature of which is specific for the sensory signals, the implemented controller, and the current state of the system and environment. If the mismatch is due to erroneous predictions about object properties, memory representations related to these properties can be updated to improve predictive control in subsequent phases of the task and in other tasks with the same

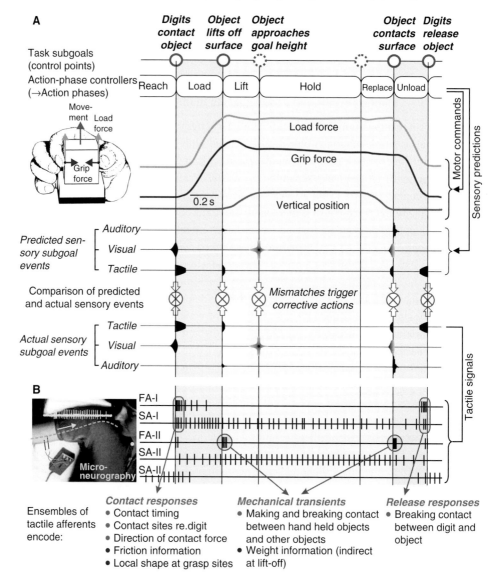

Figure 11.1. This figure is also reproduced in the color plate section. A person grasps and lifts an instrumented test object from a table, holds it in the air, and then replaces it, using the precision grip. A. The contact events shown at the top define subgoals of the task (i.e. goals of each action phase). Sequentially implemented action-phase controllers generate motor commands that bring about the required action phases. After the digits contact the object, the grip force increases in parallel with the tangential load force applied under isometric conditions. When the load force overcomes the force of gravity, the object lifts off. After the object is replaced such that it contacts the support surface, the load and grip forces decline in parallel until the object is released. In addition to issuing motor commands, the action-phase controllers predict their sensory consequences in one or more modalities (predicted sensory subgoal events). For example, when the object is replaced on the surface, the contact between the object and the surface gives rise to both predicted and actual tactile, visual and auditory sensory events. By comparing predicted and actual sensory events, the sensorimotor system can monitor task

object. This highly task- and phase-specific use of sensory information in object manipulation tasks is presumably acquired when we learn the underlying basic action-phase controllers, which occurs gradually during ontogeny until about 10 years of age (Forssberg *et al.*, 1991; Forssberg *et al.*, 1992; Gordon *et al.*, 1992; Eliasson *et al.*, 1995; Forssberg *et al.*, 1995; Paré & Dugas, 1999) (see also Chapter 17).

Although we have emphasized that manipulation tasks are organized as a sequence of discrete action phases, it is important to note that sensory predictions generated in one action phase are used in the next. Specifically, sensory predictions about the state of the motor system and environment at the termination of one action phase also provide initial state information for the subsequent action-phase controller. This information allows parameterization of the next action phase in advance, which is necessary for smooth transitions between the component actions of the task. In the absence of such anticipatory control, stuttering phase transitions would occur because the brain would have to rely on peripheral afferent signals to obtain this state information. Such a process would be time consuming due to long time-delays in sensorimotor control loops associated with receptor transduction and encoding, neural conduction, central processing and muscle activation. For example, it takes approximately 100 ms before signals from tactile sensors in the digits can bring about a significant change in fingertip actions. Even longer delays, in excess of 200 ms, are usually required to transform visual events into purposeful fingertip actions. Because of these long time delays in sensorimotor control loops operating on the hand, anticipatory control policies govern the swift and smooth transitions characteristic of dexterous manipulation. In the remainder of this chapter, we will discuss experimental results from manipulation tasks that illustrate and further develop the general principles described above.

Predictions and control points in the tactile modality

Control points for reaches for objects

The goal of the first action phase in most manipulation tasks is to bring the hand to the object in order to grasp it, or more precisely to position the digits on the object in locations that will

Figure 11.1. (cont.)
progression and detect mismatches used to bring about corrective actions tailored to the action phase. B. Schematic illustration of signals in four types of tactile afferents innervating the human fingertips as recorded from the median nerve at the level of the upper arm using the technique of microneurography (Vallbo & Hagbarth, 1968). At four points corresponding to subgoal events of the task, tactile afferents show distinct burst discharges: (1) contact responses preferentially in fast-adapting type I (FA-I; Meissner) and slowly adapting type I (SA-I; Merkel) afferents when the object is first contacted, (2) responses in the fast-adapting type II (FA-II; Pacinian) afferents related to the mechanical transients at lift-off, and (3) when objects contact the support surface, and (4) responses primarily in FA-I afferents when the object is released (goal of the unloading phase). In addition to these event-related responses, slowly adapting type II (SA-II; Ruffini) afferents and many SA-I afferents show ongoing impulse activity when forces are applied to the object. Some spontaneously active SA-II units are unloaded during the lift and cease firing. (Compiled from data presented in Westling & Johansson, 1987).

allow the development of a stable grasp in the context of the actions that will be performed with the object. In many manipulation tasks, it is likewise important that the fingertips contact the object at around the same time and that the fingertip force vectors sum to zero (Burstedt *et al.*, 1997; Flanagan *et al.*, 1999; Reilmann *et al.*, 2001). This is particularly important when initially contacting a light object that might otherwise be displaced or rotated. Signals in ensembles of tactile afferents robustly and rapidly encode these and other features of the contacts between digits and objects (Johansson & Westling, 1987; Westling & Johansson, 1987; Macefield *et al.*, 1996; Birznieks *et al.*, 2001; Jenmalm *et al.*, 2003; Johansson & Birznieks, 2004). Thus, the contact events between the digits and objects provide control points for reach phases where tactile signals can be compared with predicted tactile input concerning contact timing, geometry and forces, and errors can be assessed regarding the outcome of the executed action phase-controller. Not surprisingly, weak single pulse transcranial magnetic brain stimulation (TMS) delivered to the hand area of the contralateral primary sensorimotor cortex just before the instance of contact can interfere with this process (Lemon *et al.*, 1995). That is, it causes a disruption of the transition from the reach phase to the subsequent load phase and results in a significant and variable delay of the onset of the load phase. This effect might result from the TMS influencing the motor output causing a mismatch between the actual and predicted spatiotemporal pattern of afferent information related to contact, disruption of the sensory prediction and/or disturbed processing of tactile afferent information. Various lines of evidence indicate that tactile contact information is important both for calibration and upholding of accuracy of reach commands (Gordon & Soechting, 1995; Gentilucci *et al.*, 1997; Lackner & DiZio, 2000; Rao & Gordon, 2001; Rabin & Gordon, 2004; Säfström & Edin, 2004).

Tactile contact responses

Signals in ensembles of afferents from the entire distal phalanx can contribute to the encoding of tactile information in natural object manipulation tasks because the interaction between the fingertips and objects typically causes widespread distributions of complex stresses and strains throughout the engaged fingertips, including in the skin (Birznieks *et al.*, 2001; Jenmalm *et al.*, 2003). The non-linear deformation properties of the fingertip, with stiffness increasing with the contact force (Westling & Johansson, 1987; Pawluk & Howe, 1999), imply that it deforms quite briskly when an object is initially contacted. This deformation causes clear contact responses in SA-I and often in FA-II afferents, but most distinctly in FA-I afferents (Westling & Johansson, 1987). The spatial centroid of the afferent population response has been proposed to represent the primary contact site on the finger (Wheat *et al.*, 1995), while the recruitment of afferents and their firing rates reflect force intensity (magnitude and rate) (Knibestöl, 1973, 1975; Johansson & Vallbo, 1976; Macefield *et al.*, 1996; Goodwin & Wheat, 2004). For force direction, firing rates of individual tactile afferents distributed over the entire fingertips are tuned broadly to a preferred direction of fingertip force and this preferred direction varies amongst afferents such that ensembles of afferents can encode force direction (Figure 11.2A–C) (Birznieks

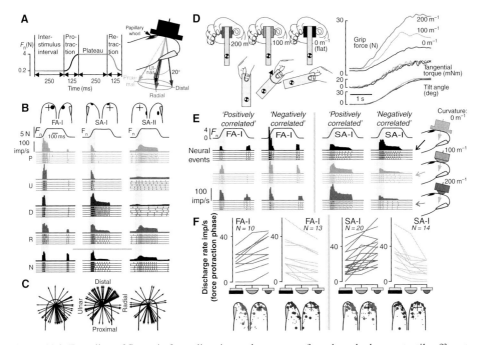

Figure 11.2. Encoding of fingertip force direction and contact surface shape by human tactile afferents under conditions representative for object manipulation tasks. A. Superimposed on a 0.2 N normal force (F_n), force was applied to the fingertip in the normal direction only (N), and together with tangential components in the proximal (P), ulnar (U), distal (D) or radial (R) directions. Each stimulus consisted of a force protraction phase (125 ms), a plateau phase (4 N force), and a retraction phase (125 ms) and was applied with either a flat or a spherically curved contact surface at a standard site on the fingertip that serves as a primary target for object contact in grasping and manipulation of small objects. B. Impulse ensembles exemplifying responses in single highly responsive FA-I, SA-I and SA-II afferents to repeated force stimuli ($n = 5$) applied in each force direction (P, U, D, R and N) with the flat contact surface. The top trace in each set shows the instantaneous discharge frequency averaged over five trials (bottom five traces). Top traces show the normal force component (F_n) superimposed for all trials. Circles on the finger indicate the location of the afferents termination and the crosses indicate the primary site of stimulation. C. Distributions of preferred directions of tangential force components for 68 SA-I, 53 FA-I and 32 SA-II afferents from the fingertip shown as unit vectors (arrows) with reference to the primary site of stimulation. These afferents terminate at various locations on a terminal phalanx. Preferred directions were estimated by vector summation of the mean firing rates during the force protraction phase (gray zone in B) obtained with different directions of the tangential force component. D. An already lifted object is tilted by 65° around the grip axis, which caused tangential torques at the grasp surfaces. The three superimposed curves (color coded) in each of the right-hand panels illustrate trials with two curved surfaces (5 and 10 mm radius) and a flat surface (curvature: 200, 100 and 0 m, respectively). Curves show the grip force, tangential torque and tilt angle against time. Note the effect of surface curvature on the coordination between grip force and tangential torque. E. Impulse ensembles show responses to repeated stimuli ($n = 5$) of two single FA-I and SA-I afferents with forces applied in the normal direction with each of the three surface curvatures used in D. Traces as in B. Left and right panels for each afferent type represent afferents for which response intensity increased ("positively correlated") and decreased ("negatively correlated") with an increase in curvature. F. Left and right panels show, for each type of afferent, afferents with responses positively and negatively correlated with surface curvature, respectively; response is represented as the mean number of impulses evoked during the protraction phase (gray zone in E) with each curvature. Circles on the fingertip as in B. (A–C adapted from Birznieks *et al.*, 2001; D, from Goodwin *et al.*, 1998; and E–F from Jenmalm *et al.*, 2003).

et al., 2001). Directional preferences of individual afferents of specific types could, for example, be combined in population models such as the vector model of direction proposed for neurons in the motor cortex (Georgopoulos *et al.*, 1986).

Control points supporting grasp stability

Practically all manipulation tasks require application of forces tangential to the contacted surfaces (load forces). For example, to lift an object with the digits contacting the sides, vertical load forces must be applied to overcome the weight of the object (Figure 11.1A). In many cases, twist forces (torques) tangential to the grasped surfaces are also applied. For example, if we lift a bar from one end, in addition to the vertical load force, we need to apply tangential torque to prevent the bar from rotating as we lift it. These tangential loads destabilize the grasp and to prevent the object from slipping (either linearly or rotationally) application of forces normal to the surface (grip forces) is required to create stabilizing frictional forces. To that end, action-phase controllers used in manipulation support grasp stability by automatically generating grip forces normal to the grasped surface that are synchronous with, and proportional to, the applied tangential loads (Figure 11.1A; see also Figure 11.3) (Johansson & Westling, 1984; Westling & Johansson, 1984; see also Chapter 1). This grip-load force coordination supports grasp stability in virtually all maneuvers that we naturally perform with objects in unimanual (Flanagan & Tresilian, 1994; Flanagan & Wing, 1995; Goodwin *et al.*, 1998; Wing & Lederman, 1998; Johansson *et al.*, 1999; Flanagan *et al.*, 1999; Santello & Soechting, 2000; LaMotte, 2000) and bimanual tasks (Johansson & Westling, 1988b; Flanagan & Tresilian, 1994; Burstedt *et al.*, 1997; Flanagan *et al.*, 1999; Witney *et al.*, 1999; Bracewell *et al.*, 2003; Gysin *et al.*, 2003; Witney & Wolpert, 2007). Thus the implemented action-phase controllers predict continuously the consequences of arm and hand motor commands regarding tangential loads acting on the object so that grip force can be suitably adjusted (see also Chapter 9).

Control of grasp stability requires, however, that the balance between the grip and load forces be adapted to the properties of contacted surfaces. The friction between digits and object surfaces determines the minimum ratio between grip and load forces required to prevent slip. Accordingly, people parametrically adapt grip-to-load force ratios for different frictional conditions, using greater ratios with more slippery surfaces (Johansson & Westling, 1984; Westling & Johansson, 1984; Flanagan & Wing, 1995; Cadoret & Smith, 1996). In fact, the local frictional condition can tailor the grip-to-load force ratios employed at individual digits within limits imposed by the overall force requirements for maintaining object equilibrium (Edin *et al.*, 1992; Birznieks *et al.*, 1998; Burstedt *et al.*, 1999; Quaney & Cole, 2004; Niu *et al.*, 2007). In the same vein, people parametrically scale the balance between the grip and load forces to the shape of the contacted surface. For example, the greater the curvature of a spherically curved grasp surface, the larger the grip force required to generate a given tangential torque (Figure 11.2D) (Goodwin *et al.*, 1998; Jenmalm *et al.*, 2000). Similarly, when lifting tapered objects, a greater grip-to-load force ratio is required

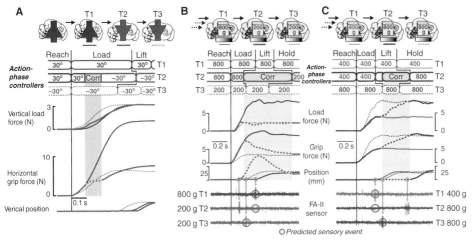

Figure 11.3. Adaptation of fingertip forces to changes in object shape and weight. A. Vertical load force and horizontal grip force in trials from a lift series in which the angle of the grasped surfaces was unpredictably varied between lifts without useful visual cues. Blue curves refer to a trial with 30° upward tapered grasp surfaces (T1) preceded by a trial with 30° upward taper. The solid red curves show the next trial (T2) performed with 30° downward tapered grasp surfaces and thus illustrate adjustments to a change in shape. The force output was initially tailored for the object shape in the previous lift before a corrective action was elicited (yellow-dashed segment in the grip force curve) that adjusted the balance between grip and load forces to better suit the 30° downward tapered surfaces. The thin red dashed curves show the following trial (T3) performed without a change in surface taper. The top diagram represents the status of the sequentially implemented action-phase controllers. Throughout T1 and in the beginning of T2 they are parameterized for the 30° upward taper. In T2 a corrective action ("Corr") is triggered about 100 ms after contact based on a detected mismatch between predicted and actual tactile information obtained at contact. This corrective action, inserted during the load phase, updates the controllers to the 30° downward taper for the remainder of the trial. In T3, the controllers remain updated to the new shape. B–C. Single unit tactile afferent responses and adjustments in force to unexpected changes in object weight based on data from single lifts. Gray circles and vertical lines indicate the instance of lift-off for each trial and the arrowheads point at the signals generated by the lift-off in a FA-II (Pacinian) afferent. The circles behind the nerve traces indicate the corresponding predicted sensory events. B. Three successive trials (T1–T3) in which the subject lifted an 800 g object (blue curves), a 200 g object (red solid curves) and then the 200 g object again (red dashed curves). The forces exerted in the first lift were adequately programmed because the participant had previously lifted the 800 g object. The forces were erroneously programmed in the first 200 g lift (T2), i.e. they were tailored for the heavier 800 g object lifted in the previous trial. The sensory information about the start of movement occurs earlier than expected which initiates a corrective action (yellow-dashed red curves) that terminates the strong force drive and brings the object back to the intended position after the marked overshoot in the vertical position. C. An adequately programmed lift with a 400 g weight (T1, green curves) was followed by a lift with 800 g (T2, blue solid curves) that was erroneously programmed for the lighter 400 g weight lifted in the previous trial. The absence of lift-off responses in FA-II afferents at the predicted point for the erroneously programmed 800 g trial elicited a corrective action (yellow-dashed blue curves) that involved abortion of the lift-phase command followed by triggering of a second load phase command that slowly and discontinuously increased grip and load forces until terminated by sensory input signaling lift-off. In the subsequent trial (T3, blue dashed curves), the participant again lifted the 800 g object. The top diagrams in B and C represent the status of the sequentially implemented action-phase controllers. In T1 they were parameterized for the 800 g (B) and 400 g (C) weight throughout. In T2, a corrective action ("Corr") was triggered about 100 ms after the occurrence of the mismatch between predicted and actual sensory information related to object lift-off. This action involved abortion of the operation of the current action-phase controller and the implementation of corrective action patterns that allow the task to continue. The corrective action was linked to an updating of the subsequently implemented controllers for the new weight. In T3, the controllers remain updated to this weight. (A, compiled from data presented in Jenmalm & Johansson, 1997; B–C, developed from Johansson & Cole, 1992).

when the grip surfaces are tapered upwards as compared to downwards (see Figure 11.3A) (Jenmalm & Johansson, 1997). These parametric adaptations to contact surface friction and shape typically result in grip forces that exceed the minimum required to prevent slips by a safety margin of 10–40% of the applied grip force.

Tactile contact responses

Tactile sensibility is critical for adaptation of grip-to-load force ratios to object surface properties (Johansson & Westling, 1984; Westling & Johansson, 1984; Jenmalm & Johansson, 1997; Jenmalm *et al.*, 2000; Monzée *et al.*, 2003; Nowak & Hermsdörfer, 2003; Cole *et al.*, 2003; Nowak *et al.*, 2004; Schenker *et al.*, 2006). In addition to forces and contact sites on the digits, the contact responses in ensembles of tactile afferents – primarily in FA-Is – rapidly convey information related to object surface properties, including friction (Johansson & Westling, 1987) and local shape (Figures 11.1 and 11.2E–F) (Jenmalm *et al.*, 2003; Johansson & Birznieks, 2004). For example, changes in the curvature of contacted surfaces, which markedly influence the grip forces in tasks involving tangential torque loads (Figure 11.2D) (Goodwin *et al.*, 1998; Jenmalm *et al.*, 2000), robustly influence firing rates in the majority of responsive tactile afferents. Roughly, one half of the afferents for which response intensity correlates with curvature show a positive correlation and half a negative correlation (Figure 11.2E–F); responsive afferents terminating at the sides and end of the fingertip tend to show negative correlations. Consequently, there is a curvature contrast signal within the population of tactile afferents.

Traditionally, it is posited that afferent information is coded by firing rates. However, in manipulation, typically, the brain quickly extracts information from discrete tactile events and expresses this information in fingertip actions faster than can be readily explained by rate codes. That is, based on the delays in sensorimotor control loops (see above) and the firing rates of tactile afferents in manipulation tasks, it can be deduced that tactile events typically influence fingertip actions when most afferents recruited have had time to fire only one impulse (Johansson & Birznieks, 2004). Recent findings in humans indicate that the relative timing of impulses from ensembles of individual afferents conveys information about important contact parameters faster than the fastest possible rate code and fast enough to account for the use of tactile signals in natural manipulation tasks (Johansson & Birznieks, 2004). Specifically, the sequence in which different afferents initially discharge in response to discrete fingertip events provides information about the shape of the contacted surface and the direction of fingertip force. The relative timing of the first spikes contains information about object shape and force direction because changes in either of these parameters differentially influenced the first-spike latency of individual afferents rather than having systematic effects on the latencies within an afferent population. For example, when the fingertip contacts a surface with a given curvature, the responsive afferents will be recruited in a particular order. With another curvature, the order will be different because some afferents are recruited earlier, and others later. Presumably, the order of recruitment of members of the populations of tactile afferents can code other contact parameters used in the control of manual actions as well, such as the friction between fingertips and contacted surfaces.

A mismatch between predicted and actual contact responses triggers a corrective action commencing ~100 ms after contact that is accompanied by an updating of the representation of the object used to control forces in future interactions with the object. Figure 11.3A illustrates this process when repeatedly lifting objects with tapered grasp surfaces where the tapering was changed between trials in an unpredictable order. First, for all trials the tapering (and hence force requirements) in the previous trial determines the initial increase in grip force, indicating that predictions based on knowledge obtained in previous trials specify the grip-load force coordination. After an unpredicted change in tapering, the grip force output is modified about 100 ms after contact with the object and tuned for the actual object properties (Figure 11.3A). By the second trial after the change, the force coordination is appropriately adapted right from the onset of force application. Knowledge about object surface properties remains critical for controlling grip forces for grasp stability when transporting held objects and using them as tools to impose forces on other environmental objects.

Under favorable conditions, visual geometric cues about object shape can provide state information for predictive parameterization of fingertip forces such that the grip-to-load force coordination is adapted to the prevailing shape right from the beginning of the force application (Jenmalm & Johansson, 1997; Jenmalm *et al.*, 2000). However, once the object is contacted, tactile signals also provide state information about object shape that can override visual predictions if necessary. With regard to friction between the hand and an object, it appears that vision is unhelpful for feedforward adaptation of force coordination. Presumably, this is because friction depends not only on the object surface but also on sweating rate and the greasiness and wetness of the skin (and objects). Thus, predictions of frictional conditions are based on memory of recent haptic experiences with the same or similar objects.

Accidental slips

Occasionally, the updating of frictional and shape representations that occurs at initial contact is inadequate and may result in an accidental slip later in the task. Such a slip usually results in a transitory and partial unloading at one digit (the slipping digit) and this increases the loads on the other digits engaged. Such transient shifts in tangential forces, reliably signaled by FA-I afferents (Johansson & Westling, 1987; Macefield *et al.*, 1996), trigger a corrective action (onset latency 70–90 ms) that results in an updating of grip-to-load force ratios and an increase in the safety margin, primarily at the slipping digit (Edin *et al.*, 1992; Burstedt *et al.*, 1997). This updated force coordination is maintained in subsequent phases of the same trial and in subsequent trials with the same object. While an increase in grip force accounts for the adjustment of the grip-to-load force ratio triggered by slip events during the hold phase in lifting trials, during the load phase it is implemented by a slowing down of the subsequent increase in load force (Johansson & Westling, 1984).

Hence, different action-phase controllers are associated with different smart corrective reflex mechanisms that support grasp stability and enable the task to progress.

Control points for object motion

The goal of many action phases in object manipulation (including tool use) is to move a held object to form or break contact with another object. The held object transmits various features of these contact events to the hand that tactile afferents can signal, including mechanical transients. For example, when we lift an object from a support surface, ensembles of FA-II afferents terminating throughout the hand and wrist signal the incidence and dynamic aspects of the lift-off event (Figure 11.1B) (Westling & Johansson, 1987). Because no sensory information is available about object weight until lift-off, a smooth and critically damped lifting motion requires that the load (lift) force drive at lift-off, which accelerates the object, be scaled predictively to object weight. People regularly form such predictions based on sensorimotor memory of the object derived from previous lifts (Johansson & Westling, 1988a). Familiar objects can be identified visually (or by haptic exploration) for retrieval of weight-related predictions, and size–weight associations can be used to predict the weights of categories of familiar objects where the items can vary in size (e.g. cups, books, loaves of bread) (Gordon *et al.*, 1991, 1993) (see also Chapter 12). In a similar vein, visual cues about object geometry can be used for anticipatory tuning of fingertip forces to the mass distribution of the object (Wing & Lederman, 1998; Salimi *et al.*, 2003).

However, if such predictions are erroneous, compensatory control processes programmed to correct for performance errors and reduce future errors are automatically elicited (Johansson & Westling, 1988a). For example, when a lifted object is lighter than predicted, the lift movement becomes faster and higher than intended (Figure 11.3B, T2). As a result, a mismatch is registered at the control point for the load phase controller because sensory events related to lift-off occur before the predicted time. This error automatically triggers a compensatory process that involves abortion of the implemented action-phase controller and execution of a corrective action program that generates motor commands that bring the object back to the intended position. However, because of delays in the sensorimotor control loops (~100 ms), the corrective action kicks in too late to avoid an overshoot in the lifting movement (see position signal in Figure 11.3B, T2). Conversely, if the object is heavier than expected, the increase in load force generated by the load phase controller finishes without giving rise to sensory events signaling lift-off (Figure 11.3C, T2). This absence of predicted sensory events triggers a corrective action program that generates slow, probing increases in fingertip forces until terminated reactively by sensory events signaling lift-off. Thus, the sensorimotor system reacts to both the presence of an unexpected sensory event and the absence of an expected sensory event and various corrective action programs can be associated with a given controller and executed depending on the characteristics of the sensory mismatch. Importantly, in addition to triggering corrective action programs, these sensory mismatches update weight-related memory for anticipatory parametric control of subsequent action phases and tasks that engage the same object (see T3 in Figure 11.3B and

C). With natural objects, usually a single lift efficiently brings about such updating (Johansson & Westling, 1988a; Gordon *et al.*, 1993) while in the presence of misleading cues or unfamiliar objects, repeated interactions with the object are usually required for establishing adequate internal representations of objects' mass and mass distributions (Gordon *et al.*, 1991, 1993; Flanagan & Beltzner, 2000; Salimi *et al.*, 2000, 2003).

When transporting and holding an object, knowledge about object weight, mass distribution and surface properties remains critical for controlling action and maintaining grasp stability. When replacing the object on a surface, the sensory goal is to produce an afferent signature signifying contact. This contact event, which represents a sensorimotor control point, is signaled by FA-II afferents that encode the timing and nature of the event (Figure 11.1B). The contact event is followed by an unloading phase where grip and load forces decrease in parallel, maintaining a grip-to-load force ratio providing grasp stability. Sensory events, especially in ensembles of FA-I afferents, related to the breaking of contact between the digits and the surface of the object represent the sensory goal of the unload phase (see "release responses" in Figure 12.1B; see also responses in FA-I afferents to the retraction phase in Figure 11.2B and E).

Predictions and control points in the visual modality

Studies of eye movements in object manipulation indicate that contact events that demarcate action phases also can be predicted in the visual modality. People use saccadic eye movements to direct their gaze to successive contact locations as they gain salience during task progression (Ballard *et al.*, 1992; Land *et al.*, 1999; Johansson *et al.*, 2001). For example, when people pick up a bar, move the bar to contact a target switch, and then replace the bar, gaze is successively directed to the grasp site on the bar, the target, and the landing surface where the bar is replaced (Figure 11.4A–B) (Johansson *et al.*, 2001). Furthermore, people may direct fixations to points where contact must be avoided, including obstacles that must be circumnavigated with the hand or by an object moved by the hand (Figure 11.4A–B). Notably, people almost never fixate their hand or objects being moved by the hand. Thus, when people direct actions towards visible objects, the implemented action-phase controllers appear to provide instructions for task- and phase-specific eye movements so as to acquire visual information optimized for guidance of the hand (Land & Furneaux, 1997; Flanagan & Johansson, 2003).

The spatiotemporal coordination of gaze and hand movements emphasizes the segmentation of manipulation tasks into distinct action phases (Figure 11.4C). At the start of most action phases, congruent hand and eye movements are launched concurrently to the contact location representing the goal of the current phase. Thus, both hand and eye movements are specified based on peripheral vision about the contact location (or on memorized landmark locations). Because eye movements are quick, gaze reaches the contact location well before the hand (or object in hand), which enables optimal use of vision for guiding the hand (Paillard, 1996; Land *et al.*, 1999; Prablanc *et al.*, 2003; Saunders & Knill, 2004). Gaze typically remains at the contact location until around the time of goal completion (e.g. until

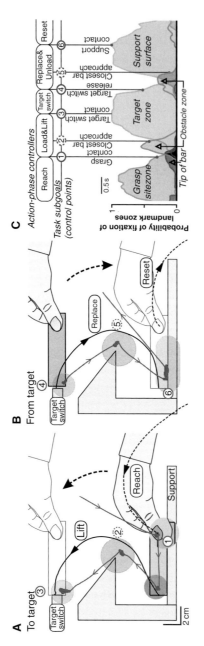

Figure 11.4. Predictions and control points in the visual modality. A–B. Gaze and hand movements for a single trial of a target contact task where the participant reached for and grasped a bar, moved it to press a target switch (A) and moved it away from the target switch and replaced it on the support surface (B). A triangular obstacle was located between the bar and target. Dashed black lines represent the path of the tip of the index finger during the reach for the bar and during the reset phase when the hand was transported away from the bar after it was replaced and released. The solid black lines represent the path of the tip of the bar. The red lines indicate the position of gaze; the thin segments with the arrowheads represent saccadic eye movements and the thick patches represent gaze fixations. The colored zones represent landmark zones that captured 90% of the fixations recorded during several trials by 10 participants. These zones are centered on the grasp site (green), tip of the bar (purple), the protruding part of the obstacle (orange), target (blue) and the support surface (pink). C. Spatiotemporal coordination of gaze and manipulatory actions shown as time-varying instantaneous probability of fixations within the landmark zones indicated in A and B derived from data pooled across 10 participants, each performing four trials with the triangular obstacle. The red circles and vertical lines mark contact events demarcating phase transitions in the task and the spatial locations of these events are schematically indicated by the location of the correspondingly numbered circle in A and B. The common time base has been normalized such that each phase of each trial has been scaled to the median duration of that phase. (Adapted from Johansson *et al.*, 2001.)

the grasp is established, the target switch is released, or the bar is replaced in the target contact task) or remains at a location where a contact should be avoided until the time of the potential contact (e.g. when the tip of the bar passed closest to the obstacle in the target contact task) (Figure 11.4). Thus, the gaze shift to the contact location associated with the next action phase occurs around the predicted time of goal completion. In fact, in most cases, both gaze and hand movement commands are initiated in anticipation of goal completion and are not delayed until sensory feedback is obtained verifying goal completion. With the latter strategy, smooth transition between successive phases of the manipulation task would not be possible because of the substantial time-delays in sensorimotor control loops. Hence, contact events that demarcate action phases can be predicted in both the tactile modality and the visual modality.

Although there is no question that tactile feedback related to control points is essential for skilled object manipulation, there are control points that do not give rise to tactile events. For example, when we drop a ball onto a surface, we typically direct our gaze to the predicted contact point at the surface. Here, sensory feedback related to contact is only available through vision and/or audition. Thus, the visual, auditory, and tactile systems can play complementary roles in predicting contact events. Prediction of contact events in the visual modality without the tactile modality engaged is evident from studies of eye movements in people who observe an actor performing familiar manipulation tasks (Flanagan & Johansson, 2003). In this situation, the gaze of both the actor and observer predicts forth-coming contact sites (e.g. where blocks are grasped and replaced in a predictable block stacking task) and gaze is maintained at each contact site until around the time of goal completion (grasp contact and block landing). By comparing actual and predicted visual feedback related to contact events, observers (and actors) may be able to obtain valuable information about outcomes of actions that can be exploited by the sensorimotor system when learning, planning and controlling future actions. These findings also support the notion that understanding of observed actions performed by others involves a mechanism that maps observed actions onto sensorimotor representations in the observer's brain implemented in real time (Rizzolatti *et al.*, 2001; Flanagan & Johansson, 2003; Rotman *et al.*, 2006).

Conclusions

Dexterity in object manipulation tasks depends on anticipatory control policies that rely on knowledge about movement–effect relationships when interacting with environmental objects. The tactile modality plays a pivotal role in gaining such knowledge because signals from tactile afferents provide direct information about mechanical interactions between the body and objects in the environment. The usefulness of visual, auditory and proprioceptive mechanisms in planning and control of object-oriented manual actions depends on learned associations between visual, auditory and proprioceptive cues and their mechanical mean-ing, primarily derived from tactile mechanisms. Signals in ensembles of tactile afferents of different types convey complementary information related to both the timing and the physical nature of contact events that represent the outcomes of motor commands to the hand. Furthermore, populations of tactile afferents encode information related to surface properties of contacted objects such as the shape and texture of contacted surfaces and the frictional conditions between these surfaces and the skin (see also Chapter 14).

Manipulatory tasks involve a sensory plan that specifies the sequence of task subgoals in terms of specific afferent neural signatures in the tactile and other modalities. This plan provides a scaffold for the selection and shaping of the action-phase controllers implemented for achieving the sensory subgoals, which generally corresponds to distinct contact events. When a given contact event gives rise to afferent signals that are adequately predicted by the implemented controllers, the task runs in a pre-defined way based on knowledge of object properties derived from internal representations gained in previous interactions with objects (and representations related to the current state of the sensorimotor system). Invalid internal representations result in mismatches between predicted and actual signals that trigger learned corrective actions – the nature of which depends on the task and its action phase and the characteristics of the error – along with updating of representations of object properties. Prediction of the terminal sensorimotor state of an active action-phase controller can also be used as a prediction of the initial state by the controller responsible for the next action phase. If the brain regularly relied on peripheral afferent information to obtain this state information, stuttering phase transitions would occur because of sensorimotor delays.

We suggest that the brain encodes, in multiple sensory modalities, planned contact events that represent sensorimotor control points where predicted and actual sensory signals can be compared. Multimodal encoding of sensorimotor control points likely allows the sensorimotor system to simultaneously monitor multiple aspects of task performance and, if prediction errors are detected, respond to the pattern of errors observed in different modalities. Furthermore, because contact events give rise to salient sensory signals from multiple modalities that are linked in time and space, they provide an opportunity for sensorimotor integration and intermodal alignment helpful for learning and upholding multimodal sensorimotor correlations that support prediction of purposeful motor commands.

Acknowledgments

The Swedish Research Council (project 08667), the sixth Framework Program of the EU (project IST-001917, IST-028056) and the Canadian Institutes of Health Research supported this work.

References

Ballard, D. H., Hayhoe, M. M., Li, F. & Whitehead, S. D. (1992). Hand-eye coordination during sequential tasks. *Philos Trans R Soc Lond B, Biol Sci*, **337**, 331–338.
Birznieks, I., Burstedt, M. K. O., Edin, B. B. & Johansson, R. S. (1998). Mechanisms for force adjustments to unpredictable frictional changes at individual digits during two-fingered manipulation. *J Neurophysiol*, **80**, 1989–2002.

Birznieks, I., Jenmalm, P., Goodwin, A. & Johansson, R. (2001). Encoding of direction of fingertip forces by human tactile afferents. *J Neurosci*, **21**, 8222–8237.

Bracewell, R. M., Wing, A. M., Soper, H. M. & Clark, K. G. (2003). Predictive and reactive co-ordination of grip and load forces in bimanual lifting in man. *Eur J Neurosci*, **18**, 2396–2402.

Burstedt, M. K. O., Edin, B. B. & Johansson, R. S. (1997). Coordination of fingertip forces during human manipulation can emerge from independent neural networks controlling each engaged digit. *Exp Brain Res*, **117**, 67–79.

Burstedt, M. K. O., Flanagan, R. & Johansson, R. S. (1999). Control of grasp stability in humans under different frictional conditions during multi-digit manipulation. *J Neurophysiol*, **82**, 2393–2405.

Cadoret, G. & Smith, A. M. (1996). Friction, not texture, dictates grip forces used during object manipulation. *J Neurophysiol*, **75**, 1963–1969.

Cohen, R. G. & Rosenbaum, D. A. (2004). Where grasps are made reveals how grasps are planned: generation and recall of motor plans. *Exp Brain Res*, **157**, 486–495.

Cole, K. J., Steyers, C. M. & Graybill, E. K. (2003). The effects of graded compression of the median nerve in the carpal canal on grip force. *Exp Brain Res*, **148**, 150–157.

Cuijpers, R. H., Smeets, J. B. & Brenner, E. (2004). On the relation between object shape and grasping kinematics. *J Neurophysiol*, **91**, 2598–2606.

Edin, B. B., Westling, G. & Johansson, R. S. (1992). Independent control of fingertip forces at individual digits during precision lifting in humans. *J Physiol*, **450**, 547–564.

Eliasson, A. C., Forssberg, H., Ikuta, K. *et al.* (1995). Development of human precision grip V. Anticipatory and triggered grip actions during sudden loading. *Exp Brain Res*, **106**, 425–433.

Flanagan, J. R. & Tresilian, J. R. (1994). Grip load force coupling: a general control strategy for transporting objects. *J Exp Psychol Hum Percept Perform*, **20**, 944–957.

Flanagan, J. R. & Wing, A. M. (1995). The stability of precision grip forces during cyclic arm movements with a hand-held load. *Exp Brain Res*, **105**, 455–464.

Flanagan, J. R. & Beltzner, M. A. (2000). Independence of perceptual and sensorimotor predictions in the size–weight illusion. *Nat Neurosci*, **3**, 737–41.

Flanagan, J. R. & Johansson, R. S. (2003). Action plans used in action observation. *Nature*, **424**, 769–771.

Flanagan, J. R., Burstedt, M. K. O. & Johansson, R. S. (1999). Control of fingertip forces in multi-digit manipulation. *J Neurophysiol*, **81**, 1706–1717.

Flanagan, J. R., Bowman, M. C. & Johansson, R. S. (2006). Control strategies in object manipulation tasks. *Curr Opin Neurobiol*, **16**, 650–659.

Forssberg, H., Eliasson, A. C., Kinoshita, H., Johansson, R. S. & Westling, G. (1991). Development of human precision grip. I: Basic coordination of force. *Exp Brain Res*, **85**, 451–457.

Forssberg, H., Kinoshita, H., Eliasson, A. C. *et al.* (1992). Development of human precision grip. 2. Anticipatory control of isometric forces targeted for objects weight. *Exp Brain Res*, **90**, 393–398.

Forssberg, H., Eliasson, A. C., Kinoshita, H., Westling, G. & Johansson, R. S. (1995). Development of human precision grip. IV. Tactile adaptation of isometric finger forces to the frictional condition. *Exp Brain Res*, **104**, 323–330.

Gentilucci, M., Toni, I., Daprati, E. & Gangitano, M. (1997). Tactile input of the hand and the control of reaching to grasp movements. *Exp Brain Res*, **114**, 130–137.

Georgopoulos, A. P., Schwartz, A. B. & Kettner, R. E. (1986). Neuronal population coding of movement direction. *Science*, **233**, 1416–1419.

Goodale, M. A., Meenan, J. P., Bülthoff, H. H. *et al*. (1994). Separate neural pathways for the visual analysis of object shape in perception and prehension. *Curr Biol*, **4**, 604–610.

Goodwin, A. W. & Wheat, H. E. (2004). Sensory signals in neural populations underlying tactile perception and manipulation. *Ann Rev Neurosci*, **27**, 53–77.

Goodwin, A. W., Jenmalm, P. & Johansson, R. S. (1998). Control of grip force when tilting objects: effect of curvature of grasped surfaces and of applied tangential torque. *J Neurosci*, **18**, 10724–10734.

Gordon, A. M. & Soechting, J. F. (1995). Use of tactile afferent information in sequential finger movements. *Exp Brain Res*, **107**, 281–292.

Gordon, A. M., Forssberg, H., Johansson, R. S. & Westling, G. (1991). Integration of sensory information during the programming of precision grip: comments on the contributions of size cues. *Exp Brain Res*, **85**, 226–229.

Gordon, A. M., Forssberg, H., Johansson, R. S., Eliasson, A. C. & Westling, G. (1992). Development of human precision grip. 3. Integration of visual size cues during the programming of isometric forces. *Exp Brain Res*, **90**, 399–403.

Gordon, A. M., Westling, G., Cole, K. J. & Johansson, R. S. (1993). Memory representations underlying motor commands used during manipulation of common and novel objects. *J Neurophysiol*, **69**, 1789–1796.

Gysin, P., Kaminski, T. R. & Gordon, A. M. (2003). Coordination of fingertip forces in object transport during locomotion. *Exp Brain Res*, **149**, 371–379.

Häger-Ross, C. & Johansson, R. S. (1996). Non-digital afferent input in reactive control of fingertip forces during precision grip. *Exp Brain Res*, **110**, 131–141.

Jenmalm, P. & Johansson, R. S. (1997). Visual and somatosensory information about object shape control manipulative finger tip forces. *J Neurosci*, **17**, 4486–4499.

Jenmalm, P., Dahlstedt, S. & Johansson, R. S. (2000). Visual and tactile information about object curvature control fingertip forces and grasp kinematics in human dexterous manipulation. *J Neurophysiol*, **84**, 2984–2997.

Jenmalm, P., Birznieks, I., Goodwin, A. W. & Johansson, R. S. (2003). Influences of object shape on responses in human tactile afferents under conditions characteristic for manipulation. *Eur J Neurosci*, **18**, 164–176.

Johansson, R. S. & Vallbo, Å.B. (1976). Skin mechanoreceptors in the human hand: an inference of some population properties. In Y. Zotterman (Ed.), *Sensory Functions of the Skin in Primates, with Special Reference to Man* (pp. 171–184). Oxford, UK: Pergamon Press Ltd.

Johansson, R. S. & Vallbo, A. B. (1979). Tactile sensibility in the human hand: relative and absolute densities of four types of mechanoreceptive units in glabrous skin. *J Physiol*, **286**, 283–300.

Johansson, R. S. & Vallbo, Å. B. (1983). Tactile sensory coding in the glabrous skin of the human hand. *Trends Neurosci*, **6**, 27–31.

Johansson, R. S. & Westling, G. (1984). Roles of glabrous skin receptors and sensorimotor memory in automatic control of precision grip when lifting rougher or more slippery objects. *Exp Brain Res*, **56**, 550–564.

Johansson, R. S. & Westling, G. (1987). Signals in tactile afferents from the fingers eliciting adaptive motor responses during precision grip. *Exp Brain Res*, **66**, 141–154.

Johansson, R. S. & Westling, G. (1988a). Coordinated isometric muscle commands adequately and erroneously programmed for the weight during lifting task with precision grip. *Exp Brain Res*, **71**, 59–71.

Johansson, R. S. & Westling, G. (1988b). Programmed and triggered actions to rapid load changes during precision grip. *Exp Brain Res*, **71**, 72–86.

Johansson, R. S. & Cole, K. J. (1992). Sensory-motor coordination during grasping and manipulative actions. *Curr Opin Neurobiol*, **2**, 815–823.

Johansson, R. S. & Birznieks, I. (2004). First spikes in ensembles of human tactile afferents code complex spatial fingertip events. *Nat Neurosci*, **7**, 170–177.

Johansson, R. S., Backlin, J. L. & Burstedt, M. K. O. (1999). Control of grasp stability during pronation and supination movements. *Exp Brain Res*, **128**, 20–30.

Johansson, R. S., Westling, G., Bäckström, A. & Flanagan, J. R. (2001). Eye-hand coordination in object manipulation. *J Neurosci*, **21**, 6917–6932.

Knibestöl, M. (1973). Stimulus-response functions of rapidly adapting mechanoreceptors in human glabrous skin area. *J Physiol*, **232**, 427–452.

Knibestöl, M. (1975). Stimulus-response functions of slowly adapting mechanoreceptors in the human glabrous skin area. *J Physiol*, **245**, 63–80.

Lackner, J. R. & DiZio, P. A. (2000). Aspects of body self-calibration. *Trends Cogn Sci*, **4**, 279–288.

LaMotte, R. H. (2000). Softness discrimination with a tool. *J Neurophysiol*, **83**, 1777–1786.

Land, M. F. & Furneaux, S. (1997). The knowledge base of the oculomotor system. *Philos Trans R Soc Lond B, Biol Sci*, **352**, 1231–1239.

Land, M., Mennie, N. & Rusted, J. (1999). The roles of vision and eye movements in the control of activities of daily living. *Perception*, **28**, 1311–1328.

Lemon, R. N., Johansson, R. S. & Westling, G. (1995). Corticospinal control during reach, grasp and precision lift in man. *J Neurosci*, **15**, 6145–6156.

Lukos, J., Ansuini, C. & Santello, M. (2007). Choice of contact points during multidigit grasping: effect of predictability of object center of mass location. *J Neurosci*, **27**, 3894–3903.

Macefield, V. G. & Johansson, R. S. (1996). Control of grip force during restraint of an object held between finger and thumb: responses of muscle and joint afferents from the digits. *Exp Brain Res*, **108**, 172–184.

Macefield, V. G., Häger-Ross, C. & Johansson, R. S. (1996). Control of grip force during restraint of an object held between finger and thumb: responses of cutaneous afferents from the digits. *Exp Brain Res*, **108**, 155–171.

Monzée, J., Lamarre, Y. & Smith, A. M. (2003). The effects of digital anesthesia on force control using a precision grip. *J Neurophysiol*, **89**, 672–683.

Niu, X., Latash, M. L. & Zatsiorsky, V. M. (2007). Prehension synergies in the grasps with complex friction patterns: local versus synergic effects and the template control. *J Neurophysiol*, **98**, 16–28.

Nowak, D. A. & Hermsdörfer, J. (2003). Digit cooling influences grasp efficiency during manipulative tasks. *Eur J Appl Physiol*, **89**, 127–133.

Nowak, D. A., Glasauer, S. & Hermsdorfer, J. (2004). How predictive is grip force control in the complete absence of somatosensory feedback? *Brain*, **127**, 182–192.

Paillard, J. (1996). Fast and slow feedback loops for the visual correction of spatial errors in a pointing task: a reappraisal. *Can J Physiol Pharmacol*, **74**, 401–417.

Paré, M. & Dugas, C. (1999). Developmental changes in prehension during childhood. *Exp Brain Res*, **125**, 239–247.

Pawluk, D. T. & Howe, R. D. (1999). Dynamic lumped element response of the human fingerpad. *J Biomech Eng*, **121**, 178–183.

Prablanc, C., Pélisson, D. & Goodale, M. A. (1986). Visual control of reaching movements without vision of the limb. I. Role of retinal feedback of target position in guiding the hand. *Exp Brain Res*, **62**, 293–302.

Prablanc, C., Desmurget, M. & Gréa, H. (2003). Neural control of on-line guidance of hand reaching movements. *Progr Brain Res*, **142**, 155–170.

Quaney, B. M. & Cole, K. J. (2004). Distributing vertical forces between the digits during gripping and lifting: the effects of rotating the hand versus rotating the object. *Exp Brain Res*, **155**, 145–155.

Rabin, E. & Gordon, A. M. (2004). Tactile feedback contributes to consistency of finger movements during typing. *Exp Brain Res*, **155**, 362–369.

Rao, A. K. & Gordon, A. M. (2001). Contribution of tactile information to accuracy in pointing movements. *Exp Brain Res*, **138**, 438–445.

Reilmann, R., Gordon, A. M. & Henningsen, H. (2001). Initiation and development of fingertip forces during whole-hand grasping. *Exp Brain Res*, **140**, 443–452.

Rizzolatti, G., Fogassi, L. & Gallese, V. (2001). Neurophysiological mechanisms underlying the understanding and imitation of action. *Nat Rev Neurosci*, **2**, 661–670.

Rotman, G., Troje, N. F., Johansson, R. S. & Flanagan, J. R. (2006). Eye movements when observing predictable and unpredictable actions. *J Neurophysiol*, **96**, 1358–1369.

Säfström, D. & Edin, B. B. (2004). Task requirements influence sensory integration during grasping in humans. *Learn Mem*, **11**, 356–363.

Salimi, I., Hollender, I., Frazier, W. & Gordon, A. M. (2000). Specificity of internal representations underlying grasping. *J Neurophysiol*, **84**, 2390–2397.

Salimi, I., Frazier, W., Reilmann, R. & Gordon, A. M. (2003). Selective use of visual information signaling objects' center of mass for anticipatory control of manipulative fingertip forces. *Exp Brain Res*, **150**, 9–18.

Santello, M. & Soechting, J. F. (1998). Gradual molding of the hand to object contours. *J Neurophysiol*, **79**, 1307–1320.

Santello, M. & Soechting, J. F. (2000). Force synergies for multifingered grasping. *Exp Brain Res*, **133**, 457–467.

Saunders, J. A. & Knill, D. C. (2004). Visual feedback control of hand movements. *J Neurosci*, **24**, 3223–3234.

Schenker, M., Burstedt, M. K., Wiberg, M. & Johansson, R. S. (2006). Precision grip function after hand replantation and digital nerve injury. *J Plast Reconstr Aesthet Surg*, **59**, 706–716.

Vallbo, Å. B. & Hagbarth, K.-E. (1968). Activity from skin mechanoreceptors recorded percutaneously in awake human subjects. *Exp Neurol*, **21**, 270–289.

Vallbo, A. B. & Johansson, R. S. (1984). Properties of cutaneous mechanoreceptors in the human hand related to touch sensation. *Hum Neurobiol*, **3**, 3–14.

Westling, G. & Johansson, R. S. (1984). Factors influencing the force control during precision grip. *Exp Brain Res*, **53**, 277–284.

Westling, G. & Johansson, R. S. (1987). Responses in glabrous skin mechanoreceptors during precision grip in humans. *Exp Brain Res*, **66**, 128–140.

Wheat, H. E., Goodwin, A. W. & Browning, A. S. (1995). Tactile resolution: peripheral neural mechanisms underlying the human capacity to determine positions of objects contacting the fingerpad. *J Neurosci*, **15**, 5582–5595.

Wing, A. M. & Lederman, S. J. (1998). Anticipating load torques produced by voluntary movements. *J Exp Psychol Hum Percept Perform*, **24**, 1571–1581.

Witney, A. G. & Wolpert, D. M. (2007). The effect of externally generated loading on predictive grip force modulation. *Neurosci Lett*, **414**, 10–15.

Witney, A. G., Goodbody, S. J. & Wolpert, D. M. (1999). Predictive motor learning of temporal delays. *J Neurophysiol*, **82**, 2039–2048.

12

Predictive mechanisms and object representations used in object manipulation

J. RANDALL FLANAGAN, KYLE MERRITT
AND ROLAND S. JOHANSSON

Summary

Skilled object manipulation requires the ability to estimate, in advance, the motor commands needed to achieve desired sensory outcomes and the ability to predict the sensory consequences of the motor commands. Because the mapping between motor commands and sensory outcomes depends on the physical properties of grasped objects, the motor system may store and access internal models of objects in order to estimate motor commands and predict sensory consequences. In this chapter, we outline evidence for internal models and discuss their role in object manipulation tasks. We also consider the relationship between internal models of objects employed by the sensorimotor system and representations of the same objects used by the perceptual system to make judgements about objects.

Introduction

Although we have designed computers that can beat grand masters at chess, we have yet to design robots that can manipulate chess pieces with anything like the dexterity of a 5-year-old child. What makes humans so good at object manipulation in comparison to robots? There is no question that the anatomy of the human hand is well adapted for manipulation. On the sensory side, the hand is richly endowed with tactile sensors that provide exquisitely precise information about mechanical interactions between the skin and objects. On the motor side, the numerous kinematic degrees of freedom of the hand enable it to grasp objects of all shapes and sizes. These sensory and motor capabilities provide the building blocks; however, it is the way in which manual tasks are organized and controlled by the nervous system that enables flexible and dexterous object manipulation. Skilled object manipulation requires the ability to generate motor commands tailored to the goals of the task and the physical properties of the manipulated objects. This involves both feedforward control, based on prediction, and feedback control that is shaped to the demands of the task. This chapter will focus on predictive mechanisms in the control of object manipulation tasks and on memory representations that support such prediction.

Sensorimotor Control of Grasping: Physiology and Pathophysiology, ed. Dennis A. Nowak and Joachim Hermsdörfer. Published by Cambridge University Press. © Cambridge University Press 2009.

Prediction and internal models

Skilled object manipulation requires the ability to estimate, in advance, the motor commands needed to achieve desired sensory outcomes. For example, when lifting objects, people scale the rate at which they increase vertical load force to the expected weight of the object and often begin to attenuate the increase in load force prior to lift-off (Johansson & Westling, 1988a). This ensures that objects are lifted smoothly and quickly regardless of their weight. In addition, when using a precision grip with the tips of the thumb and index finger on either side of the object, people scale horizontal grip forces to both the predicted load force and the expected friction between the digits and object (Johansson & Westling, 1984). This ensures that grip forces are large enough to prevent slip but not so large as to cause fatigue or damage to the hand or object (see also Chapters 1 and 11).

Skilled object manipulation also requires the ability to predict the sensory consequences of motor commands. By comparing predicted and actual sensory feedback, the motor system can monitor the progress of the task, adjust motor commands if a mismatch occurs so that the goals of the task can be achieved, and update predictive mechanisms so as to reduce future mismatches. A key feature of object manipulation tasks is that they are composed of a series of actions or phases that are often bounded by mechanical events that represent subgoals of the task. These events involve either the making or breaking of contact between the fingertips and an object or between a grasped object and another object or surface. For example, the task of picking up and replacing an object on a table top involves three mechanical events: contact between the digits and object at the end of the reach phase, the breaking of contact between the object and table top at the end of the load phase, and contact between the object and table top at the end of the replacement phase (see Chapter 11).

Although sensory feedback may be continuously predicted and monitored throughout all action phases, tactile signals associated with mechanical contact events play an essential role in the control of object manipulation tasks (see also Chapters 11 and 19). For example, when the fingertip contacts an object, ensembles of tactile afferents provide rich information about the timing, magnitude, direction and spatial distribution of forces, the shape of the contact site, and the friction between the skin and the object (Johansson & Westling, 1984; Jenmalm & Johansson, 1997; Goodwin *et al.*, 1998; Jenmalm *et al.*, 1998, 2000; Birznieks *et al.*, 2001; Johansson & Birznieks, 2004). Thus, tactile signals not only confirm successful completion of the current action phase, they also provide critical information for controlling subsequent phases. By comparing actual and predicted sensory feedback associated with contact events, the motor system can detect mismatches and respond intelligently. For example, when lifting an object, the motor system predicts the time at which it will receive tactile signals indicating that the object has lifted off the surface. If an object is lighter than expected, lift-off will occur before the predicted time and the resulting sensory mismatch will trigger a decrease in load force and grip force. Conversely, if the object is heavier than expected, lift-off will not occur at the expected time and the resulting mismatch will trigger and increase in grip force and load force (Johansson & Westling, 1987, 1988a).

When moving a hand-held object, the mapping between motor commands and sensory outcomes depends on the dynamics of the object; i.e. the relationship between forces applied to the object and its motion. Therefore, in order to accurately predict the sensory consequences of our actions, we need to take the dynamics of the object into account. In other words, the motor system must have an internal representation, or internal model, that captures the mechanical behavior of the object (see also Chapter 9). For example, to accurately predict the time of lift-off when lifting an object, the motor system must know the weight of the object. Similarly, if we lift and replace an object, attached to a table top by a spring, we need to know the stiffness of the spring to accurately predict the time of contact as the object is replaced. With information about intended arm motor commands (i.e. efference copy) and an estimate of the current state of the arm and object, an internal model of the object can be used to predict or simulate the consequences of actual motor commands (Kawato, 1999; Wolpert & Ghahramani, 2000; Flanagan *et al.*, 2006) (see also Chapter 9).

The control of grip force in manipulation tasks may also be based on predictive mechanisms that make use of internal models of object dynamics (Johansson & Westling, 1988a; Flanagan & Wing, 1997). When lifting and moving familiar objects with the grip axis normal to the plane of object motion, grip force is adjusted in phase with, and thus anticipates, movement-dependent modulations in force and torques acting tangential to the grasp surfaces (Flanagan & Wing, 1993, 1995; Flanagan & Tresilian, 1994; Blakemore *et al.*, 1998; Goodwin *et al.*, 1998; Wing & Lederman, 1998). Moreover, people can learn to generate anticipatory grip-force adjustments for a variety of loads that depend on different kinematic parameters of the movement (Flanagan & Wing, 1997; Flanagan *et al.*, 2003). Because the mapping between arm motor commands and load force depends on object dynamics, the motor system cannot rely on a set mapping between arm and grip-force motor commands to modulate grip force in phase with load force. Instead, we have argued that motor system predicts load force (and hence the required grip force) using an internal model of the dynamics of the object (Flanagan & Wing, 1997; Wolpert & Flanagan, 2001; Flanagan *et al.*, 2003). Our ability to independently modulate arm movement motor commands and grip-force motor commands has been nicely demonstrated by Danion and colleagues who showed that predictive adjustments in grip force are sensitive to loads applied to the object but not to equivalent loads applied to the arm (Danion, 2004; Descoins *et al.*, 2006). Because the transformation from arm motor commands to fingertip load forces depends on both arm and object dynamics, accurate prediction of load forces acting on the fingertips when moving a grasped object requires knowledge of arm dynamics in addition to knowledge of object dynamics. Thus, grip force can be used to examine the structure of internal models of arm dynamics in addition to object dynamics (Flanagan & Lolley, 2001).

The term forward model refers to an internal model that is used to predict the consequences of motor commands whereas the term inverse model refers to an internal model that is used to estimate the motor commands required to achieve desired sensory outcomes (see Chapter 9). Shadmehr & Mussa-Ivaldi (1994) examined the acquisition of inverse models of object dynamics using a task in which participants moved a handle attached to a robotic device that could generate complex movement-dependent loads (or force-fields). Although the force-field initially perturbed the trajectory of the hand, participants adapted such that they could move the handle directly to targets in much the same way as they did before the force-field was turned on. Importantly, when the force-field was turned off following adaptation, hand trajectories were again perturbed. This indicates that participants did not simply stiffen the limb to compensate for the force-field but, instead, learned an inverse model of the dynamics of the handle (Shadmehr & Mussa-Ivaldi, 1994). Note that after-effects are not observed following adaptation if the object in hand is released but are seen if participants re-grasp the object (Lackner & DiZio, 2005; Cothros *et al.*, 2006). This indicates that internal models can be recruited and de-recruited when grasping and releasing objects. Indeed, the fact that people can seamlessly lift myriad familiar objects shows that we can rapidly recruit and de-recruit appropriate internal models as we grasp and release objects.

As noted above, anticipatory adjustments in grip force could be generated using a forward model of the object to predict the load forces resulting from arm motor commands. However, it is also possible that grip forces could be generated using an inverse model. If one assumes that motion planning involves specifying a desired object trajectory, an inverse model could be used to transform the desired trajectory into the load forces required to move the object and hence the grip forces required to stabilize the object. However, studies showing that people can accurately predict grip forces even when they are unable to control the movement of the objects (Flanagan *et al.*, 2003; see also Flanagan & Lolley, 2001) suggest that grip forces are likely predicted based on a forward model of the object. In any case, what is clear is that knowledge of object dynamics is crucial for the accurate prediction of required grip forces in manipulation tasks.

It is important to emphasize that internal models of objects, as defined by most researchers in the field, are not necessarily complete or veridical representations of the actual dynamics of the object. Indeed, most studies examining how people adapt to unusual and novel loads applied to the hand or arm have shown that learning is action- and context-specific (e.g. Wang & Sainburg, 2004; Nozaki *et al.*, 2006). For example, studies examining adaptation of reaching movements to novel loads applied to the hand have shown limited transfer of learning when the object (or force-field linked to the object) is rotated relative to the arm (Shadmehr & Mussa-Ivaldi, 1994; Malfait *et al.*, 2002). These results suggest that when adapting to novel and unusual loads, people do not learn the full dynamics of the object mapping motion to applied force. Instead, they appear to learn a mapping between object motion and context- and action-specific motor commands (Shadmehr & Moussavi, 2000; Mah & Mussa-Ivaldi, 2003). It is an open question whether, with sufficient practice manipulating an object with novel dynamics, people form a single internal model that approximates the true dynamics of the object or a set of internal models tailored to specific contexts and actions.

In contrast to the unusual and novel loads often employed in studies of motor learning, the loads experienced in most natural manipulation tasks are familiar. For example, many of the objects we lift and move on a daily basis are standard inertial loads where the applied force varies with acceleration. When lifting familiar objects, the motor system does not have to learn a new class of dynamics; rather the challenge is to predict the load parameters. Thus, when lifting objects with inertial loads, the motor system generally attempts to predict the mass (or weight) of the object. People are very good at using information about object size and shape, obtained through vision or touch, to predict the weight (Johansson & Westling, 1988a; Gordon *et al.*, 1991a, 1991b; Mon-Williams & Murray, 2000) and weight distribution (Jenmalm *et al.*, 1998; Wing & Lederman, 1998; Johansson *et al.*, 1999; Salimi *et al.*, 2003) of objects. Although we know of no experiments that have examined the control of fingertip forces when lifting similar objects composed of different materials, it seems likely that people also use visual (and perhaps haptic) information about object material to estimate weight. Gordon and colleagues (1993) have shown that people predictively scale their fingertip forces for familiar objects of varying size, shape and material (e.g. a glass candle holder versus a box of crispbread) and it is possible that they may be using information about material, in addition to object identification, to predict object weight. Although visual (and haptic) information about object size and shape often leads to good predictions about weight and weight distribution, such prediction is based on correlations and can be erroneous. Ultimately, it is not until the object is lifted, and tactile feedback received, that the weight and weight distribution of the objects can be determined. Similarly, the friction between the digits and the contact surfaces can only be accurately determined from tactile feedback arising when the digits contact the object.

When prediction of object physical properties based on visual or haptic cues goes awry, reflex-mediated corrections of force output are observed (see above). At the same time, memory representations are updated such that, if the object is lifted a second time, prediction improves. Johansson & Westling (1984) coined the term sensorimotor memory to refer to knowledge of object properties gained from previous lifts. In the absence of useful visual cues, sensorimotor memory typically dominates fingertip force control after a single lift. For example, when repeatedly lifting a test object, the weight of which is occasionally and unexpectedly altered, people update their force output within a single trial following a weight change (Johansson & Westling, 1988a). In the presence of misleading visual cues, several trials may be required before sensorimotor memory dominates (Gordon *et al.*, 1991c; Flanagan & Beltzner, 2000; Grandy & Westwood, 2006). Sensorimotor memory is closely related to the notion of an internal model. In a study using objects with misleading size cues about weight, we have shown that sensorimotor memory for weight can be long-lasting (Flanagan *et al.*, 2001). In particular, participants who lifted a small high-density cube and a large low-density cube several times on one day exhibited accurate prediction of required fingertip forces when lifting the same objects a day later. Such persistent sensorimotor memory is tantamount to an internal model.

The idea that sensorimotor memories, or internal models, encode object mechanical properties has been challenged by Quaney *et al.* (2003). These authors showed that pinching a force transducer before lifting an object influences the grip force used during the lift. Based on this observation they argued that sensorimotor memory is based on recent fingertip actions rather than object properties. However, Cole *et al.* (2006) recently demonstrated that this finding does not extend to load force; the generation of vertical load forces at the fingertips prior to lifting an object does not influence load force development during the lift. Cole and coworkers suggested that separate memory representations may be used for grip and load force control (see also Quaney *et al.*, 2005). Because the load force required to lift the object depends solely on the physical properties of the object (i.e. object weight), we suggest that an internal model of the object is used to control load force. In contrast, grip force depends not only on weight but is also influenced by the frictional conditions at the contact surfaces and the grip safety margin selected by the individual to guard against slip. Given that grip force depends on factors that are independent of the object (e.g. the dryness of the skin), it seems reasonable that the control of grip force may involve memory mechanisms that are distinct from the internal model of the object and that can be influenced by actions, involving the fingertips, on other objects.

To test the idea that people might remember motor commands or actions rather than object properties, we conducted an experiment in which we asked participants to lift an object, instrumented with force sensors (Figure 12.1A), to different heights within a prescribed time period (Merritt & Flanagan, 2004). The object was attached to a manipulandum that could simulate different loads including an inertial and a viscous load. For each load, participants first lifted the object 20 times to a target height of 7 cm and then lifted the object another 20 times to a target height of 14 cm. In all lifts, participants were asked to lift the object to the target within a 200 ms time window. Figure 12.1B shows kinematic and force records for three lifts of the inertial (or mass) load performed by a representative participant. The solid black, solid gray and dashed black curves illustrate the last 7 cm lift, the first 14 cm lift and the second 14 cm lift. Note that reasonably good generalization was observed in the first lift to the 14 cm target as the participant increased both grip and load force. In all three lifts, grip force was modulated in phase with load force indicating good prediction of the load. Figure 12.1C shows corresponding records for three lifts with the viscous load performed by another representative participant. Although this load is somewhat unusual, the participant adapted well to the load. Specifically, in the last lift to the 7 cm target, the participant accurately and smoothly reached the target and grip force was modulated in phase with velocity-dependent load force. When first lifting the viscous load to the 14 cm target, the object undershot the target. However, partial generalization of learning was observed in that the participant appropriately increased both grip force and load force and continued to modulate grip force in phase with load force. These results indicate that when lifting both familiar and novel loads, participants acquire knowledge of dynamics that is linked to the object. (Note that it is not clear how memory representations based on motor commands could support generalization across lifts of varying height and speed that require different motor commands.)

Neural basis of anticipatory grip force adjustments

Recently, Pilon *et al.* (2007) have argued that the modulation of grip force with changes in load force, observed in many manipulation tasks, results from biomechanical rather than

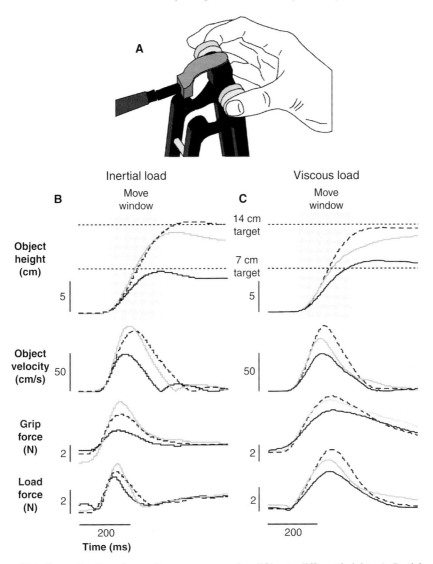

Figure 12.1. Generalization of sensorimotor memory when lifting to different heights. A. Participants grasped and lifted an object instrumented with sensors that measured the forces applied by the tips of the index finger and thumb. The object was attached to the tip of a robot manipulandum that could simulate different loads. The contact surfaces mounted on each force sensor could freely spin so that the object could effectively rotate about the grip axis. In addition, the joint between the object and manipulandum allowed rotation of the object about any axis orthogonal to the grip axis. B and C. Kinematic and force records when lifting an inertial (B) and a viscous (C) load. The solid black curves are from the last of 20 lifts to the initial 7 cm target and the solid gray curves are from the next trial that was the first lift to the 14 cm target. The dashed black curves represent the second lift to the 14 cm target. The gray boxes represent the 200 ms time window in which participants were instructed to lift the object to the target.

neural mechanisms. Specifically, these authors suggested that once an object is grasped, tangential load forces compress and move the finger pads and that this leads to changes in grip force that are proportional to the load force. Although this idea may seem attractive in terms of simplifying grip-force control, there is an abundance of evidence demonstrating that anticipatory grip forces result from neural control processes and that load and grip forces are not, in fact, mechanically coupled as assumed (but not tested) by Pilon and coworkers. In this section of the chapter, we will describe some of this evidence.

Clear decoupling between changes in grip force and changes in load force can be observed in precision-grip lifting when an object weight is unexpectedly decreased (Johansson & Westling, 1988a). When a participant expects to lift an 800 g object but actually lifts a 200 g weight, the object lifts off earlier than expected. Due to biomechanical factors (e.g. muscle shortening), there is a rapid cessation of load force increase at the moment of lift-off (see Figure 2 from Johansson & Westling, 1988a). However, grip force continues to increase for some 100 ms after lift-off. (After 100 ms, grip force starts to decrease due to a reflex-mediated mechanism triggered by the earlier than expected lift-off.) For the first 100 ms after lift-off, the grip-force profile is indistinguishable from the profile observed when the participant both expects and receives the 800 g object. Thus, for a critical 100 ms window there is a strong dissociation between changes in grip force and changes in load force and grip force is unaffected by dramatic changes in load force. This result clearly demonstrates that load force and grip force are not mechanically coupled and shows that both anticipatory and reactive changes in grip force are achieved through neural control mechanisms.

Blakemore *et al.* (1998) have shown that when a participant holds an object to which a cyclical load force is externally applied, grip force is modulated but lags behind changes in load force by some 100 ms. This indicates that participants could not predict the external load and instead relied on reactive mechanisms to modulate grip force. This decoupling between grip and load changes provides another example where grip and load forces are not mechanically coupled. Pilon *et al.* (2007) cite the Blakemore study but argue that because participants employed high overall grip forces (around 10 N on average), the finger pads would not have moved much and that therefore grip force was not modulated in phase with load force. However, Flanagan & Wing (1995) showed that when moving an object in a cyclic fashion, grip force can vary between 10 and 25 N and still be modulated in phase with load force (see Figure 2 from Flanagan & Wing, 1995). Thus, if it is true that the finger pads do not move when grip forces are large (i.e. around 10 N or higher), then the mechanical explanation for grip–load coupling posited by Pilon and colleagues cannot explain the coupling observed by Flanagan & Wing (1995). (It also cannot explain the coupling observed when lifting heavy objects where large grip forces are observed, e.g. Johansson & Westling, 1988a). Conversely, if the finger pads do move even with large grip forces, then the decoupling between grip and load force observed by Blakemore and colleagues demonstrates that finger pad motion does not lead to grip–load coupling.

Using a task similar to the one employed by Blakemore *et al.* (1998), Hermsdörfer & Blankenfeld (2008) observed that when the externally applied sinusoidal load force is suddenly and unexpectedly turned off (in catch trials), grip force continues to be modulated (in the same way as in non-catch trials) for 100 ms. If the changes in grip force were due to mechanical interactions, this continued modulation would not be observed. Witney *et al.* (1999) examined the control of grip force in a bimanual task in which participants grasped the top and bottom of a virtual object with the left and right hand. Participants were instructed to pull up with the left and, at the same time, prevent the object from moving. This required the generation of equal and opposite load forces with the two hands as well as grip forces with both hands. On unexpected catch trials, the two objects could be "unlinked" so that no load force was delivered to the right hand when the participant pulled up with the left hand. In these catch trials, participants generated increases in grip force with the unloaded right hand that closely matched those produced in normal "linked" trials. This result indicates that the modulation of grip force (in both linked and unlinked trials) was not caused by mechanical coupling with load force.

Several studies have shown that grip force is modulated in anticipation of contact forces that occur when the hand-held object strikes another object (Johansson & Westling, 1988b; Turrell *et al.*, 1999; Delevoye-Turrell & Wing, 2003). For example, when participants drop a ball into a cup held in a precision grip, they increase grip force shortly before the anticipated contact and grip force is scaled to the predicted contact force (Johansson & Westling, 1988b). Obviously, these predictive changes in grip force are achieved by neural control mechanisms. Johansson & Westling (1988b) also examined reflex-mediated changes in grip force when the experimenter dropped the ball and the participant was unaware that the ball had been dropped. The sudden increase in load force due to contact resulted in very small mechanically induced changes in grip force that were orders of magnitude smaller than the subsequent reflex-induced increase in grip force observed when the mechanical effects of the perturbation are no longer present (see Johansson *et al.*, 1992; Flanagan & Wing, 1993; Cole & Abbs, 1988 for similar results). In other words, the load force at the fingertip did not result in significant mechanical changes in grip force.

We have shown that grip force is modulated in phase with changes in load force when participants grasp objects, such as a cup, by inserting their index finger and thumb inside and pushing outwards with the finger nails (Flanagan & Tresilian, 1994). Thus, anticipatory coupling between grip force and load force is observed when the fingertip pads are not employed in gripping. Indeed, predictive coupling between normal and load forces is observed even when participants lift and move objects held between the teeth (unpublished observations from both the Johansson and Flanagan labs) or use the hand to push and pull on an object held between the teeth (Westberg *et al.*, 2001; Figure 12.2). Figure 12.2 shows results from an experiment in which participants held a bar between their teeth and hand and were instructed to either push the bar toward the mouth or pull it away from the mouth. As shown by the individual records shown in Figure 12.2B, these pushes and pulls created load forces acting at the teeth and participants modulated their bite force in anticipation of the load. Figure 12.2C shows averaged bite and load force traces for pushes. On average, bite force increased 44 ms ahead of load force and was therefore predictive. Clearly, motion of the finger pads cannot explain the anticipatory modulation of bite force with load force.

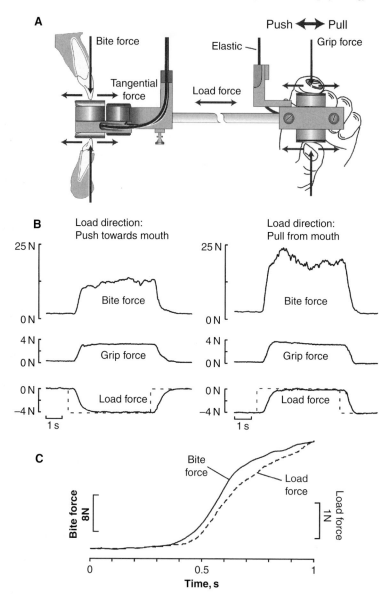

Figure 12.2. Predictive coupling between bite force and load force. A. The participant held a bar with the teeth and fingertips at either end and was instructed to use the hand to push and pull the bar while preventing it from moving with the teeth. In each trial a target load force was displayed on a monitor. Forces normal and tangential to the contact surfaces were recorded using six-axis force-torque transducers built into the bar. The load force was defined as the magnitude of the vector sum of tangential forces at the teeth. B. Examples of single trial forces when the bar was pushed and pulled by the hand. The dashed line shows the target load force. C. Averaged bite force (solid trace) and load force (dashed trace) records for pushes. On average, bite force increased 44 ms ahead of the load force. Adapted from Westberg *et al.* (2001).

Finally, Häger-Ross & Johansson (1996) examined the mechanical properties of the fingertips when subjected to load forces while gripping an object. These authors showed that grip force changes only slightly due to mechanics. Moreover, they found that grip force could increase or decrease depending on the direction of loading. Thus, fingertip mechanics simply cannot account for large grip force modulation observed during movement. All of the results described above (as well as a number of other results that we have left out) clearly show that anticipatory grip force adjustments observed in manipulation tasks are based on predictive neural control mechanisms and cannot be explained by the mechanics of the fingertip pads. Indeed, these results rule out any mechanical explanation (e.g. based on finger tendons) for grip–load coupling.

Independent object representations in action and perception

When people lift a large object and a similar but smaller object of equal weight, they typically judge the smaller of the two objects to be substantially heavier. This size-weight illusion, first described well over 100 years ago (Charpentier, 1891; Murray *et al.*, 1999), is experienced by almost all healthy people (Ross, 1969; Davis & Roberts, 1976), including children as young as 2 years of age (Robinson, 1964; Pick & Pick, 1967), and is not weakened when participants are verbally informed that the objects are equally weighted (Flourney, 1894; Nyssen & Bourdon, 1955; Flanagan & Beltzner, 2000). The size–weight illusion is powerful when only visual cues about size are available, as when lifting viewed objects by strings, but is strongest when haptic cues about object size are available, as when the hand grasps the objects directly (Ellis & Lederman, 1993).

Recently, we ruled out the hypothesis (Ross, 1969; Granit, 1972; Davis & Roberts, 1976) that the size–weight illusion arises from a mismatch between actual and expected sensory feedback related to lifting (Flanagan & Beltzner, 2000). We asked participants to repeatedly lift a small cube and an equally weighted large cube, in alternation, for a total of 40 lifts (Figure 12.3A). As expected, when lifting the two cubes for the first time, participants generated erroneous predictions about object weight based on size. That is, they over-estimated the fingertip forces required to lift the large object and underestimated the forces required to lift the small object. Figure 12.3B shows fingertip forces and force rates for the first lifts of the large and small cubes performed by a representative participant. During the first lift of the small cube, the initial rise in grip force and load force was too small and lift-off did not occur when expected. The resulting mismatch between expected and actual tactile feedback gave rise to a reflex-mediated increase in both grip and load force and lift-off then occurred. During the first lift of the large cube, overshoots of the grip and load forces were observed and lift-off occurs earlier than expected. In this case, the mismatch between expected and actual tactile feedback triggered a decrease in force output approximately 100 ms later.

Although participants were initially fooled by the misleading size cues, they adapted their force output to the true weights of the objects within 5–10 pairs of lifts. Figure 12.3C shows fingertip forces and force rates for the eighth lifts of the large and small cubes (lifts 15 and

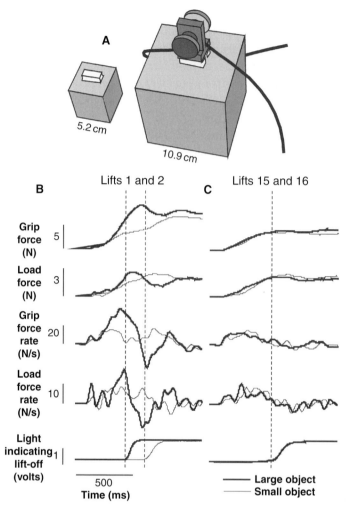

Figure 12.3. Sensorimotor adaptation when lifting size–weight stimuli. A. In alternate trials, participants lifted either a large or a small cube by grasping a handle, mounted on top of the cube, with the tips of the index finger and thumb on either side. The handle could be moved quickly between cubes and was instrumented with two force-torque sensors with circular contact surfaces (3 cm in diameter) covered in sandpaper. B and C. Grip force, load force, grip and load force rates, and the reading from a light-sensitive diode signaling lift-off recorded in the first two lifts (B) and lifts 15 and 16 (C) performed by a representative participant. This participant lifted the large object (thick traces) and then the small object (thin traces) in each pair of lifts. The trials are temporally aligned to the time at which load force started to increase. The vertical dashed lines mark lift-off times. In the eighth trial, the lift-off times for the large and small cubes were indistinguishable. Adapted from Flanagan & Beltzner (2000).

16) performed by the same participant shown in Figure 12.3B. In these lifts, the force and force rate functions for the small and large cubes were very similar and lift-off occurred at about the same time for both cubes. Importantly, grip force and load force neither overshot nor undershot their final levels and no corrective adjustments in force were observed. Thus, the participant scaled their force output appropriately for the two cubes and also generated accurate sensory predictions about the timing of lift-off. This pattern of results, observed in all participants, indicates that the sensorimotor system acquired accurate representations of the weights of the two cubes (see also Davidson & Wolpert, 2004; Grandy & Westwood, 2006). Using two new groups of participants, we assessed the strength of the size–weight illusion after a single lift of each cube and after 20 lifts of each cube (Flanagan & Beltzner, 2000). This involved asking participants to assign numbers corresponding to the weights of the two cubes after lifting them. We found that the strength of the illusion was as strong after 20 pairs of lifts as it was after the first lift.

Taken together, these results indicate that the brain maintains two independent representations of object weight: a perceptual representation that is influenced by the size of objects (as revealed by the size–weight illusion) and a sensorimotor representation that is not. The results indicate that the size–weight illusion does not arise from a mismatch between actual weight and the sensorimotor representation of weight. Instead, we suggest that the illusion stems from a mismatch between the actual weight of the object and the perceptual representation of object weight that continues to be influenced by object size even when the object is lifted a number of times and the sensorimotor representation of weight is updated (Flanagan & Beltzner, 2000).

The finding that the brain maintains separate sensorimotor and perceptual representations for object weight can be related to the growing body of work demonstrating that sensory information is processed differently (and in different brain regions) depending on whether the information is used for the guidance of action or for perceptual tasks. For example, Goodale and his colleagues have provided evidence from behavioral, neuropsychological and neuroimaging studies that visual information about object size, shape and orientation is processed in distinct neural pathways depending on whether the information is used to control grasping or to make perceptual judgements about the objects (Goodale *et al.*, 1991; Milner & Goodale, 1995; Hu & Goodale, 2000; Culham *et al.*, 2003; Ganel & Goodale, 2003).

We have suggested that the smaller of two equally weighted objects is judged to be heavier because it is heavier than would be expected based on size. Such expectations are based on the statistical relationship between size and weight learned through experience manipulating myriad objects. In addition to influencing perception, this statistical knowledge is extremely valuable in guiding our actions. Although motor commands based on such expectations will sometimes be inappropriate (as when lifting size–weight stimuli), they enable the motor system to make good guesses most of the time. Our results (Flanagan & Beltzner, 2000; Flanagan *et al.*, 2001) show that when we encounter an object that is heavier or lighter than expected, the sensorimotor system can acquire a new (and long-lasting) representation of the object without affecting the perceptual representation. Presumably, the perceptual representation is unaffected because lifting one or two objects with abnormal density does not appreciably affect the learned correlation between size and weight. It is unclear what happens when the weight of a new object, but one that belongs to a given family or type of objects, closely matches the expected weight. In principle, if predictions based on the size and type of object are accurate, it would not be necessary to form a sensorimotor representation (or internal model) of that specific object (see Chapter 9). However, this question has not been investigated and more work needs to be done to understand the relationship between correlative knowledge about the properties of families of objects (e.g. how weight scales with size for a given object family) and knowledge about the properties of individual objects in the control of object manipulation tasks.

References

Aglioti, S., DeSouza, J. F. X. & Goodale, M. A. (1995). Size-contrast illusions deceive the eye but not the hand. *Curr Biol*, **5**, 679–685.

Blakemore, S. J., Goodbody, S. J. & Wolpert, D. M. (1998). Predicting the consequences of our own actions: the role of context sensorimotor context estimation. *J Neurosci*, **18**, 7511–7518.

Birznieks, I., Jenmalm, P., Goodwin, A. W. & Johansson, R. S. (2001). Encoding of direction of fingertip forces by human tactile afferents. *J Neurosci*, **21**, 8222–8237.

Culham, J. C., Danckert, S. L., DeSouza, J. F. *et al.* (2003). Visually guided grasping produces fMRI activation in dorsal but not ventral stream brain areas. *Exp Brain Res*, **153**, 180–189.

Charpentier, A. (1891). Analyse experimentale quelques elements de la sensation de poids [Experimental study of some aspects of weight perception]. *Archiv Physiol Normales Pathologiq*, **3**, 122–135.

Cole, K. J. & Abbs, J. H. (1988). Grip force adjustments evoked by load force perturbations of a grasped object. *J Neurophysiol*, **60**, 1513–1522.

Cole, K. J., Potash, M. & Peterson, C. (2006). 'Sensorimotor' memory affects the lift and grip forces differently in a simple grip and lift task. *Soc Neurosci Abstr*, Poster 655.1.

Cothros, N., Wong, J. D. & Gribble, P. L. (2006). Are there distinct neural representations of object and limb dynamics? *Exp Brain Res*, **173**, 689–697.

Danion, F. (2004). How dependent are grip force and arm actions during holding an object? *Exp Brain Research*, **158**, 109–119.

Davidson, P. R. & Wolpert, D. M. (2004). Internal models underlying grasp can be additively combined. *Exp Brain Res*, **155**, 334–340.

Davis, C. M. & Roberts, W. (1976). Lifting movements in the size-weight illusion. *Percept Psychophys*, **20**, 33–36.

Delevoye-Turrell, Y. N. & Wing, A. M. (2003). Efficiency of grip force adjustments for impulsive loading during imposed and actively produced collisions. *Quart J Exp Psychol*, **56A**, 1113–1128.

Descoins, M., Danion, F. & Bootsma, R. J. (2006). Predictive control of grip force when moving object with an elastic load applied on the arm. *Exp Brain Res*, **172**, 331–342.

Ellis, R. R. & Lederman, S. J. (1993). The role of haptic versus visual volume cues in the size-weight illusion. *Percept Psychophys*, **53**, 315–324.

Flanagan, J. R. & Tresilian, J. (1994). Grip-load force coupling: a general control strategy for transporting objects. *J Exp Psychol Hum Percept Perform*, **20**, 944–957.

Flanagan, J. R. & Wing, A. M. (1993). Modulation of grip force with load force during point-to-point arm movements. *Exp Brain Res*, **95**, 131–143.

Flanagan, J. R. & Wing, A. M. (1995). The stability of precision grip forces during cyclic arm movements with a hand-held load. *Exp Brain Res*, **105**, 455–464.

Flanagan, J. R. & Wing, A. M. (1997). The role of internal models in motion planning and control: evidence from grip force adjustments during movements of hand-held loads. *J Neurosci*, **17**, 1519–1528.

Flanagan, J. R. & Beltzner, M. A. (2000). Independence of perceptual and sensorimotor predictions in the size-weight illusion. *Nat Neurosci*, **3**, 737–741.

Flanagan, J. R. & Lolley, S. (2001). The inertial anisotropy of the arm is accurately predicted during movement planning. *J Neurosci*, **21**, 1361–1369.

Flanagan, J. R., King, S., Wolpert, D. M. & Johansson, R. S. (2001). Sensorimotor prediction and memory in object manipulation. *Can J Exp Psychol*, **55**, 89–97.

Flanagan, J. R., Vetter, P., Johansson, R. S. & Wolpert, D. M. (2003). Prediction precedes control in motor learning. *Curr Biol*, **13**, 146–150.

Flanagan, J. R., Bowman, M. C. & Johansson, R. S. (2006). Control strategies in object manipulation tasks. *Curr Opin Neurobiol*, **16**, 650–659.

Flourney, T. (1894). De l'influence de la perception visuelle des corps sur leur poids apparrent [The influence of visual perception on the apparent weight of objects]. *L'Année Psychologiq*, **1**, 198–208.

Ganel, T. & Goodale, M. A. (2003). Visual control of action but not perception requires analytical processing of object shape. *Nature*, **426**, 664–667.

Goodale, M. A., Milner, A. D., Jakobson, L. S. & Carey, D. P. (1991). A neurological dissociation between perceiving objects and grasping them. *Nature*, **349**, 154–156.

Goodwin, A. W., Jenmalm, P. & Johansson, R. S. (1998). Control of grip force when tilting objects: effect of curvature of grasped surfaces and applied tangential torque. *J Neurosci*, **18**, 10724–10734.

Gordon, A. M., Forssberg, H., Johansson, R. S. & Westling, G. (1991a). Visual size cues in the programming of manipulative forces during precision grip. *Exp Brain Res*, **83**, 477–482.

Gordon, A. M., Forssberg, H., Johansson, R. S. & Westling, G. (1991b). The integration of haptically acquired size information in the programming of precision grip. *Exp Brain Res*, **83**, 483–488.

Gordon, A. M., Forssberg, H., Johansson, R. S. & Westling, G. (1991c). Integration of sensory information during the programming of precision grip: comments on the contributions of size cues. *Exp Brain Res*, **85**, 226–229.

Gordon, A. M., Westling, G., Cole, K. J. & Johansson, R. S. (1993). Memory representations underlying motor commands used during manipulation of common and novel objects. *J Neurophysiol*, **69**, 1789–1796.

Grandy, M. S. & Westwood, D. A. (2006). Opposite perceptual and sensorimotor responses to a size-weight illusion. *J Neurophysiol*, **95**, 3887–3892.

Granit, R. (1972). Constant errors in the execution and appreciation of movement. *Brain*, **95**, 451–460.

Häger-Ross, C. & Johansson, R. S. (1996). Nondigital afferent input in reactive control of fingertip forces during precision grip. *Exp Brain Res*, **110**, 131–141.

Hermsdörfer, J. & Blankenfeld, H. (2008). Grip force control of predictable external loads. *Exp Brain Res*, DOI 10.1007/s00221-007-1195-6.

Hu, Y. & Goodale, M. A. (2000). Grasping after a delay shifts size-scaling from absolute to relative metrics. *J Cogn Neurosci*, **12**, 856–868.

Jenmalm, P. & Johansson, R. S. (1997). Visual and somatosensory information about object shape control manipulative fingertip forces. *J Neurosci*, **17**, 4486–4499.

Jenmalm, P., Goodwin, A. W. & Johansson, R. S. (1998). Control of grasp stability when humans lift objects with different surface curvatures. *J Neurophysiol*, **79**, 1643–1652.

Jenmalm, P., Dahlstedt, S. & Johansson, R. S. (2000). Visual and tactile information about object-curvature control fingertip forces and grasp kinematics in human dexterous manipulation. *J Neurophysiol*, **84**, 2984–2997.

Johansson, R. S. & Birznieks, I. (2004). First spikes in ensembles of human tactile afferents code complex spatial fingertip events. *Nat Neurosci*, **7**, 170–177.

Johansson, R. S. & Westling, G. (1984). Roles of glabrous skin receptors and sensorimotor memory in automatic-control of precision grip when lifting rougher or more slippery objects. *Exp Brain Res*, **56**, 550–564.

Johansson, R. S. & Westling, G. (1987). Signals in tactile afferents from the fingers eliciting adaptive motor-responses during precision grip. *Exp Brain Res*, **66**, 141–154.

Johansson, R. S. & Westling, G. (1988a). Coordinated isometric muscle commands adequately and erroneously programmed for the weight during lifting task with precision grip. *Exp Brain Res*, **71**, 59–71.

Johansson, R. S. & Westling, G. (1988b). Programmed and triggered actions to rapid load changes during precision grip. *Exp Brain Res*, **71**, 72–86.

Johansson, R. S., Häger, C. & Riso, R. (1992). Somatosensory control of precision grip during unpredictable pulling forces. II: Changes in load force rate. *Exp Brain Res*, **89**, 192–203.

Johansson, R. S., Backlin, J. L. & Burstedt, M. K. (1999). Control of grasp stability during pronation and supination movements. *Exp Brain Res*, **128**, 20–30.

Kawato, M. (1999). Internal models for motor control and trajectory planning. *Curr Opin Neurobiol*, **9**, 718–727.

Lackner, J. R. & DiZio, P. (2005). Motor control and learning in altered dynamic environments. *Curr Opin Neurobiol*, **15**, 653–659.

Mah, C. D. & Mussa-Ivaldi, F. A. (2003). Generalization of object manipulation skills learned without limb motion. *J Neurosci*, **23**, 4821–4825.

Malfait, N., Shiller, D. M. & Ostry, D. J. (2002). Transfer of motor learning across arm configurations. *J Neurosci*, **22**, 9656–9660.

Merritt, K. & Flanagan, J. R. (2004). Internal models of object dynamics in skilled manipulation. Poster presented at the Seventh Annual Meeting For Health Sciences Research Trainees, Queen's University, May 2004.

Milner, A. D. & Goodale, M. A. (1995). *The Visual Brain in Action*. Oxford, UK: Oxford University Press.

Mon-Williams, M. & Murray, A. H. (2000). The size of the visual size cue used for programming manipulative forces during precision grip. *Exp Brain Res*, **135**, 405–410.

Murray, D. J., Ellis, R. R., Bandomir, C. A. & Ross, H. E. (1999). Charpentier (1891) on the size-weight illusion. *Percept Psychophys*, **61**, 1681–1685.

Nozaki, D., Kurtzer, I. & Scott, S. H. (2006). Limited transfer of learning between unimanual and bimanual skills within the same limb. *Nat Neurosci*, **9**, 1364–1366.

Nyssen, R. & Bourdon, J. (1955). Contribution to the study of the size-weight illusion by the method of P. Koseleff. *Acta Psychologia*, **11**, 467–474.

Pick, H. L. & Pick, A. D. (1967). A developmental and analytic study of the size-weight illusion. *J Exp Child Psychol*, **5**, 362–371.

Pilon, J.-F., De Serres, S. J. & Feldman, A. G. (2007). Threshold position control of arm movement with anticipatory increase in grip force. *Exp Brain Res*, **181**, 49–67.

Quaney, B. M., Rotella, D. L., Peterson, C. & Cole, K. J. (2003). Sensorimotor memory for fingertip forces: evidence for a task-independent motor memory. *J Neurosci*, **23**, 1981–1986.

Quaney, B. M., Nudo, R. J. & Cole, K. J. (2005). Can internal models of objects be utilized for different prehension tasks? *J Neurophysiol*, **93**, 2021–2027.

Robinson, H. B. (1964). An experimental examination of the size-weight illusion in young children. *Child Develop*, **35**, 91–107.

Ross, H. E. (1969). When is a weight not illusory? *Quart J Exp Psychol*, **21**, 346–355.

Salimi, I., Frazier, W., Reilmann, R. & Gordon, A. M. (2003). Selective use of visual information signaling objects' center of mass for anticipatory control of manipulative fingertip forces. *Exp Brain Res*, **150**, 9–18.

Shadmehr, R. & Mussa-Ivaldi, F. A. (1994). Adaptive representation of dynamics during learning of a motor task. *J Neurosci*, **14**, 3208–3224.

Shadmehr, R. & Moussavi, Z. M. K. (2000). Spatial generalization from learning dynamics of reaching movements. *J Neurosci*, **20**, 7807–7815.

Turrell, Y. N., Li, F.-X. & Wing, A. M. (1999). Grip force dynamics in the approach to a collision. *Exp Brain Res*, **128**, 86–91.

Wang, J. & Sainburg, R. L. (2004). Interlimb transfer of novel inertial dynamics is asymmetrical. *Exp Brain Res*, **92**, 349–360.

Westberg, K.-G., Trulsson, M. & Johansson, R. S. (2001). Oro-manual force coordination in humans. Poster presented at the Neural Control of Movement meeting, March 25–31, 2001, Seville, Spain.

Wing, A. M. & Lederman, S. J. (1998). Anticipating load torques produced by voluntary movements. *J Exp Psychol Hum Percept Perform*, **24**, 1571–1581.

Witney, A., Goodbody, S. J., & Wolpert, D. M. (1999). Predictive motor learning of temporal delays. *J Neurophysiol*, **82**, 2039–2048.

Wolpert, D. M. & Flanagan, J. R. (2001). Motor prediction. *Curr Biol*, **11**, R729–R732.

Wolpert, D. M. & Ghahramani, Z. (2000). Computational principles of movement neuroscience. *Nat Neurosci*, **3** Suppl., 1212–1217.

13

The neurohaptic control of the hand

ALLAN M. SMITH

Summary

Our knowledge about an object small enough to be grasped with the hand usually begins first with a visual appreciation of its size and shape. However, in the dark or when searching a deep pocket or purse, vision is impossible. Consequently a haptic exploration procedure is the only course of action and scanning an object's surface with the fingertips provides information about friction, shape, compliance, temperature and friction that is unattainable by visual inspection. This initial information is of particular importance to subsequent object manipulation and dexterous handling. Both exploratory hand movements and object manipulation make efficient use of specialized low-threshold mechanoreceptors in the skin which are selectively sensitive to both normal and tangential (shearing) forces as well as slip on the skin. This cutaneous feedback guides the exploratory movements and provides a signal of when a tactile target is encountered. These primary afferent signals are subsequently transformed by cell assemblies in the somatosensory cortex to generate central representations or internal models of the object's salient physical features. Neuronal signals encoding the internal model of shape, friction and center of mass are then relayed directly by cortico-cortical projections from the somatosensory cortex to motor cortex. The subsequent dexterous object manipulation is driven by anticipatory motor control strategies based on the internal model of the object's features which are used to direct grip forces and finger positions.

Introduction

The two most striking features of the biological order of primates are the expansion of the cortical gray matter and the significant structural changes in the upper extremity. The latter include an increased mobility about the shoulder and the pentadactyl hand which are obvious adaptations to an arboreal environment. The significant characteristics of the pentadactyl hand are the enhanced mobility of the digits, especially the thumb, the replacement of claws by flattened nails, and most especially, the development of highly sensitive tactile pads at the tips of the digits. The representation of the distal extremity in the brain is proportionately enlarged to rival that of the face within both the cerebral cortex and the

Sensorimotor Control of Grasping: Physiology and Pathophysiology, ed. Dennis A. Nowak and Joachim Hermsdörfer. Published by Cambridge University Press. © Cambridge University Press 2009.

cerebellum. Not surprisingly, this expansion is accompanied by functional changes in which the forelimb is no longer dominated by its locomotor function and as a result the hand is employed for a variety of voluntary manipulatory functions as well as for purely exploratory purposes.

Two broad classes of hand movements; prehension and exploration

Although the hand is capable of a wide variety of complex movements involving many joints and a large number of muscles, from a purely functional point of view, hand movements can be roughly divided into two broad categories: prehensile grips and exploratory movements. Prehension, particularly the power grip, in which all the fingers flex in against the palm, evolved in primates as part of an arboreal adaptation. The increased mobility of the thumb allowed for more delicate object manipulation in which the thumb is placed in controlled opposition to one of the other digit tips, usually the index, and has been called the precision grip (Napier, 1962, 1976). The precise control over finger pressure derives from the highly sensitive tactile pads at the tips of the digits. These same receptors play an equally important role in exploratory movements. Object manipulation requires continuous cutaneous feedback to assist the motor dexterity of the hand, whereas with tactile exploration the hand movements are organized to optimize the tactile sensations generated, adjusting the contact (i.e. normal) and sliding (i.e. tangential) forces as a function of the nature of the object and its surface features. With prehensile movements, touch sensation in the glabrous skin is used to achieve a motor objective such as secure manipulation without any slip whatsoever. In contrast, exploratory movements of the hand are directed to acquiring a particular tactile objective but surface slip on the skin is critical to the perception of friction and shape. Both processes are often seamlessly interwoven as, for example, the simple act of rummaging among articles in a deep pocket in search of an object, which is first identified (i.e. stereognosis), then securely grasped for extraction. With exploratory movements, slip is maximized as the relevant features are evaluated by scanning the fingers over the object surface. Once the desired object is identified and grasped, grip forces are applied with a sufficient safety margin to prevent slip. Both processes require continuous cortically mediated peripheral feedback about the normal and tangential forces on the skin. For the past several years our research has focused on the brain's control of prehension, and more recently on a related behavior, tactile exploration.

Prehension and the sense of touch

In 1984, Johansson and Westling (Johansson & Westling, 1984; Westling & Johansson, 1984) published two influential papers indicating that in grasping, the grip force is influenced by three important factors, namely (1) the weight of the object, (2) the friction between the object and the skin and (3) the safety margin set by the individual based on prior experience (see also Chapters 1 and 11). Anesthetizing the fingertips disrupted the coordination between grip and lifting forces adapted to the friction between the object and the skin, thereby demonstrating the importance of cutaneous afferents and implying that proprioceptive afferents alone could not compensate for the loss of tactile input. To this we would add two additionally relevant observations. The first is that free-standing objects held between the thumb and index finger are frequently dropped when the fingers are anesthetized (Augurelle *et al.*, 2003) because of a failure to recognize the unstable force vector alignment of the two fingers (Monzée *et al.*, 2003) (see also Chapters 11 and 19). The second observation is that anesthetizing the fingers reduces maximum pinch force by 25–30% despite a subjective feeling to the contrary, suggesting that with the exception of nociceptors, the skin afferents exert a strong positive feedback over the motoneurons involved in increased grip force (Augurelle *et al.*, 2003).

In Johansson and Westling's original study, friction was associated with different surface substrates (Johansson & Westling, 1984; Westling & Johansson, 1984), and although the sandpaper, suede and silk surfaces used in these studies had different coefficients of friction, it was unclear whether texture (i.e. the surface substrate) or friction (i.e. the ratio of the tangential to normal force; Bowden & Tabor, 1982) was the more important parameter. Cadoret & Smith (1996) addressed this issue by comparing textured and smooth surfaces coated with adhesives or lubricating films that altered the friction of the same surface. Friction, not texture, was the parameter of primary importance in determining the grip force and safety margin employed in object manipulation.

The precision grip and the co-contraction of antagonist muscles

Dexterous movements of the hand are achieved by a large number of muscles which produce a similarly great variety of voluntary movements. However, as Napier (1962, 1976) has pointed out the diversity of hand movements is more apparent than real and arises mainly from the variety of objects the hand is used to grasping and manipulating. The movements themselves can be classified fairly simply as prehensibility, opposability, finger abduction and finger adduction. Structurally the muscles of the hand are categorized as either extrinsic (proximal to the wrist) or intrinsic (distal to the wrist). The long tendons of the forearm flexors and extensors span the carpal joints and the metacarpal and 1st and 2nd interphalangeal joints to insert on the proximal middle and distal phalanges.

Although many voluntary movements involve the reciprocal activation of the antagonist muscles, pinching, where force is applied at the fingertips, appears to be a common exception (see also Chapter 5) Some time ago we conducted an extensive analysis of intrinsic and extrinsic muscle activity in monkeys performing an isometric precision pinch (Smith, 1981) and found that virtually all of the intrinsic and extrinsic muscles of the hand co-contract. Maier & Hepp-Reymond (1995a, 1995b) found similar antagonist co-contraction in human subjects performing a precision grip. Functionally this co-contraction would appear to be essential in order to stiffen the wrist and finger joints to ensure that force is adequately transmitted to the tips of the fingers. The radial nerve innervates the forearm extensors of the wrist and finger that are antagonists to the flexor muscles of prehension.

Paradoxically however, radial nerve palsy results in an inability to exert a pinch force to any useful degree. We will return to the question of how the brain manages the reciprocal and coactive control of antagonist muscles later.

The skin and skin receptors: structure and implicit function

Two cognitive psychologists, Susan Lederman and Roberta Klatsky, focused an extensive research program on studying purposeful exploratory hand movements in object perception and recognition (Lederman & Klatsky, 1987, 1990; Klatsky & Lederman, 1993). They showed that when subjects were asked to handle various test objects to extract certain features the subjects tended to perform stereotyped hand movements that differed depending on the features to be extracted. For example, roughness discrimination was performed by a lateral, back-and-forth movement of the finger tip over the object surface. Hard–soft (i.e. surface compliance) discrimination was achieved by probing the object with the finger tip to generate forces normal to the skin surface. Edge orientation was extracted by contour following with the finger tip. These exploratory procedures were clear functional indications of how the brain uses its motor control functions to ensure that the skin area of maximum receptor sensitivity is efficiently brought in contact with the test object. Subsequent exploratory movement of the finger over the object enhances the tactile feedback and the brain's motor control system optimizes the contact and lateral forces and the speed of tactile exploration.

Paré *et al.* (2001, 2002a, 2002b) provided a detailed description of the myelinated and unmyelinated innervation of Meissner corpuscles and Merkel ending complexes and their distribution in the fingertips of the monkey using immuno-histochemical staining techniques. From this research emerges a picture of the highly geometric distribution of receptors in the dermal-epidermal papilla. Meissner corpuscles were found in the apexes of the papilla with a density between $47/mm^2$ and $60/mm^2$ (see Figure 13.1). It was estimated that Meissner corpuscles covered about 4% of the epidermal ridges whereas, in contrast, the Merkel ending complexes found in the fundus of dermal-epidermal papilla cover about 15% of the total overlying skin surface. Although a single axon can branch as many as 18 times this density is considerably greater than the $\sim 1.0/mm^2$ proposed by Yoshioka *et al.* (2000) based on results from Johansson & Vallbo (1979). Each Merkel ending complex is surrounded on either side by a sweat gland duct. Sweat ejected on the skin surfaces in moderate quantities acts as an adhesive which affects the coefficient of friction during grasping and tactile exploration (Smith & Scott, 1996; Smith *et al.*, 1997). The geometrical arrangement of Meissner corpuscles and Merkel ending complexes would appear to be of considerable functional importance and is further supported by the demonstration of Wang & Hayward (2007) that the dermal elasticity across the fingerprint is far greater than along the ridges where the skin is more rigid. Consequently, it seems likely that the Meissner corpuscles and the Merkel ending complexes are sensitive to stretching and compression strains due to shear forces applied to the overlying skin. Paré *et al.* (2001) also noted two types of Merkel ending complexes they called *clumps* and *chains*. The clumps were compact

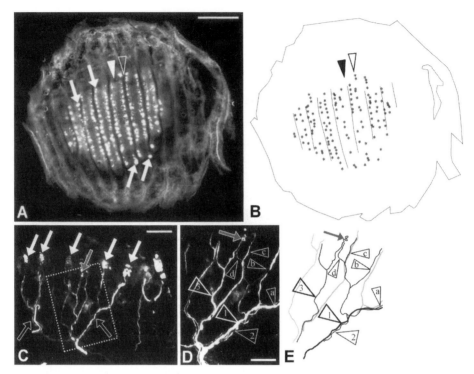

Figure 13.1. A. The Meissner cell innervation of the thumb cut parallel to the surface along the distal volar pads shown with immunofluorescence micrographs labeled with anti PGP. The solid and hollow arrow heads indicate epidermal and papillary ridges, respectively. B. A Neuro Lucida tracing of the contour and epidermal ridges (internal lines) of the section in A. The locations of the Meissner cells are mapped with asterisks.

C–E. Confocal laser imaging (C, D) of Meissner cells (solid arrows) and terminal nerve branches (hollow arrows) labeled with anti-NF in a 100 μm section cut perpendicular to the surface of the index finger. The area within the dotted rectangle is enlarged in D, showing individual axons and their branches within the nerves. In E, some of the axons (large arrowheads # 1–3) and branches (small arrowheads) were reconstructed from D. Axon # 1 innervates one Meissner cell. In this section axon # 2 gives rise to four branches (a–d) that each innervate a different MC. The Meissner cell indicated by the gray arrow in C–E receives innervation from more than one axon. Scale bars 500 μm in A and 50 μm in C and D. Figure reproduced from Paré *et al.*, 2002a.

endings located between two sweat gland ducts approximately 30 μm in width whereas the chains were more than 10 times as long extending over several hundred microns in a particular direction. This study failed to find any Ruffini-like endings and instead suggested that the chain-like terminations may account for the shear forces sensitivity of what had been described as the physiological behavior of the SA-II afferent (see also Chapter 11).

The micro-deformation of a fingerprint sliding over a smooth surface containing a smooth bump has been graphically characterized and mathematically modeled by Levesque & Hayward (2003) and illustrates beautifully the complex pattern of stick–slip stretching and compressing of the fingertip skin during tactile exploration. The qualitative differences detected among smooth, isothermic surfaces of different materials such as wood, glass, plastic or metal are almost certainly derived from the different stick–slip patterns generated on the fingertip. The sensations arising from these tangential forces are readily perceived and discriminated (Biggs & Srinivasan, 2002; Paré *et al.*, 2002a) and consequently available to provide feedback control for grasping and tactile exploration (Goodwin & Wheat, 2004).

The motor cortex and hand movements

The critical importance of the primary motor cortex in dexterous hand movements has been recognized for more than a century. The revelation of direct connections between the motor cortex and the motoneurons of the extrinsic and intrinsic hand muscles by Kuypers (Lawrence & Kuypers, 1968) and Phillips (1969) focused considerable attention on the role the motor cortex might play in the control of independent finger movements. The correlation between the discharge frequency of some motor cortical cells and pinch force and its rate of application has reinforced the notion that the motor cortex plays a particularly significant role in controlling the precision grip (Smith *et al.*, 1975). Strick & Preston (1982) found two distinct zones within the motor cortex of Cebus monkeys; a rostral zone dominated by deep tissue proprioceptive afferents and a caudal zone predominately receiving cutaneous afferents. Recently, Rathelot & Strick (2006) discovered that many of the cells of origin of the corticomotoneuronal projections lie deep within the rostral wall of the central sulcus which would suggest that the majority of corticomotoneurons receive cutaneous afferents from the skin of the fingers and palm. Recording the activity of single motor cortical neurons in monkeys trained to perform a semi-naturalistic grasp, lift and hold task, Picard & Smith (1992) found that the discharge of motor cortical cells was strongly influenced by the surface texture of the grasped object. In spite of the simple inverse relationship between friction and the grip forces used to lift and hold an object against gravity, the relationship between single neuron activity and object texture and/or friction is rather complex. That is, some motor cortex cells produce an increase in activity for low friction objects that one would expect since more grip force is required to lift and hold slippery objects. However, other neurons were more active when the fingers contacted objects with rough, high-friction surfaces. From these studies we know that neuronal discharge in the motor cortex is influenced by shear forces, slip and surface texture on the skin. It remains to be shown just how each of these is decoded to enable the dexterous handling under a variety of different conditions.

Other cortical motor areas and hand movements

Recent anatomical research, particularly from Strick's laboratory, has demonstrated that there are many other regions of the cerebral cortex that are directly involved in motor control that were either unidentified or considered "premotor." Dum & Strick (1991) showed that these areas have substantial direct projections to the spinal cord that terminate in the motor regions of the intermediate zone. Moreover, the ventral premotor area (PMv) and the dorsal premotor area (PMd) have individual digit representations, identified by movements evoked by micro-stimulation, and both regions are densely interconnected with M1. Surprisingly, the projections from the supplementary medial area (SMA) were stronger to the PMd and PMv than those to the primary motor cortex. In addition, our group found clear hand representations in all of these areas as well an additional separate area in the ventral cingulate sulcus (CMAv) on the medial wall beneath the SMA (Cadoret & Smith, 1997). Functionally, the neurons in the SMA and CMAv seemed very similar regarding the onset of activity preceding hand movements, the type and size of receptive fields and the response latencies to perturbations applied to objects held in the hand. The neurons in the dorsal and ventral premotor areas had responses similar to the SMA and CMAv during grasping and holding movements of the hand with one significant exception. The neurons in the PMv have clear large visual receptive fields that appeared to be sensitive to the direction of motion in the visual field (Boudreau *et al.*, 2001). Unfortunately the exact retinotopic size and location of these fields could not be ascertained and awaits further investigation.

Cerebellum and hand movements

Based on both clinical and experimental evidence the cerebellum has a long association with the control of voluntary movements and motor skill acquisition (see also Chapter 26). Lesions of the cerebellum of sufficient size produce striking deficits in movement coordination. The ataxic movements are associated with the anatomical organization of the cerebellum with oculomotor control related to the floccular nodular region, axial and proximal muscle represented in the midline vermis and the more distal limb muscles represented laterally in the paravermal area. The cerebellum is a phylogenetically old structure and it appears in the early evolution of chordates. Despite its somato-topographic organization, the cellular composition and connectivity of the cerebellar cortex and nuclei are remarkably homogeneous throughout. Moreover, other than proportionate changes in size and cell number, the cerebellum is surprisingly similar in all mammals. The entire output of the cerebellar cortex exerts an inhibitory action on the deep cerebellar and vestibular nuclei which for the most part exert an excitatory action on the long motor pathways to the spinal cord.

Given its uniform structure in most mammalian species we entertained the conjecture that the cerebellum exercises a time-varying control over joint and limb stiffness by regulating the degree of co-contraction in antagonist muscles. Since object manipulation and pinching involve the reciprocal inhibition and co-contraction of antagonist muscles the cerebellum would appear to be particularly important for the dexterous motor skills of the hand. During co-contraction of antagonist hand muscles during pinching Purkinje cells were generally inhibited whereas with reciprocally activated wrist and finger muscles the Purkinje cells were reciprocally activated (Frysinger *et al.*, 1984). In contrast cerebellar nuclear cells were

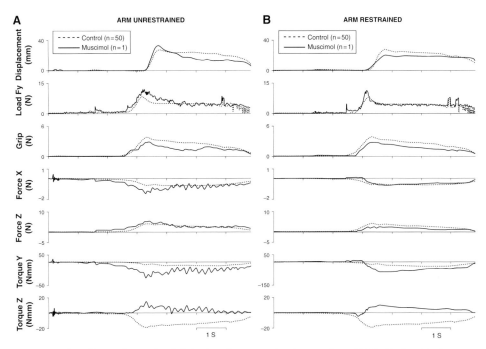

Figure 13.2. Effect of muscimol injection into anterior interpositus (thick line) with the arm unsupported at the elbow and wrist in A and supported in B. The line represents a 50-trial average without muscimol. The X, Y and Z planes correspond to the horizontal, vertical and saggital planes respectively.

generally activated during the co-contraction of antagonist hand muscles associated with pinching (Wetts *et al.*, 1985).

More recently we have studied the effects of transient inactivation by direct injection of muscimol into the cerebellar nuclei on grasping and object manipulation behavior (Monzée *et al.*, 2004). Only inactivation of the interpositus nucleus produced clear ataxia of the upper limb. Surprisingly, when the limb was supported at the elbow and wrist, object manipulation was well coordinated (Figure 13.2B). In contrast, when the arm was left unsupported a classic dysmetria and dynamic tremor appeared, particularly about the pronation and supination axis of the forearm (Figure 13.2A). A review of the clinical literature revealed several studies reporting substantial deficits in grasping and manual manipulation in patents with cerebellar lesions (Mai *et al.*, 1988; Müller & Dichgans, 1994; Bastian & Thach, 1995; see also Chapter 26). However, since it appears that in these studies the forearm was not fully constrained, therefore it would be of some interest to re-examine finger dexterity in patients with cerebellar lesions but with the forearm fully constrained. Exactly why the flexors and extensors of the wrist and fingers are unaffected by cerebellar lesions when the forearm is fully constrained remains somewhat of a mystery, but it may reflect the predominant control over the muscles of the hand by the cerebral cortex.

Somatosensory cortex and hand movements

The glabrous skin representation of the digits in the somatosensory cortex is characterized by its disproportionately large surface area, its predominantly cutaneous input and generally discrete, small receptive fields. Although this is particularly true of area 3b, it is also true for the thumb–index finger region of areas 1 and 2. It is not clear if area 5 has any glabrous skin representation of the hand at all. Using a continuously moving stimulus, a recent study of hand area 3b neurons by DiCarlo *et al.* (1998) showed that the majority of neurons have an "inhibitory" or disfacilitatory receptive field in addition to the usual excitatory one, and these fields can be located on one, two, three or four sides of the excitatory field. These fields would likely contribute to the directional sensitivity to tangential forces. DiCarlo *et al.* (1998) suggested that area 3b is an intermediate step in the ultimate elaboration of texture and form perception. We agree with this notion and further suggest that efferents of individual area 3b neurons converge on single cells in area 2 to provide information about tangential force direction, which is then projected to area 4, motor cortex for the purpose of controlling prehensile forces.

Ablation or inactivation of the somatosensory cortex produces profound deficits in all tasks requiring cutaneous feedback such as precise stimulus localization, texture and shape discrimination, and the dexterous object manipulation of small objects (Randolph & Semmes, 1974; Hikosaka *et al.*, 1985; Iwamura & Tanaka, 1991; Brochier *et al.*, 1999). Inactivation of areas 3b, 1 and 2 also abolished the monkey's ability to sustain grasping beyond a brief pinch (Brochier *et al.*, 1999), which is consistent with the common clinical observation that patients suffering from the loss of touch sensation on the hand have great difficulty in maintaining grip force and must pay close attention to avoid dropping grasped objects (see also Chapter 19). The second major effect of somatosensory cortical inactivation is a substantial increase in peak grip force because of inadequate finger pressure feedback plus an inability to sense the shape of an object and align the fingers properly for grasping, since the direction of force vectors on the skin are no longer adequately perceived (see also Chapters 11 and 19). Despite the exceptionally high peak grip force, sustained grasping is seriously debilitated. After cortical inactivation with muscimol the monkey was unable to hold a 25 g load for 2 s because the grip force decayed too rapidly.

Role of somatosensory cortex in grasping

Our initial somatosensory recording studies focused on neuronal activity in monkeys trained to grasp an instrumented object that measured the grip and lifting forces. Cells recorded in the somatosensory cortex during the manipulation of this object, which varied in weight, texture and friction (Salimi *et al.*, 1999a, 1999b, 1999c) were classified as rapidly or slowly adapting by their apparent similarity to peripheral afferents. The discharge of neurons with cutaneous receptive fields on the thumb or index finger was recorded over an area extending from 3b rostrally to area 7 in the posterior bank of the intraparietal sulcus caudally. When the object surface and weight were kept constant an effect of smooth-surface

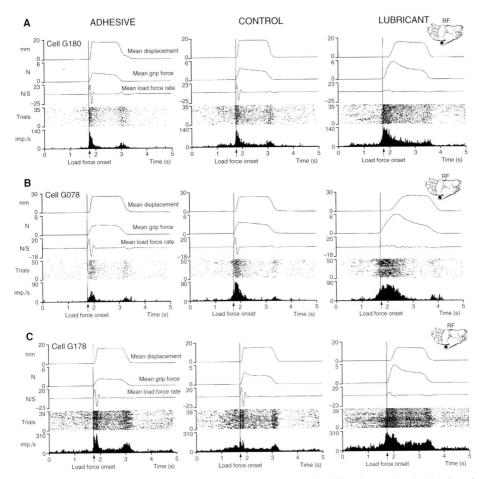

Figure 13.3. The upper panel (A) shows a cell with increased activity lifting the low-friction lubricated surface and probably related to object slip. The middle panel (B) shows a cell unrelated to either shear or slip and most likely related to grip force pressure on the skin. The lower panel (C) shows a cell related to tangential shear in the vertical plane.

friction was demonstrated by applying lubricants (e.g. petroleum jelly or talc) or adhesives (e.g. rosin or sucrose) to the grasping surface which facilitated slip and shear respectively on the skin. In general, the coatings had a very powerful effect on peak discharge frequency. One group of cells had maximal activity with lubricant coatings and decreased activity with the adhesive coating suggesting that these cells were particularly sensitive to slip (Figure 13.3A). Another category of neurons was particularly activated by quick tangential forces applied to skin coated with an adhesive and were labeled shear-force sensitive (Figure 13.3B). However, most neurons were activated by both shear and slip as shown by enhanced activity with both the lubricant and the adhesive compared with uncoated smooth metal (Figure 13.3C).

Tactile exploration

The juxtaposition of the sense of touch in grasping and tactile exploration was the major motivation for examining tactile exploratory procedures in greater detail. The subjective sensation of roughness had historically been associated with spatial attributes of textured surfaces such as groove width or inter-asperity spacing. However, studying the deployment of finger forces as subjects rated the stickiness of smooth surfaces, Smith & Scott (1996) arrived at the conclusion that subjects estimated surface friction based on the ratio of the normal contact force with the tangential sliding force. In a subsequent study, subjects were asked to rate the roughness of a variety of surfaces with embossed asperities at different spatial periods (Smith *et al.*, 2002a). The results suggested that the subjective sensation of roughness was best correlated with the fluctuations in tangential force as compared with the contact force, the coefficient of friction or the spatial geometry as measured by the inter-asperity spacing. The dissociation between roughness and spatial geometry was achieved by comparing roughness estimates of the dry surface with roughness estimates of the same surface with liquid soap added to the finger. Figure 13.4 shows a similar reduction in subjective roughness and for the root mean square of the tangential force variations as a result of the lubrication. That is, the same spatial geometry feels less rough if the tangential forces are reduced as a result of lubrication.

In an additional study, human subjects (Smith *et al.*, 2002b) performed a tactile search with the finger tips for a small asperity on a 7.0 cm flat circular surface. The search was accomplished without visual feedback, and the search surface was mounted on a force and

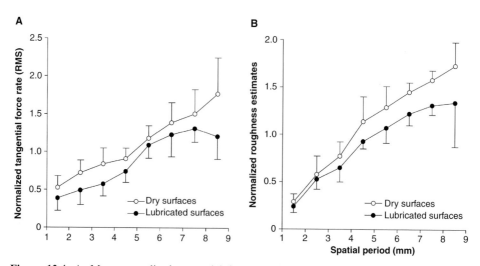

Figure 13.4. A. Mean normalized tangential force rate RMS (dFtan/dt) averaged across the six subjects (± SD) for each of the lubricated and dry surface textures. B. Mean normalized roughness estimates averaged across the six subjects (± SD) for each of the lubricated and dry surface textures. From Smith *et al.* (2002a).

torque sensory that tracked the finger movement from the first touch to final contact with the tactile target. By using a variety of rough and smooth background surfaces we were able to study the deployment of the contact force and tangential force used in the exploratory procedure. The results indicated that both humans and monkeys control the contact and sliding forces, presumably to optimize target perception.

In conclusion, the flow of tactile information from the skin of the fingers to cerebral cortex is of critical importance for both prehensile, manipulatory behaviors and for tactile exploration. However, slip of a grasped object during precision handling is an error signal whereas slip during tactile explorations is a necessary component of perception. How the somatosensory cortex and motor cortex mutually influence each other in the collaborative control of the hand is an exciting topic for future research.

Acknowledgments

The technical assistance of L. Lessard, J. Jodoin, C. Gauthier and C. Valiquette is gratefully acknowledged. This research was supported by the Canadian Institutes for Health Research (CIHR), le Fonds de Recherche en Santé du Québec (FRSQ) and the National Institute for Neurological Disorders and Stroke (NINDS).

References

Augurelle, A.-S., Smith, A. M., Lejeune, T. & Thonnard, J.-L. (2003). Importance of cutaneous feedback in maintaining the safety margin during the manipulation of hand-held objects. *J Neurophysiol*, **89**, 665–671.

Bastian, A. J. & Thach, W. T. (1995). Cerebellar outflow lesions: a comparison of movement deficits resulting from lesions at the levels of the cerebellum and thalamus. *Ann Neurol*, **38**, 881–892.

Biggs, J. & Srinivasan, M. A. (2002). Tangential versus normal displacements of skin: Relative effectiveness for producing tactile sensations. *10th International Symposium on Haptic Interfaces for Virtual Environment and Teleoperator Systems (IEEE Computer Society)*, 121–128.

Birznieks, I., Jenmalm, P., Goodwin, A. W. & Johansson, R. S. (2001). Encoding of direction of fingertip forces by human tactile afferents. *J Neurosci*, **21**, 8222–8237.

Boudreau, M.-J., Brochier, T., Paré, M. & Smith, A. M. (2001). Activity in ventral and dorsal premotor cortex in response to predictable force-pulse perturbations in a precision grip task. *J Neurophysiol*, **86**, 1067–1078.

Bowden, F. P. & Tabor, D. (1982). *Friction. An Introduction to Tribology*. Malabar, FL: Robert E. Krieger Publishing Company.

Brochier, T., Boudreau, M.-J., Paré, M. & Smith, A. M. (1999). The effects of muscimol inactivation of small regions of motor and somatosensory cortex on independent finger movements and force control in the precision grip. *Exp Brain Res*, **128**, 31–40.

Cadoret, G. & Smith, A. M. (1996). Friction, not texture dictates grip forces during object manipulation. *J Neurophysiol*, **75**, 1963–1969.

Cadoret, G. & Smith, A. M. (1997). Comparison of the neuronal activity in the SMA and the ventral cingulate cortex during prehension in the monkey. *J Neurophysiol*, **77**, 153–166.

DiCarlo, J. J., Johnson, K. O. & Hsiao, S. S. (1998). Structure of receptive fields in area 3b of primary somatosensory cortex in the alert monkey. *J Neurosci*, **18**, 2626–2645.

Dum, R. P. & Strick, P. L. (1991). Premotor areas: nodal points for parallel efferent systems involved in the central control of movement. In D. R. Humphrey & H.-J. Freund (Eds.), *Motor Control: Concepts and Issues* (pp. 383–397). Chichester, UK: John Wiley & Sons.

Frysinger, R. C., Bourbonnais, D., Kalaska, J. F. & Smith, A. M. (1984). Cerebellar cortical activity during antagonist co-contraction and reciprocal inhibition of forearm muscles. *J Neurophysiol*, **51**, 32–49.

Goodwin, A. W. & Wheat, H. E. (2004). Sensory signals in neural populations underlying tactile perception and manipulation. *Annu Rev Neurosci*, **27**, 53–77.

Hikosaka, O., Tanaka, M., Sakamoto, M. & Iwamura, Y. (1985). Deficits in manipulative behaviors induced by local injections of muscimol in the first somatosensory cortex of the conscious monkey. *Brain Res*, **325**, 375–380.

Iwamura, Y. & Tanaka, M. (1991). Organization of the first somatosensory cortex for manipulation of objects: an analysis of behavioral changes induced by muscimol injection into identified cortical loci of awake monkeys. In O. Franzen & J. Westman (Eds.), *Information Processing in the Somatosensory System. Wenner-Gren International Symposium Series* (pp. 371–380). New York, NY: Stockton Press.

Johansson, R. S. & Vallbo, A. B. (1979). Detection of tactile stimuli. Thresholds of afferent units related to psychophysical thresholds in the human hand. *J Physiol (Lond)*, **297**, 405–422.

Johansson, R. S. & Westling, G. (1984). Influences of cutaneous sensory input on the motor coordination during precision manipulation. In C. von Euler, O. Franzen, U. Lindblom & D. Otteson (Eds.), *Somatosensory Mechanisms* (pp. 249–260). London: Macmillan Press.

Klatzky, R. L. & Lederman, S. J. (1993). Toward a computational model of constraint-driven exploration and haptic object identification. *Perception*, **22**, 597–621.

Lawrence, D. G. & Kuypers, H. G. J. M. (1968). The functional organization of the motor system in the monkey. I. The effects of bilateral pyramidal lesions. *Brain*, **91**, 1–14.

Lederman, S. J. & Klatzky, R. L. (1987). Hand movements: a window into haptic object recognition. *Cogn Psychol*, **19**, 342–368.

Lederman, S. J. & Klatzky, R. L. (1990). Haptic classification of common objects; knowledge-driven exploration. *Cogn Psychol*, **22**, 421–459.

Lederman, S. J. & Klatzky, R. L. (1997). Relative availability of surface and object properties during early haptic processing. *J Exp Psychol*, **23**, 1680–1707.

Levesque, V. & Hayward, V. (2003). Experimental evidence of lateral skin strain during tactile exploration. *Proc Eurohaptics*, **2003**, 14–24.

Mai, N., Bolsinger, P., Avarello, M., Diener, H. C. & Dichgans, J. (1988). Control of isometric finger force in patients with cerebellar disease. *Brain*, **111**, 973–998.

Maier, M. A. & Hepp-Reymond, M.-C. (1995a). EMG activation patterns during force production in precision grip. I. Contribution of 15 finger muscles to isometric force. *Exp Brain Res*, **103**, 108–122.

Maier, M. A. & Hepp-Reymond, M.-C. (1995b). EMG activation patterns during force production in precision grip. II. Muscular synergies in the spatial and temporal domain. *Exp Brain Res*, **103**, 123–136.

Monzée, J., Lamarre, Y. & Smith, A. M. (2003). The effects of digital anesthesia on force control in a precision grip. *J Neurophysiol*, **89**, 672–683.

Monzée, J., Drew, T. & Smith, A. M. (2004). The effects of muscimol inactivation of the deep cerebellar nuclei on precision grip. *J Neurophysiol*, **91**, 1240–1249.

Müller, F. & Dichgans, J. (1994). Dyscoordination of pinch and lift forces during grasp in patients with cerebellar lesions. *Exp Brain Res*, **101**, 485–492.

Napier, J. R. (1962). The evolution of the hand. *Sci Amer*, **207**, 56–62.

Napier, J. R. (1976). *The Human Hand*. Burlington, NC: Carolina Biology Readers.

Paré, M., Elde, R., Mazurkiewicz, J. E., Smith, A. M. & Rice, F. L. (2001). The Meissner corpuscle revised: a multi-afferented mechanoreceptor with nociceptor immunochemical properties. *J Neurosci*, **21**, 7236–7246.

Paré, M., Smith, A. M. & Rice, F. L. (2002a). Distribution and terminal arborizations of cutaneous mechanoreceptors in the glabrous finger pads of the monkey. *J Comp Neurol*, **445**, 347–359.

Paré, M., Carnahan, H. & Smith, A. M. (2002b). Magnitude estimation of tangential force applied to the fingerpad. *Exp Brain Res*, **142**, 342–348.

Phillips, C. G. (1969). The motor apparatus of the baboon's hand. *Proc Roy Soc B*, **173**, 141–174.

Picard, N. & Smith, A. M. (1992). Primary motor cortical activity related to the weight and texture of grasped objects in the monkey. *J Neurophysiol*, **68**, 1867–1881.

Randolph, M. & Semmes, J. (1974). Behavioral consequences of selective subtotal ablation in the postcentral gyrus of *Macaca mulatta*. *Brain Res*, **70**, 55–70.

Rathelot, J. A. & Strick, P. L. (2006). Muscle representation in the macaque motor cortex: an anatomical perspective. *Proc Natl Acad Sci USA*, **103**, 8257–8262.

Salimi, I., Brochier, T. & Smith, A. M. (1999a). Neuronal activity in somatosensory cortex of monkeys using a precision grip. I. Receptive fields, and discharge patterns. *J Neurophysiol*, **81**, 825–834.

Salimi, I., Brochier, T. & Smith, A. M. (1999b). Neuronal activity in somatosensory cortex of monkeys using a precision grip. III. Responses to altered friction and perturbations. *J Neurophysiol*, **81**, 845–857.

Salimi, I., Brochier, T. & Smith, A. M. (1999c). Neuronal activity in somatosensory cortex of monkeys using a precision grip. II.Responses to object textures and weights. *J Neurophysiol*, **81**, 835–844.

Smith, A. M. (1981). The coactivation of antagonist muscles. *Can J Physiol Pharmacol*, **59**, 733–747.

Smith, A. M. & Scott, S. H. (1996). The subjective scaling of smooth surface friction. *J Neurophysiol*, **75**, 1957–1962.

Smith, A. M., Hepp-Reymond, M.-C. & Wyss, U. R. (1975). Relation of activity in precentral cortical neurons to force and rate of force change during isometric contractions of finger muscles. *Exp Brain Res*, **23**, 315–332.

Smith, A. M., Cadoret, G. & St-Amour, D. (1997). Scopolamine increases prehensile force during object manipulation by reducing palmar sweating and decreasing skin friction. *Exp Brain Res*, **114**, 578–583.

Smith, A. M., Chapman, C. E., Deslandes, M., Langlais, J.-S. & Thibodeau, M.-P. (2002a). Role of friction and tangential forces in the subjective scaling of tactile roughness. *Exp Brain Res*, **144**, 211–223.

Smith, A. M., Gosselin, G. & Houde, B. (2002b). Deployment of fingertip forces in tactile exploration. *Exp Brain Res*, **147**, 209–218.

Strick, P. L. & Preston, J. B. (1982). Two representations of the hand in area 4 of a primate. II. Somatosensory input organization. *J Neurophysiol*, **48**, 150–159.

Wang, Q. & Hayward, V. (2007). In vivo biomechanics of the fingerpad skin under local tangential traction. *J Biomech*, **40**, 851–860.

Westling, G. & Johansson, R. S. (1984). Factors influencing the force control during precision grip. *Exp Brain Res*, **53**, 277–284.

Wetts, R., Kalaska, J. F. & Smith, A. M. (1985). Cerebellar nuclear cell activity during antagonist cocontraction and reciprocal inhibition of forearm muscles. *J Neurophysiol*, **54**, 231–244.

Yoshioka, T., Gibb, B., Dorsch, A. K., Hsiao, S. S. & Johnson, K. O. (2001). Neural coding mechanisms underlying perceived roughness of finely textured surfaces. *J Neurosci*, **21**, 6905–6916.

14

Points for precision grip

ALAN M. WING AND SUSAN J. LEDERMAN

Summary

We describe constraints on grip points in reaching and lifting objects. Most objects afford a choice of points providing stable grip with thumb and index finger. We overview experiments showing how micro (surface texture determining friction) and macro (local shape for determining direction of the surface normal relative to interdigit force, and global shape for determining center of mass) geometric features affect precision grip. We summarize the roles of visual and haptic cues in selection of grip points and describe how planning takes account not only of the object but also the intended action in directing grasp to these points. We support our arguments with evidence taken from studies of normal and disordered motor behavior.

Introduction

In characterizing the sensorimotor control of grasping, other chapters in this book have emphasized coordination between the hand, which shapes to and grasps the object, and the arm, which moves the hand to the object and lifts both hand and object through space (Jones & Lederman, 2006; see also Chapters 1, 11–13). Generally, the object concerned has been provided with vertical and parallel sides, the grasp has been a precision grip, and the goal of the action has been to maintain a stable grip so that the object neither translates nor rotates relative to the hand.

In precision grip, the thumb contacts one side of the object and provides a force normal (i.e. at right angles) to the surface that opposes a normal force developed by the index finger on the other side of the object. If the forces are precisely opposed, there is no resultant force or torque to displace or twist the object, a first sense in which the grip is stable. If the normal forces are sufficient given the nature of the digit–object contact surfaces, the resulting friction allows the development of vertical lift forces tangential to the grip surfaces. This means that the object can be held securely without risk of the digits slipping under tangential load forces due, for example, to gravity (i.e. weight) in lifting the object (see Figure 14.1). This is a second sense in which the grip is stable.

Even within the limited context of using precision grip to lift a parallel-sided object such as a cuboid, there are many different places on its surfaces that afford stable grip. The cuboid

Sensorimotor Control of Grasping: Physiology and Pathophysiology, ed. Dennis A. Nowak and Joachim Hermsdörfer. Published by Cambridge University Press. © Cambridge University Press 2009.

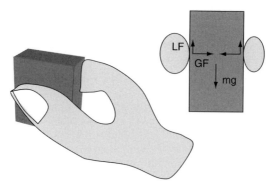

Figure 14.1. In picking up the cuboid, the frictional effects of the normal force at each digit determine the tangential lift force that can be developed to overcome object weight in lifting.

in Figure 14.1 could be picked up with thumb and finger in contact with any of a large set of pairs of contact points, as long as they allow the digits to exert opposed forces. These pairs of points could range, vertically, from top to bottom or, horizontally, from one edge to the other. If the hand is rotated 180 degrees by forearm pronation, the placement of thumb and index finger could be reversed. Or if the hand orientation is turned through 90 degrees by wrist flexion, the digits could grasp the alternate set of opposed surfaces.

What determines which surfaces, and which points on the surfaces, are selected in grasping? In this chapter we consider constraints that operate to determine people's choice of grip points, not only for regular solids such as cubes, but also for irregular solids. These constraints include the surface friction at the grip points, which determines the tendency to slip; local contours around the grip points, which can change the normal force and affect the tendency for slip; and the presence of torque around the line between grip points, which tends to cause the object to rotate. In the next section we review studies of gripping in lifting and moving objects. The following sections treat visual and haptic cues to grip points and how constraints on grip points can affect the process of selecting and forming a grasp when reaching to take hold of the object (see also Chapter 11).

On a terminological point, the terms grasp and grip are often used synonymously when applied to hand function. However, dictionary definitions usually give the sense that grip is limited to a static posture involving contact between hand and object, while grasp often involves movement towards grip. In this chapter we adopt this contrast, with grip treated as a target for grasp. Dictionary definitions also indicate that grip, but not grasp, is used to refer to designated areas for contact on an object. This use of the term grip emphasizes the nature of the hand–object interface and is also a focus of this chapter.

Gripping

Here we consider three topics relating to precision grip. Our first topic concerns the way people set normal force in anticipation of the friction between object and digits at the grip surfaces. Local object geometry at either digit determines normal and tangential forces in gripping and this is our second topic. Our third topic relates to people's anticipation of torques around the thumb–finger grip axis. In each of these cases, we propose that a corollary of anticipatory adjustment of grip force is that given free choice, people choose grip points that they anticipate will improve grip stability for a given level of force between the digits. Thus, the chosen grip points may yield improved friction, force between the digits that is better aligned with the surface normals or reduced torque about the grip axis.

Basic grip and friction

In the depiction of lifting in Figure 14.1, we referred to friction at the digit–object contact surfaces which determines the size of the tangential lift force that can be developed to overcome object weight. According to the Coulomb model of friction, the limiting tangential force is given by the product of the normal force and the coefficient of friction, which depends on the properties of the surfaces in contact (Bhavikatti & Rajashekharappa, 1994). Seminal studies of precision grip by Johansson and colleagues (Johansson & Westling, 1984; Westling & Johansson, 1984) (see also Chapters 1 and 11) showed that with experience gained from repeated lifting of a test object, normal force is modulated according to the coefficient of friction. When the friction coefficient is low with a smooth surface (e.g. suede), the normal force used is greater, whereas when friction is high with a rough surface (e.g. sandpaper), the normal force used is less. These differences in normal force are reflected in the rate of rise of normal force. Such changes in force rate begin with initial object contact, which indicates that the adjustments of normal force are made on a predictive basis. The adjustments in normal force when gripping different surface textures are not only observed in lifting, but also in anticipation of the fluctuations in inertial forces associated with moving objects (Flanagan & Wing, 1995).

The adjustments in normal force observed when using a precision grip with different friction surfaces mean that the safety margin against slip (the amount by which normal force exceeds the minimum required to prevent slip) remains rather constant. This suggests that people try to maintain normal force at relatively low levels in lifting and moving objects, and it seems reasonable to speculate that this might be on grounds of muscular efficiency. As a corollary, we have observed in our laboratory that, when people are asked to lift a bar using one of two alternate pairs of grip points, which have smooth and rough surfaces respectively, they tend to choose the rougher one that affords improved stability or lower grip force (Figure 14.2). We thought that sense of effort, perhaps linked to a perception of reduced object weight (Flanagan *et al.*, 1995), might underlie this behavior. However, in our study we observed across subjects that the choice was evident on the very first lift. This suggested to us a role of generalized knowledge, acquired prior to the experiment in the laboratory, and applied on the basis of visual input from the current trial. Nonetheless, the important point is that friction at the digit–object contact surfaces is a factor affecting choice of grasp points. This point may be seen as underpinning the established design principle that rough textured

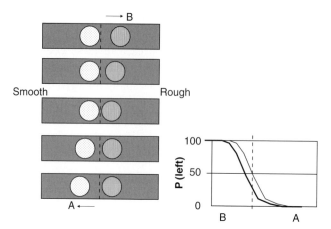

Figure 14.2. Schematic of data from study showing when participants were asked to lift and place the bar shown with smooth and rough grips to left and right of center, they tended to avoid the left (smoother) surface, even on the first lift. Thus, the thick line in the graph on the right shows that, when the two pairs of grip surfaces were centered (dashed line), the probability of using the left (smooth) grip was less than the 50% expected if their choice had been random. When the rough surface was further away from the center (B), which increased exposure to torque, the preference for the left (smooth) surface, which results in less torque, increased. The thin line in the graph shows the function obtained when the two surfaces were equally smooth.

handles are preferred to smooth handles and are particularly helpful for those, such as the elderly, with limited grip.

Local shape constraints on grip points

As noted in the introduction to this chapter, when an object with parallel sides is held in a precision grip, the forces developed at either digit are equal in magnitude and act along opposed surface normals to maintain the object in stable equilibrium. If the arm applies upward lift, the normal forces, in combination with friction, allow tangential forces to be developed at the digits to match, and so lift, the weight of the object. If the object sides are not parallel, the situation changes. For example, consider picking up a pyramid-shaped object. Common experience suggests that this is much harder than picking up a cube, as the fingers will tend to slide up the object when squeezing to try and get sufficient "purchase" to allow lifting. An experimental analysis of this situation was provided by Jenmalm & Johansson (1997) (Figure 14.3). They asked subjects to pick up an instrumented object in which two flat grip surfaces tapered in (like a pyramid) or out (like an upside-down pyramid) and there was also a parallel-sided condition.

The results showed that participants adjusted the horizontal force between the digits, up for positive taper (pyramidal object) and down for negative taper, so that the safety margin (amount by which the normal force exceeds the minimum required to prevent slip) remained

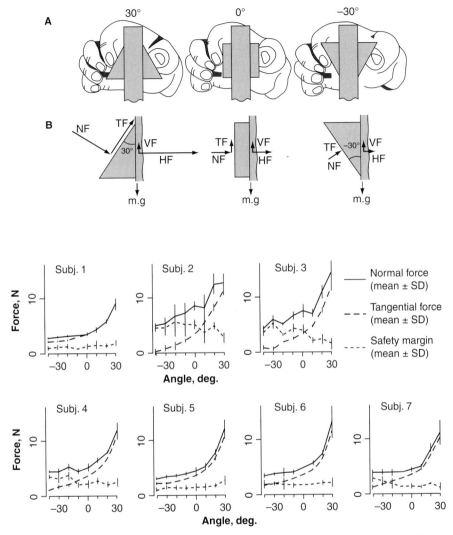

Figure 14.3. Participants lifted a manipulandum (A) in which the grip surfaces tapered with angles varying in degrees from –30 "in" (left), 0 parallel (center) to +30 "out" (right). Subjects scaled the horizontal force (B) so that the normal force was scaled to the tangential force experienced in lifting (C).

relatively constant. The adaptation of horizontal force for the different surfaces was evident from very early after contact, with higher force rates used for more positive taper angles. As long as vision was available, this was the case whether the taper angle on the current trial was novel or corresponded to the angle used in the preceding block of trials. However, when blindfolded, a change to a new taper angle resulted in the force rate initially being appropriate to the old taper angle, and only adjusting to the new angle after some delay

(70–90 ms), resulting in a longer lift phase on such trials. Such a delay is sufficient for tactile feedback to reset grip forces through supraspinal, possibly cortical, circuits. However, on the second trial at the new taper angle, the force rate was appropriate from the very first contact. This suggests a role of experience in setting force rate, based on visual information (from the current trial) or haptic information (from the previous trial). The sensitivity of grip forces to local geometry, with surface angle effectively acting as a limit on friction, might have been due to slips experienced in the paradigm, or from more general experience gained across the domain of all objects. In either case it suggests that people will choose grip surfaces that give more stable grip.

Weight distribution as a constraint on precision grip

In the studies of normal and tangential forces in lifting and moving described above, the load was designed so that the grip axis (the line joining the digit contact surfaces) was aligned with the center of mass (CoM) or, if not directly aligned, the CoM lay below the grasp axis so that the two were in line with the direction of motion. If, instead, the grip axis lies some distance from the CoM, the weight of the object in lifting, or the inertia in moving, acting though the CoM creates a torque load around the grip axis. This torque tends to cause the object to rotate around the grip axis unless the torque is matched by torsional friction generated by the normal force, which adds extra complexity to keeping the object stable in the hand. We studied lifting and moving an object in which the CoM could be displaced from the grip axis by varying distances (Wing & Lederman, 1998). In both lifting and moving, we found evidence of anticipatory adjustments in normal force proportional to the torque load (see Figure 14.4A).

Given that people are sensitive to torque load when the grip points are fixed, we hypothesized in another study (Lederman & Wing, 2003) that, when people are free to choose where they grasp an object, their selection may be influenced by a strategy of minimizing the torque, that is, by selecting a grip axis that includes the CoM. We asked participants to pick up a range of planar solids placed on a glass table top. With a video camera under the table we were able to observe their grasp points. Figure 14.4B shows how, in the majority of cases, the CoM indeed fell very close to the grip axes in support of our hypothesis. In our study we controlled weight, thus all the objects were equally heavy. Interestingly, Eastough & Edwards (2007) showed, in grasping cylinders standing on end, the heavier ones are grasped more on the diameter and this was associated with larger distance between the thumb and index finger (grasp aperture) in reaching. It seems reasonable to suppose that, in their study, participants were attempting to reduce torque in lifting the cylinders. With heavier objects, because torque is greater for a given grip axis, participants tended to grasp across the diameter of the cylinder, which included CoM and, in this way, sought to minimize torque.

Cues to grip point selection

Vision was available in the Lederman & Wing (2003) study of picking up planar objects. The importance of visual analysis in this study was demonstrated by the inclusion

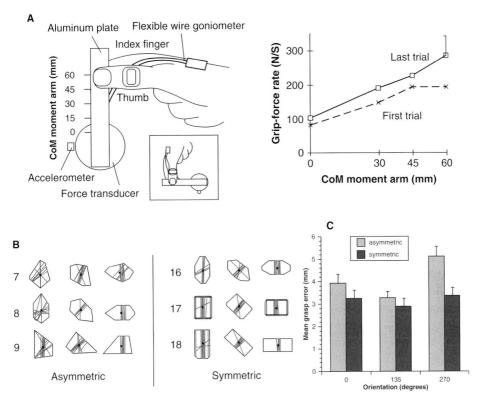

Figure 14.4. A. Apparatus for assessing torque effects on grip force (left). Peak grip force rate is scaled to torque indicating anticipation of torque (right) (Wing & Lederman, 1998). B. Grip axes superimposed on a subset of the object shapes fall close to the center of mass (CoM) shown with the black dot. C. The average distance from the grip axis to CoM was less for symmetric than asymmetric objects. (Lederman & Wing 2003).

of non-symmetric objects. In contrast to symmetric objects, where the axis or axes of symmetry can be used to estimate CoM, no such cues were available for non-symmetric objects. With these objects, the problems of perceptual analysis resulted in greater distance between grip axis and CoM. Interestingly, the characteristic of placing the grip axis so that it included CoM was absent in a neurological patient DM with impaired vision for action (Goodale *et al.*, 1994). However, her vision for perceptual judgement was apparently unaffected, as shown by her ability to discriminate the shapes she was being asked to grasp. The authors interpreted this observation as support for the dissociation of vision for perception and action (see also Goodale *et al.*, 1996). In contrast, a study using similar planar shapes with stroke patients, who exhibited visual neglect, failed to find evidence of dissociation between vision for perceptual judgment and vision for action. A right-directed bias was present, both in the patients' perceptual judgments about shape and in their selection of the grip axis relative to CoM (Marotta *et al.*, 2003).

It seems reasonable to suppose that the choice of grip points might be influenced, not only by vision, but also by knowledge of object properties derived from haptic cues i.e. information derived from tactile and proprioceptive inputs (Loomis & Lederman, 1986). Weight is an obvious property relevant to grasping, in that it determines the required lift and grip forces. Attributes particularly relevant to the choice of grip points are surface texture and shape, which may be determined by exploratory procedures such as lateral motion and contour following (Lederman & Klatzky, 1987).

A further important property in determining the grip axis (as noted above) is CoM location. In a recent study in our lab (Endo, Bracewell and Wing, in preparation) we asked participants to pick up a cuboid inside which a weight was placed (and so was not visible) either in the middle or at one end. During the trial sequence the weight was moved without informing the participants. Over trials this resulted in a reliable shift in digit placement towards the position of the weight. Interestingly, the information about favored grip points was found to be transferable (e.g. if the object was rotated or if the other hand was used) even though digit function in generating asymmetric grip and lift forces when grip points are constrained does not transfer in this way (Salimi *et al.*, 2000). One possibility is that the memory system supporting grip point position is based on a geometric, possibly visual, representation whereas digit function in grip is based on a more labile sensorimotor memory system.

Approach to gripping: Reach and grasp

So far we have considered factors that constrain grip points when lifting and moving objects, and cues that provide information on those grip points. We now turn to examine what is known about the selection of the particular grip points used from the set of all possible grip points. When presented with an object for grasping, what is known about the stages in planning a secure grip appropriate to the object and the envisaged action using the object?

Transport and grasp as two independent processes

Seminal work on reaching and grasping objects in the environment was carried out by Jeannerod (1981) (see also Chapters 2 and 10). He proposed two separate aspects of control that are mediated by separate visual streams, the first related to transport of the hand towards the target, the second concerned shaping the hand to the object. Initially these aspects were attributed to independent processing channels because transport was unaffected by unpredictable change in object size which produced a change in grasp aperture. Subsequently, with recognition of a degree of temporal and spatial interdependence, coupling between channels was included in theoretical accounts (e.g. Hoff & Arbib, 1993). Although focusing attention on the coordination of hand and arm movements, Jeannerod's approach did not address mapping of the digits onto grip points. In contrast, Smeets & Brenner (1999) proposed that the coordination of hand aperture with transport might be understood as the result of planning a goal trajectory for each digit to target grip points, which brings the work on grasping closer to that on grip point selection.

Kinematic constraints on grip posture

When applying the digits to grip points selected on considerations outlined earlier, in anatomical terms there is freedom in choosing different postures for the hand and arm. For instance, consider picking up a cylinder placed on its end with target grip points placed on a diameter half-way up. It might seem natural to use the hand with the forearm in a neutral position midway between pronation and supination. However, it is quite possible to pronate the forearm from this posture (or even to supinate it) and, in this way, to exchange the grip points mapped by the thumb and index finger.

The first study to examine factors that determine the orientation of the hand in such situations was that of Rosenbaum *et al.* (1990). Recognizing that elements of the task subsequent to grasping the object might play a role, they introduced directionality in the target object by differentiating the two ends of a cylinder by the use of black and white markings. They contrasted the effect of asking participants to lift the object from its horizontal starting position and place it vertically at another position, with either the black or white end uppermost. They found that the object's orientation in this final target position influenced whether the hand was pronated or supinated in picking up the object from the start position. On the basis of a further study which obtained comfort ratings for various hand orientations, Rosenbaum *et al.* (1990) argued that participants' choice of hand posture in picking the object was determined by an "end-state comfort" strategy. In the present context, we would argue that, if reaching to grasp involves planning for constraints due to movement goals, these might well be expected to interact with constraints on grip points reflecting constraints arising in the object. However, at present this remains a speculative point since, to date, end-state comfort studies have involved whole-hand power grip and this tends to reduce issues of grip point selection.

What might be the neuroanatomical bases underlying such grasping behavior? Hermsdörfer *et al.* (1999) compared stroke patients with left vs. right brain damage using their ipsilesional hand (since the other hand was in many cases hemiparetic) to reach and take hold of a cylindrical object presented in various orientations. They found that right-, but not left-, brain-damaged participants were slow and less accurate in their reaching. The authors suggested this may have reflected impaired visuospatial processing, normally executed by the right hemisphere. They then contrasted this task with one which required, as before, taking hold of the cylinder but now placing it with a specified end in a target hole. In this task, right brain damaged participants exhibited slowed but otherwise normal movement. However, they failed to show the end-state comfort effect in taking hold of the cylinder, which resulted in awkward postures in the targeting phase of the action. In contrast left brain damaged did show the end-state comfort effect in grasping the cylinder but used abnormal rotational movements in targeting and also in the initial reaching. The authors suggested the contrast reflects right hemisphere spatial analysis for end-state planning and left hemisphere organisation of sequential elements of movement. Given the lateralization of grasping function suggested by Hermsdörfer *et al.* (1999), in future research it would be interesting to evaluate whether object constraints on grip points interact with postural constraints and whether this interaction is more affected by spatial (right) or sequential (left) impairments following stroke (see also Chapters 21 and 29). However, in order to obtain grip point effects, it would be advisable for the research to use precision rather than power grip.

Summary and conclusions

This chapter has been about perception and action effects of object micro (texture) and macro (local and global shape) geometry as constraints on selection of precision grip points for lifting and moving objects with the thumb and index finger. We noted anticipatory adjustment of grip force for surface texture, local shape effects on surface normal, and position of CoM relative to the grip axis. We then proposed the corollary that, given freedom of choice, people select grip points that improve grip stability for a given level of force between the digits. In the examples we considered, the chosen grip points yielded improved friction, force between the digits that was better aligned with the surface normals or resulted in less torque about the grip axis. We then asked what cues contributed to the selection of grip points. We noted the use of visual attributes in this regard, but also the use of haptic information. An unresolved issue is whether the use of these cues implies abstract internal representations of object mechanics, that might, for instance, allow prediction of non-alignment of the CoM with the grip axis. Alternatively it might be that the cues allow the observer to generalize previous experience with objects providing similar cues. Finally we considered reach to grasp as the precursor to gripping which may be considered as the stage when planning for constraints due to movement goals can interact with constraints embodied in the object.

Acknowledgments

This work was supported by funding to AMW from the sixth Framework Program of the EU (project IST-2006-27141; http://www.immersence.info/) and to SJL from the Natural Sciences and Engineering Council of Canada.

References

Bhavikatti, S. S. & Rajashekharappa, K. G. (1994). *Engineering Mechanics*. Chichester, UK: John Wiley.

Eastough, D. & Edwards, M. G. (2007). Movement kinematics in prehension are affected by grasping objects of different mass. *Exp Brain Res*, **176**, 193–198.

Flanagan, J. R. & Wing, A. M. (1995). The stability of precision grip forces during cyclic arm movements with a hand-held load. *Exp Brain Res*, **105**, 455–464.

Flanagan, J. R., Wing, A. M., Allison, S. & Spenceley, A. (1995). Effects of surface texture on weight perception when lifting objects with a precision grip. *Percept Psychophys*, **57**, 282–290.

Goodale, M. A., Meenan, J. P., Bulthoff, H. H. *et al.* (1994). Separate neural pathways for the visual analysis of object shape in perception and prehension. *Curr Biol*, **1**, 604–610.

Goodale, M., Jakobson, L. & Servos, P. (1996). The visual pathways mediating perception and prehension. In A. M. Wing, P. Haggard & J. R. Flanagan (Eds.), *Hand and Brain: The Neurophysiology and Psychology of Hand Movements* (pp. 15–32). San Diego, CA: Academic Press.

Salimi, A., Hollander, I., Frazier, W. & Gordon, A. M. (2000). Specificity of internal representations underlying grasping. *J Neurophysiol*, **84**, 2390–2397.

Hermsdörfer, J., Laimgruber, K., Kerkhoff, G., Mai, N. & Goldenberg, G. (1999). Effects of unilateral brain damage on grip selection, coordination, and kinematics of ipsilesional prehension. *Exp Brain Res*, **128**, 41–51.

Hoff, B. & Arbib, M. A. (1993). Models of trajectory formation and temporal interaction of reach and grasp. *J Motor Behav*, **25**, 175–192.

Jeannerod, M. (1981). Intersegmental coordination during reaching at natural visual objects. In J. Long & A. Baddeley (Eds.), *Attention and Performance IX* (pp. 153–169). Hillsdale, NJ: Lawrence Erlbaum.

Jenmalm, P. & Johansson, R. S. (1997). Visual and somatosensory information about object shape and control of manipulative finger tip forces. *J Neurosci*, **17**, 4486–4499.

Johansson, R. S. & Westling, G. (1984). Roles of glabrous skin receptors and sensorimotor memory in automatic control of precision grip when lifting rougher or more slippery objects. *Exp Brain Res*, **56**, 550–564.

Jones, L. A. & Lederman, S. J. (2006). *Human Hand Function*. New York: Oxford University Press.

Lederman, S. J. & Klatzky, R. L. (1987). Hand movements: a window into haptic object recognition. *Cogn Psychol*, **19**, 342–368.

Lederman, S. J. & Wing, A. M. (2003). Grasp point selection, perceptual judgment and object symmetry. *Exp Brain Res*, **152**, 156–165.

Loomis, J. M. & Lederman, S. J. (1986). Tactual perception. In K. Boff, L. Kaufman & J. Thomas (Eds.), *Handbook of Perception and Human Performance* (pp. 31–1 – 31–41). New York, NY: John Wiley.

Marotta, J. J., McKeeff, T. J. & Behrmann, M. (2003). Hemispatial neglect: its effects on visual perception and visually guided grasping. *Neuropsychologia*, **41**, 1262–1271.

Rosenbaum, D. A., Marchak, F., Barnes, H. J. *et al.* (1990). Constraints for action selection: overhand versus underhand grip. In M. Jeannerod (Ed.), *Attention and Performance XIII* (pp. 321–342). Hillsdale, NJ: Lawrence Erlbaum.

Smeets, J. B. J. & Brenner, E. (1999). A new view on grasping. *Motor Control*, **3**, 237–271.

Westling, G. & Johansson, R. S. (1984). Factors influencing the force control during precision grip. *Exp Brain Res*, **53**, 277–284.

Wing, A. M. & Lederman, S. J. (1998). Anticipating load torques produced by voluntary movements. *J Exp Psychol: Hum Percept Perform*, **24**, 1571–1581.

15

Two hands in object-oriented action

SATOSHI ENDO, ALAN M. WING AND R. MARTYN BRACEWELL

Summary

This chapter examines object-oriented bimanual coordination and reviews studies which have furthered our understanding of how the central nervous system coordinates the movement of the two hands. Each section deals with different aspects of bimanual coordination and the relevant underpinning neurobiology. First, we describe how bimanual behavior is inherently constrained by the sensorimotor system which preferentially processes and executes symmetrical movements. Second, we discuss how the dynamics of the two hands are integrated to maintain the equilibrium of bimanual performance using anticipatory mechanisms. The third section deals with handedness and how the inherent laterality of our motor system influences bimanual behavior. In the final section, we show how some of the lateral preferences may be over-ridden according to the demands of certain tasks.

Introduction

One hand affords reaching, grasping and manipulation of objects of various shapes and sizes. However, two hands dramatically increase the capacity and range of human dexterity to include larger, heavier objects and to permit greater relative motions of manipulated parts. To achieve coordinated bimanual actions, the kinematics and dynamics of each hand need to be temporally and spatially orchestrated. We consider a number of bimanual manipulative tasks, both in terms of behavioral control issues and also in terms of the underlying neuro-anatomy and physiology. A key issue in object manipulation is the use of an appropriate grip force (GF) to maintain stability of the grasped object. Prior to considering bimanual coordination of grasping, we briefly review relevant work on unimanual manipulation of objects.

In order to lift an object with precision grip, frictional force arising from GF normal to the surfaces must counteract the load force (LF) of the object to avoid slippage (Flanagan *et al.*, 1999; see also Chapters 1, 11 and 12). When the load is predictable, the increase of GF slightly precedes that of LF, whilst economically keeping the GF/LF ratio relatively low (Flanagan & Wing, 1997). If the mechanical properties change unexpectedly, so that GF is inadequate, there is rapid feedback correction. In the event of inadequate GF, the slippage of

Sensorimotor Control of Grasping: Physiology and Pathophysiology, ed. Dennis A. Nowak and Joachim Hermsdörfer. Published by Cambridge University Press. © Cambridge University Press 2009.

the object through the digits is registered by the sensory receptors embedded in the skin and a reactive increase in GF is produced (Johansson & Westling, 1984). When the load of the object remains the same, GF will be scaled within a few trials of lifting the object, suggesting the central nervous system (CNS) stores information about the object's mechanical properties (Johansson & Cole, 1992).

Bimanual force production and the crosstalk model

A key issue in bimanual action is how the separate elements of motor control for each hand are coordinated by the CNS. Consider bimanual actions to separate targets; for example reaching with the left and right hands from the midline to targets at different distances on the left and right, or pressing two buttons, one with each index finger, to achieve different force levels. Some have proposed that such bimanual actions originate from a single movement plan, after observing that movement parameters such as movement amplitudes and timings in bimanual pointing tasks are highly correlated (Schmidt, 1975). Others have argued that there is a separate control plan for each hand and bimanual correlations result from intermingling bilaterally delegated information (Marteniuk *et al.*, 1984). The latter account, referred to as the crosstalk model, has been examined both in kinematic (e.g. Marteniuk *et al.*, 1984; Spijkers *et al.*, 1994) and kinetic (e.g. Rinkenauer *et al.*, 2001; Heuer *et al.*, 2002) terms.

Kinematic interference of bilateral grasping movements has been addressed in a task which required participants to make bilateral reaching and grasping movements to two independent targets (Castiello *et al.*, 1993; Dohle *et al.*, 2000). It is generally assumed that prehension involves two components: transport of the arm and scaling of grip aperture to the target size. These two components are subserved by different mechanisms to meet their respective target demands (i.e. distance to and size of the target, respectively), yet the temporal envelopes of these two components are tightly coupled (Jeannerod, 1981; see also Chapters 2 and 10). When a person makes prehensile movements to bilaterally matched targets, transport and grasp of the two hands are carried out symmetrically with strong spatial and temporal couplings. However, when the targets differ in size, grip aperture is independently scaled with ease while transport of the arms remains strongly coupled (Dohle *et al.*, 2000). Thus, bilateral coupling of arm movements appear "hardwired," whereas coupling of grip aperture is flexibly modulated.

At the kinetic level, Rinkenauer *et al.* (2001) studied coordinative constraints in concurrent bimanual force production over a range of timing and force level requirements. When participants used the index fingers to apply simultaneous, bilaterally symmetric forces, the degree of bimanual coupling, assessed by correlation, was high in both temporal and non-temporal (i.e. force level) features. When the two hands were required to produce different levels of force, participants successfully adjusted the forces to hand-specific targets without affecting the timing. Thus, specification of bilateral force levels was somewhat independent of temporal specification. When asymmetric timing was required, in contrast, bimanual decoupling was incomplete and coordination of the bilateral force levels, as

indexed by correlation, decreased slightly. The lowered correlation in force when performing with different timings was taken to suggest that the specification of timing originated at a
more global level. Thus, assimilation in bilateral force production is modulated as a function
of temporal complexity of the bimanual task.

Cortical substrates for bimanual constraints

The distal muscles of the hand are primarily under the control of the contralateral primary
motor cortex (M1). Thus, the corpus callosum, which allows reciprocal information
exchange between homologous cortical areas in the two hemispheres, has been considered
as a source of neural crosstalk (Swinnen, 2002). In fact, acallosal patients show difficulty in
learning new bimanual movements whilst well-learned movements remain relatively intact
(Franz *et al.*, 2000). These patients are unusually prone to non-temporal crosstalk, for
instance with spatially asymmetric movements, compared with those without neurological
impairment (Franz *et al.*, 1996; Eliassen *et al.*, 1999).

The neuropsychological data are consistent with a recent structural MRI study showing
that the strength of callosal connections between medial motor areas, including supplementary motor area (SMA) and caudal cingulate cortex, is positively correlated with resistance
to bimanual crosstalk in a rhythmic movement task (Johansen-Berg *et al.*, 2007). A related
point with a developmental perspective is that a stronger susceptibility to bimanual crosstalk
in children may be due to their less-developed corpora callosa (Franz & Fahey, 2007;
Muetzel *et al.*, 2008). There is an inherent tendency to produce mirror symmetric movements, even during unimanual actions; an inhibitory mechanism, likely to be mediated via
the corpus callosum, is crucial in allowing the execution of asymmetric actions
(Daffertshofer *et al.*, 2005; Duque *et al.*, 2005; Serrien, 2008).

As a possible neural source of crosstalk, the SMAs have received much attention due
to denser interhemispheric connections than those involving the M1s (Gould *et al.*, 1986;
Rouiller *et al.*, 1994). The involvement of the SMAs in bimanual coordination has been
established by a variety of methodologies. For instance, Brinkman (1981) demonstrated
that SMA lesions lead to the movements becoming symmetrical, even in a task requiring
asymmetric actions of the two arms. However, this behavior was eliminated after an
additional callosal section was performed. Moreover, disruption of SMA by repetitive
transcranial magnetic stimulation (TMS) resulted in reduced interhemispheric coupling
and degradation of bimanual performance (Serrien *et al.*, 2002). Overall, these findings
emphasize the contribution of bilateral SMAs and the corpus callosum to neural
crosstalk.

In contrast to traditional teaching, there is increasing evidence of bilateral representation
of the hands in M1 (Davare *et al.*, 2007) such that 20% of corticospinal fibers from M1
descend to the ipsilateral hand (Nathan *et al.*, 1990; Wassermann *et al.*, 1994). As a
consequence, one might infer that each hand is under the influence of both M1s. The
ipsilateral influence alters the muscular activation, and, as a result, the movement that
each arm performs becomes slightly similar to the movement of the opposite arm, giving

rise to movement assimilation (Swinnen *et al.*, 1991). Thus, neural crosstalk could take place within a single M1, though the precise role of ipsilateral control is still debated.

Predictive mechanisms and bimanual coordination

In this section, we discuss issues relating to bimanual coordination involved in bimanual support of an object. Consider placing an object on a hand (the *postural* hand) and then removing it using the other hand. As the object is supported by the postural hand, the weight of the object vertically pushes the hand downwards (Figure 15.1). This force is counteracted by an equal amount of force produced by the postural hand in order to maintain its position. When the object is removed, the loss of the downward force from the object results in an upwards movement of the postural hand because the counteracting upward force is now inappropriately large.

The unloading task has provided a good means of studying bimanual coordination (Massion, 1992). In these studies, unloading by the participant was contrasted with load removal by an external agent. When the unloading was performed by the participant's other hand, the postural hand produced little movement, indicating that the CNS adjusted the forces to maintain the stability of the postural hand (Gahery & Massion, 1981; Lum *et al.*, 1992). Electromyogram recordings from the postural arm have shown that this compensatory force adjustment occurred prior to the load offset in an anticipatory manner (Hugon *et al.*, 1982; Dufosse *et al.*, 1985). However, this predictive behavior, termed an anticipatory postural adjustment (APA), was not observed when the unloading was produced by an external agent or when the participants only had control over the *timing* of unloading by pressing a button to initiate the unloading (Dufosse *et al.*, 1985).

More recently, researchers have explored a virtual object paradigm in which two robotic arms or torque motors simulate reaction forces between the two hands (Witney *et al.*, 2000; Ohki *et al.*, 2002; Diedrichsen *et al.*, 2003). This paradigm is particularly useful in allowing measurement of the development and decay of APAs, since the reaction force can be manipulated in novel fashions to explore the potential for learning new modes of coordination between the hands.

Figure 15.1. Bimanual unloading task (adapted from Diedrichsen *et al.*, 2005). A. An object rests steadily on the postural hand. B. When the object is unloaded by another person, the postural hand moves slightly upward due to the unpredictable loss of the load force of the object. C. The postural hand maintains its stability when the object is unloaded by the other hand, hence the unloading is predictable.

In the study of Witney *et al.* (2000), for example, each hand held an object that was attached to its own torque motor. On each trial the participant was required to pull on the object held in the left hand and to maintain the position of the object held in the right hand. The torque motors were computer controlled so that the two objects could be either "linked" so that the forces on the objects were equal and opposite, acting as though they were a single object, or "unlinked," so that they acted as two independent objects. A predictive response in the restraining hand was only necessary when the objects were linked and was unnecessary in the unlinked condition where there was no risk of force applied to one object causing a force at the other which might lead to slippage.

After a few trials with linked objects, the participants successfully learned to produce the counteracting force to meet the target force with minimum deviation. To determine whether the exerted force was modulated by an anticipatory mechanism, the authors measured the after-effects of learning by inserting unpredictable unlinked trials. The analysis revealed that anticipatory grip force developed within a few trials. When the torque motors simulated an unlinked object that did not have normal physical properties, the anticipatory response was reduced. Thus, the CNS can appropriately switch between different modes of anticipatory control based on the previous experience of an object's mechanical properties.

Using a similar paradigm to that used by Witney *et al.*, Ohki and colleagues (Ohki & Johansson, 2000; Ohki *et al.*, 2002) further showed that this higher-order motor representation governs reflexive responses. Thus an unexpected perturbation to one hand led to reflexive force modulation of both hands soon after the participants experienced the "linked" trials but not during "unlinked" trials (Ohki & Johansson, 2000; Ohki *et al.*, 2002).

Whilst these studies provide strong evidence for central representation of bimanual action, the higher-order representation of bimanual action appears somewhat independent of unimanual action (Schulze *et al.*, 2002; Nozaki *et al.*, 2006; Theorin & Johansson, 2007). According to a recent motor adaptation study by Nozaki *et al.* (2006), adaptation to a novel force field while performing a reaching task unimanually did not generalize to an equivalent bimanual task, and vice versa. Hence, motor adaptation to a novel force field was largely restricted to the coordination pattern the individuals were originally exposed to. Furthermore, Iyengar *et al.* (2007) showed that lifting an object with one hand whilst lightly touching the remaining hand improved GF coordination of the lifting hand, possibly because the remaining hand acts as an extra source of information for training the predictive mechanism for the hand grasping the object. These findings suggest that bimanual coordination is not a simple addition of two unimanual movements.

Subcortical structures in bimanual coordination

Although a number of studies have indicated that links between areas in the two cerebral cortices are a source of bimanual crosstalk, the contribution of subcortical structures should not be ignored. For example, the cerebellum is an integral source of anticipatory actions (Wolpert *et al.*, 1998). A study by Diedrichsen *et al.* (2005) suggested that the general characteristics of APAs are present in cerebellar patients: they exhibited force adjustment of the postural hand prior to and appropriately scaled to the unloading. However, APAs occurred abnormally faster than in the neurologically healthy control group, and the patients failed to adjust their behavior to a situation where APAs were no longer needed because the mechanical link between the two hands was eliminated. These authors concluded that the cerebellum is critically involved in organizing different elements of an action in a temporally coherent manner and in short-term modification of anticipatory responses.

Another set of subcortical structures implicated in bimanual coordination are the basal ganglia (Kraft *et al.*, 2007). The basal ganglia are strongly connected to SMA (Alexander *et al.*, 1990) and their abnormal activity in, for example, Parkinson's disease (PD) has been associated with reduced excitability of SMA as revealed by functional imaging studies (Sabatini *et al.*, 2000). Patients with PD show difficulty in making bilateral asymmetric movements (Brown *et al.*, 1993), and often make an unintended "mirror movement" of a contralateral limb (Cincotta *et al.*, 2006). The similarity of "mirror movement" symptoms caused by SMA lesion and PD reflects the tight connections between SMA and the basal ganglia, and further supports the notion that coordination requires interactions between subcortical and cortical structures.

Assigning two hands to different subgoals

In bimanual tasks the two hands need not take equivalent roles, and a number of authors have suggested that each hand's function maps onto specific manipulative and postural sub-goals (Guiard, 1987; Macneilage *et al.*, 1987; Weiss *et al.*, 2000; Wiesendanger & Serrien, 2004; Johansson *et al.*, 2006). In general, the postural role refers to the hand which defines a global coordinate for a given action, and the manipulative role to the hand that completes the action with finer movements. Guiard (1987) related these subgoals to handedness after observing that people typically assign their dominant hand to a manipulative role, such as opening a lid of a box, whilst the non-dominant hand supports the box. His account conceptualized bimanual action as "a kinematic chain" of two hands and proposed that the action of the non-dominant hand forms a reference frame for action by the dominant hand.

A similar subgoal perspective was taken by Wiesendanger and colleagues. In a bimanual drawer-opening task, participants were trained to reach and open a drawer to pick up an object inside the drawer. The drawer was spring-loaded so the one hand had to hold the drawer open to obtain a reward with the other hand. Kinematic analyses revealed very consistent timing, with the arrival of the pull-hand (i.e. postural role) leading the pick-hand (i.e. manipulative role) by 10–50 ms, despite large temporal variability in the single limb components prior to this point (Serrien & Wiesendanger, 2000). This temporal synchronization of the two hands was sustained when an external load was added to the pull-hand which led to a prolonged movement time (Perrig *et al.*, 1999). In contrast, when sensory feedback from the leading arm was distorted by vibration, the movements of the hands became desynchronized, suggesting the importance of sensory feedback of the postural hand in defining the action of the manipulative hand (Kazennikov & Wiesendanger, 2005).

Similar to the observation made by Guiard (1987), all the participants in the drawer-pulling experiments assigned their dominant hand to the pick-hand (manipulative role) even though they were not given any instruction to do so (Wiesendanger & Serrien, 2004).

It is assumed that both hand dominance and functional specialization of the hands are a result of the CNS maximizing efficacy of the overall performance rather than a simple preference for using one hand over the other. In particular, it is assumed that the dominant arm is more efficient in exploiting intersegmental dynamics (i.e. torque as a consequence of the movement produced by an adjacent proximal body segment) when controlling the trajectory of the hand (Sainburg, 2005), due to a more accurate anticipatory knowledge (Serrien *et al.*, 2006; Heuer, 2007). The non-dominant hand, operating largely under feedback control, shows superiority in counteracting unexpected perturbations thus providing a more stable spatial reference for the dominant arm (Sainburg, 2005).

Whilst Sainburg originally proposed the theory of lateralized differences to account for intralimb dynamics of unimanual action, the theory has been applied to bimanual tasks such as rhythmic movement (Heuer, 2007) and GF production (Ferrand & Jaric, 2006; de Freitas *et al.*, 2007a). In a series of experiments conducted by Jaric and colleagues (Ferrand & Jaric, 2006; de Freitas *et al.*, 2007a), for example, participants grasped and applied tensile or compressive forces to fixed handles in order to match the average of the LFs exerted by the two hands with an external target. The results indicated superior GF coordination by the non-dominant hand than the dominant hand, indexed by a lower GF/LF ratio (Ferrand & Jaric, 2006) and a higher accuracy of GF and LF modulation (de Freitas *et al.*, 2007a, 2007b).

Lateralization at the cortical level

There is abundant evidence for functional and anatomical lateralization at the cortical level (Serrien *et al.*, 2006). Functional imaging studies have shown that the hemisphere contralateral to the dominant hand (i.e. the dominant hemisphere) has a superior role in motor control in general and especially during skilled movements (e.g. Grafton *et al.*, 2002; Haaland *et al.*, 2004). For example, unimanual action with the non-dominant hand leads to bilateral neural activity in M1 whereas action with the dominant hand only involves the contralateral hemisphere (Amunts *et al.*, 2000). A similar observation has been made in bimanual action such that the activation level was higher in the non-dominant hemisphere (Jancke *et al.*, 2000; Ullen *et al.*, 2003). These studies suggest that the non-dominant hemisphere is less efficient in processing movement-related information and more susceptible to neural crosstalk.

Bimanual object manipulation

So far, we have reviewed how internally derived factors can modulate bimanual behavior directed at distinct targets for each hand. However, to our knowledge, bilateral grip

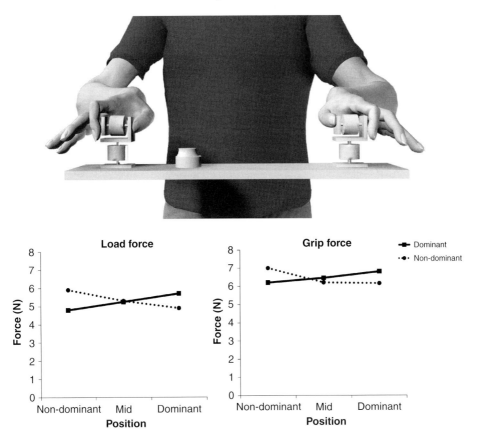

Figure 15.2. Top. A schematic illustration of the apparatus used by Bracewell *et al*. (2003). With each hand the participants grasped and vertically lifted coupled force transducers attached to a tray with an additional mass placed at varying locations. Bottom. The interaction effects on grip and load forces of hand and position of the external mass. Varying the load between the hands as shown on the left resulted in matched scaling of grip force as shown on the right.

coordination directed at a single object has only been explored in a small number of studies (Bracewell *et al*., 2003; Johansson *et al*., 2006; Scholz & Latash, 1998). These studies are important because we can study how bimanual coordination evolves from these internal factors interacting with external factors including the mechanical properties of the objects. In the study by Bracewell *et al*. (2003), participants either voluntarily lifted with two hands a manipulandum with predictable loading, or were required to hold the manipulandum steady when exposed to unpredictable additional loading (Figure 15.2). When the manipulandum was voluntarily lifted, the increase of GF was temporally coupled with LF change at each hand. Unlike GF scaling in a unimanual task, during bimanual object manipulation, GF required by each hand to support an object may vary, as long as the net GF exerted by the two hands counteracts overall LF. In fact, the sum of the bilateral GFs was scaled to the total weight of the object. Interestingly, GF of one hand increased as the location of the

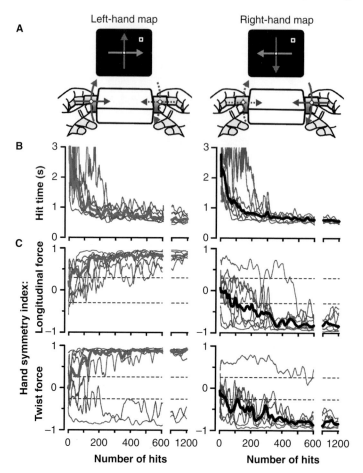

Figure 15.3. Apparatus and experimental data from Johansson *et al.* (2006). The thin lines in the graphs are data obtained from each participant, and the solid one represents the median. A. Schematic of the apparatus. The directional congruency between the applied forces and cursor movements defined the mapping rule. B. The duration decreased over successive trials. C. Hand-asymmetry indices, obtained by correlating movements of the object and the force in their respective (i.e. longitudinal and twisting) dimensions, became strongly biased as the trials were repeated. Positive (negative) values indicate that the left (right) hand served a manipulative role.

external mass was closer to it. This is contradictory to what would be expected from the crosstalk model, which suggests there is an inherent tendency for bimanual coordination (i.e. force level) to be symmetric. When the effects of the LFs on the GF were factored out, nevertheless, the bilateral GFs were found to remain correlated, indicating that there is a common underlying process in specifying GF between the hands. Furthermore, this correlation was stronger during reactive GF production in which the participants reactively increased GF to counteract the unpredictable LF change. The authors suggested that the neural crosstalk for scaling bilateral GFs occurs at a relatively low level of motor control, given there was greater assimilation of bilateral GFs in reflex-triggered GF production (Bracewell *et al.*, 2003).

Interaction of externally and internally driven factors can further extend to functional specialization of the two hands as reviewed in the previous section. In a study by Johansson *et al.* (2006), the participants bimanually held a stylus and exerted longitudinal and twist forces in order to control a visual cursor and hit a target displayed on a screen (Figure 15.3). To keep the object steady in the air, the force at one hand had to match that of the other in magnitude but with opposite direction. During the task, participants experienced two different spatial mappings to control the vertical position of the visual cursor. In one, cursor movements were compatible with the force directions produced by the left hand; in the other, with the right hand. Joanna and Colleagues hypothesized that one hand would provide a postural role for the other hand, and the directional displacement of the object would reflected the force exerted by the manipulative hand just as in opening a jar. The results revealed that the allocation of the manipulative hand was determined by the spatial congruency between direction of the twisting force and the movement of the cursor. As the participants were exposed to the task repeatedly, this tendency was increased and stabilized. Thus, functional specialization can be induced by externally driven factors and over-ride the internal ones.

Interestingly, functional specialization between two ostensibly symmetric effectors has also been reported when two people each apply force to achieve a shared goal. Reed *et al.* (2006) used a device that allowed recording of forces created by two partners. The device was a turntable with a handle attached at each side, and paired participants grasped and moved one handle each to rotate the turntable to a visual target as quickly as possible. Even though verbal communication was prevented, the participants rapidly developed a strategy such that one participant took responsibility for acceleration and the other deceleration. This division of labor resulted in improved efficiency as indexed by reduced response times.

Mirror neurons and complementary action

The study by Reed *et al.* (2006) indicates that functional specialization can be observed at an *inter*personal level of action coordination. From a neuroscience perspective, the discovery of the mirror neurons has inspired many researchers to propose accounts for the understanding of the actions (including speech) of others. The mirror neurons, found in the ventral premotor cortex and the posterior parietal cortex, are characterized by their unique property of firing when one is executing a particular action (e.g. grasping an object) as well as when perceiving another person performing the same action (Rizzolatti *et al.*, 2001). These neurons have led to wide-ranging speculation as to their function including imitation, understanding, and learning of an action in an interpersonal situation, all of which involve internally "mirroring" another person's movement. In contrast, though it may still involve mirroring another's action, a recent functional imaging study has shown that the ventral premotor area preferentially responds to planning of complementary action rather than imitation (Newman-Norlund *et al.*, 2007). In fact, there is a tendency for humans to choose complementary action when, for example, receiving an object from another person (Shibata *et al.*, 2007). Thus, the mirror neuron system might also underpin the ability of people to coordinate their actions when cooperating to achieve a goal, as in the task studied by Reed *et al.* (2006). These studies of interpersonal coordination could provide new insights into functional specialization and complementary action of two effectors found in bimanual action.

Conclusions

The coordination of the two hands in space, time and force is an essential part of bimanual object manipulation. Bimanual movements involve many unique coordination properties, and each raises interesting questions as to how the movement is organized in the CNS. For example, we have reviewed how the complexity of coordination in bimanual tasks limits our capacity to process and execute asymmetric movements of two hands. We also discussed how certain bimanual movements, such as picking out an object from a drawer opened with the other hand, can naturally elicit asymmetric, yet highly coordinated, actions of the two hands. Such asymmetric movements are a result of a complex interaction between internal "inherent" factors, such as neural crosstalk and handedness, and external factors, including the mechanical properties of the object and the task demands.

Acknowledgment

This work was supported by the sixth Framework Program of the EU (project IST-2006-27141; http://www.immersence.info/).

References

Alexander, G. E., Crutcher, M. D. & Delong, M. R. (1990). Basal ganglia-thalamocortical circuits – parallel substrates for motor, oculomotor, prefrontal and limbic functions. *Progr Brain Res*, **85**, 119–146.

Amunts, K., Jancke, L., Mohlberg, H., Steinmetz, H. & Zilles, K. (2000). Interhemispheric asymmetry of the human motor cortex related to handedness and gender. *Neuropsychologia*, **38**, 304–312.

Bracewell, R. M., Wing, A. M., Soper, H. M. & Clark, K. G. (2003). Predictive and reactive co-ordination of grip and load forces in bimanual lifting in man. *Eur J Neurosci*, **18**, 2396–2402.

Brinkman, C. (1981). Lesions in supplementary motor area interfere with a monkey's performance of a bimanual coordination task. *Neurosci Lett*, **27**, 267–270.

Brown, R. G., Jahanshahi, M. & Marsden, C. D. (1993). The execution of bimanual movements in patients with Parkinson, Huntington and cerebellar disease. *J Neurol Neurosurg Psychiatry*, **56**, 295–297.

Castiello, U., Bennett, K. M. B. & Stelmach, G. E. (1993). The bilateral reach to grasp movement. *Behav Brain Res*, **56**, 43–57.

Cincotta, M., Borgheresi, A., Balestrieri, F. *et al.* (2006). Mechanisms underlying mirror movements in Parkinson's disease: a transcranial magnetic stimulation study. *Mov Disord*, **21**, 1019–1025.

Daffertshofer, A., Peper, C. E. & Beek, P. J. (2005). Stabilization of bimanual coordination due to active interhemispheric inhibition: a dynamical account. *Biol Cybern*, **92**, 101–109.

Davare, M., Duque, J., Vandermeeren, Y., Thonnard, J. L. & Olivier, E. (2007). Role of the ipsilateral primary motor cortex in controlling the timing of hand muscle recruitment. *Cerebr Cortex*, **17**, 353–362.

de Freitas, P. B., Krishnan, V. & Jaric, S. (2007a). Force coordination in static manipulation tasks: effects of the change in direction and handedness. *Exp Brain Res*, **183**, 487–497.

de Freitas, P. B., Krishnan, V. & Jaric, S. (2007b). Elaborate force coordination of precision grip could be generalized to bimanual grasping techniques. *Neurosci Lett*, **412**, 179–184.

Diedrichsen, J., Verstynen, T., Hon, A., Lehman, S. L. & Ivry, R. B. (2003). Anticipatory adjustments in the unloading task: is an efference copy necessary for learning? *Exp Brain Res*, **148**, 272–276.

Diedrichsen, J., Verstynen, T., Lehman, S. L. & Ivry, R. B. (2005). Cerebellar involvement in anticipating the consequences of self-produced actions during bimanual movements. *J Neurophysiol*, **93**, 801–812.

Dohle, C., Ostermann, G., Hefter, H. & Freund, H. J. (2000). Different coupling for the reach and grasp components in bimanual prehension movements. *Neuroreport*, **11**, 3787–3791.

Dufosse, M., Hugon, M. & Massion, J. (1985). Postural forearm changes induced by predictable in time or voluntary triggered unloading in man. *Exp Brain Res*, **60**, 330 334.

Duque, J., Mazzocchio, R., Dambrosia, J. *et al.* (2005). Kinematically specific interhemispheric inhibition operating in the process of generation of a voluntary movement. *Cerebr Cortex*, **15**, 588–593.

Eliassen, J. C., Baynes, K. & Gazzaniga, M. S. (1999). Direction information coordinated via the posterior third of the corpus callosum during bimanual movements. *Exp Brain Res*, **128**, 573–577.

Ferrand, L. & Jaric, S. (2006). Force coordination in static bimanual manipulation: effect of handedness. *Motor Control*, **10**, 359–370.

Flanagan, J. R. & Wing, A. M. (1997). The role of internal models in motion planning and control: evidence from grip force adjustments during movements of hand-held loads. *J Neurosci*, **17**, 1519–1528.

Flanagan, J. R., Burstedt, M. K. O. & Johansson, R. S. (1999). Control of fingertip forces in multidigit manipulation. *J Neurophysiol*, **81**, 1706–1717.

Franz, E. A. & Fahey, S. (2007). Developmental change in interhemispheric communication – evidence from bimanual cost. *Psychol Sci*, **18**, 1030–1031.

Franz, E. A., Eliassen, J. C., Ivry, R. B. & Gazzaniga, M. S. (1996). Dissociation of spatial and temporal coupling in the bimanual movements of callosotomy patients. *Psychol Sci*, **7**, 306–310.

Franz, E. A., Waldie, K. E. & Smith, M. J. (2000). The effect of callosotomy on novel versus familiar bimanual actions: a neural dissociation between controlled and automatic processes? *Psychol Sci*, **11**, 82–85.

Gahery, Y. & Massion, J. (1981). Coordination between posture and movement. *Trends Neurosci*, **4**, 199–202.

Gould, H. J., Cusick, C. G., Pons, T. P. & Kaas, J. H. (1986). The relationship of corpus-callosum connections to electrical-stimulation maps of motor, supplementary motor, and the frontal eye fields in owl monkeys. *J Comp Neurol*, **247**, 297–325.

Grafton, S. T., Hazeltine, E. & Ivry, R. B. (2002). Motor sequence learning with the nondominant left hand – a PET functional imaging study. *Exp Brain Res*, **146**, 369–378.

Guiard, Y. (1987). Asymmetric division of labor in human skilled bimanual action – the kinematic chain as a model. *J Motor Behav*, **19**, 486–517.

Haaland, K. Y., Elsinger, C. L., Mayer, A. R., Durgerian, S. & Rao, S. M. (2004). Motor sequence complexity and performing hand produce differential patterns of hemispheric lateralization. *J Cogn Neurosci*, **16**, 621–636.

Heuer, H. (2007). Control of the dominant and nondominant hand: exploitation and taming of nonmuscular forces. *Exp Brain Res*, **178**, 363–373.

Heuer, H., Spijkers, W., Steglich, C. & Kleinsorge, T. (2002). Parametric coupling and generalized decoupling revealed by concurrent and successive isometric contractions of distal muscles. *Acta Psychologica*, **111**, 205–242.

Hugon, M., Massion, J. & Wiesendanger, M. (1982). Anticipatory postural changes induced by active unloading and comparison with passive unloading in man. *Pflugers Archiv-Eur J Physiol*, **393**, 292–296.

Iyengar, V., Santos, M. J. & Aruin, A. S. (2007). Does the location of the touch from the contralateral finger application affect grip force control while lifting an object? *Neurosci Lett*, **425**, 151–155.

Jancke, L., Peters, M., Himelbach, M. *et al.* (2000). fMRI study of bimanual coordination. *Neuropsychologia*, **38**, 164–174.

Jeannerod, M. (1981). Intersegmental coordination during reaching at natural visual objects. In J. Long & A. Baddeley (Eds.), *Attention and Performance IX* (pp. 153–168). Hillsdale, NJ: Lawrence Erlbaum.

Johansen-Berg, H., Della-Maggiore, V., Behrens, T. E. J., Smith, S. M. & Paus, T. (2007). Integrity of white matter in the corpus callosum correlates with bimanual co-ordination skills. *Neuroimage*, **36**, T16–T21.

Johansson, R. S. & Westling, G. (1984). Roles of glabrous skin receptors and sensorimotor memory in automatic-control of precision grip when lifting rougher or more slippery objects. *Exp Brain Res*, **56**, 550–564.

Johansson, R. S. & Cole, K. J. (1992). Sensory-motor coordination during grasping and manipulative actions. *Curr Opin Neurobiol*, **2**, 815–823.

Johansson, R. S., Theorin, A., Westling, G. *et al.* (2006). How a lateralized brain supports symmetrical bimanual tasks. *PLoS Biol*, **4**, 1025–1034.

Kazennikov, O. V. & Wiesendanger, M. (2005). Goal synchronization of bimanual skills depends on proprioception. *Neurosci Lett*, **388**, 153–156.

Kraft, E., Chen, A. W., Flaherty, A. W. *et al.* (2007). The role of the basal ganglia in bimanual coordination. *Brain Res*, **1151**, 62–73.

Lum, P. S., Reinkensmeyer, D. J., Lehman, S. L., Li, P. Y. & Stark, L. W. (1992). Feedforward stabilization in a bimanual unloading task. *Exp Brain Res*, **89**, 172–180.

Macneilage, P. F., Studdertkennedy, M. G. & Lindblom, B. (1987). Primate handedness reconsidered. *Behav Brain Sci*, **10**, 247–263.

Marteniuk, R. G., MacKenzie, C. L. & Baba, D. M. (1984). Bimanual movement control – information-processing and interaction effects. *Q J Exp Psychol A*, **36**, 335–365.

Massion, J. (1992). Movement, posture and equilibrium – interaction and coordination. *Progr Neurobiol*, **38**, 35–56.

Muetzel, R. L., Collins, P. F., Mueller, B. A. *et al.* (2008). The development of corpus callosum microstructure and associations with bimanual task performance in healthy adolescents. *Neuroimage*, **39**, 1918–1925.

Nathan, P. W., Smith, M. C. & Deacon, P. (1990). The corticospinal tracts in man – course and location of fibers at different segmental levels. *Brain*, **113**, 303–324.

Newman-Norlund, R. D., van Schie, H. T., van Zuijlen, A. M. J. & Bekkering, H. (2007). The mirror neuron system is more active during complementary compared with imitative action. *Nat Neurosci*, **10**, 817–818.

Nozaki, D., Kurtzer, I. & Scott, S. H. (2006). Limited transfer of learning between unimanual and bimanual skills within the same limb. *Nat Neurosci*, **9**, 1364–1366.

Ohki, Y. & Johansson, R. S. (2000). Reactive finger responses are influenced by grasp motor set prepared for objects' behaviour. *Eur J Neurosci*, **12**, 197.

Ohki, Y., Edin, B. B. & Johansson, R. S. (2002). Predictions specify reactive control of individual digits in manipulation. *J Neurosci*, **22**, 600–610.

Perrig, S., Kazennikov, O. & Wiesendanger, M. (1999). Time structure of a goal-directed bimanual skill and its dependence on task constraints. *Behav Brain Res*, **103**, 95–104.

Reed, K., Peshkin, M., Hartmann, M. J. *et al.* (2006). Haptically linked dyads – are two motor-control systems better than one? *Psychol Sci*, **17**, 365–366.

Rinkenauer, G., Ulrich, R. & Wing, A. M. (2001). Brief bimanual force pulses: correlations between the hands in force and time. *J Exp Psychol Hum Percept Perform*, **27**, 1485–1497.

Rizzolatti, G., Fogassi, L. & Gallese, V. (2001). Neurophysiological mechanisms underlying the understanding and imitation of action. *Nat Rev Neurosci*, **2**, 661–670.

Rouiller, E. M., Babalian, A., Kazennikov, O. *et al.* (1994). Transcallosal connections of the distal forelimb representations of the primary and supplementary motor cortical areas in macaque monkeys. *Exp Brain Res*, **102**, 227–243.

Sabatini, U., Boulanouar, K., Fabre, N. *et al.* (2000). Cortical motor reorganization in akinetic patients with Parkinson's disease – a functional MRI study. *Brain*, **123**, 394–403.

Sainburg, R. L. (2005). Handedness: differential specializations for control of trajectory and position. *Exercise Sport Sci Rev*, **33**, 206–213.

Schmidt, R. A. (1975). Schema theory of discrete motor skill learning. *Psychol Rev*, **82**, 225–260.

Scholz, J. P. & Latash, M. L. (1998). A study of a bimanual synergy associated with holding an object. *Hum Move Sci*, **17**, 753–779.

Schulze, K., Luders, E. & Jancke, L. (2002). Intermanual transfer in a simple motor task. *Cortex*, **38**, 805–815.

Serrien, D. J. (2008). Coordination constraints during bimanual versus unimanual performance conditions. *Neuropsychologia*, **46**, 419–425.

Serrien, D. J. & Wiesendanger, M. (2000). Temporal control of a bimanual task in patients with cerebellar dysfunction. *Neuropsychologia*, **38**, 558–565.

Serrien, D. J., Strens, L. H. A., Oliviero, A. & Brown, P. (2002). Repetitive transcranial magnetic stimulation of the supplementary motor area (SMA) degrades bimanual movement control in humans. *Neurosci Lett*, **328**, 89–92.

Serrien, D. J., Ivry, R. B. & Swinnen, S. P. (2006). Dynamics of hemispheric specialization and integration in the context of motor control. *Nat Rev Neurosci*, **7**, 160–167.

Shibata, H., Suzuki, M. & Gyoba, J. (2007). Cortical activity during the recognition of cooperative actions. *Neuroreport*, **18**, 697–701.

Spijkers, W., Tachmatzidis, K., Debus, G., Fischer, M. & Kausche, I. (1994). Temporal coordination of alternative and simultaneous aiming movements of constrained timing structure. *Psychol Res Psychologische Forsch*, **57**, 20–29.

Swinnen, S. P. (2002). Intermanual coordination: from behavioural principles to neural-network interactions. *Nat Rev Neurosci*, **3**, 350–361.

Swinnen, S. P., Young, D. E., Walter, C. B. & Serrien, D. J. (1991). Control of asymmetrical bimanual movements. *Exp Brain Res*, **85**, 163–173.

Theorin, A. & Johansson, R. S. (2007). Zones of bimanual and unimanual preference within human primary sensorimotor cortex during object manipulation. *Neuroimage*, **36**, T2–T15.

Ullen, F., Forssberg, H. & Ehrsson, H. H. (2003). Neural networks for the coordination of the hands in time. *J Neurophysiol*, **89**, 1126–1135.

Wassermann, E. M., Pascualleone, A. & Hallett, M. (1994). Cortical motor representation of the ipsilateral hand and arm. *Exp Brain Res*, **100**, 121–132.

Weiss, P. H., Jeannerod, M., Paulignan, Y. & Freund, H. J. (2000). Is the organisation of goal-directed action modality specific? A common temporal structure. *Neuropsychologia*, **38**, 1136–1147.

Wiesendanger, M. & Serrien, D. J. (2004). The quest to understand bimanual coordination. *Brain Mech Integr Posture Move*, **143**, 491–505.

Witney, A. G., Goodbody, S. J. & Wolpert, D. M. (2000). Learning and decay of prediction in object manipulation. *J Neurophysiol*, **84**, 334–343.

Wolpert, D. M., Miall, R. C. & Kawato, M. (1998). Internal models in the cerebellum. *Trends Cogn Sci*, **2**, 338–347.

16

Dynamic grasp control during gait

PRISKA GYSIN, TERRY R. KAMINSKI AND ANDREW M. GORDON

Summary

When transporting an object during locomotion, the inertial forces that are indirectly generated through the motion of multiple body parts must be taken into account to prevent object slippage. The grip–inertial force coupling that maintains a secure grasp on a hand-held object is preserved across a variety of locomotor tasks that include variations in velocity and precision demands (e.g. transporting a cup of water). When the locomotor pattern is altered by changing the step length or stepping over an obstacle, the grip–inertial force coupling continues to be under anticipatory control. However, the coupling is less robust and can be explained by increased attention demands. Furthermore, the fine motor grasping functions and gross motor locomotor functions are precisely coordinated across multiple limb segments to ensure successful performance right from the onset of gait initiation. These findings support the notion that grip force is based on moment-to-moment predictions of inertial forces acting on the object at gait initiation and throughout predictable variations in the gait cycle. Internal representations of the interactions between body segments through which inertia is transferred to the object–digit interface are proposed to provide the basis for this anticipatory grip force control.

Introduction

A central question in the study of systems motor control is how simultaneous tasks involving multiple body segments are coordinated. For example, during voluntary movements with a hand-held object, grip (normal) force is coupled to the object's load as well as to the motion-induced inertial (tangential) force in an anticipatory manner to prevent slippage. This coupling is observed during arm movements varying in rate and direction, and while using different grasp configurations (Flanagan & Wing 1993, 1995; Flanagan *et al.*, 1993; Flanagan & Tresilian, 1994; Kinoshita *et al.*, 1996; see also Chapters 1, 12 and 14). Internal models of the dynamic interactions between an object and performer's actions are thought to underlie these anticipatory control processes (see Chapter 9) (Flanagan & Wing, 1997; Hermsdörfer *et al.*, 2000; Flanagan & Lolley, 2001; Flanagan *et al.*, 2003; Nowak *et al.*, 2004; White *et al.*, 2005). Coordinating grip with inertial force variations

Sensorimotor Control of Grasping: Physiology and Pathophysiology, ed. Dennis A. Nowak and Joachim Hermsdörfer. Published by Cambridge University Press. © Cambridge University Press 2009.

occurring indirectly through a chain of motions originating further away from the hand-held object during tasks such as jumping (Flanagan & Tresilian, 1994) or walking (Gysin *et al.*, 2003, 2008) requires estimating the end product of multiple effectors. The present chapter will outline the nature of grasp control in the context of such whole-body actions, focusing particularly on transporting an object during locomotion with varying task demands.

Grip force coordination during regular rhythmic gait

Walking involves a repetitive sequence of lower limb motions to propel the body forward. Typically, during these stepping actions the trunk oscillates in the sagittal, frontal and transverse planes (Saunders *et al.*, 1953; Murray *et al.*, 1964, 1970; Waters *et al.*, 1973; Thorstensson *et al.*, 1984). If a person holds an object while walking, forward propulsion combined with the cyclical vertical motions of the body and arm affects the object's inertia in a cyclical manner. The vertical motions of the trunk and a hand-held object during self-paced gait (1.3 m/s; 1.9 steps/s) are illustrated across four steps (Figure 16.1). As seen in the

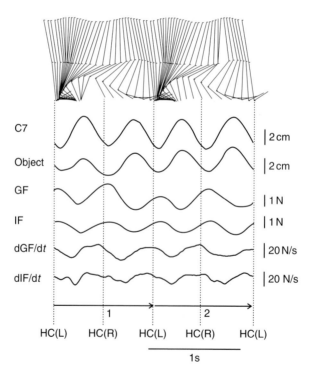

Figure 16.1. Stick figure depicting hip, knee, and ankle joint motions (left leg); pattern of the vertical displacement of the trunk (C7) and hand-held object; and pattern of the grip (GF) and inertial (IF) forces and their derivatives (dGF/d*t* and dIF/d*t*) of a subject during self-paced gait, across two full gait cycles (see arrows 1 and 2). Vertical dashed lines indicate right (R) and left (L) heel contacts (HC). The object was held with a precision grip (thumb opposing index finger) of the right hand and transported anterior to the body with the grip surfaces oriented parallel to the plane of motion.

figure, the body (as represented by movement of C7) and the object move in a synchronized manner with two upward and two downward sinusoidal oscillations per gait cycle. Vertical displacement of the object is less than the body, indicating that the viscoelastic properties of the arm musculature are incorporated into the gait cycle to dampen the effects of inertial forces acting at the hand (Gysin *et al.*, 2003).

The points where maximum and minimum inertial forces occur are similar when cyclical vertical motion of an object is generated directly by arm movement or indirectly by locomotion; both occur at the lower and upper reversal points of the object's path (see Figure 16.1). The lowest vertical position of the object and highest inertial forces occur shortly after heel contact. The highest position of the object and lowest inertial forces are seen during mid-stance when the single stance leg is vertically aligned with the trunk. Generally, inertial force fluctuations on a hand-held object during normal cyclic gait are quite low, averaging approximately 1 N between minima and maxima. These values correspond to about 20–30% of those observed during slow-paced cyclical vertical arm movements (Flanagan & Wing, 1995; Serrien & Wiesendanger, 2001), or to about 10% of those observed during cyclical jumping (Flanagan & Tresilian, 1994).

Three strategies could potentially be used to adjust to the oscillating inertial force during locomotion. First, given the small inertial force oscillations during normal walking, simply grasping harder with a force slightly higher than the maximum generated inertial force would appear to be a viable strategy. Second, subjects could develop grip force in reaction to sensory feedback arising from tactile stimulation associated with fluctuating inertial force. However, this would result in temporal grip-force delays since feedback responses take a minimum of 60 ms to be initiated (e.g. Johansson & Westling, 1984, 1987; Cole & Abbs, 1988) (see also Chapters 9 and 12). Finally, subjects could oscillate grip force in anticipation of the locomotor-induced inertial force oscillations (i.e. couple the two forces), similar to tasks with larger inertial force fluctuations as described above.

As illustrated in Figure 16.1, when transporting an object during regular rhythmic gait, this latter strategy is employed, whereby grip force is continuously modulated and thereby precisely coupled in time and magnitude with the arising inertial force. This coupling is preserved across different gait velocities and precision demands (i.e. transporting a cup of water), showing cross-correlation coefficients between the force rates that are consistently high (mean $r \geq 0.8$) with grip force lags close to zero (Gysin *et al.*, 2003). An energy efficiency level similar to that observed during stance (mean grip–inertial force ratio ~2.5) is maintained during these walking tasks (Gysin *et al.*, 2003). Nevertheless, the coupling of the two forces appears to be more robust during faster gait. This may reflect an increased need for more precise coordination between grip and inertial forces because the risk for slippage is higher when inertial forces are greater (Gysin *et al.*, 2003). Consequently, the precise control of grip force observed during locomotion indicates that subjects anticipate variations in the inertial forces acting on the hand and plan their grasping forces accordingly. Therefore, subjects seem to access an internal representation of the grasping demands and efficiently maintain an appropriate grip force on the object. Presumably, this representation takes into account the interactions between body segments through which inertia is transferred to the object–digit interface (see also Chapter 9).

Methodological considerations when analyzing grip force during locomotion

During locomotion the trunk, and to a lesser extent the transported object, normally deviate from a straight line of progression (alternating toward the side of the supporting limb). During self-paced locomotion, this sinusoidal, medial–lateral displacement averages ~4 cm at the trunk (as represented by C7) while a hand-held object deviates ~2.5 cm. The path consists of a single sinusoid for each gait cycle (compared with two sinusoids in the vertical direction) with midline positions at heel contact and maximum deviations during midstance. In order to assess whether the measurement of grasping force is confounded by such lateral shifts, we asked subjects to walk with the object rotated 90° so that the grip surfaces were parallel to the frontal plane. This allowed us to separate the lateral forces associated with medial–lateral movement of the body (acting tangential to the grip surfaces) from the grip forces (acting perpendicular to the grip surfaces). In this condition, lateral force fluctuation magnitudes were < 5% of the grip force maxima and were out of phase with grip force oscillations. Furthermore, during locomotion with a forward-oriented object, the individual grip forces applied with the thumb and index finger, respectively, were in phase, indicating that these force increases and decreases occurred simultaneously. In contrast, when shaking an object from side-to-side, out-of-phase fluctuations occur at each digit (Kinoshita *et al.*, 1996). In addition, the two grip forces observed during locomotion were perfectly correlated (Pearson's $r = 1$) and comparable in amplitude (< 4% difference in maximum grip force which was similar to the difference measured during quiet stance). This strong similarity between the two grip forces indicates that the medial–lateral forces generated from lateral shifts of the body or other out-of-plane motions have a minimal effect on the coupling between grip and inertial force.

During all tasks in which the hand-held object is not mechanically constrained, out-of-plane movements (e.g. rotations in the frontal or transverse planes) could potentially create load or inertial forces that are falsely measured as grip forces. To determine whether these out-of-plane forces confound our results, we attempted to quantify their effects. If an object with a symmetrical mass distribution is held between two fingers and moved with no out-of-plane translations and rotations (parallel to the gravitational force vector and the primary plane of motion), then the grasping forces on either side of the object are equal and opposite (Johansson & Westling, 1984). Therefore, force asymmetries measured at the finger and thumb would be the result of out-of-plane deviations in object orientation (grip force ± gravitational and inertial forces acting at each transducer). To adjust for this out-of-plane orientation, we calculated the force difference between the grip force measures on each side of the object and subtracted this difference from total grip force as follows: GFadjusted = ((absGF1 + absGF2) – (absGF1 – absGF2))/2. Comparison of the raw grip force with the adjusted grip force during self-paced (~1 m/s) and fast-paced (~1.8 m/s) gait indicated marginal magnitude differences between the two forces at their maxima (< 2% and similar to the differences during quiet stance). Furthermore, the oscillating pattern observed during locomotion did not change and the coupling between grip and inertial force was maintained. These results suggest that the tertiary forces generated through out-of-plane motions of the hand-held object do not significantly influence the oscillations in the grip force associated with locomotion.

Grip-force coordination during predictable gait variations

Given the regularity of inertial force fluctuations in evenly paced cyclical object manipulations (e.g. Flanagan *et al.*, 1993; Flanagan & Tresilian, 1994; Flanagan & Wing, 1995; Kinoshita *et al.*, 1996; Serrien & Wiesendanger, 2001; Gysin *et al.*, 2003), the question arose whether anticipatory grip force during locomotion can be maintained equally well when the regularity of the gait cycle is altered (Gysin *et al.*, 2008). To explore this issue, subjects were required to deviate from their regular rhythmic gait pattern under three predictable conditions. These gait alterations led to changes in step frequency and body accelerations which resulted in temporal and magnitude deviations in the inertial force acting on the hand-held object. Specifically, subjects altered their regular gait cycles by taking a single short step, long step (marked on the ground) or by stepping on and over an obstacle (~15 cm high). To standardize step length, tapelines guided foot placement for the regularly spaced and altered step length along the walkway. Grip force remained tightly coupled to the inertial force with mean time lags below the 60 ms threshold needed for feedback modifications across the three step variations (see example of a subject in Figure 16.2A and mean values in Figure 16.2B). Thus, subjects were able to anticipate the variations in inertial force oscillations (Gysin *et al.*, 2008). In addition, at times of maximum inertial force, safety margins were equal to or higher (long step) than those observed during normal locomotion (Gysin *et al.*, 2008). These larger safety margins may be related to less stable body positions in a manner similar to safety margin increases found when an object is manipulated in an unfamiliar physical environment (Nowak *et al.*, 2001) or when two objects are accelerated in an asymmetric pattern during the performance of a bimanual task (Serrien & Wiesendanger, 2001).

There are two possibilities as to how variations in inertial force could be estimated accurately during locomotion. First, the location of the stepping targets and dimensions of the obstacle may cue retrieval of memory representations (Gordon *et al.*, 1993) acquired during previous locomotor experiences, allowing subjects to predict the upcoming inertial force changes. Alternatively, subjects might interpolate or extrapolate estimates of inertial force fluctuations, based on the perceived difference between the present circumstances and familiar step distances and heights. Such a process would ensure appropriate estimates of grip force without the need of practicing each specific step variation. This latter explanation is in line with the central nervous system's capacity for adjusting grip force to novel load torques (Wing & Lederman, 1998).

Interestingly, variations in the walking pattern influence the temporal precision of grip force prior to and/or during the altered steps (Gysin *et al.*, 2008). Figure 16.2B shows that at

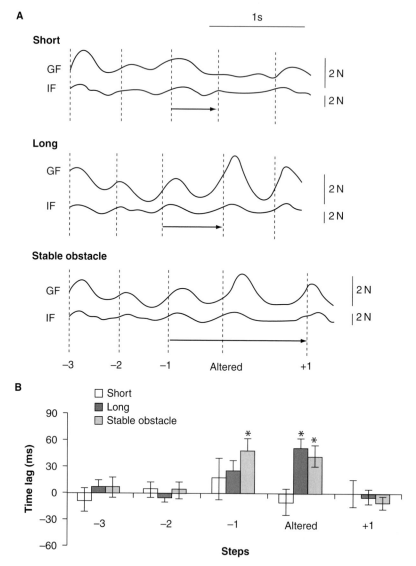

Figure 16.2. A. Pattern of the grip (GF) and inertial (IF) forces during three conditions with predictable gait alterations (short step, long step, and stepping on and over a stable obstacle; marked with arrows in the figure). The steps prior to (−3, −2 and −1) and after (+1) the altered steps are unaltered. Grip and inertial force maxima occurred shortly after heel contacts (indicated by vertical dashed lines). B. Mean (± SEM) time lags (ms) of the maximum grip force relative to the maximum inertial force (positive values indicate that grip force lagged inertial force) during the three conditions with predictable gait alterations (short step, long step and stepping on and over a stable obstacle). * Steps that differ ($P < 0.05$) from steps −3, −2 and +1.

these instances there were relatively longer (and significantly different from preceding and following steps) mean time lags in peak grip force within the 60 ms time window of anticipatory force control, during the obstacle and long step tasks. Initiating a second motor task while performing an initial one requires a shift in attention that may affect the continuity of the initial movement (Sternad *et al.*, 2007). Similarly, performance decrements may occur in dual-task situations that involve the combination of a motor and a cognitive task (e.g. Chen *et al.*, 1996; Müller *et al.*, 2004; Singhal *et al.*, 2007; Siu *et al.*, 2008). Therefore, we speculate that visuo-motor processes related to the transfer of gaze to the obstacle prior to lifting the foot onto or over it (Patla & Vickers, 1997; Di Fabio *et al.*, 2003a, 2003b) might interfere with the grip–inertial force coupling prior to stepping onto the obstacle. Similar to reaching movements of the arm (e.g. Desmurget *et al.*, 1999; Pisella *et al.*, 2000) additional planning processes related to gait trajectory modifications (McFadyen & Carnahan, 1997; Ivanenko *et al.*, 2005) may account for decrements in temporal grip-force precision of the altered steps themselves.

Grip-force coordination during gait initiation

The previous studies focused on grip–load force coupling that was present during ongoing activities. The way in which this coupling unfolds during the early phases of a task may differ from what is observed once the rhythmical pattern is established. When lifting, pushing or pulling an object during stance, grip force is precisely tuned to the associated postural adjustments (i.e. center of pressure displacement, ground reaction forces and torques), indicating the use of a common internal reference (Wing *et al.*, 1997; Forssberg *et al.*, 1999). In these tasks, subjects initiated their actions with the upper extremity. Less is known about grip-force coordination when motion of a hand-held object is initiated indirectly by movement of body parts further away from the arm. Consequently, we used gait initiation while holding an object as an opportunity to investigate the emergence of grip–inertial force coupling while transitioning from stance to the cyclical activity of walking (Diermayr *et al.*, 2008).

Gait initiation is referred to as a sequence of postural shifts that culminate in a forward step (Elble *et al.*, 1994). Typically, prior to foot off, asymmetric vertical ground reaction forces between the two legs can be measured with an initial weight shift towards the swing limb and subsequent weight shift to the stance limb (e.g. Nissan & Whittle, 1990; Brunt *et al.*, 1991; Jian *et al.*, 1993; Brunt *et al.*, 1999, 2005) (see Figure 16.3A). When initiating gait while holding an object, grip force started to increase ~65 ms prior to inertial force onset and ~30 ms after gait onset (Diermayr *et al.*, 2008) (see Figure 16.3). Onset of grip force prior to inertial force implies anticipation of the start of the object's motion. Such anticipation is similar to grip force increases described in object motions that were initiated through direct actions of the arm (e.g. Flanagan & Wing, 1993; Wing *et al.*, 1997; Forssberg *et al.*, 1999; Delevoye-Turrell *et al.*, 2003). Thus, regardless of the origins of the inertial forces acting on the hand-held object, the CNS is able to anticipate the resulting demands for increasing grip force (see Chapter 9).

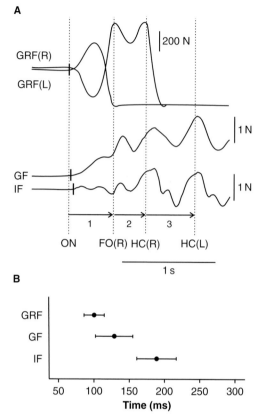

Figure 16.3. A. Bilateral force plate recordings of the vertical ground reaction forces (GRF) of the right swing leg (R) (starting leg) and left stance leg (L) and pattern of the grip (GF) and inertial forces (IF) of a subject during gait initiation (paced as fast as possible). Vertical dashed lines indicate onset of gait (ON), foot off (FO) of the right swing leg, and heel contacts (HC) of the right swing leg and left stance leg. Arrows (1, 2 and 3) show the first, second and third phase of gait initiation. Small vertical solid lines indicate onset of GRF (corresponds to the onset of gait), GF and IF. B. Mean (± SEM) onset times (relative to a predictable auditory starting signal) of ground reaction (GRF), grip (GF) and inertial (IF) forces when initiating gait (as fast as possible).

The example in Figure 16.3A illustrates that prior to foot off, an increase in grip force occurs before fluctuating synchronously with the inertial force. Only a moderate correlation was observed between grip and inertial force rates (mean $r = 0.6$ across the group; grip force preceding inertial force by ~20 ms) during this initial phase of gait initiation (Diermayr *et al.*, 2008). This relatively low correlation may be due to the object's low velocity (mean ~0.2 m/s across the group) and relatively lower inertial force magnitudes, an effect that was also observed during slower gait velocities (Gysin *et al.*, 2003). In the subsequent phase between foot off and heel contact of the swing leg (starting leg), the object's velocity increased to approximately 0.7 m/s and the correlation coefficients became similar to those seen during steady state gait (mean $r = \sim 0.8$ across phases and group; time lags close to zero). The proximity of the grip force onset to the onset of gait, and its close temporal linkage with inertial force, together suggest anticipatory grip force planning from the onset of gait initiation (Diermayr *et al.*, 2008). This clearly demonstrates that fine motor grasping functions and gross motor functions across multiple limbs can be precisely coordinated to ensure a successful performance right from the beginning of a multicomponent task.

Neuro-motor control processes in object transport during locomotion

Internal feed-forward models are widely used (see Chapter 9) to explain the above-illustrated parallel grip–inertial force control (Flanagan & Wing, 1997; Blakemore *et al.*, 1998; Wing & Lederman, 1998; Salimi *et al.*, 2000; Flanagan & Lolley, 2001; Flanagan *et al.*, 2001, 2003; Kawato *et al.*, 2003; Nowak *et al.*, 2004; White *et al.*, 2005). An internal forward model predicts the consequences of a given action in the context of a given state (Jordan & Rumelhart, 1992). In the framework of grip–load force coupling, this means that the CNS builds an internal feed-forward model of the prevailing object load, based on an efference copy of the motor commands for a load-force producing activity (Kawato, 1999). This model incorporates the dynamic properties of the object as well as those of the performer's actions (Flanagan & Wing, 1997) (see also Chapter 12). Proponents of a "threshold position control paradigm" suggest that the grip–inertial force coupling observed during object manipulation does not rely on internal models, but instead can be explained by intrinsic anticipatory properties of the system's dynamics (Pilon *et al.*, 2007). According to this theory, parallel grip-load force control could emerge as an inevitable byproduct of the mechanical interactions between the grasping fingers and the applied load. Based on these principles and on empirical data, Pilon *et al.* created a grip-force model where force began to increase slightly prior to or simultaneously with the onset of rapid point-to-point horizontal movement of a hand-held object. Although accommodating changes in load force without programming grip force appears to be an appealing solution, it does not take into account the fact that the grip–inertial force relationship is task-dependent and not mutually synchronous (see Chapter 12). In particular, under comparable mechanical constraints, marked differences in the timing and amplitudes of grip force relative to load force can be observed when an object is manipulated voluntarily versus externally (Johansson & Westling, 1988; Blakemore *et al.*, 1998; Witney *et al.*, 1999; Delevoye-Turrell *et al.*, 2003; Nowak, 2004). For example, grip force increased in parallel with load force and adequately scaled to the load impact during voluntary arm movements (Nowak, 2004). In contrast, excessive grip force with maxima that lagged load force by approximately 100 ms was used when movements were externally guided. This example illustrates that the temporal relationship between the two forces is not fixed, but is sensitive to the predictability of a task. Furthermore, during reactive, feedback-based tasks, grip force time lags vary between 60 and 600 ms, depending upon the load force rate, type of stimulus (loading versus unloading), direction of impact, phase of a task (dynamic versus static) and

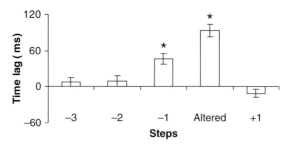

Figure 16.4. Mean (±SEM) time lags (ms) of the maximum grip force relative to the maximum inertial force (positive values indicate that grip force lagged inertial force) during a condition with an unpredictable gait alteration (stepping on and over an unstable obstacle). Values across five steps (peaks occurred shortly after heel contacts) are shown. The altered step consists of the step *on* the unstable obstacle. The steps prior to (−3, −2, −1) and after (+1) are unaltered. * Steps that differ ($P < 0.05$) from steps −3, −2 and +1.

integrity of digital sensibility and cerebellum (Cole & Abbs, 1988; Johansson *et al.*, 1992a, 1992b, 1992c; Häger-Ross *et al.*, 1996; Serrien & Wiesendanger, 1999; Mrotek *et al.*, 2004).

Given the temporal variability described above, predictability of the task seemingly is what permits parallel grip–inertial forces to be maintained when walking unimpeded with an object in the hand. To test this possibility, we altered the stability of the stance foot during walking by asking subjects to step on and then over a pliant obstacle (~15 cm high) that induced postural perturbation when stepping on it (Gysin *et al.*, 2008). Subjects were not able to anticipate the changes in ground reaction forces and related inertial forces transmitted to the object, as evidenced by a 95 ms lag in peak grip force relative to peak inertial force (Figure 16.4). This is in marked contrast to the precise temporal coupling of the two forces during unimpeded walking described above. Thus, during locomotion with a hand-held object, efficient and well-timed grip force does not simply occur as a result of intrinsic mechanical properties within the grasping hand. Rather, given the high degree of coupling between grip and inertial forces during the performance of predictable tasks, there is strong evidence that the CNS relies on memory representations of relevant and predictable features of the action. This information is likely to increase the precision of a feed-forward internal model used to specify grip-force timing and amplitude.

The cerebellum is crucial in anticipatory control of grasping (see Chapter 26). Studies of individuals with cerebellar dysfunctions have found that these subjects had impaired ability in predictive grip-force planning (Babin-Ratté *et al.*, 1999; Nowak *et al.*, 2002). Brain-imaging studies have shown the cerebellum is the primary CNS structure responsible for the coupling between grip and inertial forces seen during force-oscillating activities (Kawato *et al.*, 2003; Boecker *et al.*, 2005) (see also Chapter 26). Thus, cerebellar regions contain the neural substrate of the internal feed-forward representations for the coordination of grip and inertial forces. However, to retrieve memory representations and to transform them into motor actions, additional interactions between the cerebellum, task-planning and task-execution areas in the CNS are necessary. These areas are likely to vary, depending upon task complexity. For example, the posterior parietal cortex (PPC) is active during tasks that require higher-level cognitive functions, such as the formation of intentions during early stages of movement planning (for review see Andersen & Buneo, 2002). Recent work in non-primates also suggests a role of the PPC in visually guided gait modifications, i.e. when stepping over an obstacle (Lajoie & Drew, 2007). If a task involves variation in limb postures, as opposed to repeating postural sequences, the superior parietal cortex appears to play a key role in continuously updating a representation of the body schema (Pellijeff *et al.*, 2006). This may be particularly important when grip force needs to be specified in tasks that require variations of limb postures, such as when walking with an object in a variable environment or during the transition from stance to gait. In addition, higher-level planning areas in the parietal and prefrontal cortices that are activated during single isolated actions, but not repetitive automated ones (Schaal *et al.*, 2004) may also be essential in cyclical gait with single alterations. On a temporal dimension, the same brain areas, known to be activated in "cognitively controlled" timing (Lewis & Miall, 2003) may contribute to estimating changes in the duration of a step (e.g. when varying the step length or stepping over an obstacle) during locomotion.

The basal ganglion is another main structure for force regulation and planning. For example, posterior basal ganglia nuclei (globus pallidus interna, posterior putamen and substantia nigra) regulate basic aspects of dynamic force pulse production (i.e. increased activation during force pulses, compared to steady-state force) whereas anterior ones (caudate nucleus) are involved in the selection of different force amplitudes (Vaillancourt *et al.*, 2007) (see also Chapters 8 and 22). In addition, the globus pallidus externa and the anterior putamen increase activation during the production of similar force pulses (compared to steady-state force) and further increase activation during the selection of different force pulses (Vaillancourt *et al.*, 2007). Furthermore, as shown by impaired performances of individuals with Parkinson's disease, basal ganglia structures appear to be crucial in CNS processes that involve temporal coordination of multiple effectors (see also Chapter 22), including multi-digit force applications (Poizner *et al.*, 2000; van den Berg *et al.*, 2000; Schettino *et al.*, 2003; Baltadjieva *et al.*, 2006; Muratori *et al.*, 2008). Thus, the integrity of basal ganglia structures may be particularly crucial when coordinating grasping forces on a hand-held object during locomotion where inertial forces are transmitted indirectly across multiple effectors and where variations of inertial forces may occur when the gait pattern is altered.

Conclusions

Transporting a hand-held object via whole body actions requires that grasping forces take into account inertial forces that are the indirect result of the motions of multiple body parts. The goal of walking with a hand-held object is to progress safely without risking slippage while simultaneously minimizing the energy costs of the task. This requires that gross motor postural and dynamic functions of the limbs and fine motor grasping functions of the hand all strive towards this common goal. The presented studies demonstrate that a tight grip–inertial force coupling can be maintained across a variety of locomotor demands, including gait initiation. The coupling strength is higher during faster paced movements and unimpeded gait, indicating that grip force planning is sensitive to the consequences of velocity-related object slippage and attention demands of a task. Anticipatory control in object transport during locomotion across a variety of rhythms and changes in inertia suggests that grip force is based on moment-to-moment estimates of the resulting inertial forces acting on the object. Internal representations located in the cerebellum, supported by a network of CNS structures, are proposed to provide the basis for this kind of stable relationship between grip and inertial forces. The study of grip-force strategies used during locomotion is aimed at enhancing our understanding of actions across multiple effectors. We aim to further explore the role of the grasping forces during whole body actions in challenging environments and conditions where disruptive forces on the object and/or body stability demand rapid anticipatory or reactive adjustments to avoid falls or object slippage.

References

Andersen, R. A. & Buneo, C. A. (2002). Intentional maps in posterior parietal cortex. *Annu Rev Neurosci*, **25**, 189–220.

Babin-Ratté, S., Sirigu, A., Gilles, M. & Wing, A. (1999). Impaired anticipatory finger grip-force adjustments in a case of cerebellar degeneration. *Exp Brain Res*, **128**, 81–85.

Baltadjieva, R., Giladi, N., Gruendlinger, L., Peretz, C. & Hausdorff, J. M. (2006). Marked alterations in the gait timing and rhythmicity of patients with de novo Parkinson's disease. *Eur J Neurosci*, **24**, 1815–1820.

Blakemore, S. J., Goodbody, S. J. & Wolpert, D. M. (1998). Predicting the consequences of our own actions: the role of sensorimotor context estimation. *J Neurosci*, **18**, 7511–7518.

Boecker, H., Lee, A., Mühlau, M. *et al.* (2005). Force level independent representations of predictive grip force-load force coupling: a PET activation study. *Neuroimage*, **25**, 243–252.

Brunt, D., Lafferty, M. J., McKeon, A. *et al.* (1991). Invariant characteristics of gait initiation. *Am J Phys Med Rehabil*, **70**, 206–212.

Brunt, D., Liu, S. M., Trimble, M., Bauer, J. & Short, M. (1999). Principles underlying the organization of movement initiation from quiet stance. *Gait Posture*, **10**, 121–128.

Brunt, D., Santos, V., Kim, H. D., Light, K. & Levy, C. (2005). Initiation of movement from quiet stance: comparison of gait and stepping in elderly subjects of different levels of functional ability. *Gait Posture*, **21**, 297–302.

Chen, H. C., Schultz, A. B., Ashton-Miller, J. A. (1996). Stepping over obstacles: dividing attention impairs performance of old more than young adults. *J Gerontol A Biol Sci Med Sci*, **51**, M116–M122.

Cole, K. J. & Abbs, J. H. (1988). Grip force adjustments evoked by load force perturbations of a grasped object. *J Neurophysiol*, **60**, 1513–1522.

Delevoye-Turrell, Y. N., Li, F. X. & Wing, A. M. (2003). Efficiency of grip force adjustments for impulsive loading during imposed and actively produced collisions. *Q J Exp Psychol A*, **56**, 1113–1128.

Desmurget, M., Epstein, C. M., Turner, R. S. *et al.* (1999). Role of the posterior parietal cortex in updating reaching movements to a visual target. *Nat Neurosci*, **2**, 563–567.

Di Fabio, R. P., Greany, J. F. & Zampieri, C. (2003a). Saccade-stepping interactions revise the motor plan for obstacle avoidance. *J Motor Behav*, **35**, 383–397.

Di Fabio, R. P., Zampieri, C. & Greany, J. F. (2003b). Aging and saccade-stepping interactions in humans. *Neurosci Lett*, **339**, 179–182.

Diermayr, G., Gysin, P., Hass, C. J. & Gordon, A. M. (2008). Grip force control during gait initiation with a hand-held object. *Exp Brain Res*, **190**, 337–345.

Elble, R. J., Moody, C., Leffler, K. & Sinha, R. (1994). The initiation of normal walking. *Mov Disord*, **9**, 139–146.

Flanagan, J. R. & Wing, A. M. (1993). Modulation of grip force with load force during point-to-point arm movements. *Exp Brain Res*, **95**, 131–143.

Flanagan, J. R. & Tresilian, J. R. (1994). Grip-load force coupling: a general control strategy for transporting objects. *J Exp Psychol Hum Percept Perform*, **20**, 944–957.

Flanagan, J. R. & Wing, A. M. (1995). The stability of precision grip forces during cyclic arm movements with a hand-held load. *Exp Brain Res*, **105**, 455–464.

Flanagan, J. R. & Wing, A. M. (1997). The role of internal models in motion planning and control: evidence from grip force adjustments during movements of hand-held loads. *J Neurosci*, **17**, 1519–1528.

Flanagan, J. R. & Lolley, S. (2001). The inertial anisotropy of the arm is accurately predicted during movement planning. *J Neurosci*, **21**, 1361–1369.

Flanagan, J. R., Tresilian, J. & Wing, A. M. (1993). Coupling of grip force and load force during arm movements with grasped objects. *Neurosci Lett*, **152**, 53–56.

Flanagan, J. R., King, S., Wolpert, D. M. & Johansson, R. S. (2001). Sensorimotor prediction and memory in object manipulation. *Can J Exp Psychol*, **55**, 87–95.

Flanagan, J. R., Vetter, P., Johansson, R. S. & Wolpert, D. M. (2003). Prediction precedes control in motor learning. *Curr Biol*, **13**, 146–150.

Forssberg, H., Jucaite, A. & Hadders-Algra, M. (1999). Shared memory representations for programming of lifting movements and associated whole body postural adjustments in humans. *Neurosci Lett*, **273**, 9–12.

Gordon, A. M., Westling, G., Cole, K. J. & Johansson, R. S. (1993). Memory representations underlying motor commands used during manipulation of common and novel objects. *J Neurophysiol*, **69**, 1789–1796.

Gysin, P., Kaminski, T. R., & Gordon, A. M. (2003). Coordination of fingertip forces in object transport during locomotion. *Exp Brain Res*, **149**, 371–379.

Gysin, P., Kaminski, T. R., Hass, C. J. & Gordon, A. M. (2008). Effects of gait variations on grip force coordination during object transport. *J Neurophysiol*, **100**, 2477–2485.

Häger-Ross, C., Cole, K. J. & Johansson, R. S. (1996). Grip-force responses to unanticipated object loading: load direction reveals body- and gravity-referenced intrinsic task variables. *Exp Brain Res*, **110**, 142–150.

Hermsdörfer, J., Marquardt, C., Philipp, J. *et al.* (2000). Moving weightless objects. Grip force control during microgravity. *Exp Brain Res*, **132**, 52–64.

Ivanenko, Y. P., Cappellini, G., Dominici, N., Poppele, R. E. & Lacquaniti, F. (2005). Coordination of locomotion with voluntary movements in humans. *J Neurosci*, **25**, 7238–7253.

Jian, Y., Winter, D. A., Ishac, M. G. & Gilchrist, L. (1993). Trajectory of the body COG and COP during initiation and termination of gait. *Gait Posture*, **1**, 9–22.

Johansson, R. S. & Westling, G. (1984). Roles of glabrous skin receptors and sensorimotor memory in automatic control of precision grip when lifting rougher or more slippery objects. *Exp Brain Res*, **56**, 550–564.

Johansson, R. S. & Westling, G. (1987). Signals in tactile afferents from the fingers eliciting adaptive motor responses during precision grip. *Exp Brain Res*, **66**, 141–154.

Johansson, R. S. & Westling, G. (1988). Programmed and triggered actions to rapid load changes during precision grip. *Exp Brain Res*, **71**, 72–86.

Johansson, R. S., Häger, C. & Bäckström, L. (1992a). Somatosensory control of precision grip during unpredictable pulling loads. III. Impairments during digital anesthesia. *Exp Brain Res*, **89**, 204–213.

Johansson, R. S., Häger, C. & Riso, R. (1992b). Somatosensory control of precision grip during unpredictable pulling loads. II. Changes in load force rate. *Exp Brain Res*, **89**, 192–203.

Johansson, R. S., Riso, R., Häger, C. & Bäckström, L. (1992c). Somatosensory control of precision grip during unpredictable pulling loads. I. Changes in load force amplitude. *Exp Brain Res*, **89**, 181–191.

Jordan, M. I. & Rumelhart, D. E. (1992). Forward models: supervised learning with a distal teacher. *Cogn Sci*, **16**, 307–354.

Kawato, M. (1999). Internal models for motor control and trajectory planning. *Curr Opin Neurobiol*, **9**, 718–727.

Kawato, M., Kuroda, T., Imamizu, H. *et al*. (2003). Internal forward models in the cerebellum: fMRI study on grip force and load force coupling. *Prog Brain Res*, **142**, 171–188.

Kinoshita, H., Kawai, S., Ikuta, K. & Teraoka, T. (1996). Individual finger forces acting on a grasped object during shaking actions. *Ergonomics*, **39**, 243–256.

Lajoie, K. & Drew, T. (2007). Lesions of area 5 of the posterior parietal cortex in the cat produce errors in the accuracy of paw placement during visually guided locomotion. *J Neurophysiol*, **97**, 2339–2354.

Lewis, P. A., & Miall R. C. (2003). Distinct systems for automatic and cognitively controlled time measurement: evidence from neuroimaging. *Curr Opin Neurobiol*, **13**, 250–255.

McFadyen, B. J. & Carnahan, H. (1997). Anticipatory locomotor adjustments for accommodating versus avoiding level changes in humans. *Exp Brain Res*, **114**, 500–506.

Mrotek, L. A., Hart, B. A., Schot, P. K. & Fennigkoh, L. (2004). Grip responses to object load perturbations are stimulus and phase sensitive. *Exp Brain Res*, **155**, 413–420.

Müller, M. L., Jennings, J. R., Redfern, M. S. & Furman, J. M. (2004). Effect of preparation on dual-task performance in postural control. *J Motor Behav*, **36**, 137–146.

Muratori, L. M., McIsaac, T. L., Gordon, A. M. & Santello, M. (2008). Impaired anticipatory control of force sharing patterns during whole-hand grasping in Parkinson's disease. *Exp Brain Res*, **185**, 41–52.

Murray, M. P., Drought, A. B. & Kory, R. C. (1964). Walking pattern of normal men. *J Bone Joint Surg*, **46A**, 335–360.

Murray, M. P., Kory, R. C. & Sepic, S. B. (1970). Walking pattern of normal women. *Arch Phys Med Rehabil*, **51**, 637–650.

Nissan, M. & Whittle, M. W. (1990). Initiation of gait in normal subjects: a preliminary study. *J Biomed Eng*, **12**, 165–171.

Nowak, D. A. (2004). Different modes of grip force control: voluntary and externally guided arm movements with a hand-held load. *Clin Neurophysiol*, **115**, 839–848.

Nowak, D. A., Hermsdörfer, J., Philipp, J. *et al.* (2001). Effects of changing gravity on anticipatory grip force control during point-to-point movements of a hand-held object. *Motor Control*, **5**, 231–253.

Nowak, D. A., Hermsdörfer, J., Marquardt, C. & Fuchs, H. H. (2002). Grip and load force coupling during discrete vertical arm movements with a grasped object in cerebellar atrophy. *Exp Brain Res*, **145**, 28–39.

Nowak, D. A., Hermsdörfer, J., Schneider, E. & Glasauer, S. (2004). Moving objects in a rotating environment: rapid prediction of Coriolis and centrifugal force perturbations. *Exp Brain Res*, **157**, 241–254.

Patla, A. E. & Vickers, J. N. (1997). Where and when do we look as we approach and step over an obstacle in the travel path? *Neuroreport*, **8**, 3661–3665.

Pellijeff, A., Bonilha, L., Morgan, P. S., McKenzie, K. & Jackson, S. R. (2006). Parietal updating of limb posture: an event-related fMRI study. *Neuropsychologia*, **44**, 2685–2690.

Pilon, J. F., De Serres, S. J. & Feldman, A. G. (2007). Threshold position control of arm movement with anticipatory increase in grip force. *Exp Brain Res*, **181**, 49–67.

Pisella, L., Grea, H., Tilikete, C. *et al.* (2000). An 'automatic pilot' for the hand in human posterior parietal cortex: toward reinterpreting optic ataxia. *Nat Neurosci*, **3**, 729–736.

Poizner, H., Feldman, A. G., Levin, M. F. *et al.* (2000). The timing of arm-trunk coordination is deficient and vision-dependent in Parkinson's patients during reaching movements. *Exp Brain Res*, **133**, 279–292.

Salimi, I., Hollender, I., Frazier, W. & Gordon, A. M. (2000). Specificity of internal representations underlying grasping. *J Neurophysiol*, **84**, 2390–2397.

Saunders, J. B., Inman, V. T. & Eberhart, H. D. (1953). The major determinants in normal and pathological gait. *J Bone Joint Surg Am*, **35-A**, 543–558.

Schaal, S., Sternad, D., Osu, R. & Kawato, M. (2004). Rhythmic arm movement is not discrete. *Nat Neurosci*, **7**, 1136–1143.

Schettino, L. F., Rajaraman, V., Jack, D. *et al.* (2003). Deficits in the evolution of hand preshaping in Parkinson's disease. *Neuropsychologia*, **42**, 82–94.

Serrien, D. J. & Wiesendanger, M. (1999). Role of the cerebellum in tuning anticipatory and reactive grip force responses. *J Cogn Neurosci*, **11**, 672–681.

Serrien, D. J. & Wiesendanger, M. (2001). Regulation of grasping forces during bimanual in-phase and anti-phase coordination. *Neuropsychologia*, **39**, 1379–1384.

Singhal, A., Culham, J. C., Chinellato, E. & Goodale, M. A. (2007). Dual-task interference is greater in delayed grasping than in visually guided grasping. *J Vis*, **7**, 5 1–12.

Siu, K. C., Catena, R. D., Chou, L. S., van Donkelaar, P. & Woollacott, M. H. (2008). Effects of a secondary task on obstacle avoidance in healthy young adults. *Exp Brain Res*, **184**, 115–120.

Sternad, D., Wei, K., Diedrichsen, J. & Ivry, R. B. (2007). Intermanual interactions during initiation and production of rhythmic and discrete movements in individuals lacking a corpus callosum. *Exp Brain Res*, **176**, 559–574.

Thorstensson, A., Nilsson, J., Carlson, H. & Zomlefer, M. R. (1984). Trunk movements in human locomotion. *Acta Physiol Scand*, **121**, 9–22.

Vaillancourt, D. E., Yu, H., Mayka, M. A. & Corcos, D. M. (2007). Role of the basal ganglia and frontal cortex in selecting and producing internally guided force pulses. *Neuroimage*, **36**, 793–803.

van den Berg, C., Beek, P. J., Wagenaar, R. C. & van Wieringen, P. C. (2000). Coordination disorders in patients with Parkinson's disease: a study of paced rhythmic forearm movements. *Exp Brain Res*, **134**, 174–186.

Waters, R. L., Morris, J. & Perry, J. (1973). Translational motion of the head and trunk during normal walking. *J Biomech*, **6**, 167–172.

White, O., McIntyre, J., Augurelle, A. S. & Thonnard, J. L. (2005). Do novel gravitational environments alter the grip-force/load-force coupling at the fingertips? *Exp Brain Res*, **163**, 324–334.

Wing, A. M. & Lederman, S. J. (1998). Anticipating load torques produced by voluntary movements. *J Exp Psychol Hum Percept Perform*, **24**, 1571–1581.

Wing, A. M., Flanagan, J. R. & Richardson, J. (1997). Anticipatory postural adjustments in stance and grip. *Exp Brain Res*, **116**, 122–130.

Witney, A. G., Goodbody, S. J. & Wolpert, D. M. (1999). Predictive motor learning of temporal delays. *J Neurophysiol*, **82**, 2039–2048.

17

Development of grasping and object manipulation

BRIGITTE VOLLMER AND HANS FORSSBERG

Summary

The development of skilled hand movements such as grasping and object manipulation is of fundamental importance to the ability to perform everyday life activities. The purpose of this chapter is to provide an overview on development of grasping and object manipulation. The first part describes developmental characteristics of prehension, i.e. reaching and grasping. The second part deals with the development of independent finger movements, which is an important prerequisite for both grip formation and object manipulation. In the third part, aspects of manipulation of unstable and stable objects are discussed. This includes discussion of the development of sensory control mechanisms, i.e. adaptation to friction and weight of the manipulated object. Finally, the concept of neuronal group selection, a concept that implies that development is the result of complex interaction between genes and environment, is described.

Introduction

Prehension (i.e. reaching and grasping) and manipulation of objects are motor skills that are fundamentally important for exploration and interaction with the environment. Although human infants can grasp from an early age, it takes several years before children are able to perform these skills in a mature pattern. The development of these skilled hand movements and underlying neural mechanisms is the focus of this chapter.

Prehension involves two main components, i.e. reaching and grasping. First, the hand has to be moved to the location of the object. Second, the grip must be adapted to size, shape, orientation and the intended use of the object (see also Chapters 1 and 12). It is well established that reaching and grasping are coordinated by two interacting motor programs in which the hand grip is shaped during transportation of the hand to the object. The direction, amplitude and temporal parameters of both the reaching and grasping movements are programmed in advance based on visual information. The velocity profiles have typical "bell-shaped" trajectories with an initial rapid acceleration phase followed by a slower phase of deceleration. Corrections, if required, are made at the end of the movement, which results in multiple peaks in the velocity trajectory.

Sensorimotor Control of Grasping: Physiology and Pathophysiology, ed. Dennis A. Nowak and Joachim Hermsdörfer. Published by Cambridge University Press. © Cambridge University Press 2009.

A prerequisite for successful object manipulation is the ability to move fingers independently of each other. Independent finger movements are required for both forming various grips and manipulation of an object. Independent control of the fingers, i.e. breaking innate synergies in which all fingers are moved together, appears to be dependent on an intact motor cortex and intact cortico-spinal tracts (Lawrence & Kuypers, 1968a, 1968b; Schieber, 1990). During object manipulation, e.g. when rotating a coin between the fingers, the vectors of fingertip forces applied to the object have to be controlled precisely. Most of our knowledge about the sensory-motor mechanisms involved in the control of object manipulation has been derived from lifting stable (non-deformable) objects with the precision grip – a paradigm developed by Johansson and collaborators (for details see Chapters 1, 11 and 12). Only recently, an interesting paradigm for investigation of manipulation of unstable objects has been developed (Valero-Cuevas *et al.*, 2003; see below).

Development of prehension

Even in the newborn infant, a basic neuro-muscular infrastructure for reaching and grasping is present. When an object is placed in the palm of a newborn infant, the tactile stimulation triggers a grasp reaction in which all digits are flexed around the object. Similarly, in newborns, reaching movements aimed towards objects within the centre of the visual field are present (Von Hofsten, 1982). However, the first goal-directed reaching and grasping movements are observed at around 4 months of age (Von Hofsten, 1991). The subsequent development of prehension is a dynamic process in which reaching and grasping movements become more integrated and efficient over time. There are three indispensable aspects: (1) refinement of the reaching movement (which requires correct perception of distance and location), (2) improvement of grip formation linked to better coordination with reaching and (3) development of the ability to reach and grasp without visual guidance (which requires memorizing an object's intrinsic and extrinsic characteristics).

Kinematic studies of infant reaching (Konzcak & Dichgans, 1997) show in very young infants (age 5 months) curved movement paths with high intra-individual variability. Subsequently, these paths become quickly straighter up until 9 months of age, followed by slower improvement. At 3 years of age, there are still some differences in the kinematic profile compared with an adult pattern. Figure 17.1 (from Kuhtz-Buschbeck *et al.*, 1998) illustrates the development of kinematic profiles of prehension in three groups of children aged 4, 7 and 12 years respectively. It can be seen that over the age range shown, variability decreases and kinematic profiles become smoother with increasing age.

In adults, the hand grip is shaped during transportation of the hand to the object (preshaping). The hand starts to close before reaching the object, with the maximum grip aperture occurring at 60–80% of the total movement time, and with the precise timing depending on the size of the object (Jeannerod, 1984; Paillard 1987; Jakobson & Goodale, 1991; Jeannerod *et al.*, 1998) (see also Chapters 2 and 10). In infants, the reaching and grasping movements are not yet integrated into one action but are more sequentially organized. Von Hofsten & Rönnqvist (1988) investigated the coordination between reaching

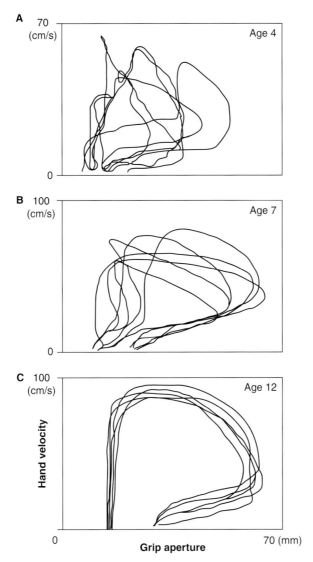

Figure 17.1. A–C. Kinematic profiles of prehension at three ages (4, 7, 12 years) with hand velocity plotted against grip aperture in three children.

and grasping under visual guidance in infants aged 6, 9 and 13 months. In all infant groups, hand closure started before touching the object, while only the 13-month-old infants had a pattern similar to adults. In none of the infant groups was timing dependent on the size of the object, and adjustment of grip aperture to object size was only seen in the 9- and 13-month-olds. A mature pattern of integration of reaching and grasping does not develop until about 12 years of age (e.g. Kuhtz-Buschbeck *et al.*, 1998; Schneiberg *et al.*, 2002). At 6 years of age, the reach-to-grasp coordination is still variable and unstable and there is a non-linear improvement with age, with a critical period at age 8 years (Fetters & Todd, 1987; Konczak *et al.*, 1995; Olivier *et al.*, 2007). Kuhtz-Buschbeck *et al.* (1998) investigated grip formation during visually guided reaching in children aged 4–12 years. The 4-year-olds exhibited anticipatory pre-shaping of hand aperture appropriate to object size, but used a wider hand opening than older children or adults. Over the investigated age range of 4–12 years, grip formation improved and developed towards a mature uniform adult pattern that was well integrated with the transport of the hand.

Visual information about the extrinsic and intrinsic properties of an object plays a fundamental role in programming of reaching and grasping movements before the movements are initiated. Visual guidance and feedback seems to be especially important for endpoint accuracy. Without having sight of the hand and the object during the movement, adults open the hand wider, reach the maximum of hand aperture earlier, and prolong the final phase of the movement. Several studies indicate that young children cannot yet use the visual information for accurate advance programming of the reaching and grasping movements, but that they need on-line vision to guide the hand correctly (Kuhtz-Buschbeck *et al.*, 1998). The ability to program movements in advance based on visual information seems to develop at around 5–6 years. A study by Smyth *et al.* (2004) in children showed that at the age of 5–6 years, the absence of sight of the hand during the movement has no effect on the timing and grip aperture. At 9–10 years of age, children performed the movement without sight of the hand in a similar way to adults.

Development of independent finger movements

The ability to move fingers independently of each other is a prerequisite for grip formation and object manipulation. This ability evolves gradually during infancy. Initially, the finger movements seem to be coupled to each other and fingers flex and extend in synergy. In newborns, reflexive closure of the hand is present in the grasp reflex (i.e. all fingers flex in synergy). When voluntary grasping develops at around 3 months of age, all fingers are still flexed in synergy and objects are held with a power grip (or palmar grip). Subsequently, the flexor synergy is broken up, and thumb, index finger and the ulnar fingers can be moved more independently of each other. From about age 10 months, the index finger and the thumb can be isolated from the other fingers and small objects can be picked up in a precision grip, i.e. between the tip of the index finger and the thumb (pincer grip). However, until about 3 years of age it is difficult for the child to perform more sophisticated isolated finger movements (e.g. sequential opposition of the fingers to the thumb). Further refinement of finger coordination and hand shaping takes place throughout the first decade of life.

The neural control of independent finger movements can be studied by measuring of forces that are produced unintentionally by fingers not involved in a voluntary task that requires maximal voluntary force production in one finger ("enslaving"). Shim *et al.* (2007) found that over the age range 6–10 years, finger inter-dependency ("enslaving") decreased with age. The level of finger independency in flexion in the 10-year-old children was similar to that of adults, whereas in extension, the level of independency at age 10 years was not. These findings indicate that independent finger movements in flexion develop at a faster rate than in extension movements.

Development of object manipulation

Manipulation of unstable objects

For successful manipulation of objects, the force vectors of the fingertip forces need to be controlled precisely. Recently, a paradigm for investigation of finger force coordination when manipulating unstable objects has been developed in adults by Valero-Cuevas *et al.* (the Strength-Dexterity Test, Valero-Cuevas *et al.*, 2003). This test consists of a series of springs that have to be compressed to their solid length (Figure 17.2A). The individual springs in the test set differ with regard to dexterity and force requirements. Each individual spring is characterized by a strength index, which refers to the force required to compress a spring to its solid length, and a dexterity index, which measures the tendency of a spring to buckle during compression. This test has recently been adapted for use in children and, in a first study in a population of typically developing children and adolescents aged 5–16 years, it was found that over the observed age range, performance on the test improved in a linear way (own unpublished data, see Figure 17.2B). Furthermore, performance on this test correlated with both performance on a test that assesses gross manual dexterity (the Box and Block test) and pinch strength measured with a pinch meter.

Manipulation and lifting of stable objects

When lifting a stable (non-deformable) object, the grip force (normal to the contact surface) is automatically initiated prior to the lift force (vertical tangential force) and thereafter increased in parallel with the lift force increase in order to provide grasp stability and prevent the object from slipping out of the hand (Westling & Johansson, 1984) (see also Chapters 1 and 12). A similar automatic predictive enhancement of the grip force (normal force) is generated when an object is moved through the air in order to resist the inertial forces (tangential) generated by the acceleration and deceleration of the object (Wing, 1996) (see also Chapters 1 and 12). This type of invariant, task-related motor behavior is believed to represent the neural output from functional motor programs (or synergies) that simplify the demands on the nervous system by reducing the degrees of freedom that have to be controlled (Bernstein, 1967). The grip force and lift force are automatically coupled together by the motor program and are not controlled independently. The basic coupling of the forces is a prerequisite for the sensory motor mechanisms that control adaptation of the fingertip forces to the physical properties of the object (see also Chapter 12).

This basic grip–lift synergy is not present when infants start to use their hands for grasping. When the precision grip emerges around 10 months of age, children use a

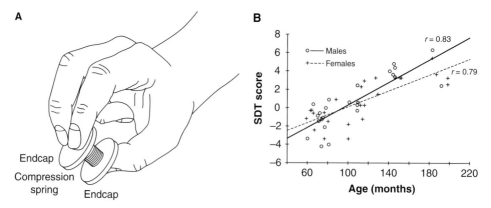

Figures 17.2. A. A compression spring from the Strength–Dexterity test (SD-test) kit and correct finger positions for task performance. B. The scatter plot and regression lines of SD-test participant ability measures versus age, plotted separately for males and females. SDT score = measure of performance on the SD-test.

sequential order in which they first establish the grip force and thereafter initiate the lift force (Forssberg *et al.*, 1991, see Figure 17.3). In young infants, the initial gripping of the object is often combined with a negative lift force where the object is pressed down against the surface. With this strategy children produce a firm grasp before lifting. During the second half of the second year, children start to develop automatized coupling between the grip force and load force. The coupling develops rapidly over subsequent years, but is then followed by a long phase of fine tuning that is not completed until adolescence. A major characteristic of the precision grip in children is a large intra-individual variability of the fingertip force coordination in lifts, in particular at the youngest ages.

The long-term development of a mature grip–lift synergy suggests a similarly slow development of the motor program that controls fingertip force output. It also raises the question whether this is an innate behavior (program) that matures slowly, or a learned motor activity without any predetermined neural structure. The large intra-subject variation of the force patterns between subsequent lifts in young children, in contrast to the invariant pattern seen in adults, reflects some type of learning process. At the same time well-coordinated lifts are mixed with less mature patterns. This variation might indicate that at an early age several prototypic motor programs are present. The large variation in young children might allow the central nervous system to evaluate the outcome of the various motor patterns (programs), then select the most efficient and discard the least efficient patterns (see "Neuronal group selection theory" later in this chapter). Interestingly, in a recent study in children with hypoplastic digits who had surgery at ages 1–13 years with transfer of the big toe to the hand, age did not seem to be critical for developing the proper grip–lift force coordination during lifting with the new digit (Schenker *et al.*, 2007). This is contradictory to earlier thoughts and indicates that there is no critical period, but that the capacity to develop appropriate grip–lift synergy remains during all of childhood.

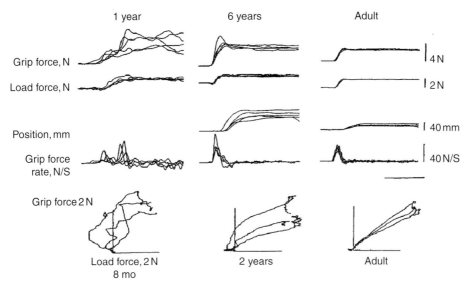

Figure 17.3. Recordings of isometric fingertip forces while subjects are lifting an instrumented object 5 cm above the table surface and keep it stationary in the air. The object was lifted by a precision grip between the pulps of the thumb and index finger. Force transducers measured the grip force (normal to the surface), load force (vertical tangential force) and the vertical position. Five trials are superimposed from three subjects at different ages. Grip force rate indicates the first-time derivative of the grip force. The grip force is plotted against the load force (abscissa) in the panels below. Note (i) the large variation between trials in the youngest subject and the invariant pattern in the adult, indicating an automatized control; (ii) the parallel increase of grip force and load force in adults versus the sequential increase in children (leading grip force); (iii) the single-peaked grip force rate in the adult indicating a programmed force generation versus multiple peaks in the 1-year-old indicating a feedback-controlled probing strategy (modified from Forssberg *et al.*, 1991)

Sensory control mechanisms

When a fragile object is grasped and manipulated (e.g. when picking a berry), the amplitude of the fingertip forces has to be adapted precisely so as not to crush or drop the object. Similarly, when a glass of water is lifted, the movement has to be smooth in order not to spill the water. These examples illustrate that the amplitude of force development in the automatized grip–lift synergy has to be adapted to the physical properties of the object, i.e. the friction of the object's surface and weight of the object.

Adaptation to friction – tactile adjustments

The adaptation of the grip force to the friction of the object's contact surface is conveyed by the cutaneous mechano-receptors at the fingertips, which allow detection of small *micro slips* between the skin and the object (Johansson & Westling, 1984) (see also Chapters 11 and 13). This tactile information is used to automatically adjust grip force to be just above the level when the object starts to slip. If the grip force is small and closely above the slip level, micro slips will occur and induce fast corrections, upgrading the grip force well above the slip level. In order to avoid dropping the object, an object with a *slippery surface* requires a larger grip force than an object with a *rough surface*. This is particularly important during the load phase before the object is lifted. Depending on the friction of the object, the grip force increases. This is automatically generated so as to provide grasp stability and is set to increase at different rates, i.e. faster for a slippery than for a rough surface. Two mechanisms are responsible for this adaptation. First, there is anticipatory programming of the grip-force slope based on stored information about the frictions obtained during previous lifts of the same object. Second, directly after contact, there is a fast upgrading based on information on the actual friction followed by adjustment of the grip force (Edin *et al.*, 1997).

Young children are able to adapt the grip force to friction of the object early in life and long before the grip–lift synergy is properly developed. This suggests that both sensory and motor mechanisms needed for recording the friction between skin and object, as well as the processing of this information to adaptive grip-force motor commands, develop faster than the grip–lift synergy (see Figure 17.4; Forssberg *et al.*, 1992). However, young children generate excessive grip force, which results in a large safety margin. This could be a "safety strategy" indicating that the sensory processes are not fully developed after all. A similar "safety strategy" is used in adults with local finger anesthesia that blocks the tactile information from the fingertips (see also Chapters 11, 13 and 19). The excessive grip forces could also be a compensatory mechanism to avoid slips, since the grip force in young children is not stable but oscillates with minima close to the slip level that jeopardizes grasp stability.

Adaptation to weight – predictive parameterisation of force

In order to achieve smooth and dampened vertical lifting movements, the lift force increase must be decreased to just above zero at lift-off. If it is zero or below, the object will not lift. If it is too high, the object will be lifted briskly, since excessive lift force increase is transformed into acceleration. A well-adjusted lift force at lift-off could be achieved by a "probing strategy" with stepwise force increase until the object lifts (feedback control). However, such a strategy would be time consuming. Instead, in adults, the load force is programmed in one pulse adapted to the weight of the object (see also Chapter 12). This results in a "bell-shaped" force rate trajectory (i.e. the first-time derivative of the force) in which the lift force rate peaks in the middle of the loading phase and is decreased just above zero at lift-off (Johansson & Westling, 1988). Since explicit information on an object's weight cannot be acquired before lift-off, the adaptation of the force output has to rely on stored information. The existence of such information becomes obvious when the weight of the object is changed without knowledge of the subject who lifts the object. If the object is lighter than expected, the object will briskly lift earlier than expected, while a heavier object will not lift until additional force pulses are generated. This suggests complex sensory-motor processes in which information about an object's weight is stored from previous experience in memory representations, which are used in order to program the load force increase in advance such that it targets the weight of the object.

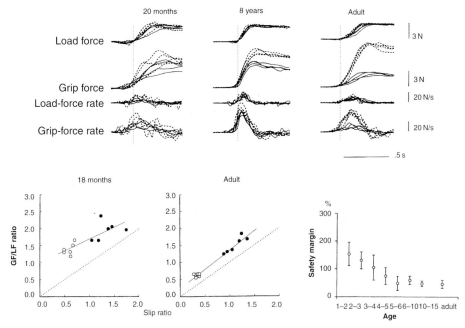

Figure 17.4. The same precision grip–lift paradigm as described in Figure 17.3. The friction at the contact surfaces of the object was changed by using sandpaper (dashed lines, open circles – high friction) or silk (continuous lines, filled circles – low friction). Single force trajectories from four "sandpaper" and "silk" lifts, respectively, are superimposed in three subjects of different ages in the top panels. Note the clear separation of the grip force trajectories in "sandpaper" and "silk" lifts at older ages, being much smaller in the youngest child. In the lower left panels the ratio between the grip force and the load force during the static phase of the lift (when the object is held in the air) is plotted against the slip ratio (abscissa), i.e. the ratio between grip force and load force when the object starts to slip out of the hand. Six lifts with "sandpaper" and "silk", respectively, are plotted. The dotted line represents the slip ratio. Note that already the youngest child adapted the grip force to the friction of the contact surfaces, but had a much larger safety margin to the slip ratio than the older subjects. In the lower right panel the safety margin for "sandpaper" is plotted expressed as a percentage of the slip ratio (modified from Forssberg *et al.*, 1995).

When infants start to lift objects they use a "probing strategy" with stepwise force increase (Forssberg *et al.*, 1992, 1995). Since the "anticipatory processes" of adults have not yet developed, there is a dependency on a feedback strategy instead. Towards the end of the second year, the force rate profiles become more "uni-modal." However, it is not until age 6–8 years that a "bell-shaped" force rate profile which indicates that the entire force output is programmed in advance is achieved. During the same period, children develop the ability to adjust the amplitude of the force rate pulse that targets the weight of the object. At an early age, there are only minor differences in the first force pulse depending on weight and children have to produce several force pulses before the object is lifted. Not before the age of 10 years of age have children developed a force pulse control that is as efficient as that seen in adults. This development of lift behavior clearly shows that young children use a feedback controlled movement in the beginning. Gradually, they develop an anticipatory control and start to program force output in advance. This transfer, from feedback control to anticipatory programming, is, however, not completely developed before 8–10 years of age. In parallel with the development of the motor commands producing "bell-shaped" force trajectories, neural systems develop that store critical parameters in internal memory representations and use them for programming of the "bell-shaped" force rate.

The internal representation of an object's weight is used to set parameters of both the load force and the grip force. In addition, as mentioned previously, the amplitude of the grip force has to be set separately in the start phase of the lift in order to avoid slips of the object. This results in faster grip force increase for objects with slippery surfaces. Already in children as young as 2 years of age, a difference in the initial phase of the lift can be detected when objects with different frictional characteristics are lifted. This indicates that children very early are able to store tactile information on the object's surface in internal memory representations and use them to program the grip-force increase in relation to load-force increase.

Visual information is not only used for prehension, but also for prediction of the physical properties of the object and for anticipatory programming of force output. This has been demonstrated by studying subjects who lift objects of similar weight, but of different sizes. The forces were targeted to a higher weight when the large object was lifted (Gordon *et al.*, 1991b, 1992). Similar results were obtained when blindfolded subjects used their hands to estimate the size of the object. This indicates that visual (or haptic; Gordon *et al.*, 1991a) information that provides cues about the size of an object is transformed into weight estimates that are then used to program the force parameters before movement is initiated. While children start to use information about weight and friction for programming of the force output during their second year, they are unable to use visual size cues before the third year of life (Gordon *et al.*, 1992) (see also Chapter 11). This suggests that circuits used to transform object size into weight parameters develop later than neural mechanisms used for storing information about weight and friction. It is therefore likely that this transformation from one modality to another requires more complex cortical processes.

Triggered corrections – reflexes involved in manipulation

Grasp reflexes induced by tactile stimulation are present during the first months of life but diminish as goal-directed grasping develops. However, reflex-triggered grip adjustments during manipulation are still important in adults. A micro slip (see above), induced by a too-low grip force, will trigger a brisk grip-force increase, followed by adaptation of the grip force to a higher level. Similarly, a sudden load increase will trigger brisk reflex increases in both grip force and load force (Johansson & Westling, 1988) (see also Chapter 11). These triggered corrections are fast, with latencies around 65 ms, which allows supraspinal transmission. A similar reflex response to sudden weight perturbation is already present in children during the second year of life. However, it is less efficient, i.e. the latency is longer (100 ms) and the amplitude smaller (see Figure 17.5; Eliasson *et al.*, 1995). During

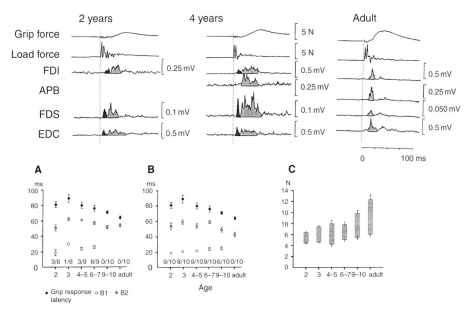

Figure 17.5. Averaged force and EMG recordings from an experiment in which the subjects held an instrumented receptacle with the precision grip, while the experimenter dropped a 75 g disc 12 times. At the top, the grip force and load force of three subjects at various ages together with the EMG signals from four muscles are shown: first dorsal interosseus (FDI), abductor pollicis brevis (APB), flexor digitorum superficialis (FDS) and extensor digitorum communis (EDC). The recordings were synchronized to the impact of the disc on the receptacle (vertical dotted line). The EMG responses were divided into two bursts, depending on the latency after impact: less than 30 ms (dark shading), more than 40 ms (light shading). In the lower row A and B show the latencies from the impact to the grip force response (filled circles), and the first (open circle) and second (shaded circle) electromyogram (EMG) burst from FDI (A) and FDS (B). The numbers beneath the symbols indicate the number of subjects that exhibited the early EMG burst in relation to subjects tested in each age group. Note the diminishing number of subjects with a first burst with increasing age, in conjunction with shorter latencies of the second burst and of the grip-force response. In C the amplitude of the grip response is plotted for each age group. The lower end of the vertical bars indicates the baseline grip force level prior to impact. The upper end marks show the peak grip force after impact, i.e. the bars represent the actual grip response. Note the increasing amplitude with age (modified from Eliasson *et al.*, 1995).

development, the grip response latency decreases and the grip force amplitude increases to adult levels around 9–10 years of age. The development of the grip-force responses is accompanied by changes in the EMG pattern reflecting various reflex pathways. Cutaneo-muscular reflex pathways, probably used during manipulation, have been studied by weak electrical stimulation of the digits (Jenner & Stephens, 1982; Evans 1997). The cutaneous stimulation evokes first a short latency reflex burst (E1) followed by an inhibitory component (I1). The short latencies indicate that these effects are mediated through spinal pathways only. They are followed by a long latency reflex. Several studies have indicated that these effects are transmitted through a transcortical loop (Jenner & Stephens, 1979). At birth, only the spinal excitatory component (E1) is elicited. During the first year of life, the E1 decreases in conjunction with the appearance of I1. The long-latency E2 component does not develop until the second year of life. At 2 years of age children have developed the characteristic triphasic pattern. In grasping, the improved grip-force response, triggered by sudden weight perturbations, is accompanied by similar characteristic EMG changes. A reflex burst corresponding to E2 (latency 40–55 ms) is enhanced, while a short-latency burst (latency 20 ms) diminishes with age. This indicates a dramatic reorganization of the cutaneous reflex pathways, from stereotyped reflex patterns, organized at the spinal level, to more flexible and functional supra-spinal reactions that are well integrated in motor skills. Most likely inter-neurons in the spinal pathways are inhibited by supra-spinal circuits that block the spinal reflexes, while at the same time transcortical pathways that can be adapted to the ongoing motor activity are developed.

Neuronal group selection theory

One principal property of normal development is variation (Touwen, 1978). Variation is present in practically all developmental parameters, such as motor performance, developmental sequence, or the duration of developmental stages. The development of the motor coordination of the precision grip follows a similar pattern. When young children start to grasp, the coordination of grip force and load force varies substantially. Several different coordination patterns are used, amongst them patterns in which the forces increase in parallel. With increasing age, the parallel coordination pattern is used more and more frequently, while other patterns are used less frequently. Eventually, the child has established a consistent parallel force pattern in a "bell-shaped" profile.

Recently, Sporns & Edelman (1993) have put forward a theory that may end the old "nature–nurture" debate (see Forssberg, 1999). The "neuronal group selection" theory is based on the use-dependent selection of synapses and neurons, and describes how active neurons survive and develop, while less active neurons are discarded through a dynamic epigenetic regulation of synaptic connectivity, cell division, migration and death, neurite extension and retraction. The theory implies that there is a repertoire of genetically determined neural networks at the onset of development that can induce a broad repertoire of movement patterns. The large variation of movement patterns reflects the variation of neural networks. Development proceeds with selection of movement patterns, i.e. neural circuits, on the basis of afferent information produced by the behavior and the environment. The selection is accomplished by a "re-entry signaling" process, in which the most favorable patterns (neuronal networks, motor programs) are retained. The nature of this process is not known, and neither is it known how favorable patterns are discriminated from less favorable patterns. The process results in selection of one pattern and a reduction of the behavioral variation. The theory of neuronal group selection is appealing because it implies that development is the result of complex interaction between genes and environment.

References

Bernstein, N. (1967). *The Coordination and Regulation of Movements*. Oxford, UK: Pergamon.

Edin, B. B., Westling, G. & Johansson, R. S. (1997). Independent control of human finger tip forces at individual digits during precision lifting. *J Physiol*, **450**, 547–564.

Eliasson, A. C., Forssberg H., Ikuta K. *et al.* (1995). Development of human precision grip. V. Anticipatory and triggered grip actions during sudden loading. *Exp Brain Res*, **106**, 425–433.

Evans, A. L. (1997). Development and function of cutaneomuscular reflexes and their pathophysiology in cerebral palsy. In K. J. Connolly & H. Forssberg (Eds.), *Neurophysiology and Neuropsychology of Motor Development* (pp. 145–161). London: MacKeith Press.

Fetters, L. & Todd, J. (1987). Quantitative assessment of infant reaching movements. *J Motor Behav*, **19**, 147–166.

Forssberg, H. (1999). Neural control of human motor development. *Curr Opin Neurobiol*, **9**, 676–682.

Forssberg, H., Eliasson, A. C., Kinoshita, H., Johansson, R. S. & Westling, G. (1991). Development of human precision grip I: Basic coordination of force. *Exp Brain Res*, **85**, 451–457.

Forssberg, H., Kinoshita, H., Eliasson, A. C. *et al.* (1992). Development of human precision grip II. Anticipatory control of isometric forces targeted for object's weight. *Exp Brain Res*, **90**, 393–398.

Forssberg, H., Eliasson, A. C., Kinoshita, H., Johansson, R. S. & Westling, G. (1995). Development of human precision grip IV: Tactile adaptation of isometric finger forces to the frictional condition. *Exp Brain Res*, **104**, 323–330.

Gordon, A. M., Forssberg, H., Johansson, R. S. & Westling, G. (1991a). The integration of haptically acquired size information in the programming of precision grip. *Exp Brain Res*, **83**, 483–488.

Gordon, A. M., Forssberg, H., Johansson, R. S. & Westling, G. (1991b). Visual size cues in the programming of manipulative forces during precision grip. *Exp Brain Res* **83**, 477–482.

Gordon, A. M., Forssberg, H., Johansson, R. S., Eliasson, A. C. & Westling, G. (1992). Development of human precision grip. III. Integration of visual size cues during the programming of isometric forces. *Exp Brain Res*, **90**, 399–403.

Jakobson, L. S. & Goodale, M. A. (1991). Factors affecting higher-order movement planning: a kinematic analysis of human prehension. *Exp Brain Res*, **86**, 199–208.

Jeannerod, M. (1984). The timing of natural prehension movements. *J Motor Behav*, **16**, 235–254.

Jeannerod, M. (1991). The interaction of visual and proprioceptive cues in controlling reaching movements. In D. R. Humphrey & H. J. Freund (Eds.), *Motor control: Concepts and Issues* (pp. 277–291). Chichester, UK: John Wiley & Sons Ltd.

Jeannerod, M., Paulignan, Y. & Weiss, P. (1998). Grasping an object: one movement, several components. *Novartis Foundation Symposium*, **218**: 5–16; discussion pp. 16–20.

Jenner, J. R. & Stephens, J. A. (1979). Evidence for a transcortical cutaneous reflex response in man. *J Physiol*, **293**, 39P–40P.

Jenner, J. R. & Stephens, J. A. (1982). Cutaneous reflex responses and their central nervous pathways studied in man. *J Physiol*, **333**, 405–419.

Johansson, R. S. & Westling, G. (1984). Roles of glabrous skin receptors and sensorimotor memory in automatic control of precision grip when lifting rougher or more slippery objects. *Exp Brain Res*, **56**, 550–564.

Johansson, R. S. & Westling, G. (1988). Coordinated isometric muscle commands adequately and erroneously programmed for the weight during lifting task with precision grip. *Exp Brain Res*, **71**, 59–71.

Konczak, J. & Dichgans, J. (1997). The development toward stereotypic arm kinematics during reaching in the first 3 years of life. *Exp Brain Res*, **117**, 346–354.

Konczak, J., Borutta, M., Topka, H. & Dichgans, J. (1995). The development of goal-directed reaching in infants: hand trajectory formation and joint torque control. *Exp Brain Res*, **106**, 156–168.

Kuhtz-Buschbeck, J. P., Stolze, H., Jöhnk, K., Boczed-Funcke, A. & Ilert, M. (1998). Development of prehension movements in children: a kinematic study. *Exp Brain Res*, **122**, 424–432.

Lawrence, D. G. & Kuypers, H. G. (1968a). The functional organization of the motor system in the monkey. I. The effects of bilateral pyramidal lesions. *Brain*, **91**, 1–14.

Lawrence, D. G. & Kuypers, H. G. (1968b). The functional organization of the motor system in the monkey. II. The effects of lesions of the descending brain-stem pathways. *Brain*, **91**, 15–36.

Olivier, I., Hay, L., Bard, C. & Fleury, M. (2007). Age-related differences in the reaching and grasping coordination in children: unimanual and bimanual tasks. *Exp Brain Res*, **179**, 17–27.

Paillard, J. (1987). Cognitive versus sensorimotor encoding of spatial information. In P. Ellen & C. Thinus-Blanc (Eds.), *Cognitive Processes and Spatial Orientation in Animal and Man* (pp. 43–77). Dordrecht, the Netherlands: Martinus Nijhoff.

Schenker, M., Wiberg, M., Kay, S. P. & Johansson, R. S. (2007). Precision grip function after free toe transfer in children with hypoplastic digits. *J Plastic Reconstr Aesth Surg*, **60**, 13–23.

Schieber, M. H. (1990). How might the motor cortex individuate movements? *Trends Neurosci*, **13**, 440–445.

Schneiberg S., Svestrup H., McFadyen B., McKinley, P. & Levin, M. (2002). The development of coordination for reach-to grasp movements in children. *Exp Brain Res*, **146**, 142–156.

Shim, K. S., Oliveira, M. A., Hsu, J. *et al.* (2007). Hand digit control in children: age-related changes in hand digit force interactions during maximum flexion and extension force production tasks. *Exp Brain Res*, **176**, 374–386.

Smyth, M. M., Katamba, J. & Peacock, K. A. (2004). Development of prehension between 5 and 10 years of age: distance scaling, grip aperture, and sight of hand. *J Motor Behav*, **36**, 91–103.

Sporns, O. & Edelman G. M. (1993). Solving Bernstein's problem: a proposal for the development of coordinated movement by selection. *Child Dev*, **64**, 960–981.

Touwen, B. C. (1978). Variability and stereotypy in normal and deviant development. In B. C. Touwen (Ed.), *Neurological Development in Infancy* (pp. 99–110). London: SIMP and Heinemann Medical Books.

Valero-Cuevas, F. J., Smaby, N., Venkadesan, M., Peterson, M. & Wright, T. (2003). The strength-dexterity test as a measure of dynamic pinch performance. *J Biomech*, **36**, 265–270.

Von Hofsten, C. (1982). Eye-hand coordination in the newborn. *Dev Psychol*, **18**, 450–461.

Von Hofsten, C. (1991). Structuring of early reaching movements: a longitudinal study. *J Motor Behav*, **23**, 280–292.

Von Hofsten, C. & Rönnqvist, L. (1988). Preparation for grasping an object: a developmental study. *J Exp Psychol Hum Percept Perform*, **14**, 610–621.

Westling, G. & Johannsson, R. S. (1984). Factors influencing the force control during precision grip. *Exp Brain Res*, **53**, 277–284.

Wing, A. M. (1996). Anticipatory control of grip force in rapid arm movement. In A. M. Wing, P. Haggard & J. R. Flanagan (Eds.), *Hand and Brain: The Neurophysiology and Psychology of Hand Movements* (pp. 301–324). San Diego, CA: Academic Press.

18

The effects of aging on sensorimotor control of the hand

KELLY J. COLE

Summary

Aging-related decline in hand function is ubiquitous and inexorable, beginning at about age 60 years. This decline disproportionately impacts dexterous grasp and manipulation. Many potential explanations for this decline have been offered, but the causes remain poorly understood. Here we report observations from our laboratory on timed tasks demonstrating that the forces and kinematics of dexterous grasp and manipulation in old adults differ from young adults, even at "comfortable" performance speeds. These observations support recent suggestions that controlling the moments of force applied to grasped objects is a fundamental problem in old age. Possible explanations lead to a review of contemporary issues related to the sensorimotor control of the aging hand. These topics include: sensory deterioration (both peripheral and central); reduced ability to coordinate muscle forces to control force vectors at the fingertip of a single digit and across digits; increased moment-to-moment force fluctuations; and loss of independent control of the right and left hands. On an optimistic note, training and practice appears to slow or reverse declining hand function in healthy aging.

Introduction

Hand function deteriorates unequivocally in healthy aging. Well-known decreases in muscle strength, mostly from sarcopenia (Holloszy, 1995; Hughes et al., 2001; Doherty, 2003), can account for increased difficulty in accomplishing some daily living skills, such as opening containers (Sperling, 1960; Shiffman, 1992). Manual dexterity also declines in old age, but in a manner that is dissociated from decreasing muscle strength. Impaired manual dexterity is significantly self-reported by healthy elderly persons (Falconer et al., 1991; Hughes et al., 1991) and is identified as a correlate of functional dependency in old age independent of cognitive status (Ostwald et al., 1989; Falconer et al., 1991) and with predictive value (Scholer et al., 1990). This decline in manual dexterity appears to be ubiquitous and inexorable in healthy individuals 60 years and older. In a rare longitudinal study, significant declines were reported for manual dexterity (and particularly fine manual dexterity) over a 3-year period in a large sample (n = 264) of healthy, community-dwelling individuals

Sensorimotor Control of Grasping: Physiology and Pathophysiology, ed. Dennis A. Nowak and Joachim Hermsdörfer. Published by Cambridge University Press. © Cambridge University Press 2009.

60 years and older, *regardless of age* (Desrosiers *et al.*, 1999). Exceptions to this decline were rare. The nature of this age-related decline in manual dexterity is the focus of this chapter, with a particular emphasis on contemporary reports that address potential mechanisms of declining fine dexterity in aging.

Aging-related changes in the nervous system and muscle are legion and thus there is no shortage of candidate explanations for declining manual function (for reviews see Katzman & Terry, 1983; Schaumberg *et al.*, 1983; Carmeli *et al.*, 2003; Chance, 2006; Dinse, 2006; Drachman, 2006; Raz & Rodrigue, 2006; Raz *et al.*, 2007). The continuing difficulty lies in causally linking motor behavioral changes with specific anatomical or physiological changes of the neural and/or muscular systems. Attempts at linking specific changes in neuro-muscular function with motor behavior often involve studies employing correlation analysis. For example, the availability of dopamine receptors (type D2) declines in healthy old age (in caudate and putamen, but not cerebellum), which correlates with the speed of finger tapping and cognitive tests of frontal brain function, regardless of age (e.g. Volkow *et al.*, 1998, 2000). Similarly, atrophy in the substantia nigra correlates with performance on the Purdue Pegboard test (Pujol *et al.*, 1992) as does reduced performance on cognitive tests that assess frontal lobe function (Jodar & Junque, 1998). These studies strongly support a long-standing theory that impaired dopaminergic systems in healthy old age contribute significantly to aging-related changes in motor and cognitive behavior. As another example, tactile spatial resolution at the fingertips correlates with performance on the Grooved Pegboard Test (Tremblay *et al.*, 2003), strengthening another long-standing theory that declining sensory functioning contributes to impaired hand function in old age. In contrast to these correlation approaches, Kornatz *et al.* (2005) demonstrated that training with light loads in healthy old adults (but not heavy-load training) reduced age-related fluctuations in acceleration of the index finger during precision lengthening and shortening actions of the first dorsal interosseous muscle, and improved performance on the Purdue Pegboard test. This latter study, though lacking the rigor of a well-designed clinical trial, strengthens the theory that the age-related decreases in force steadiness during skeletal muscle contractions contribute to impaired hand function (Enoka *et al.*, 2003), at least for actions performed at relatively low forces (Barry *et al.*, 2007).

As our knowledge of the causes of age-related deterioration of manual dexterity increases, so should our ability to intervene. On an encouraging note, there are additional reports that the rate of decline of manual dexterity can be slowed or even reversed through exercise or intensive manual practice (Salthouse, 1984; Keen *et al.*, 1994; Krampe & Ericsson, 1996; Ranganathan *et al.*, 2001a, 2001b; Krampe, 2002; Kornatz *et al.*, 2005; Buchman *et al.*, 2006; Adamo *et al.*, 2007). The mechanisms behind these improvements are unknown, but changes in central neuronal structure and function seem likely considering the recent explosion in our knowledge of the structural and functional adaptations of the central nervous system following intense, repetitive practice (cf. Nudo & Milliken, 1996; Nudo, 2006). Indeed, a new perspective on the functional losses in old age is emerging, which contends that aging causes brain plasticity processes with negative consequences that reinforce a downward spiral of function and adaptation (Dinse, 2006; Mahneke *et al.*,

2006). Based on this perspective, Mahneke and colleagues (2006) designed and tested, in a preliminary fashion, an intervention for memory function via appropriately designed behavioral training paradigms. Significant improvements in memory were reported. Even more remarkably, new motor memories formed after practice of novel tasks involving fine manual dexterity were "pristinely" retained over a 2-year study period *without rehearsal* in healthy aged adults, even into their 10th decade (Smith *et al.*, 2005).

Behavioral slowing

Aging-related decline of manual dexterity has been measured by a variety of instruments, but the most frequent and longest-standing reports have focused on the increased time to perform manual tasks (e.g. Welford, 1958; Jebsen *et al.*, 1969; Welford *et al.*, 1969; Kellor *et al.*, 1971). Recent studies have demonstrated that advancing age affects fine motor tasks of the hand in particular (Smith *et al.*, 1999, 2005). Smith and colleagues (1999) studied the time healthy adults needed to remove a close-fitting nut from various shaped smooth rods (straight, single curve and double-curve) by sliding the nut over the rod (see Figure 18.1 for a similar apparatus). A fourth task involved simply lifting the nut from a platform. Regression analysis of performance time across age revealed a slight linear increase in the time required to lift the nut from the platform. In stark contrast, the double-curve rod showed a precipitous increase in performance time beginning around age 60 years. In an extensive study of 497 cognitively and neurologically normal adults from 18–95 years of age, Smith and colleagues (2005) again demonstrated that the performance time for the double-curve rod underwent a marked increase in rate of slowing and variance around age 62 years. Their regression

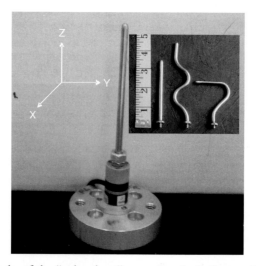

Figure 18.1. Photographs of the "rod and nut" apparatus. A six-degree-of-freedom force–torque transducer acts as the base into which the various rods were inserted. The inset shows the three rods and a size reference.

analysis revealed a 12 ms/year rate of slowing on this task for individuals between 30 and 62 years of age, and a 60 ms/year rate of slowing for individuals older than 62 years of age.

Likewise, Desrosiers and colleagues (1999) in their longitudinal study of healthy adults older than 60 years documented greater declines after 3 years for dexterous manipulation of objects compared to simple grasp and lift tasks (cf. Table 2, page 399 of their report). For example, the time required to grasp and move a jar with the right hand increased by ~ 4% at the 3-year measurement point, while the time to handle coins with the right hand increased by ~17%. Potential mechanisms for manual slowing must explain the disproportionate effects on tasks that demand greater dexterity.

Beyond behavioral slowing

Studies of increased time to manipulate objects raise the obvious question of whether dexterous use of the hand in old age is simply slower than in young adults, or if it is different in other ways as well. Seminal work by Welford (1958, 1977, 1981) and work in the early 20th century demonstrated that old adults' motor performance is slower and more variable than young adults'. This early work focused mostly on temporal variability and movement accuracy in rapid movements like finger tapping. Over the past two decades observations of kinematic, kinetic and muscle activity in old age clearly demonstrate that hand actions in healthy old adults differ from young adults in more ways than simply slowing.

We (Darling, Hynes and Cole, unpublished observations) recently re-examined the task of sliding a nut over smooth rods of different shapes, as employed by Smith and colleagues (1999, 2005). We sought to determine if age-related slowing was accompanied by increased forces of the nut against the rod. Slowing without impaired ability to orient the nut in space should yield little difference in forces against the rod for young versus old adults. We also examined performance on a second task; placing keyed pegs in a slotted pegboard using a modified Grooved Pegboard task (Lafayette Instruments, Lafayette, Indiana, USA). Peg and hand kinematics were monitored in three dimensions via small electromagnetic sensors (Flock-of-Birds, Ascension Technologies, North Carolina, USA). These tasks (nut/rod, Grooved Pegboard) are of considerable interest relative to the study of declining fine motor control in old age because rapid performance of these tasks fundamentally depends upon the ability to dynamically orient small hand-held objects. The results of these investigations indicate that age-related slowing was accompanied by forces and motions that were different than those in young adults, even when the tasks were performed at self-selected "comfortable" speeds. The ensuing discussion will serve as a vehicle for reviewing the contemporary literature on the effects of aging on sensorimotor control of the hand during dexterous grasp and manipulation.

We modified the apparatus of Smith and colleagues (1999, 2005) by incorporating a 6-degree-of-freedom force/torque transducer (ATI Nano 17, North Carolina) into the base that received the straight and curved rods (Figure 18.1). Thus, we monitored the forces applied to the rod along perpendicular axes (X, Y) in the horizontal plane, and the forces along the vertical axis (Z). Eighteen healthy, community-dwelling adults were studied consisting of

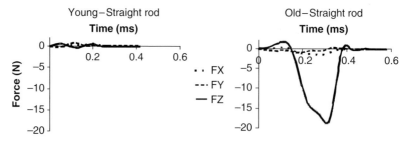

Figure 18.2. Examples of force signals (amplitude × time) in the X, Y, and Z directions (cf. Figure 18.1) obtained from a young (left panel) and old subject (right panel) each of who lifted the nut from the straight rod under the instruction to remove the nut as fast as possible (negative force in the Z direction corresponds to force upward in the vertical direction).

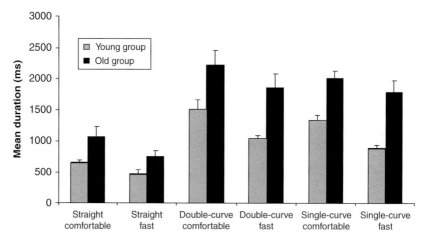

Figure 18.3. Average duration ($\pm$ standard error) for the old and young groups to remove the nut from each rod, performed under the "fast" and "comfortable" speed conditions. All pairs of bars are different at $P < 0.015$.

10 young adults (5 females, 5 males; 21–22 years of age) and 8 old adults (7 females and 1 male; 65–83 years of age; average age 73.8 years). All subjects began with the straight rod and removed the nut five times at a self-selected "comfortable" speed, followed by five repetitions in which they were instructed to remove the nut "as quickly as possible." This order was repeated for the double-curve rod, and then the single-curve rod.

The time required to execute the tasks (Figures 18.2 and 18.3) replicated the findings of Smith and colleagues (1999). The old adults required disproportionately longer durations to remove the nut from the two curved rods compared to the straight rod (significant Age Group × Rod Shape interaction; repeated-measures ANOVA; $F_{2,32} = 5.59$; $P < 0.0083$). Inspection of the forces applied against the rods revealed a striking difference between the young and old adults. All old subjects produced large vertical forces against the rods, even

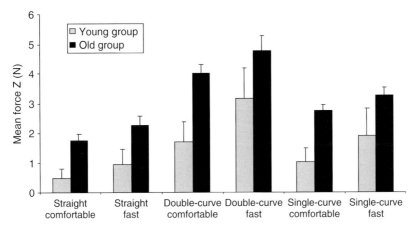

Figure 18.4. Mean force in the vertical direction over the trial duration, averaged for each group (+ standard error), performed under the "fast" and "comfortable" speed conditions. All vertical forces were in the upward direction. All pairs of bars are different at $P < 0.001$ except Double-curve fast and Single-curve fast.

when lifting the nut from the *straight* rod, but only slightly larger forces in the horizontal plane (Figures 18.2 and 18.4). Moreover, old subjects were slower and produced substantially larger and more variable vertical forces than the young subjects even when they moved at their self-selected "comfortable" (slower) speed. This latter finding is consistent with suggestions that age-related slowing of hand/arm movements (in a line-drawing task) does not represent a strategy to minimize error, but is due to a decline in "motor coordination" (Morgan *et al.*, 1994; Bellgrove *et al.*, 1998).

Why did large vertical forces occur for the old subjects on the straight rod, while young adults produced relatively small vertical forces? Lifting the nut with relatively small vertical force on the rod is accomplished by minimizing contact (friction) between the nut and the rod. The task of lifting the nut quickly could be accomplished with less precision (i.e. significant contact between the nut and rod) if one simply generates large lifting forces on the nut to overcome frictional forces. Only one of our young subjects used this latter strategy, while the remaining nine young subjects lifted the nut quickly with little evidence of contact with the rod. In contrast, all old subjects demonstrated large vertical forces on the rod *and* slower lifting speeds. The relatively small horizontal forces that were observed imply that the frictional forces were not from reduced precision in positioning the nut in the horizontal plane. Instead, we believe that the nut was tilted by the older adults as they lifted it, causing it to bind slightly and transfer vertical lifting forces to the rod. Indeed, analysis of hand "roll" indicated that old adults averaged nearly 15 degrees more roll into forearm pronation compared with young adults on the straight rod task ($P < 0.015$), although we cannot determine if object roll was from forearm or finger actions.

Our observations of the kinematics of a keyed peg as it is inserted into a slot of matching shape also indicate that controlling the orientation of grasped objects is problematic in old age. The subjects described previously participated in this experiment using a modified version of the Grooved Pegboard Test. The keyed peg was rigidly affixed to a wooden disk (30 mm diameter × 22 mm thick) so that it protruded about 22 mm from the center of the disk. Vision of the keyed peg and slot were not obscured by the wooden disk. An electromagnetic sensor was inserted into the top of the disk, with the cable supported to minimize forces on the disk. The object was positioned in a vertical slot at the start of each trial. The starting position and orientation of the hand, forearm and arm were held constant across trials and subjects, and structured so that the disk was grasped similarly across subjects and trials. Specifically, subjects grasped the object as if they were lifting a drinking glass; the thumb was on one side of the disk with two fingers positioned in opposition to the thumb, and the forearm was parallel to the ground in intermediate pronosupination. Subjects grasped the object, lifted it from the starting slot, and moved it a few inches to slide it into the target slot. This task was performed three times at a self-selected "comfortable" speed and again three times "as fast as possible." This was repeated for three different slot orientations.

The old subjects averaged about 150 ms slower than the young subjects when inserting the pegs, regardless of whether they performed at their comfortable speed or more quickly (main effect of Group; $F_{1,15} = 12.963$; $P < 0.0026$), which is consistent with our observations from the nut/rod task. We observed that old subjects performed the task with significantly greater "roll" of the object than the young subjects, averaged from start to completion of each trial (main effect of Group; $F_{1,15} = 5.1$; $P < 0.04$). Object "roll" in this experiment corresponds to a peg orientation away from vertical that would result, for example, from pronation/supination movements of the forearm, or from similar moments of force on the object from the thumb and fingers. Whereas the young adults maintained nearly a perfectly vertical orientation, the old subjects averaged about 12 degrees of roll. The old subjects also demonstrated greater variations in roll angle over the duration of a trial, as measured by the standard deviation over each trial (significant main effect of Group; $F_{1,15} = 30.5$; $P < 0.001$). It is crucial to note that subjects did not need to "roll" the object to visualize the peg/slot orientations, nor did the old subjects appear to do so for this reason.

Controlling total external moments on objects – a fundamental problem in aging?

Our observations, although preliminary, are consistent with recent speculation that declining control over the external moments applied to objects during grasp contributes fundamentally to declining fine dexterity in old age (Shim *et al.*, 2004; Cole, 2006). Additional kinematic data are required to address the extent to which this declining control of the external moment results from the misapplication of force at the fingertips (cf. Shim *et al.*, 2004; Cole, 2006), versus forearm pronation/supination, versus movements of more proximal joints. For example, in a study of the relationship between manual dexterity, assessed via the Grooved Pegboard Test, and tactile spatial acuity, Tremblay and colleagues (2003) reported anecdotal observations that old adults appeared to increase their wrist and shoulder movements "… when they tried to compensate for their inability to manipulate the pegs …" (p. 131).

The ability of old adults to control the external moments they apply to objects during grasp can be considered first at the level of the individual digit, and second, during grasp with multiple digits. We recently reported that old adults tend to direct their index fingertip force differently than do young adults when simply pressing the index finger against a rigid plate (Cole, 2006). Young and old subjects were instructed to match a target force ranging from 2.5–15 N with the perpendicular component of their total fingertip force. Subjects viewed a display of this perpendicular force whereas forces tangential to the force plate were not displayed. After reaching the target force the display was turned off. Young adults produced fingertip forces that were nearly perpendicular to the plate at all target forces while old adults produced fingertip forces that deviated constantly in the proximal ("flexion") direction of the horizontal plane, averaging between 10 and 15 degrees across the various force targets (Figure 18.5). Old adults also demonstrated an ulnar direction to their fingertip force that averaged between 5 and 10 degrees across the force targets, whereas young adults demonstrated force vectors that deviated less than 5 degrees. These observations may be important because small biases in fingertip force direction, if not compensated by other digits during grasp, can impart unwanted rotation to objects, and particularly to lightweight objects. The need to correct "misdirected" forces of the digits as they unfold may contribute to the characteristic slowing of manual performance in old age.

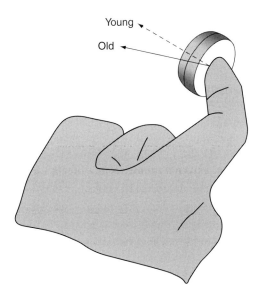

Figure 18.5. Illustration showing a more proximally directed fingertip force for old individuals while pressing against a rigid plate, compared to a more perpendicularly directed force for young adults (cf. Cole, 2006).

In support of this view are studies by Latash, Zatsiorsky and colleagues concerning age-related changes in finger coordination and their effects on the moment of force (Shim *et al.*, 2004; Olafsdottir *et al.*, 2007) (see also Chapter 3). In a four-finger isometric pressing task, subjects attempted to produce a trapezoidally shaped "total" moment profile in time, where the pivot point for the total moment was located between the middle and ring fingers along the long axis of the hand (Olafsdottir *et al.*, 2007). Pressing forces were independently transduced at the four digits, and forearm rotation was prevented. Old subjects exerted larger and more variable total pressing forces, larger variance of the total moment of force, and larger forces by the fingers that produced a moment of force that opposed the required moment direction. Old subjects failed to demonstrate finger force coordination patterns ("synergies") that would stabilize the total moment of force, in contrast to the demonstration of such synergies by young adult subjects. These data may reflect "an impairment of rotational hand actions ..." which "... may contribute to failure at a variety of everyday tasks relying on rotational hand action, including spilling the contents of a mug, failing to turn the key to open the door lock, producing poorly legible handwriting, etc." (Olafsdottir *et al.*, 2007, p. 1498). Similarly, Shim and colleagues (2004) studied the force coordination patterns while old and young subjects held a vertical handle in a prismatic grasp (thumb opposed by the four remaining digits, forearm secured mid-way in pronosupination) (see also Chapter 3). Subjects generated fingertip forces to produce the total moment required to hold the eccentrically weighted object without tilting. In another task they were required to hold the object without tilting it while producing ramp increases in grip force. These tasks required the accurate and simultaneous production of the total force and moment. As expected the old subjects were less accurate and more variable in these tasks, but they also showed worse coordination; that is, they were less able to produce the needed covariation of the finger forces and of the components that made up the total moment.

Shim and colleagues (2004) also suggested that observations of large grip forces in old adults (e.g. Cole, 1991; Kinoshita & Francis, 1996; Cole *et al.*, 1999) may result from the need to compensate for the higher moments produced by the "antagonist" fingers (those fingers that produced a moment opposite to that of the required moment direction). They suggested that these higher antagonist moments also may increase the passive resistance of the hand to unexpected variations in applied torque and thus may be adaptive to the decline in the control of finger forces and moments. However, exceedingly large finger-pressing force also has been reported when old adults attempted to stabilize a small paddle against unexpected perturbations (Cole & Rotella, 2001). In this experiment a single digit pressed against the paddle. Thus, there may be multiple explanations for high grip forces in old adults, depending on the task and conditions. For example, the size of grip force applied by old adults varies depending on whether relevant object properties (friction; weight) can be anticipated (Cole *et al.*, 1999).

Does impaired sensory processing contribute to declining fine dexterity?

Sensory deterioration at the peripheral and central levels remains a viable explanation for declining fine dexterity, and should be considered in relation to the reduced ability to orient grasped objects. The transduction, transmission, and central processing of tactile sensory information deteriorate in old age (Verrillo, 1979; Kenshalo, 1986; Cerella, 1990; Schmidt *et al.*, 1990; Stevens, 1992; Gescheider *et al.*, 1994a, 1994b, 1996; Stevens & Patterson, 1995; Stevens *et al.*, 1998; Tremblay *et al.*, 2003, 2005; Manning & Tremblay, 2006). It is likely that muscle spindles of the hand and forearm change as well. The number of intrafusal fibers in muscle spindles declines in old age (chain fibers, in particular), along with increased thickening of the capsule (Swash & Fox, 1972; Liu *et al.*, 2005). These changes may contribute to observations of reduced static position sense at the digits and elbow in old age (Ferrell *et al.*, 1992), although aging-related changes in central processing of sensory information are likely to contribute as well (Adamo *et al.*, 2007).

Attempts at linking reduced tactile sensory function in old age to motor behavioral changes indicate that the impact of declining sensory function depends on the motor task. The profound changes in the histology of tactile mechanoreceptors associated with fast-adapting tactile afferents (Dickens *et al.*, 1963; Cauna, 1965; Bolton *et al.*, 1966; Bruce & Sinclair, 1980) provoked us to test whether old age impairs the ability to adjust fingertip forces when faced with unpredictable changes in the friction of objects (Cole *et al.*, 1999) and unpredictable pulling loads on an object restrained in grasp (Cole & Rotella, 2001). The rationale for these studies was the important role that fast-adapting tactile afferents play in encoding object friction and tangential skin loading during grasp (Johansson & Westling, 1987; Westling & Johansson, 1987; Macefield & Johansson, 1996; Macefield *et al.*, 1996; see also Chapters 11 and 19). Indeed, old adults were impaired in their ability to respond rapidly via adaptive changes in fingertip forces to unpredictable changes in the friction of objects, or to loads on the object that would pull the object from the hand. By contrast, we observed that old adults, when grasping and lifting small spheres without vision, did not slow compared with performing the task with vision any more than did young adults (Cole *et al.*, 1998). We interpreted these latter results to indicate that tactile deterioration in old age does not contribute significantly to grasping and lifting small objects. However, discounting the role of tactile deterioration in the aging-related decline of manual function on the basis of the single task of lifting small spheres was short-sighted. Tactile deterioration may well affect dexterous manipulation of a more demanding nature, even when friction is constant and tangential loads are predictable. For example, Tremblay and colleagues (2003) reported substantial aging-related deterioration of tactile spatial acuity using grating orientation detection thresholds (Van Boven *et al.*, 1989) that correlated strongly with scores on the Grooved Pegboard Test. Note that performance on the Grooved Pegboard Test requires precise orientation of the peg in relation to the slot, in contrast to securing a sphere in grasp.

It is worthwhile to consider aging-related sensory decline in relation to our findings on the nut/rod task and our observations of excessive object "roll" on the pegboard task. First, the findings of Cole & Rotella (2001) may be important in relation to our observations of large vertical forces when old adults removed a nut from a straight rod. We reported that the old subjects' response latencies to *rapid* pulling loads did not differ from that of young adults, but old adults' response latencies became increasingly longer compared with young adults for progressively slower ramp loads (Cole & Rotella, 2001). Regression analyses indicated that the old adults' threshold for responding to the pulling loads was twice that of the young adults. This reduced sensitivity to tangential loads at the skin of the fingertips may have contributed to the large vertical forces seen in the nut/rod experiment, because the old adults would have to experience greater drag on the rod (and greater tangential forces on the skin at the contact patches with the nut) before responding with changes in nut orientation. However, it is not clear that old adults responded to nut/rod contact by changing orientation of the nut (see Figure 18.2).

Although an increased threshold for detecting vertical loading of the hand during contact between the nut and straight rod may affect performance on that task, it does not explain our preliminary observations of excessive object roll while handling the disk/peg in our modified version of the Grooved Pegboard test. Perhaps a more fundamental question is whether or not aging is associated with a reduced ability to encode/decode sensory (peripheral afferent or efferent-copy) information related to the *orientation* of grasped objects. We are not aware of any studies that have directly addressed this issue in aging, either with regard to the orientation of objects relative to the hand, or relative to any other coordinate frame. Likewise, we are unaware of studies of the perception of the direction of applied force to the skin in old age. This may be relevant with regard to old adults' ability to control the total external moment of force on an object during grasp.

It is reasonable to expect that old adults may be impaired in their ability to detect changes in the orientation of the hand (and thus grasped objects), based on studies of deteriorating static joint position sense as previously noted. One might also predict an age-related impairment of the ability to detect changes in the orientation of grasped objects within the hand. This ability most likely depends on encoding/decoding signals related to the magnitude and direction of forces at the skin contact patches with the object, and changing object edges/contours at contact patches. Multiple afferent sources are required for these functions. For example, the brain is informed about the directions of forces at the glabrous skin by signals from at least three types of mechano-receptive tactile afferents (fast adapting with small receptive fields – FA-I, slow-adapting with small receptive fields – SA-I, and slow-adapting with large receptive fields – SA-II; Birznieks *et al.*, 2001). Deteriorating function of FA-I and SA-I afferents in old age is well known from psychophysical studies of vibration sense and spatial acuity (cf. Stevens *et al.*, 1998; Tremblay *et al.*, 2003). Manning & Tremblay (2006) also have documented aging-related deterioration of tactile pattern recognition at the fingertip (letter recognition) that they suggest is the result of changes in peripheral signals *and* central processing.

Determining the orientation of objects within grasp may depend upon the nervous system's ability to monitor dynamic changes in the locations of edges and curves against the skin. This ability in particular may depend on central nervous system mechanisms. Objects with curvatures typical of those that are grasped and manipulated, when applied to the skin of the fingertip with contact forces typical of grasp, evoked responses in SA-I, SA-II and FA-I afferents over vast regions of the terminal phalanx, with discharge frequencies that correlated with curvature (Jenmalm *et al.*, 2003). More importantly, curvature and force direction had interactive effects on afferent discharge rate. Based on these interactive effects the authors concluded that "... the CNS must possess mechanisms that disentangle inter-actions between these and other parameters of stimuli on the fingertips" (Jenmalm *et al.*, 2003). Thus, well-known findings of reduced density of tactile receptors at the hand in old age (Cauna, 1965) are expected to impair the ability to determine the spatial relationship between the hand and object to the extent that these relationships depend upon the direction and magnitude of forces between the object and hand, and on the resolution of changing edges and curvatures of objects. Old adults may be additionally impaired in their ability to detect or control object orientation within grasp by changes in central nervous system mechanisms that "disentangle" the complex stimulus interactions during grasp.

Finally, Johansson & Birznieks (2004) reported that the relative timing of the first impulses from ensembles of afferents conveys information about the direction of fingertip force and the shape of the surface that contacted the fingertip. These first impulses provide information faster than rate coding, and provide a mechanism that accounts for observations of extremely rapid adaptation of fingertip forces to mechanical events during grasp and manipulation. Loss of tactile receptor density and function in old age surely must compro-mise this early detection system and negatively impact fine, dexterous manipulation. We must expand our knowledge concerning how old age affects our ability to detect and respond to changes in object orientation within grasp.

Force variability and declining dexterity

Old adults demonstrate a reduced ability to maintain steady forces with the digits at low force levels (Galganski *et al.*, 1993; Laidlaw *et al.*, 2000; Ranganathan *et al.*, 2001a, 2001b; Kornatz *et al.*, 2005; Tracy *et al.*, 2005; Sosnoff & Newell, 2006; Barry *et al.*, 2007). The impact of this reduced steadiness on dexterous grasp and manipulation can be inferred from studies demonstrating that practice on skilled finger movements improved force steadiness and scores on tests of dexterity (Ranganathan *et al.*, 2001b; Kornatz *et al.*, 2005). Reduced force steadiness in old age has been studied at the level of a single muscle, joint and digit in attempts to understand the cause of increased force fluctuations (cf. Enoka *et al.*, 2003). The impact on steadiness of the position of the hand in space has been studied as well (Ranganthan *et al.*, 2001b), but we are unaware of any studies concerning how force unsteadiness affects control of the orientation of grasped objects. Studies of this nature may reveal the strength and nature of the relationship between the phenomenon of aging-related force unsteadiness at the level of muscles and joints, to declining dexterous grasp and manipulation.

Bilateral hand function

Aging-related changes in bilateral object manipulation have received little attention. However, many skills of daily living important to old adults involve manipulating an object with both hands, and sharing or transferring object loads between hands. Shinohara *et al.*

(2003) reported the provocative result that old adults have a reduced ability to suppress unintended activity in contralateral hand muscles during tasks that were performed with the opposite hand. This area of study would also appear to require greater attention in the future.

Conclusions

Declining hand function in aging is manifest as slowing of manual tasks even at self-selected "comfortable" speeds, reduced precision (or increased clumsiness) and increased variability in fine manipulation. Recent work indicates that reduced control of external moments on objects, and therefore increased errors in object orientation during grasp and manipulation, may be an important characteristic of aging-related declines in manual dexterity. The potential causes of this reduced control generate a list of the "usual suspects," including changes in sensory functioning, sarcopenia, declining function of dopaminergic systems, increased force unsteadiness, along with newer candidates like impaired finger coordination synergies. The challenge for scientists is to design experiments that better address causality using these more specific characteristics of declining fine dexterity. Possible designs include intervention studies, and longitudinal studies that track changes in fine dexterity beyond slowing, along with changes in measures that reflect candidate mechanisms.

Acknowledgments

Warren G. Darling, PhD and Stephanie M. Hynes, Department of Integrative Physiology, The University of Iowa, made substantial and significant contributions to all phases of the original projects that were summarized in this chapter. While cited in the text as "unpublished observations," these projects are ongoing.

References

Adamo, D. E., Martin, B. J. & Brown, S. H. (2007). Age-related differences in upper limb proprioceptive acuity. *Percept Mot Skills*, **104**, 1297–1309.

Barry, B., Pascoe, M., Jesunathadas, M. & Enoka, R. (2007). Rate coding is compressed but variability is unaltered for motor units in a hand muscle of old adults. *J Neurophysiol*, **97**, 3206–3218.

Bellgrove, M., Phillips, J., Bradshaw, J. & Gallucci, R. (1998). Response (re-)programming in aging: a kinematic analysis. *J Gerontol A Biol Sci Med Sci*, **53**, M222–M227.

Birznieks, I., Jenmalm, P., Goodwin, A. & Johansson, R. (2001). Encoding of direction of fingertip forces by human tactile afferents. *J Neurosci*, **21**, 8222–8237.

Bolton, C. F., Winkelmann, R. K. & Dyck, M. D. (1966). A quantitative study of Meissner's corpuscles in man. *Neurology*, **16**, 1–9.

Bruce, M. & Sinclair, D. (1980). The relationship between tactile thresholds and histology in the human finger. *J Neurol Neurosurg Psychiatry*, **43**, 235–242.

Buchman, A., Boyle, P., Wilson, R., Bienias, J. & Bennett, D. (2006). Physical activity and motor decline in older persons. *Muscle Nerve*, **35**, 354–362.

Carmeli, E., Patish, H. & Coleman, R. (2003). The aging hand. *J Gerontol Med Sci*, **58A**, 146–152.

Cauna, N. (1965). The effects of aging on the receptor organs of the human dermis. In W. Montagna (Ed.), *Advances in Biology of the Skin, Vol. VI* (pp. 63–96). New York, NY: Pergamon.

Cerella, J. (1990). Information processing rates in the elderly. In J. E. Birren, K. W. Schaie (Eds.), *Handbook of the Psychology of Aging, 3rd edition* (pp. 201–221). New York, NY: Academic Press.

Chance, S. A. (2006). Subtle changes in the ageing human brain. *Nutr Health*, **18**, 217–224.

Cole, K. J. (1991). Grasp force control in older adults. *J Motor Behav*, **23**, 251–258.

Cole, K. (2006). Age-related directional bias of fingertip force. *Exp Brain Res*, **175**, 285–291.

Cole, K. & Rotella, D. (2001). Old age affects fingertip forces when restraining an unpredictably loaded object. *Exp Brain Res*, **136**, 535–542.

Cole, K. J., Rotella, D. L. & Harper, J. G. (1998). Tactile impairments cannot explain the effect of age on a grasp and lift task. *Exp Brain Res*, **121**, 263–269.

Cole, K., Rotella, D. & Harper, J. (1999). Mechanisms of age-related changes in fingertip forces during precision gripping and lifting in adults. *J Neuroscience*, **19**, 3238–3247.

Desrosiers, J., Hebert, R., Bravo, G. & Rochette, A. (1999). Age-related changes in upper extremity performance of elderly people: a longitudinal study. *Exp Gerontol*, **34**, 393–405.

Dickens, W. N., Winkelmann, R. K. & Mulder, D. W. (1963). Cholinesterase demonstration of dermal nerve endings in patients with impaired sensation: a clinical and pathological study of 41 patients and 27 control subjects. *Neurology*, **13**, 91–100.

Dinse, H. R. (2006). Cortical reorganization in the aging brain. *Prog Brain Res*, **157**, 57–80.

Doherty, T. J. (2003). Invited review: aging and sarcopenia. *J Appl Physiol*, **95**, 1717–1727.

Drachman, D. A. (2006). Aging of the brain, entropy, and Alzheimer disease. *Neurology*, **67**, 1340–1352.

Enoka, R., Christou, E., Hunter, S. *et al.* (2003). Mechanisms that contribute to differences in motor performance between young and old adults. *J Electromyogr Kinesiol*, **13**, 1–12.

Falconer, J., Hughes, S. L., Naughton, B. J. *et al.* (1991). Self report and performance-based hand function tests as correlates of dependency in the elderly. *J Am Geriatr Soc*, **39**, 695–699.

Ferrell, W., Crighton, A. & Sturrock, R. (1992). Age-dependent changes in position sense in human proximal interphalangeal joints. *Neuroreport*, **3**, 259–261.

Galganski, M. E., Fuglevand, A. J. & Enoka, R. M. (1993). Reduced control of motor output in a human hand muscle of elderly subjects during submaximal contractions. *J Neurophysiol*, **69**, 2108–2115.

Gescheider, G. A., Beiles, E. J., Checkosky, C. M., Bolanowski, S. J. & Verrillo, R. T. (1994a). The effects of aging on information-processing channels in the sense of touch. 2. Temporal summation in the P channel. *Somatosensory Motor Res*, **11**, 359–365.

Gescheider, G. A., Bolanowski, S. J., Hall, K. L., Hoffman, K. E. & Verrillo, R. T. (1994b). The effects of aging on information-processing channels in the sense of touch. 1. Absolute sensitivity. *Somatosensory Motor Res*, **11**, 345–357.

Gescheider, G. A., Edwards, R. R., Lackner, E. A., Bolanowski, S. J. & Verrillo, R. T. (1996). The effects of aging on information-processing channels in the sense of touch. 3. Differential sensitivity to changes in stimulus intensity. *Somatosensory Motor Res*, **13**, 73–80.

Holloszy, J. O. (Ed.) (1995). Workshop on sarcopenia: muscle atrophy in old age. Airlie, Virginia, September 19–21, Proceedings. *J Gerontol A Biol Sci Med Sci*, **50A**, 1–161.

Hughes, S. L., Edelman, P., Chang, R. W., Singer, R. H. & Schuette, P. (1991). The GERI-AIMS. Reliability and validity of the arthritis impact measurement scales adapted for elderly respondents. *Arthritis Rheumat*, **34**, 856–865.

Hughes, V., Frontera, W., Wood, M. *et al.* (2001). Longitudinal muscle strength changes in older adults: influence of muscle mass, physical activity, and health. *J Gerontol A Biol Sci Med Sci*, **56**, B209–B217.

Jebsen, R. H., Taylor, N., Trieschmann, R. B., Trotter, M. J. & Howard, L. A. (1969). An objective and standardized test of hand function. *Arch Phys Med Rehab*, **50**, 311–319.

Jenmalm, P., Birznieks, I., Goodwin, A. & Johansson, R. (2003). Influence of object shape on responses of human tactile afferents under conditions characteristic of manipulation. *Eur J Neurosci*, **18**, 164–176.

Jodar, M. & Junque, C. (1998). Frontal functions in normal aging and the performance in Purdue Pegboard test. In V. Vellas, J. Fitten & G. Frisoni (Eds.), *Research and Practice in Alzheimer's Disease* (pp. 151–162). New York, NY: Springer.

Johansson, R. S. & Westling, G. (1987). Signals in tactile afferents from the fingers eliciting adaptive motor-responses during precision grip. *Exp Brain Res*, **66**, 141–154.

Johansson, R. & Birznieks, I. (2004). First spikes in ensembles of human tactile afferents code complex spatial fingertip events. *Nat Neurosci*, **7**, 170–177.

Katzman, R. & Terry, R. (1983). Normal aging of the nervous system. In R. Katzman (Ed.), *The Neurology of Aging* (pp. 15–50). Philadelphia, PA: F.A. Davis.

Keen, D. A., Yue, G. H. & Enoka, R. M. (1994). Training-related enhancement in the control of motor output in elderly humans. *J Appl Physiol*, **77**, 2648–2658.

Kellor, M., Frost, J., Silberberg, B., Iverson, I. & Cummings, R. (1971). Hand strength and dexterity. *Am J Occup Ther*, **25**, 77–83.

Kenshalo, D. R. (1986). Somesthetic sensitivity in young and elderly humans. *J Gerontol*, **41**, 732–742.

Kinoshita, H. & Francis, P. R. (1996). A comparison of prehension force control in young and elderly individuals. *Eur J Appl Physiol*, **74**, 450–460.

Kornatz, K., Christou, E. & Enoka, R. (2005). Practice reduces motor unit discharge variability in a hand muscle and improves manual dexterity in old adults. *J Appl Physiol*, **98**, 2072–2080.

Krampe, R. (2002). Aging, expertise and fine motor movement. *Neurosci Biobehav Rev*, **26**, 769–776.

Krampe, R. & Ericsson, K. (1996). Maintaining excellence: deliberate practice and elite performance in young and older pianists. *J Exp Psychol Gen*, **125**, 331–359.

Laidlaw, D., Bilodeau, M. & Enoka, R. (2000). Steadiness is reduced and motor unit discharge is more variable in old adults. *Muscle Nerve*, **23**, 600–612.

Liu, J., Eriksson, L., Thornell, L. & Pedrosa-Domellof, F. (2005). Fiber content and myosin heavy chain composition of muscle spindles in aged human biceps brachii. *J Histochem Cytochem*, **53**, 445–454.

Macefield, V. G. & Johansson, R. S. (1996). Control of grip force during restraint of an object held between finger and thumb: responses of muscle and joint afferents from the digits. *Exp Brain Res*, **108**, 172–184.

Macefield, V. G., Rothwell, J. C. & Day, B. L. (1996). The contribution of transcortical pathways to long-latency stretch and tactile reflexes in human hand muscles. *Exp Brain Res*, **108**, 147–154.

Mahneke, H., Bronstone, A. & Merzenich, M. (2006). Brain plasticity and functional losses in the aged: scientific bases for a novel intervention. *Prog Brain Res*, **157**, 81–109.

Manning, H. & Tremblay, F. (2006). Age differences in tactile pattern recognition at the fingertip. *Somatosens Mot Res*, **23**, 147–155.

Morgan, M., Phillips, J. G., Bradshaw, J. L. *et al.* (1994). Age-related motor slowness – simply strategic? *J Gerontol*, **49**, M133–M139.

Nudo, R. (2006). Mechanisms for recovery of motor function following cortical damage. *Curr Opin Neurobiol*, **16**, 638–644.

Nudo, R. J. & Milliken, G. W. (1996). Reorganization of movement representations in primary motor cortex following focal ischemic infarcts in adult squirrel monkeys. *J Neurophysiol*, **75**, 2144–2149.

Olafsdottir, H., Zhang, W., Zatsiorsky, V. & Latash, M. (2007). Age-related changes in multifinger synergies in accurate moment of force production tasks. *J Appl Physiol*, **102**, 1490–1501.

Ostwald, S. K., Snodown, D. A., Rysavy, S. D. M., Keeana, N. L. & Kane, R. L. (1989). Manual dexterity as a correlate of dependency in the elderly. *J Am Geriatr Soc*, **37**, 963–969.

Pujol, J., Junque, C., Vendrell, P., Grau, J. & Capdevila, A. (1992). Reduction of the substantia nigra width and motor decline in aging and Parkinson's disease. *Arch Neurol*, **49**, 1119–1122.

Ranganathan, V., Siemionow, V., Sahgal, V. & Gue, G. (2001a). Effects of aging on hand function. *J Am Geriatr Soc*, **49**, 1478–1484.

Ranganathan, V., Siemionow, V., Sahgal, V., Liu, J. & Gue, G. (2001b). Skilled finger movement exercise improves hand function. *J Gerontol A Biol Sci Med Sci*, **56**, M518–M522.

Raz, N. & Rodrigue, K. (2006). Differential aging of the brain: patterns, cognitive correlates and modifiers. *Neurosci Biobehav Rev*, **30**, 730–748.

Raz, N., Rodrigue, K. & Haacke, E. (2007). Brain aging and its modifiers: insights from in vivo neuromorphometry and susceptibility weighted imaging. *Ann NY Acad Sci*, **1097**, 84–93.

Salthouse, T. (1984). Effects of age and skill in typing. *J Exp Psychol Gen*, **113**, 345–371.

Schaumberg, M. D., Spencer, P. S. & Ochoa, J. (1983). The aging human peripheral nervous system. In R. Katzman (Ed.), *The Neurology of Aging* (pp. 111–122). Philadelphia, PA: F. A. Davis.

Schmidt, R. F., Wahren, L. K. & Hagbarth, K. E. (1990). Multiunit neural responses to strong finger pulp vibration. I. Relationship to age. *Acta Physiol Scand*, **140**, 1–10.

Scholer, S. G., Potter, J. F. & Burke, W. J. (1990). Does the Williams Manual Test predict service use among subjects undergoing geriatric assessment? *J Am Geriatr Soc*, **38**, 767–772.

Shiffman, L. M. (1992). Effects of aging on adult hand function. *Am J Occup Ther*, **46**, 785–792.

Shim, J., Lay, B., Zatsiorsky, V. & Latash, M. (2004). Age-related changes in finger coordination in static prehension tasks. *J Appl Physiol*, **97**, 213–224.

Shinohara, M., Keenan, K. & Enoka, R. (2003). Contralateral activity in homologous hand muscle during voluntary contractions is greater in old adults. *J Appl Physiol*, **94**, 373–384.

Smith, C., Umberger, G., Manning, B. *et al.* (1999). Critical decline in fine motor hand movements in human aging. *Neurology*, **53**, 1458–1461.

Smith, C., Walton, A., Loveland, A. *et al.* (2005). Memories that last in old age: motor skill learning and memory preservation. *Neurobiol Aging*, **26**, 883–890.

Sosnoff, J. & Newell, K. (2006). Are age-related increases in force variability due to decrements in strength? *Exp Brain Res*, **174**, 86–94.

Sperling, G. (1960). The information available in brief visual presentations. *Psychol Monog Gen Applied*, **74**, 1–30.

Stevens, J. (1992). Aging and spatial acuity of touch. *J Gerontol: Psychol Sci*, **47**, 35–40.

Stevens, J. C. & Patterson, M. Q. (1995). Dimensions of spatial acuity in the touch sense: Changes over the life span. *Somatosensory Motor Res*, **12**, 29–47.

Stevens, J. C., Cruz, L. A., Marks, L. E. & Lakatos, S. (1998). A multimodal assessment of sensory thresholds in aging. *J Gerontol Ser B Psychol Sci*, **53**, P263–P272.

Swash, M. & Fox, K. (1972). The effect of age on human skeletal muscle. Studies of the morphology and innervation of muscle spindles. *J Neurol Sci*, **16**, 417–432.

Tracy, B., Maluf, K., Stephenson, J., Hunter, S. & Enoka, R. (2005). Variability of motor unit discharge and force fluctuations across a range of muscle forces in older adults. *Muscle Nerve*, **32**, 533–540.

Tremblay, F., Wong, K., Sanderson, R. & Cote, L. (2003). Tactile spatial acuity in elderly persons: assessment with grating domes and relationship with manual dexterity. *Somatosensory Motor Res*, **20**, 127–132.

Tremblay, F., Mireault, A., Dessurault, L., Manning, H. & Sveistrup, H. (2005). Postural stabilization from fingertip contact: II. Relationships between age, tactile sensibility and magnitude of contact forces. *Exp Brain Res*, **164**, 155–164.

Van Boven, R., Tilghman, D. & Johnson, K. (1989). Oral tactile spatial resolution: a pyschophysical study. *J Dent Res*, **68**, 329.

Verrillo, R. T. (1979). Change in vibrotactile thresholds as a function of age. *Sensory Proc*, **3**, 49–59.

Volkow, N., Gur, R., Wang, G. *et al.* (1998). Association between decline in brain dopamine activity with age and cognitive and motor impairment in healthy individuals. *Am J Psychiatry*, **155**, 344–349.

Volkow, N., Logan, J., Fowler, J. *et al.* (2000). Association between age-related decline in brain dopamine activity and impairment in frontal and cingulate metabolism. *Am J Psychiatry*, **157**, 75–80.

Welford, A. (1958). *Aging and Human Skill*. Oxford, UK: Oxford University Press for the Nuffield Foundation.

Welford, A. (1981). Signal, noise, performance, and age. *Hum Factors*, **23**, 97–109.

Welford, A. T. (1977). Motor performance. In J. E. Birren & K. W. Schaie (Eds.), *Handbook of the Psychology of Aging* (pp. 450–496). New York, NY: Van Nostrand Reinhold.

Welford, A. T., Norris, A. H. & Shock, N. W. (1969). Speed and accuracy of movement and their changes with age. *Acta Psychol*, **30**, 3–15.

Westling, G. & Johansson, R. S. (1987). Responses in glabrous skin mechanoreceptors during precision grip in humans. *Exp Brain Res*, **66**, 128–140.

Part III

The pathophysiology of grasping

19

Disorders of the somatosensory system

JOACHIM HERMSDÖRFER AND DENNIS A. NOWAK

Summary

This chapter reviews impairments of grasping and other fine motor tasks following disorders of the somatosensory system. The first part reports findings from transient anesthesia induced experimentally in healthy human subjects. The second part summarizes studies on the effects of lesions to the peripheral sensory system. Findings in patients with sensory deficits following polyneuropathy or carpal tunnel syndrome are differentiated from chronic complete somatosensory deafferentation. The latter group of very rare subjects provides the unique possibility of investigating the function of the motor system deprived of sensory input. The last part summarizes the effects of central lesions due to stroke or cerebral palsy that frequently affect the somatosensory system. The results for various motor tasks including prehensile movements are reported. Specific emphasis is placed on analyses of grip-force control during object manipulation since somatosensory feedback is particularly important for these activities and ample research has been performed during the last few years, enabling comparisons between patient groups.

Introduction

Clarifying the role of sensory information in the control of voluntary movement and force production is one of the most essential questions in sensorimotor research. The most obvious way to investigate this question is to study the effects of damage to the sensory system on movement execution. Indeed, there was controversy about the effects of a complete lack of sensory information at the beginning of the 20th century. Some authors reported that monkeys with sections of the dorsal root of the hand afferents discontinued using the impaired hands in daily life, indicating a mandatory role of sensory input to each aspect of motor control (Mott & Sherrington, 1895). This observation was contradicted by reports of profound recovery of function after the same surgical procedure when the section was bilateral or the use of the non-impaired hand was restricted after unilateral surgery (Munk 1909; Knapp et al., 1963). From this observation the necessity of active use of the deafferented hand to regain function was emphasized.

Sensorimotor Control of Grasping: Physiology and Pathophysiology, ed. Dennis A. Nowak and Joachim Hermsdörfer. Published by Cambridge University Press. © Cambridge University Press 2009.

This chapter concentrates on the effects of disturbances of the somatosensory system on manipulative hand functions in humans. Results of transient perturbations in healthy subjects and of chronic impairments in patients with damage to the peripheral or central nervous system are presented.

The nervous system is equipped with various receptors recording information about movement and forces. Proprioception subserved by muscle spindles, tendon and joint receptors is usually distinguished from exteroception conveyed by receptors in the skin. Cutaneous receptors are of particular importance for skilled object manipulation (Johansson, 1996; Macefield & Johansson, 1996; Flanagan & Johansson, 2002). However, muscle spindles, joint and tendon receptors signal forces and movements and can compensate for deficits of cutaneous receptor function (Phillips, 1986; Häger-Ross & Johansson, 1996; Jones & Lederman, 2006) (see below). A profound review of receptor types and their specific functions during object grasping and manipulation is provided in Chapters 11 and 13.

Transient perturbation of somatosensory afferents in healthy subjects

Several methods to transiently interrupt sensory information from the hand have been applied in studies of healthy subjects. One simple manipulation is the use of gloves that prevent the direct contact of the skin with the grasped object. As a consequence, information about surface properties and shear forces at the fingertips is much less precise (Kinoshita, 1999). Information from the skin can also be selectively disturbed by cooling with sprays or gels (Nowak & Hermsdörfer, 2003a). Stronger and longer-lasting deterioration of afferents can be achieved by injections of local anesthetics. Typically a ring-block was applied to the base of individual fingers anesthetizing primarily cutaneous afferents from the fingers (Nowak *et al.*, 2001; Augurelle *et al.*, 2003; Monzee *et al.*, 2003). Anesthesia of the whole hand was achieved by inflating pressure cuffs around the forearm that render the distal body parts ischemic (Goodwin *et al.*, 1972; Phillips, 1986). Such an intervention, however, affects not only afferents of skin, muscles and joints, but also efferent motor signals. Similarly, mechanical compression of a nerve deteriorates afferent and efferent signals (Cole *et al.*, 2003).

An almost invariable effect of these manipulations was an increase of the grip force applied against a grasped object. For example, the safety margin, calculated as the difference between actual grip force and slip force (minimal force to prevent slippage; see Chapters 1, 3 and 11), increased when an object was held with gloves as compared with barehanded holding and the grip-force increase was related to the thickness of the gloves (Kinoshita, 1999). Large increases of the grip force were observed after injection of anesthetics into the proximal phalanx of the grasping fingers. During stationary holding, increases of the ratio between grip force and load of 50–300% were reported (Johansson & Westling, 1984; Augurelle *et al.*, 2003; Monzee *et al.*, 2003). The effect of cooling of the fingers on the grip force was tested during dynamic movements of a grasped object (see Chapter 1). The ratio between grip force and load at the lower and upper turning point of a cyclic vertical movement increased by 43–69% (Nowak & Hermsdörfer, 2003a). Figure 19.1 shows examples of single up- and downward movements of a grasped object with and without anesthetic ring block of the fingers. Despite similar time courses of the load, grip force is substantially increased during anesthesia. This and comparable experiments found large increases of the grip force:load force ratio between about 50% and more than 300% during dynamic load production (Nowak *et al.*, 2001, 2002a; Augurelle *et al.*, 2003). Paradoxically, while the grip force during object manipulation increases, the maximum grip force decreases by up to 30% during digital anesthesia (Duque *et al.*, 2005).

One logical primary reason for the grip-force increases is a strategic response to ensure against slippage of the object despite lack of appropriate sensory information. However, loss of the object was nevertheless frequently reported in these studies. Various reasons seem to be responsible for this finding. While grasping an object with anesthetized fingers the initial increase of the grip force has been shown to attenuate over time (Augurelle *et al.*, 2003). During object manipulations with time-varying loads, the increase of the grip-force level may not suffice to compensate transient load peaks, for example high loads at the lower turning point of cyclic vertical movements (Augurelle *et al.*, 2003). In addition, the direction of force exertion and the placement of the fingers is less accurate during anesthesia (Monzee *et al.*, 2003). Placement of the fingers away from the center of gravity causes torques that can destabilize the object and increase the effective load (see Chapter 1). Thus, grip force increases following anesthesia can be considered as a combination of increased safety margin and consequence of increased total load.

In addition to increases of the grip force, various other aspects of skilled object manipulation deteriorate during anesthesia of the fingers. Thus, the different phases of a grasping and lifting movement (see Chapter 11) are prolonged, the coordination between the grip force and the load force is less precise, and the adaptation of the grip force to changing object properties such as weight and surface structure is inaccurate or even lacking (Johansson & Westling, 1984; Jenmalm & Johansson, 1997; Augurelle *et al.*, 2003; Monzee *et al.*, 2003). It is important, however, that the feedforward mechanism of grip force control during self-produced loads is preserved despite impaired sensory input from the fingertips. For example it is obvious from Figure 19.1 that grip force rises and falls near simultaneously with the load force and both forces peak at similar time points irrespective of anesthesia that affects the level of the grip force (Nowak *et al.*, 2001, 2002a; see also Augurelle *et al.*, 2003). From this observation it was hypothesized that, despite decreases of precision, the function of internal models may not be critically dependent on sensory input (Nowak *et al.*, 2001; Hermsdörfer *et al.*, 2005) (see Chapters 9 and 12).

Apart from impaired force control during object contact, anesthesia of the fingers also deteriorates the kinematics of prehensile movements (see the paradigm in Chapters 2 and 10). When subjects with finger anesthesia grasped a seen object without visual feedback of the hand, the whole movement slowed down and the hand path (transport movement) became more variable (Gentilucci *et al.*, 1997). In addition, the time to open the fingers was prolonged and maximum finger aperture increased. The authors hypothesized that precise information about the time of object contact is necessary to optimize all components

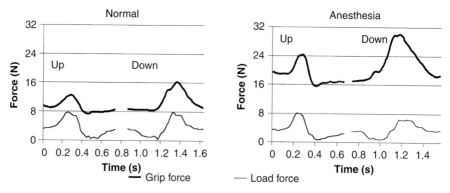

Figure 19.1. Grip force (thick line) and load force (thin line) during single up- and downward movements of a grasped object. Performance of a healthy subject before (left) and during anesthesia of the grasping fingers (right). Modified from Nowak *et al.* (2001).

of the prehensile movement. In addition, a strategic response due to uncertainty might also be important. Adaptation of the movement to variations in amplitude and object size was however not impaired by anesthesia of the fingers.

The response of healthy subjects to transient impairments of sensory information provides a schema of altered behavior that may also be expected in patients with deficits of the somatosensory system. As outlined below, examinations of patients show similarities but also discrepancies between both groups.

Lesions of the peripheral sensory system

Polyneuropathy and carpal tunnel syndrome

Damage to the peripheral sensory system that affects the function of the hand can originate from a variety of neurological diseases. Polyneuropathies of inflammatory or non-inflammatory origin and nerve compression syndromes such as carpal tunnel syndrome (CTS) are the most common of these diseases. They affect the motor and the sensory systems to varying degrees. This chapter concentrates on variants with primarily sensory deficits. Chapter 20 provides a more profound consideration of grip-force control in CTS patients particularly during multi-finger tasks.

Interestingly, a number of studies on grip-force control in patients with mild to moderate symptoms following peripheral lesions failed to reveal notable impairments (Thonnard *et al.*, 1999; Nowak *et al.*, 2003b). Thus, the grip force during stationary holding of an object as well as during vertical cyclic movements of an object were similar in a combined group of patients with polyneuropathies and CTS and a group of control subjects, despite all patients revealing mild to moderate deficits in sensory tests such as perception of touch and vibration or 2-point discrimination (Nowak *et al.*, 2003b). Other studies in similar patient populations found a moderate increase of grip force in similar tasks (Lowe & Freivalds,

1999; Nowak & Hermsdörfer, 2003b). These findings were supported by simulations of CTS in healthy subjects. Compression of the median nerve at a milder degree deteriorated touch perception but did not change the grip force while holding an object. Strong compression, however, increased the grip force although this increase saturated at the highest compression strengths (Cole *et al.*, 2003).

However, the synchrony between the grip force and self-generated load was preserved in all studies. Thus feedforward grip-force control seems to be generally preserved after mild to moderate damage to the peripheral sensory system.

Traumatic nerve injury

Apart from neurologists, hand surgeons have long been interested in the consequences of damage to peripheral nerves transmitting sensory information from the hand. Moberg (1962, 1991) emphasized the tight linkage between sensory and motor aspects in impaired manual dexterity and complained about the lack of a functionally relevant measure of sensory loss.

Recently, the effects of replantation of the hand after injury and of sutured single digital nerves on grip-force control was investigated (Schenker *et al.*, 2006). All patients had impaired two-point discrimination of varying degrees. The grip force during lifting of objects with different weights was increased in about half of the patients. This increase resulted partially from an increase of the safety margin. In addition, an increase of the grip force was necessary to compensate for load increases due to misalignment of the fingers and inaccurate force production (as discussed previously in healthy subjects with anesthesia). Patients adapted their grip force to different object weights but frequently failed to adapt to varying object shapes. Measures of impaired force control correlated only reasonably with the clinical test of sensibility (see below in stroke patients).

Complete sensory deafferentation

As already outlined at the beginning of the chapter, particular insight into the role of sensory information in motor control was expected from studying motor systems completely deprived of somatosensory input. Such a constellation exists not only in animals with experimental lesions but also in humans with pure sensory neuropathies. In these rare cases, the disease selectively destroys the large myelinated sensory fibers while the motor fibers remain intact.

Despite their severe deficits these patients usually are able to live independently and to perform daily motor tasks. However, manual skills necessary for eating, drinking or dressing are typically reported to be impaired (Rothwell *et al.*, 1982; Sanes *et al.*, 1985; Cole & Paillard, 1995). Handwriting is very difficult and slowed. In general, these patients report being highly dependent on visual feedback in all of their motor actions. Figure 19.2 shows two characteristic alterations in everyday motor actions in one patient (GL). One image shows an imprecise grip with only the middle and little finger contacting a sheet of wrapping paper and the second image shows exaggerated hand opening while grasping a glass.

Figure 19.2. Snapshots from video recordings of two manual activities in a patient with severe somatosensory deafferentation of the limbs and the trunk (GL). Left: unwrapping a present; right: grasping a glass.

In a number of these patients, elementary motor functions of the wrist and fingers were studied. Some of these tests revealed surprisingly good performance. Rothwell *et al.* (1982) showed in a single case that the hand could be shaped according to verbal command even without visual feedback. The subject could produce movements of varying amplitudes and velocity quite precisely with and without visual feedback (although deficits in a related task were reported in a larger patient sample; see Sanes & Jennings, 1984; Sanes *et al.*, 1985). Similarly, another single case tested for force production in a precision grip was able to produce fractions of her maximum force and to reproduce a force with the other hand with near-normal precision (Teasdale *et al.*, 1993; Lafargue *et al.*, 2003). These experiments suggest that deafferented patients are able to learn, maintain and control the production of motor output despite lack of somatosensory information. Deficits were however obvious as variability and instability increased over time in the absence of visual feedback. In particular, perturbations applied against a wrist or finger movement were not compensated so that the movements missed the target and were not corrected (Rothwell *et al.*, 1982; Sanes & Jennings, 1984).

A study of the transport and grasp component of a prehensile movement in another single case revealed further performance aspects that were preserved despite somatosensory deafferentation (Gentilucci *et al.*, 1994). While movement time was prolonged and variability increased, the velocity and the duration of the transport component were scaled to the reaching distance and the size of the grip aperture was scaled to the size of the object. Only in the final phase of the movement during object approach were abnormal hesitations and re-openings of the hand observed. Another study interested in bimanual interaction reported preserved performance in a passing task when one hand moved and passed the target to the other hand in the absence of visual feedback (Simoneau *et al.*, 1999). Transport of one hand and grip formation with the other hand was well coordinated in the deafferented subject. Both components were however uncoordinated during unimanual prehension without visual feedback. The authors concluded that especially in the absence of vision, proprioceptive feedback is necessary to coordinate the transport and grasp component within one hand.

The control of grip forces following somatosensory deafferentation is of particular interest because no or only very small movements arise during the generation of grip forces applied against solid objects. Thus, force production cannot be effectively controlled by vision so that the primary compensatory input channel for deafferented subjects is ineffective.

We had the possibility to study GL, a woman in her mid-50s at the time of the experiments, with complete deafferentation of all limbs and the trunk up to the level of the nose for more than 25 years. She was unable to perceive any movement of her limbs or any pressure exerted against the skin. The performance examples in Figure 19.2 were recorded from GL. Figure 19.3 shows her and a control subject's performance during vertical cyclic movements of a grasped object (Hermsdörfer *et al.*, 2008). Such movements produce oscillating acceleration and load profiles that are accompanied by synchronous oscillations of the grip force in healthy subjects (see Chapter 1 – Figure 1.3). In the experiment, the frequency of the movement and the weight of the object were varied in order to produce different time courses and levels of load. Figure 19.3 shows the resulting load profiles for two trials with different loads in both subjects. While GL was able to move her arm in the prescribed fashion, the anticipatory oscillations of the grip force were largely missing; her grip-force profile was irregular and unrelated to the load profile. This is particularly obvious from the plots of grip force versus load that vary with a near linear relationship in the control subject but seem unrelated in GL. From this finding it was concluded that an internal forward model of the dynamics of the object and of the own motor system was absent in GL (see also Gordon *et al.*, 1995; Sainburg *et al.*, 1995; Sarlegna *et al.*, 2003 for aiming movements). Such models, which are established during development, probably must be at least intermittently updated by somatosensory input (see Chapters 1 and 9; Miall & Wolpert, 1996; Wolpert & Kawato, 1998). The absence of anticipatory grip-force coupling with self-generated loads was also obvious during discrete upward and downward movements tested in GL (Nowak *et al.*, 2004; Hermsdörfer *et al.*, 2004).

Other important aspects of altered grip-force control in GL are obvious from Figure 19.3. Thus, grip force was clearly increased compared with the control subject. This increase was expected from corresponding findings in healthy subjects with transient sensory impairments and other patients with sensory loss (see above). Increased grip force was a recurrent result of all the studies we conducted in GL and another patient with a comparable deafferentation (Nowak *et al.*, 2003a, 2004; Hermsdörfer *et al.*, 2004, 2008; Nowak & Hermsdörfer, 2006). However, despite the increase, GL scaled the grip-force level to the average load level. This is obvious in Figure 19.3 from the plot of averaged grip force maxima versus load maxima for single trials. A range of loads was generated by combinations of two different movement frequencies and two object weights. Despite high variability, the grip force maxima of GL increased with the load, the slope being even steeper than in the control subject. Such preserved performance indicates that GL was able to infer the

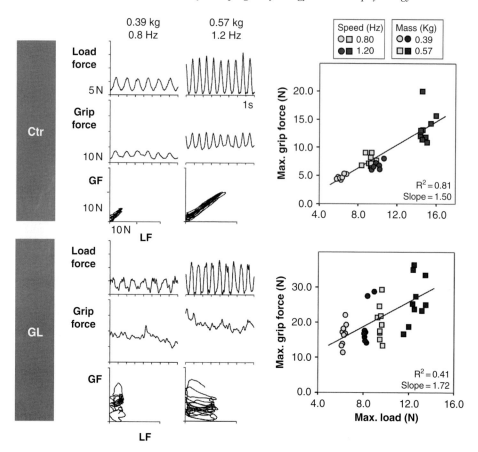

Figure 19.3. Grip-force control of movement-induced loads in a subject with complete somatosensory deafferentation (GL) and a control subject (Ctr). Left side: Load force (LF), grip force (GF) and GF/LF-plot during vertical cyclic movements with two different frequencies and with two objects of different weight in both participants. Right side: Maximum grip force in dependency of the actual maximum load generated during movements with combinations of two different frequencies and two different object weights. Modified from Hermsdörfer *et al.* (2008).

load variations and emit a crude anticipatory response either on the basis of her motor command for the arm movement (efferent copy) or on the basis of alternative and indirect load information. Such mechanisms were probably responsible for other demonstrations of surprisingly good performance in deafferented subjects during weight matching (Fleury *et al.*, 1995) and object lifting (Nowak *et al.*, 2004).

Interestingly, a second patient with somatosensory deafferentation of trunk and limbs (IW) showed a very similar pattern of crude grip-force anticipation of the load level (Hermsdörfer *et al.*, 2008). The grip-force–load-force coupling was however less deficient in this patient (although not normal), suggesting that in the case of such severe deafferentation the motor system may find different solutions for compensation and may exploit remaining signals in idiosyncratic ways.

In conclusion, studies in deafferented subjects emphasize the importance of somatosensory feedback for many aspects of motor performance, particularly if an action lasts for a longer time, if visual control is not possible or is difficult, and if unpredictable perturbations occur. The studies have however also shown that the subjects have many preserved motor abilities and are extremely proficient in integrating alternative information into their motor programs.

Somatosensory deficits following central lesions

Somatosensory disturbances are a frequent consequence of cerebral lesions. Estimates of the frequency of affected patients vary in a relatively wide range between about 25% up to 85% in unselected samples of patients with brain lesions (Hermsdörfer *et al.*, 1994; Kim & Choi-Kwon, 1996; Smania *et al.*, 2003; Blennerhassett *et al.*, 2007). In a survey in 272 patients with stroke and traumatic brain injury, disturbances in common clinical tests of sensibility were manifested in 30% of the patients (Hermsdörfer *et al.*, 1994). The most frequently affected tasks included kinesthesia (reproduction of passive movement), localization, two-point discrimination, stereognosia and mild touch, in descending order. The high frequency of 85% was reported in a study of similar tasks (plus tactile discrimination) using refined psychophysical methods and a relatively narrow definition of the normal range (1 SD) (Kim & Choi-Kwon, 1996).

Stroke

In patients with stroke, sensory deficits are typically combined with motor deficits (see Chapter 21). There exist however pure sensory strokes with lesions confined to the somatosensory areas of the cortex or to sensory structures of the thalamus (Jeannerod *et al.*, 1984; Smania *et al.*, 2003). In one such case with a unilateral cortical lesion of sensory areas (postcentral gyrus extending into supramarginal gyrus), Jeannerod *et al.* (1984) conducted a thorough investigation of prehensile movements. Although the severe sensory deficit was confined to the contralesional hand and the shoulder was only mildly affected, the patient's performance deficits were comparable to or even worse than those in patients with complete sensory deafferentation (see above; Rothwell *et al.*, 1982; Gentilucci *et al.*, 1994). While the first phase of the transport component from movement start to maximum velocity was near normal, the second phase was prolonged and characterized by an irregular movement path and velocity profile. The grasp component was imprecise or missing. As in the deafferented patients, the severity of deficit depended critically on the availability of visual feedback, with complete failure when the hand was not visible, improvement for vision during the terminal phase, and further improvement but remaining deficits when full vision was available. Other tasks such as maintaining a constant grip force, tracking a sinusoidal force target (compare Chapter 1) or functional tasks such as crumpling a sheet of paper were clearly impaired. This and other studies in comparable patients showed that extensive lesions in primary central sensory areas severely affect hand function and cannot be compensated by alternative processing networks.

A powerful method to evaluate the impact of sensory deficits on hand function is correlation analyses in patient samples with sensory deficits of varying severity. In one study, Blennerhassett *et al.* (2007) assessed the functional performance of 45 patients during the grasping and lifting of objects with varying surface structure and weight. Performance was compared with two tasks that ostensibly assessed sensory performance aspects relevant for the lifting task, namely tactile surface discrimination and weight discrimination assessed with psychophysical methods. Correlation between the scores derived from the sensory tests and two scores characterizing the temporal aspect and force parameters of the lifting task revealed a coefficient of 0.34 for the relationship between the tactile discrimination score and both lifting scores and no significant correlation with weight perception. Thus, explicit surface discrimination was significantly related to deficits in lifting performance ($P < 0.03$) although only a small amount of variation could be explained.

We analyzed the control of grip force during three manipulation tasks in 19 chronic stroke patients and compared outcome parameters with the ability to resist an external perturbation without visual feedback (Hermsdörfer *et al.*, 2003). The perturbation consisted of a moderately fast increase of the force acting against a precision grip (2.5 to 7.5 N in 1 s). This increase caused an invariable opening of the grip that had to be resisted by the subject. The amount of displacement before the opening was stopped was considered as a measure of the effectiveness of sensory processing. The displacement correlated well with clinical tests of sensibility (proprioception: Spearman $r = 0.60$, tactile discrimination: $r = 0.68$).

Figure 19.4 shows the performance for both hands in one patient with severe sensory deficits of the right hand during the perturbation task and during cyclic vertical movements of a grasped object. The load perturbation caused a much larger displacement of the fingers in the right than in the left hand. During cyclic movements with the grasped object, the acceleration and load profile were similar for both hands; grip-force level was however substantially increased in the involved hand. In the patient sample (right side of Figure 19.4) the grip force was higher with higher rank orders and a corresponding larger displacement in the perturbation task. The correlation between grip force (relative to load) and displacement was high (Spearman $r = 0.77$). In addition, the maximum coefficient of cross-correlation, which is a measure for the precision of coupling between grip force and load force, decreased with increasing inability to resist the perturbation ($r = -0.62$). The time-lag between grip force and load was however near zero and independent of rank order in all but two patients, indicating that feedforward processing was not affected in most of the patients.

High correlations with the displacement were also detected in the other manipulation tasks for the amount of grip force (object hold: $r = 0.80$, object transport: $r = 0.70$) and temporal parameters (time to reach grip force during transport: $r = 0.71$). A study of grip-force control in acute stroke patients, all of whom had sensory deficits, found very similar

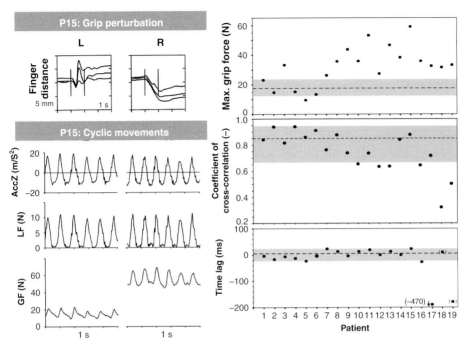

Figure 19.4. Left side: Performance of a stroke patient following infarction of the left medial cerebral artery causing contralateral sensory loss without paresis. Examples for two tasks are shown for both hands: Displacement of the fingers in a precision grip during unpredictable increases of the load (2.5 to 7.5 N, between vertical lines) (above). Vertical acceleration (AccZ), load force (LF) and grip force (GF) during vertical cyclic movements of a grasped object (below). Right side: Average maximum grip force, maximum coefficient of cross-correlation and corresponding time lag during cyclic movements of a grasped object in 19 patients with stroke. Patients are ordered by increasing displacements of fingers in the grip perturbation task. The shaded areas indicate the range of healthy subjects. Modified from Hermsdörfer *et al.* (2003).

performance patterns during object grasping and lifting and during discrete vertical movements (Nowak *et al.*, 2003c).

The ability to resist an unexpected load perturbation during grasping therefore emerged as a sensitive test to predict deficits during object manipulation. Exaggerated grip forces can be considered as a strategic response to a lack of somatosensory information in order to protect against involuntary loss of the object (Hermsdörfer *et al.*, 2004; Nowak & Hermsdörfer, 2005). Sensory deficits, however, not only caused uneconomical force production but also induced slowness and uncoordination. The fact that the correlations were higher for the perturbation tasks than for the psychophysically assessed discrimination tasks may indicate that implicit processing of sensory input shares more common features with grip force control during object manipulation than explicit processing of physical object features (see Chapter 12).

Dexterous manipulation of objects and tools is just one aspect of skilled hand function. Another aspect is the explorative function of the hand in what has been termed active touch or active haptic sensing (Phillips, 1986; Lederman & Klatzky, 1987; Jones & Lederman, 2006) (see Chapter 13). Active touch requires precise and fast integration of sensory input and motor output to identify physical characteristics and identities of objects in the hand. Binkofski *et al.* (2001) showed in stroke patients that contralesional exploratory finger movements can be disturbed even without or with only mild sensory and motor deficits. The finger movements during object exploration were altered in a characteristic way; fingers moved more slowly, more irregularly and the movement path covered a much larger space (see also Pause *et al.*, 1989). In contrast, finger-tapping was largely normal. The deficit was closely related to astereognosia, that is, inability to identify objects by active touch without sight. Lesions of the posterior part of the parietal lobe seem to be particularly related to the combination of deficits.

Cerebral palsy

Patients with cerebral palsy exhibit various deficits of dexterity and force control outlined in detail in Chapter 31. The contribution of frequent sensory deficits in cerebral palsy to performance in functional tasks was studied by Gordon & Duff (1999) using correlation analyses similar to those conducted in stroke patients (see above). The ability to adapt and anticipate grip force according to different surface structures of a grasped object was compared with the performance in clinical tests of sensibility and motor performance in 15 children with cerebral palsy. Among the sensory tests, two-point discrimination correlated most strongly with the difference in static grip forces for sandpaper and rayon surface ($r = -0.68$), emphasizing the role of tactile sensibility and indicating that children with low tactile acuity did not economically regulate grip force according to surface friction. The study also suggested that different aspects of the grasping and lifting task are associated with specific elementary sensory and motor abilities. Thus, the correlation of the two-point discrimination with performance measures was strongest during the static phase and weaker during the earlier dynamic phase.

Conclusion

Data from studies in healthy subjects with experimentally induced sensory deficits and from patients with sensory loss due to peripheral nerve damage or central brain lesions clearly emphasize the central role of somatosensory feedback for the manipulative function of the hand. In studies that assessed grip forces during object manipulation, an increase of grip force was the most frequent consequence of sensory impairments irrespective of the cause. However, as suggested by preserved performance in patients with moderate peripheral sensory deficits, grip-force increases are not a mandatory consequence of sensory loss, nor are sensory deficits the only cause of grip-force increases. Indeed, excessive grip forces have been reported in numerous neurological diseases that are not associated with prominent impairments of sensory perception such as stroke of the internal capsule (Wenzelburger *et al.*, 2005), amyotrophic lateral sclerosis (Nowak & Hermsdörfer, 2002; Nowak *et al.*, 2003d; Hermsdörfer *et al.*, 2004), or degenerative cerebellar diseases (Fellows *et al.*, 2001; Nowak *et al.*, 2002b; Rost *et al.*, 2005) (see Chapter 26). As outlined above in detail, sensory deficits affect a high number of other aspects of motor performance such as precision, speed, regularity, repeatability, coordination and endurance. It is however remarkable that feedforward control seems not to be vulnerable to sensory loss with the exception of severe and long-lasting deafferentation.

References

Augurelle, A. S., Smith, A. M., Lejeune, T. & Thonnard, J. L. (2003). Importance of cutaneous feedback in maintaining a secure grip during manipulation of hand-held objects. *J Neurophysiol*, **89**, 665–671.

Binkofski, F., Kunesch, E., Classen, J., Seitz, R. J. & Freund, H. J. (2001). Tactile apraxia. Unimodal apractic disorder of tactile object exploration associated with parietal lobe lesions. *Brain*, **124**, 132–144.

Blennerhassett, J. M., Matyas, T. A. & Carey, L. M. (2007). Impaired discrimination of surface friction contributes to pinch grip deficit after stroke. *Neurorehabil Neural Repair*, **21**, 263–272.

Cole, J. & Paillard, J. (1995). Living without touch and peripheral information about body position and movement: studies with deafferented subjects. In J. L. Bermudez, A. Marcel & N. Eilan (Eds.), *The Body and the Self* (pp. 245–266). Cambridge, MA: MIT Press.

Cole, K. J., Steyers, C. M. & Graybill, E. K. (2003). The effects of graded compression of the median nerve in the carpal canal on grip force. *Exp Brain Res*, **148**, 150–157.

Duque, J., Vandermeeren, Y., Lejeune, T. M. *et al.* (2005). Paradoxical effect of digital anaesthesia on force and corticospinal excitability. *Neuroreport*, **16**, 259–262.

Fellows, S. J., Ernst, J., Schwarz, M., Töpper, R. & Noth, J. (2001). Precision grip deficits in cerebellar disorders in man. *Clin Neurophysiol*, **112**, 1793–1802.

Flanagan, J. R. & Johansson, R. S. (2002). Hand movements. In V. S. Ramachandran (Ed.), *Encyclopedia of the Human Brain, Vol. 2* (pp. 399–414). San Diego, CA: Academic Press.

Fleury, M., Bard, C., Teasdale, N. *et al.* (1995). Weight judgment: the discrimination capacity of a deafferented subject. *Brain*, **118**, 1149–1156.

Gentilucci, M., Toni, I., Chieffi, S. & Pavesi, G. (1994). The role of proprioception in the control of prehension movements: a kinematic study in a peripherally deafferented patient and in normal subjects. *Exp Brain Res*, **99**, 483–500.

Gentilucci, M., Toni, I., Daprati, E. & Gangitano, M. (1997). Tactile input of the hand and the control of reaching to grasp movements. *Exp Brain Res*, **114**, 130–137.

Goodwin, G. M., McCloskey, D. I. & Matthews, P. B. (1972). The contribution of muscle afferents to kinaesthesia shown by vibration induced illusions of movement and by the effects of paralysing joint afferents. *Brain*, **95**, 705–748.

Gordon, A. M. & Duff, S. V. (1999). Relation between clinical measures and fine manipulative control in children with hemiplegic cerebral-palsy. *Dev Med Child Neurol*, **41**, 586–591.

Gordon, J., Ghilardi, M. F. & Ghez, C. (1995). Impairments of reaching movements in patients without proprioception. 1. Spatial errors. *J Neurophysiol*, **73**, 347–360.

Häger-Ross, C. & Johansson, R. S. (1996). Nondigital afferent input in reactive control of fingertip forces during precision grip. *Exp Brain Res*, **110**, 131–141.

Hermsdörfer, J., Mai, N., Rudroff, G. & Münßinger, M. (1994). *Untersuchung zerebraler Handfunktionsstörungen. Ein Vorschlag zur standardisierten Durchführung.* Dortmund, Germany: Borgmann.

Hermsdörfer, J., Hagl, E., Nowak, D. A. & Marquardt, C. (2003). Grip force control during object manipulation in cerebral stroke. *Clin Neurophysiol*, **114**, 915–929.

Hermsdörfer, J., Hagl, E. & Nowak, D. A. (2004). Deficits of anticipatory grip force control after damage to peripheral and central sensorimotor systems. *Hum Mov Sci*, **23**, 643–662.

Hermsdörfer, J., Nowak, D. A., Lee, A. *et al.* (2005). The representation of predictive force control and internal forward models: evidence from lesion studies and brain imaging. *Cogn Proc Int Quart Cogn Sci*, **6**, 48–58.

Hermsdörfer, J., Elias, Z., Cole, J. D., Quaney, B. M. & Nowak, D. A. (2008). Preserved and impaired aspects of feedforward grip force control after chronic somatosensory deafferentation. *Neurorehabil Neural Repair*, **22**, 374–384.

Jeannerod, M., Michel, F. & Prablanc, C. (1984). The control of hand movements in a case of hemianaesthesia following a parietal lesion. *Brain*, **107**, 899–920.

Jenmalm, P. & Johansson, R. S. (1997). Visual and somatosensory information about object shape control manipulative fingertip forces. *J Neurosci*, **17**, 4486–4499.

Johansson, R. S. (1996). Sensory control of dexterous manipulation in humans. In A. M. Wing, P. Haggard & J. R. Flanagan (Eds.), *Hand and Brain* (pp. 381–414). San Diego, CA: Academic Press.

Johansson, R. S. & Westling, G. (1984). Roles of glabrous skin receptors and sensorimotor memory control of precision grip when lifting rougher or more slippery objects. *Exp Brain Res*, **56**, 550–564.

Jones, L. A. & Lederman, S. J. (2006). *Human Hand Function*. Oxford, UK: Oxford University Press.

Kim, J. S. & Choi-Kwon, S. (1996). Discriminative sensory dysfunction after unilateral stroke. *Stroke*, **27**, 677–682.

Kinoshita, H. (1999). Effect of gloves on prehensile forces during lifting and holding tasks. *Ergonomics*, **42**, 1372–1385.

Knapp, H. D., Taub, E. & Berman, A. J. (1963). Movements in monkeys with deafferented forelimbs. *Exp Neurol*, **7**, 315.

Lafargue, G., Paillard, J., Lamarre, Y. & Sirigu, A. (2003). Production and perception of grip force without proprioception: is there a sense of effort in deafferented subjects? *Eur J Neurosci*, **17**, 2741–2749.

Lederman, S. J. & Klatzky, R. L. (1987). Hand movements: a window into haptic object recognition. *Cogn Psychol*, **19**, 342–368.

Lowe, B. D. & Freivalds, A. (1999). Effect of carpal-tunnel syndrome on grip force coordination on hand tools. *Ergonomics*, **42**, 550–564.

Macefield, V. G. & Johansson, R. S. (1996). Control of grip force during restraint of an object held between finger and thumb: responses of muscle and joint afferents from the digits. *Exp Brain Res*, **108**, 172–184.

Miall, R. C. & Wolpert, D. M. (1996). Forward models for physiological motor control. *Neural Networks*, **9**, 1265–1279.

Moberg, E. (1962). Criticism and study of methods for examining sensibility in the hand. *Neurology*, **12**, 8–19.

Moberg, E. (1991). The unsolved problem – how to test the functional value of hand sensibility. *J Hand Ther*, **4**, 105–110.

Monzee, J., Lamarre, Y. & Smith, A. M. (2003). The effects of digital anesthesia on force control using a precision grip. *J Neurophysiol*, **89**, 672–683.

Mott, F. W. & Sherrington, C. S. (1895). Experiments upon the influence of sensory nerves upon movement and nutrition of the limbs. *Proc R Soc Lond B*, **57**, 488.

Munk, H. (1909). Über die Folge des Sensibilitätsverlustes der Extremität für deren Motilität. Über die Funktion von Hirn und Rückenmark. In *Gesammelte Mitteilungen* (pp. 247–285). Berlin: Hirschwald.

Nowak, D. A. & Hermsdörfer, J. (2002). Impaired coordination between grip force and load force in amyotrophic lateral sclerosis: a case-control study. *Amyotrophic Lat Sclerosis*, **3**, 199–207.

Nowak, D. A. & Hermsdörfer, J. (2003a). Digit cooling influences grasp efficiency during manipulative tasks. *Eur J Appl Physiol*, **89**, 127–133.

Nowak, D. A. & Hermsdörfer, J. (2003b). Selective deficits of grip force control during object manipulation in patients with reduced sensibility of the grasping digits. *Neurosci Res*, **47**, 65–72.

Nowak, D. A. & Hermsdörfer, J. (2005). Grip force behavior during object manipulation in neurological disorders: toward an objective evaluation of manual performance deficits. *Mov Disord*, **20**, 11–25.

Nowak, D. A. & Hermsdörfer, J. (2006). Predictive and reactive control of grasping forces: on the role of the basal ganglia and sensory feedback. *Exp Brain Res*, **173**, 650–660.

Nowak, D. A., Hermsdörfer, J., Glasauer, S. *et al.* (2001). The effects of digital anaesthesia on predictive grip force adjustments during vertical movements of a grasped object. *Eur J Neurosci*, **14**, 756–762.

Nowak, D. A., Glasauer, S., Meyer, L., Mai, N. & Hermsdörfer, J. (2002a). The role of cutaneous feedback for anticipatory grip force adjustments during object movements and externally imposed variation of the direction of gravity. *Somatosensory Mot Res*, **19**, 49–60.

Nowak, D. A., Hermsdörfer, J., Marquardt, C. & Fuchs, H. H. (2002b). Load force coupling during discrete vertical movements in patients with cerebellar atrophy. *Exp Brain Res*, **145**, 28–39.

Nowak, D. A., Glasauer, S. & Hermsdörfer, J. (2003a). Grip force efficiency in long-term deprivation of somatosensory feedback. *Neuroreport*, **14**, 1803–1807.

Nowak, D. A., Hermsdörfer, J., Marquardt, C. & Topka, H. (2003b). Moving objects with clumsy fingers: how predictive is grip force control in patients with impaired manual sensibility? *Clin Neurophysiol*, **114**, 472–487.

Nowak, D. A., Hermsdörfer, J. & Topka, H. (2003c). Deficits of predictive grip force control during object manipulation in acute stroke. *J Neurol*, **250**, 850–860.

Nowak, D. A., Hermsdörfer, J. & Topka, H. (2003d). When motor execution is selectively impaired: control of manipulative finger forces in amyotrophic lateral sclerosis. *Mot Contr*, **7**, 304–320.

Nowak, D. A., Glasauer, S. & Hermsdörfer, J. (2004). How predictive is grip force control in the complete absence of somatosensory feedback? *Brain*, **127**, 182–192.

Pause, M., Kunesch, E., Binkofski, F. & Freund, H. J. (1989). Sensorimotor disturbances in patients with brain lesions of the parietal cortex. *Brain*, **112**, 1599–1625.

Phillips, C. G. (1986). Movements of the hand. *Sherrington Lecture*, **17**.

Rost, K. R., Nowak, D. A., Timman, D. T. & Hermsdörfer, J. (2005). Preserved and impaired aspects of predictive grip force control in cerebellar patients. *Clin Neurophysiol*, **116**, 1405–1414.

Rothwell, J. C., Traub, M. M., Day, B. L. *et al*. (1982). Manual motor performance in a deafferented man. *Brain*, **105**, 515–542.

Sainburg, R. L., Ghilardi, M. F., Poizner, H. & Ghez, C. (1995). Control of limb dynamics in normal subjects and patients without proprioception. *J Neurophysiol*, **73**, 820–835.

Sanes, J. N. & Jennings, V. A. (1984). Centrally programmed patterns of muscle activity in voluntary motor behavior of humans. *Exp Brain Res*, **54**, 23–32.

Sanes, J. N., Mauritz, K.-H., Dalakas, M. C. & Evarts, E. V. (1985). Motor control in humans with large-fiber sensory neuropathy. *Hum Neurobiol*, **4**, 101–114.

Sarlegna, F., Blouin, J., Bresciani, J. P. *et al*. (2003). Target and hand position information in the online control of goal-directed arm movements. *Exp Brain Res*, **151**, 524–535.

Schenker, M., Burstedt, M. K. O., Wiberg, M. & Johansson, R. S. (2006). Precision grip function after hand replantation and digital nerve injury. *J Plast Reconstr Aesth Surg*, **59**, 706–716.

Simoneau, M., Paillard, J., Bard, C. *et al*. (1999). Role of the feedforward command and reafferent information in the coordination of a passing prehension task. *Exp Brain Res*, **128**, 236–242.

Smania, N., Montagnana, B., Faccioli, S., Fiaschi, A. & Aglioti, S. M. (2003). Rehabilitation of somatic sensation and related deficit of motor control in patients with pure sensory stroke. *Arch Phys Med Rehabil*, **84**, 1692–1702.

Teasdale, N., Forget, R., Bard, C. *et al*. (1993). The role of proprioceptive information for the production of isometric forces and for handwriting tasks. *Acta Psychol*, **82**, 179–191.

Thonnard, J. L., Saels, P., Vandenbergh, P. & Lejeune, T. (1999). Effects of chronic median nerve compression at the wrist on sensation and manual skills. *Exp Brain Res*, **128**, 61–64.

Wenzelburger, R., Kopper, F., Frenzel, A. *et al*. (2005). Hand coordination following capsular stroke. *Brain*, **128**, 64–74.

Wolpert, D. M. & Kawato, M. (1998). Multiple paired forward and inverse models for motor control. *Neural Networks*, **11**, 1317–1329.

20

Multi-digit grasping and manipulation: effect of carpal tunnel syndrome on force coordination

JAMIE A. JOHNSTON AND MARCO SANTELLO

Summary

Skilled manipulatory behaviors require complex spatial and temporal coordination of the digits. In healthy individuals, visual and somatosensory feedback is processed and integrated with motor commands thus ensuring successful interactions with objects. This process can be disrupted by a number of neuromuscular diseases. One of the most severely debilitating diseases of hand function is carpal tunnel syndrome (CTS), a compression neuropathy of the median nerve resulting in (1) somatosensory deficits in the thumb, index, middle and ring fingers (lateral half) and, in severe cases, (2) motor deficits in the thumb. Most studies that have investigated the effect of CTS on grasp control have focused on force coordination between the affected digits only. For patients with CTS, control of whole-hand grasping poses the additional challenge of coordinating all digits, a subset of which is characterized by deficits in sensorimotor capabilities. Our research on five-digit grasping shows that CTS affects patients' ability to create accurate sensorimotor memories of multi-digit forces for dexterous manipulation. This knowledge significantly extends and complements the information provided by existing clinical tools to diagnose and monitor CTS, with potential to improve the efficacy of clinical interventions such as physical rehabilitation and hand surgery.

Introduction

The coordination of digit forces during manipulatory behaviors relies on the ability to effectively integrate somatosensory and visual feedback with motor commands responsible for modulating forces at individual digits. Successful sensorimotor integration leads to the formation of "sensorimotor memories" or "internal models" (Edin *et al.*, 1992; Flanagan & Wing, 1993, 1997; Gordon *et al.*, 1993; Jenmalm & Johansson, 1997; Jenmalm *et al.*, 2000; Salimi *et al.*, 2000; 2003; Aoki *et al.*, 2006) of object properties and the forces necessary to manipulate the object. Proper scaling of digit forces implemented through these processes relies on repeated exposure to an object whose properties are stable over time and are predictable with a high degree of certainty, i.e. lifting and holding an object whose weight or center of mass are known to remain the same across trials (Gordon *et al.*, 1993; Salimi *et al.*,

Sensorimotor Control of Grasping: Physiology and Pathophysiology, ed. Dennis A. Nowak and Joachim Hermsdörfer. Published by Cambridge University Press. © Cambridge University Press 2009.

2000, 2003; Santello & Soechting, 2000; Reilmann *et al.*, 2001; Rearick *et al.*, 2002; see Chapters 9 and 12 for more details). Evidence for the use of sensorimotor memories has been revealed by studying force modulation occurring from the time the digit contacts the object to the onset of object lift (force rise phase), i.e. before object properties can be ascertained through use of tactile information. For instance, during the force rise phase, subjects tend to (1) use normal-to-tangential force ratios successfully employed during previous manipulations of the same object (see Chapters 9 and 12) and (2) exhibit a high degree of linear covariation between digit pairs, each reflecting anticipatory grip-force modulation (e.g. Gordon *et al.*, 1993).

Sensory information is also known to play a crucial role in upgrading erroneously planned fingertip forces (Macefield *et al.*, 1996; Augurelle *et al.*, 2003; Monzee *et al.*, 2003; see Chapter 11). Specifically, modulation of forces, either within- or across-digits, that occur *during* object manipulation as a result of perturbations or unexpected changes in object properties relies on responses triggered by tactile feedback. Therefore, in addition to the role of somatosensory feedback for the formation of sensorimotor memories described above, tactile feedback in particular has been found to be responsible for triggering short-latency force adjustments when digit forces are erroneously planned, i.e. when there is a mismatch between expected and actual object properties. Under digital anesthesia, these short-latency force responses are absent or delayed and subsequent manipulatory behaviors are characterized by the production of excessively large forces (Johansson & Westling, 1984a, 1984b; Edin *et al.*, 1992; Monzee *et al.*, 2003). In these circumstances, excessive grip forces found during object hold are developed prior to object lift, i.e. during the force rise phase. This suggests that subjects attempt to compensate for loss of tactile information by using excessive grip forces to minimize the risk of object slip (Cole *et al.*, 2003). Tactile input has at least two advantages over the use of on-line visual information during object manipulation. First, the former is characterized by significantly shorter latencies, i.e. tactile afferents are able to generate corrective force responses at ~35 ms (Johansson & Westling, 1987) vs. ~150–180 ms for visuomotor latencies (Welford, 1972). The implications are that when tactile afferents cannot provide veridical information about a mechanical stimulus (e.g. transient changes in shear forces on the fingerpads due to object slipping or tilting), vision might partially compensate but the motor responses might occur too late to prevent the object from falling. Second, tactile feedback can provide more precise information about certain object properties, such as texture, which may not be readily available to the visual system. In such cases, the role of vision in modulating force responses is likely minimal.

We start to appreciate the fundamental significance of somatosensory feedback and sensorimotor integration on manipulatory behavior when the use of somatosensory information from the digits is obstructed by a neuropathy or neurological disorder. Such is the case in carpal tunnel syndrome (CTS), which results in a constellation of symptoms including aching and burning, tingling, numbness, weakness and clumsiness in the affected hand. It is estimated that 400 000 CTS surgeries are performed in the USA each year (Mondelli *et al.*, 2004). Because of the high prevalence of this disorder and its potential for disability, it is imperative that effective techniques quantifying complex aspects of grasping and manipulatory behaviors be made available to clinicians to help improve diagnosis and determine the effectiveness of clinical interventions. Additionally, improving our knowledge of sensorimotor integration in patients affected by nerve compression can significantly contribute to understanding fundamental concepts in motor control and neuro-science such as neural plasticity and control of multiple degrees of freedom, i.e. how the central nervous system adapts the spatio-temporal coordination of multi-digit forces to somatosensory deficits that selectively impair a subset of digits of the hand.

Carpal tunnel syndrome

Carpal tunnel syndrome is a compression neuropathy of the median nerve. The median nerve originates in the brachial plexus and runs along the upper arm and forearm. It enters the hand through the carpal tunnel which is formed by the distal carpal bones (trapezoid, trapezium, capitate and hamate) and the more distal portion of the flexor retinaculum, i.e. the transverse carpal ligament. The median nerve shares the carpal tunnel with the m. flexor pollicis longus tendon, as well as the four tendons of both m. flexor digitorum superficialis and profundus (Levangie & Norkin, 2001). Thickening of the synovial lining of these tendons, due to repetitive hand activity, may increase fluid pressure in the tunnel (Armstrong *et al.*, 1984; Werner *et al.*, 1997) leading to a combination of stretching, shearing and/or compressive forces applied to the nerve (Werner & Andary, 2002). Prolonged mechanical compression of the nerve results in ischemic damage and/or changes in the myelination of the nerve, and in severe cases axonal loss can occur (Nora *et al.*, 2004). The electrophysiological consequences of this damage include slowing in axonal conduction velocity and nerve block.

The median nerve is a mixed nerve comprised of both sensory and motor axons innervating most extrinsic hand flexor muscles and some intrinsic muscles. It also relays sensory information from the palmar aspect of the thumb, index, middle and ring finger (lateral half). Sensory information from mechanoreceptors of tendons and joints of the thumb also travels through the carpal tunnel; however skin sensory information from the palmar aspect of the hand over the thenar eminence, conveyed by the palmar cutaneous branch, travels outside of the carpal tunnel and thus is not affected by CTS. Median nerve compression leads to sensorimotor impairments in the hand that begin with deficits in sensation in the thumb, index, middle and lateral half of the ring finger (palmar and the most distal dorsal aspect of these digits) and progresses to include motor deficits predominantly in the thumb. Note that the extrinsic digit flexors are unaffected by CTS because innervation of these muscles occurs proximal to the site of nerve compression. Thus, digit force production can still occur through activation of the extrinsic muscles, while force production from intrinsic muscles – primarily innervating the thumb – would be impaired. Since both intrinsic and extrinsic muscles innervate each digit, this constitutes a complex control problem as these muscles differ in their force generation capabilities and dependence on wrist posture (Schieber & Santello, 2004; Johnston *et al.*, 2009). An additional issue is that tactile input is likely to be important for the coordination of intrinsic and extrinsic muscle activity underlying digit force modulation. Therefore, although extrinsic muscles are not directly affected by median nerve compression, sensory deficits affecting intrinsic muscles may affect the coordination of both sets of muscles. Clearly, a definition of "sensory" or "motor" deficits for CTS cannot account for subtler, yet functionally crucial, aspects of the ability of patients to grasp and manipulate objects.

Effect of carpal tunnel syndrome on grasp control

To date, research on the effects of CTS on the coordination of grip forces has been equivocal, showing that individuals with CTS either (1) produce greater normal force than or (2) exhibit no difference from healthy controls. It is important to note that these studies examined two-digit grasping only (with the exception of Nowak *et al.*, 2003, see below). Most importantly, the analysis performed by these studies, however informative, was limited to the coordination of normal and tangential forces within – rather than across – digits. For instance, Thonnard *et al.* (1999) reported that patients with CTS were able to develop normal and tangential forces *within each digit* in a quantitatively similar fashion controls, as further demonstrated by their ability to adjust both digit forces to different frictional properties, i.e. from non-slippery (brass) to slippery (talc on brass). The authors concluded that the apparent lack of correlation between hand function (dexterity and grip force coordination) and the electrodiagnostic tests (suggestive of severe CTS) may be partially due to the latter's assessment of only a minority of fibers constituting a peripheral nerve, i.e. large myelinated fibers. Nevertheless, even though the smaller fibers (Aδ and unmyelinated C fibers from nociceptors) may still be intact, their role in sensorimotor transformations for grasping is likely to be significantly smaller than that played by large myelinated fibers. Nowak *et al.* (2003) asked patients with moderately impaired tactile sensitivity due to either chronic median nerve compression or axonal or demyelinating sensory polyneuropathy to perform point-to-point arm movements with a hand-held instrumented object and compared their performance with an age- and gender-matched control group. Their main result was that the sum of all digit forces exerted during stationary hold and point-to-point movements of the object was similar to that found in the control subjects. This task was designed such that subjects used all five digits, i.e. *both CTS-affected and non-affected digits*, to hold the object. In this case, the use of digits with intact sensorimotor capabilities may have contributed to appropriately modulating overall grip force. This could not be determined, however, since forces at individual digits were not examined. In contrast to these studies, larger normal-to-tangential force ratios were observed by Lowe & Freivalds (1999) in patients with CTS compared with control subjects while performing a tool gripping task. It is unclear, however, whether task differences and/or the degree of severity of CTS accounts for the different results reported by Nowak *et al.* (2003) and Lowe & Freivalds (1999). Lastly, studies in which median nerve neuropathies were artificially induced by mechanical compression of the nerve (Cole *et al.*, 2003) or by injection of anesthesia into the carpal tunnel (Dun *et al.*, 2007) support Lowe & Freivalds (1999) findings of increased normal force amplitude relative to control conditions. It is important to note, however, that artificially induced nerve "neuropathies" reflect *acute* changes in nerve function. The behavioral manifestations of acute changes may differ from changes in nerve function due to *chronic* compression, as in CTS, as the latter may also lead to (1) a reorganization of the somatosensory hand area in the cortex (Tecchio *et al.*, 2002; Napadow *et al.*, 2006), (2) a change in the excitability of the spinal cord (Jaberzadeh & Scutter, 2006), and/or (3) behavioral changes to compensate for functional deficits due to long-term, progressive degeneration of nerve function. Regardless, an increase in normal grip force with respect to tangential force is likely to be a compensatory mechanism used by patients with CTS to prevent the object from dropping.

The study described in the following sections focuses on the effects of CTS on sensori-motor integration for multi-digit grasping. Our goal was to determine the extent to which patients with CTS can integrate tactile feedback from CTS-affected and non-affected digits and use this information to plan for and control digit forces during a five-digit grasp. Our hypothesis is that patients with CTS will be less successful than controls at using prior sensory information to adequately plan appropriate forces for object manipulation.

Effect of CTS on multi-digit grasping

We collected data from six patients with CTS and six age- and gender-matched controls (mean age: 64; range: 51–82). The patients were diagnosed with mild or moderately severe CTS based on results from electrodiagnostic tests conducted by a neurologist in consultation with a hand surgeon from The Mayo Clinic, Scottsdale. Subjects were asked to grasp, lift, hold for approximately 8 s and replace a grip manipulandum (Figure 20.1) using all digits.

The task was to maintain the object vertical, measured by a position sensor, throughout all grip phases. We changed the object's center of mass (CM) by placing a weight (0.1 kg) at one of three locations at the base of the device. The rationale for adding an external torque to

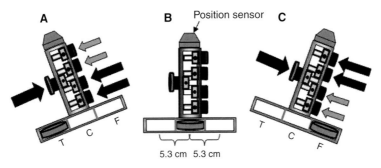

Figure 20.1. Grip manipulandum with the mass added (5.3 cm distance from the center) on the (A) thumb, (B) center and (C) finger side. Black and gray arrows indicate larger and smaller hypothetical normal forces, respectively, required to align the object with the vertical during lift.

the object was to cause object tilt, and therefore assess the patients' ability to (1) appropriately sense object tilt – primarily through changes in shear forces at the fingertips – and (2) modulate forces appropriately at individual digits to minimize the tilt.

To assess the extent to which patients with CTS could utilize sensorimotor memories to generate repeatable force coordination patterns, we changed object CM using a blocked design. Specifically, subjects performed a block of seven trials (the first two were practice) with the same CM, one block for each of the three CM locations. We used a screen to occlude vision of the added mass to prevent subjects from anticipating, using vision, the direction of the external torque. Repeated manipulation of the same object allows healthy subjects to plan the temporal evolution of manipulative forces in an anticipatory fashion. In fact, healthy subjects require only one or two practice trials to learn how to appropriately modulate digit forces to object properties such that their force coordination patterns are appropriate for use in subsequent trials (Rearick & Santello, 2002). The extent to which forces were properly planned was assessed by measuring (a) linear covariation of normal forces across digit pairs during force rise, (b) normal force magnitude and across-trial variability during object hold and (c) magnitude of object tilt during lift. Correct force planning is revealed by very small across-trial force variability and approximately linear relations between forces exerted by digit pairs that are consistent across trials. Object tilts should be minimal in the case of correctly planned forces as subjects can anticipate the net torque to be exerted through the digits before having to sense it during the lift.

Our preliminary analysis revealed that patients with CTS are less capable of integrating somatosensory information acquired through previous experience to plan appropriate grip forces for subsequent manipulations. These interpretations are supported by three main findings. First, patients with CTS exhibited greater across-trial variability (coefficient of variation) in digit forces than did controls (range across CM conditions and digits: 0.11–0.31 vs. 0.08–0.19, respectively). This indicates a reduced ability to use tactile information to plan and establish an appropriate force coordination pattern for counteracting the external torque. The difference in variability is evident in Figure 20.2, showing representative normal force traces for one patient with CTS and one control subject.

Figure 20.2 is also illustrative of the fact that patients produced significantly larger forces, at all digits, than did controls (means: 30.69 ± 9.25 N vs. 23.5 ± 8.04 N, respectively; $P < 0.001$). We also examined the linear covariation of forces across all digit pairs during the force rise phase of the grasp. As illustrated in Figure 20.3, the linear relations of forces between digit pairs were weaker and more variable across trials (as evident in the slopes) in the patients than in the controls (slope range across CM conditions, digit pairs and subjects: 0.03–1.35 vs. 0.03–0.43, respectively).

The impairments indicated by the previous findings resulted in greater object tilt ($1.71° \pm 0.86°$ vs. $1.18° \pm 0.61°$; $P < 0.001$) with greater across-trial variability (range across CM and subjects: $0.41°$–$1.12°$ vs. $0.22°$–$0.64°$) in patients with CTS than controls (for example see Figure 20.4). Analysis of the time course of object tilt during lift, however, is likely to be more informative than that of peak object tilt. Figure 20.4 shows that the patient with CTS

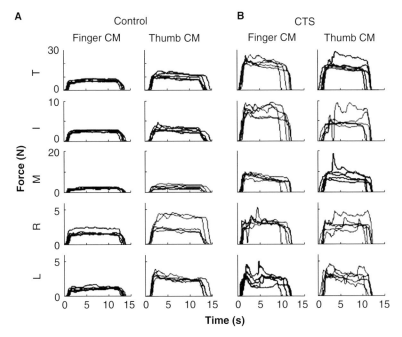

Figure 20.2. Forces for each digit (rows) over all trials for one control subject (A) and one patient with CTS (B). Left and right columns show data from finger and thumb side CM, respectively. T, I, M, R and L denote thumb, index, middle, ring and little fingers, respectively.

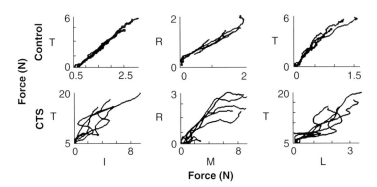

Figure 20.3. Force covariation of three digit pairs during force rise phase (CM on finger side) for the same control subject and patient with CTS (top and bottom panels, respectively) as in Figure 20.2. Digits are labeled as in Figure 20.2.

took longer and was less capable of returning the object to the appropriate orientation. Furthermore, note that the patient with CTS overcompensates for the external torque when the weight is placed on the thumb side, i.e. the tilt occurs in the opposite direction from that which is expected.

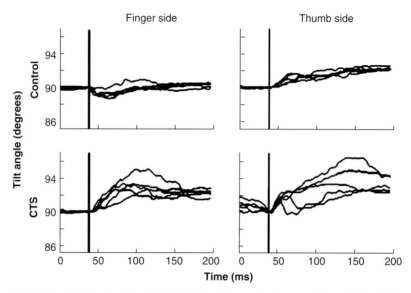

Figure 20.4. Object tilt for all trials with the CM location on the finger and thumb side (left and right columns, respectively) for the same control subject and patient with CTS (top and bottom row, respectively) as in Figure 20.2. The vertical line indicates lift onset and 90° represents vertical orientation of the object. Tilts towards the fingers and thumb are angles less than and greater than 90°, respectively.

Discussion

In agreement with literature on two-digit grasping in patients with acute or chronic median nerve compression (Lowe & Freivalds, 1999; Cole *et al.*, 2003), we found that patients with CTS produced excessive grip forces when compared with healthy controls during five-digit object hold. Excessively large forces produced by patients with CTS can be interpreted as a compensatory mechanism developed through the years in response to the patients' inability to finely regulate forces as a result of poorly sensed object properties. The inability to sense actual CM location after repeated exposure to the task and to modulate digit forces accordingly led to a strategy of "squeezing harder" with all digits to counteract the external torque. This strategy is strikingly different from that used by control subjects who modulate forces at specific digits (Figure 20.2).

We also found that excessive forces were exerted not only by the CTS-affected digits, but also by the ring and little finger. Based on median nerve function and previous research on two-digit grasping, one might expect to observe impairments in the ability to control force amplitude in the CTS-affected digits only. However, the requirements of static equilibrium dictate that (1) the normal forces of the thumb and the opposing digit(s) must be equal in magnitude and opposite in direction; (2) the sum of the tangential forces of all the digits involved must equal the weight of the object; (3) the net moment on the object is zero and (4) the normal-to-tangential force ratio at each digit must be greater than the ratio at which slips occur (see Chapter 3 for more details). Due to space limitations, the analyses presented here focused only on (1) above, from which it follows that if larger than physiological forces (control data) are exerted by a subset of digits (i.e. CTS-affected digits), they will result in larger forces also at digits with intact sensorimotor capabilities. Consequently, the larger forces (relative to the control subject) exerted by ring and little fingers suggest that they are appropriately compensating for excessive forces produced by the CTS-affected digits, i.e. to minimize or stop ongoing object tilt the non-affected digits must produce unnecessarily large forces to counteract those produced by the CTS-affected digits. Yet, the use of excessive forces is still less effective than that used by the control subject for controlling the orientation of the object (see Figure 20.4). Furthermore, this behavior is highly inefficient as it may lead to fatigue, further exacerbation of the median nerve compression and may prevent dexterous control of manipulation.

Excessive forces in patients with CTS relative to healthy controls are just one aspect of the impact of CTS on grasp function. Additionally, patients with CTS exhibited larger across-trial force variability than control subjects suggesting that they attempted to find an appropriate solution (i.e. force coordination pattern) to minimize object tilt for each object CM location, but could not, despite the fact that they lifted the device with the same CM for a total of seven consecutive trials. In contrast, the force coordination pattern used by control subjects was fairly consistent across trials, supporting the notion that they were able to form and use sensorimotor memories built through the practice trials to select and use – repeatably – appropriate multi-digit force coordination patterns in response to the changes in CM location. These interpretations are further supported by the observation that force coordination during force rise in the patients with CTS was characterized by time-varying relations between digit forces which were less consistent, both within and across trials, than in the control subjects. This is the case even for the ring and little fingers which are less and not affected, respectively, by median nerve compression. The linear covariation measure is quite revealing as weaker force covariations indicate deficits in the ability to synchronize force output at given digit pairs, a behavior which is a defining characteristic of whole-digit grasping in healthy individuals (Santello & Soechting, 2000; Rearick & Santello, 2002), but disrupted in patients affected by neurological disorders (e.g. Parkinson's disease; Rearick *et al.*, 2002).

Conclusion

Carpal tunnel syndrome is a compression neuropathy of the median nerve resulting in sensory and motor deficits in a subset of digits in the hand. By using tasks that mimic activities of daily living we have effectively detected subtle yet functionally important differences in grasp control planning and execution in patients with CTS. Reduced capability for sensorimotor integration in these patients results in digit force coordination deficits and therefore impairment in effective object manipulation. These deficits could lead to the problems in activities of daily living and patients' subjective feelings of "clumsiness," e.g. difficulties with fine manipulation, dropping objects, spilling drinks, inability to place objects in small spaces (e.g. lock and key).

Acknowledgements

This work was supported by an ASU/Mayo Seed Grant and National Institutes of Health Grant RO1 HDO57152. The authors would like to thank Drs. Scott Duncan, Anthony Smith and Mark Ross for consulting, screening and referring their patients with CTS. We would also like to thank Ms. Marianne Merritt, R. N. for scheduling our patients.

References

Aoki, T., Niu, X., Latash, M. L. & Zatsiorsky, V. M. (2006). Effects of friction at the digit-object interface on the digit forces in multi-finger prehension. *Exp Brain Res*, **172**, 425–438.

Armstrong, T. J., Castelli, W. A., Evans, F. G. & Diaz-Perez, R. (1984). Some histological changes in carpal tunnel contents and their biomechanical implications. *J Occup Med*, **26**, 197–201.

Augurelle, A. S., Smith, A. M., Lejeune, T. & Thonnard, J. L. (2003). Importance of cutaneous feedback in maintaining a secure grip during manipulation of hand-held objects. *J Neurophysiol*, **89**, 665–671.

Cole, K. J., Steyers, C. M. & Graybill, E. K. (2003). The effects of graded compression of the median nerve in the carpal canal on grip force. *Exp Brain Res*, **148**, 150–157.

Dun, S., Kaufmann, R. A. & Li, Z. -M. (2007). Lower median nerve block impairs precision grip. *J Electromyogr Kinesiol*, **17**, 348–354.

Edin, B. B., Westling, G. & Johansson, R. S. (1992). Independent control of human finger-tip forces at individual digits during precision lifting. *J Physiol*, **450**, 547–564.

Flanagan, J. R. & Wing, A. M. (1993). Modulation of grip force with load force during point-to-point arm movements. *Exp Brain Res*, **95**, 131–143.

Flanagan, J. R. & Wing, A. M. (1997). The role of internal models in motion planning and control: evidence from grip force adjustments during movements of hand-held loads. *J Neurosci*, **17**, 1519–1528.

Gordon, A. M., Westling, G., Cole, K. J. & Johansson, R. J. (1993). Memory representations underlying motor commands used during manipulation of common and novel objects. *J Neurophysiol*, **69**, 1789–1796.

Jaberzadeh, S. & Scutter, S. (2006). Flexor carpi radialis motoneuron pool in subjects with chronic carpal tunnel syndrome are more excitable than matched control subjects. *Man Ther*, **11**, 22–27.

Jenmalm, P. & Johansson, R. S. (1997). Visual and somatosensory information about object shape control manipulative fingertip forces. *J Neurosci*, **17**, 4486–4499.

Jenmalm, P., Dahlstedt, S. & Johansson, R. S. (2000). Visual and tactile information about object-curvature controls fingertip forces and grasp kinematics in human dexterous manipulation. *J Neurophysiol*, **84**, 2984–2997.

Johansson, R. S. & Westling, G. (1984a). Factors influencing force control during precision grip. *Exp Brain Res*, **53**, 277–284.

Johansson, R. S. & Westling, G. (1984b). Roles of glabrous skin receptors and sensorimotor memory in automatic control of precision grip when lifting rougher or more slippery objects. *Exp Brain Res*, **56**, 550–564.

Johansson, R. S. & Westling, G. (1987). Signals in tactile afferents from the fingers eliciting adaptive motor responses during precision grip. *Exp Brain Res*, **66**, 141–154.

Johnston, J. A., Winges, S. A. & Santello, M. (2009). Neural control of hand muscles during prehension. In D. Sternad (Ed.), *Progress in Motor Control. A Multidisciplinary Perspective* (pp. 577–596). New York: Springer.

Levangie, P. K. & Norkin, C. C. (2001). *Joint Structure and Function: A Comprehensive Analysis, 3rd edition*. Philadelphia, PA: FA Davis Co.

Lowe, B. D. & Freivalds, A. (1999). Effect of carpal tunnel syndrome on grip force coordination in hand tools. *Ergonomics*, **42**, 550–564.

Macefield, V. G., Hager-Ross, C. & Johansson, R. S. (1996). Control of grip force during restraint of an object held between finger and thumb: responses of cutaneous afferents from the digits. *Exp Brain Res*, **108**, 155–171.

Moberg, E. (1962). Criticism and study of methods for examining sensibility in the hand. *Neurology*, **12**, 8–19.

Mondelli, M., Padua, L. & Reale, F. (2004). Carpal tunnel syndrome in elderly patients: results of surgical decompression. *J Peripher Nerv Syst*, **9**, 168–176.

Monzee, J., Lamarre, Y. & Smith, A. M. (2003). The effects of digital anesthesia on force control using a precision grip. *J Neurophysiol*, **89**, 672–683.

Napadow, V., Kettner, N., Ryan, A. *et al.* (2006). Somatosensory cortical plasticity in carpal tunnel syndrome – a cross-sectional fMRI evaluation. *Neuroimage*, **31**, 520–530.

Nora, D. B., Becker, J., Ehlers, J. A. & Gomes, I. (2004). Clinical features of 1039 patients with neurophysiological diagnosis of carpal tunnel syndrome. *Clin Neurol Neurosurg*, **107**, 64–69.

Nowak, D. A., Hermsdorfer, J., Marquardt, C. & Topka, H. (2003). Moving objects with clumsy fingers: how predictive is grip force control in patients with impaired manual sensibility? *Clin Neurophysiol*, **114**, 472–487.

Rearick, M. P. & Santello, M. (2002). Force synergies for multifingered grasping. Effect of predictability in object center of mass and handedness. *Exp Brain Res*. **44**, 38–49.

Rearick, M. P., Stelmach, G. E., Leis, B. & Santello, M. (2002). Coordination and control of forces during multifingered grasping in Parkinson's disease. *Exp Neurol*, **177**, 428–442.

Reilmann, R., Gordon, A. M. & Henningsen, H. (2001). Initiation and development of fingertip forces during whole-hand grasping. *Exp Brain Res*, **140**, 443–452.

Salimi, I., Hollender, I., Frazier, W. & Gordon, A. M. (2000). Specificity of internal representations underlying grasping. *J Neurophysiol*, **84**, 2390–2397.

Salimi, I., Frazier, W., Reilmann, R. & Gordon, A. M. (2003). Selective use of visual information signaling objects' center of mass for anticipatory control of manipulative fingertip forces. *Exp Brain Res*, **150**, 9–18.

Santello, M. & Soechting, J. F. (2000). Force synergies for multifingered grasping. *Exp Brain Res*, **133**, 457–467.

Schieber, M. H. & Santello, M. (2004). Hand function: peripheral and central constraints on performance. *J Appl Physiol*, **96**, 2293–2300.

Tecchio, F., Padua, L., Aprile, I. & Rossini, P. M. (2002). Carpal tunnel syndrome modifies sensory hand cortical somatotopy: a MEG study. *Hum Brain Mapp*, **17** 28–36.

Thonnard, J.-L., Saels, P., Van den Bergh, P. & Lejeune, T. (1999). Effects of chronic median nerve compression at the wrist on sensation and manual skills. *Exp Brain Res*, **128**, 61–64.

Welford, A. T. (1972). Skill learning and performance. The obtaining and processing of information: some basic issues relating to analysing inputs and making decisions. *Res Q*, **43**, 295–311.

Werner, R. A. & Andary, M. (2002). Carpal tunnel syndrome: pathophysiology and clinical neurophysiology. *Clin Neurophysiol*, **113**, 1373–1381.

Werner, R., Armstrong, T. J., Bir, C. & Aylard, M. K. (1997). Intracarpal canal pressures: the role of finger, hand, wrist and forearm position. *Clin Biomech*, **12**, 44–51.

21

Stroke

CATHERINE E. LANG AND MARC H. SCHIEBER

Summary

Stroke results in irreversible brain damage, with the type and severity of symptoms dependent upon the location and the amount of injured brain tissue. The most common neurological impairment caused by stroke is partial weakness, called paresis, reflecting a reduced ability to voluntarily activate spinal motoneurons. In conjunction with the general reduced ability to voluntarily activate spinal motoneurons, there is often a reduced ability to selectively activate the spinal motoneuron pools, i.e. turning on some neurons while not turning on others. Together, these mechanisms result in altered movement control of many muscles, especially the contralesional hand and arm muscles used for grasping. Because of the altered muscle control, a variety of kinematic and kinetic alterations are observed during grasping in people with paresis post stroke. Impairments in grasping are related to the inability to use the hand for functional activities during daily life. In rare instances, stroke affects the posterior parietal lobe, resulting in distinct grasping deficits that are substantially different from grasping deficits seen after corticospinal system damage. Future studies investigating grasping post stroke could include the examination of both kinematic and kinetic aspects of grasping in the same subject samples, the examination of different types of grasping (e.g. palmar, precision), and the examination of different time points post stroke.

General information about stroke

Stroke is an acute neurological event that is caused by an alteration in blood flow to the brain. The alteration in blood flow can be either a deprivation of blood to the brain tissue (ischemic stroke) or a spilling of blood (hemorrhagic stroke) onto the brain tissue. Ischemic strokes account for about 85% of all strokes. Stroke is a major health problem. More than 700 000 new strokes occur each year in the USA, and stroke is the leading cause of adult long-term disability (Kelly-Hayes *et al.*, 1998).

Stroke results in irreversible brain damage, with the type and severity of symptoms dependent upon the location and the amount of injured brain tissue. The most common neurological impairment caused by stroke is partial weakness, often called paresis, reflecting a reduced ability to activate spinal motoneurons voluntarily. Total paralysis, or plegia, is the most severe form of paresis, reflecting a complete inability to activate motoneurons voluntarily. Post stroke paresis or plegia most dramatically affects the side of the body contralateral to the damaged brain tissue. Typically, the entire contralateral half of the body is weak or paralyzed (hemiparesis or hemiplegia), although occasionally smaller strokes weaken only the contralateral arm or only the leg (monoparesis or monoplegia). Note that paresis can result from a wide range of neurological diseases in addition to stroke, such as multiple sclerosis, cerebral palsy, amyotrophic lateral sclerosis, traumatic brain injury and spinal cord injury. In all cases, the disease has produced a lesion that results in a decreased ability to voluntarily activate motoneurons in the anterior horn of the spinal cord.

Around 80% of people with stroke experience acute hemiparesis, resulting in a diminished ability to use their affected extremities for purposeful movement (Granger *et al.*, 1988; Gray *et al.*, 1990). Other impairments often accompany hemiparesis, such as hemianesthesia, hemianopsia, aphasia and dysarthria (Granger *et al.*, 1988; Patel *et al.*, 2000; Han *et al.*, 2002). Only about 40% of such patients achieve full recovery (Wade & Hewer, 1987; Reding & Potes, 1988; Jorgensen *et al.*, 1995a, 1995b). The remaining 60% of stroke survivors have persistent motor and non-motor impairments that significantly disrupt their ability to participate in home and community life. Even for stroke survivors who are considered fully recovered, quality of life is decreased compared with stroke-free, community-dwelling individuals, after controlling for age and comorbidities (Lai *et al.*, 2002). As the population of the world ages and the rate of obesity increases, more individuals are expected to have strokes and to live with the disabling consequences of stroke.

Paresis post stroke

Despite the fact that stroke often injures multiple brain areas, motor deficits due to stroke in humans can be reasonably predicted by the proportion of the corticospinal system that is damaged (Pineiro *et al.*, 2000; Ward *et al.*, 2006). The corticospinal system, i.e. the primary motor cortex, the non-primary motor cortical areas, and the corticospinal tract, is distributed in a proximal to distal gradient to the cervical spinal cord, such that the motoneuron pools of the distal upper-extremity segments receive the greatest proportion of inputs (Clough *et al.*, 1968; Fetz & Cheney, 1980; Palmer & Ashby, 1992; Porter & Lemon, 1993; Dum & Strick, 1996, 2002). Consistent with the disruption of this input gradient, there is a clinical perception that the severity of hemiparesis is greatest in the distal muscles and least in the proximal muscles of the upper extremity (Colebatch & Gandevia, 1989). This clinical perception, however, is not well supported by systematic studies of hemiparetic severity across the upper extremity in larger patient populations (Bard & Hirschberg, 1965; Lang & Beebe, 2007). With respect to grasping, corticospinal system lesions disrupting corticofugal fibers within the posterior limb of the internal capsule result in chronic disabilities affecting dextrous movements, where the more posterior lesions (those most likely to damage fibers from the primary motor cortex) were strongly related to greater deficits in the kinematic and kinetic features of grasping (Wenzelburger *et al.*, 2005).

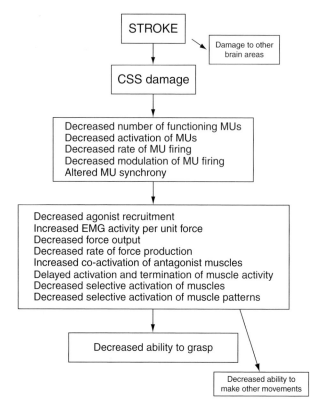

Figure 21.1. Schematic of relationships between stroke, damage to the corticospinal system (CSS), motor unit (MU) activity, muscle activity and grasping behavior. Note that damage to other brain areas results in alterations in other domains besides the motor system and that movements other than grasp are affected by this same cascade.

Loss of corticospinal system input to spinal motoneuron pools is the neural mechanism that produces paresis post stroke (Figure 21.1). Diminished corticospinal input causes a reduction in the number of motor units in the hemiparetic muscles that can be recruited voluntarily (McComas *et al.*, 1973). During strong voluntary effort, fewer motor units are recruited, and the discharge rates of recruited motor units are slower than typically found in normal muscles (Rosenfalck & Andreassen, 1980; Young & Mayer, 1982; Jakobsson *et al.*, 1991, 1992; Gemperline *et al.*, 1995; Frontera *et al.*, 1997). This may account for the observation that the amount of time required to develop peak forces in hemiparetic muscles is prolonged (Canning *et al.*, 1999). Furthermore, the ability to modulate motor unit discharge rates is impaired, such that the range of modulation and its variability are reduced (Rosenfalck & Andreassen, 1980; Dietz *et al.*, 1986; Gemperline *et al.*, 1995; Frontera *et al.*, 1997). Measurements of synchrony between motor unit pairs in hemiparetic hand muscles show either a narrowing or a broadening of the cross-correlogram peak (Farmer *et al.*, 1993), indicating decreased overall corticospinal tract input to spinal motoneurons (narrowing), or the loss of focused input (broadening). Interestingly, the re-emergence of motor unit synchrony appears to be loosely related to the functional recovery of the hemiparetic limb.

At the muscle level, loss of corticospinal input and altered activation of motor units post stroke results in a decreased ability to recruit targeted (agonist) muscles for a given task (Fellows *et al.*, 1994; Gowland *et al.*, 1992; Hammond *et al.*, 1988; Kamper & Rymer, 2001). Paretic muscles often produce more electromyogram activity (EMG activity) per unit force compared with the homologous muscle on the side unaffected by the stroke (Tang & Rymer, 1981). The decreased ability to recruit agonist muscles is frequently accompanied by co-activation of antagonist muscles (Sahrmann & Norton, 1977; Hammond *et al.*, 1988; Bourbonnais & Vanden Noven, 1989; Dewald *et al.*, 1995; Kamper & Rymer, 2001), although this is not always the case (Gowland *et al.*, 1992; Fellows *et al.*, 1994). Initiation and termination of muscle activity are delayed, i.e. the paretic muscles are slow to turn on and slow to turn off (Angel, 1975; Sahrmann & Norton, 1977; Dietz & Berger, 1984). Spatial and temporal patterns of muscle activation are disrupted across single joints (Bourbonnais *et al.*, 1989; Canning *et al.*, 2000), multiple joints (Dewald *et al.*, 1995; Dewald & Beer, 2001), and even at joints not directly involved during a given motor activity (Boissy *et al.*, 1997; Lang & Schieber, 2004).

Taken together, these findings suggest that the loss of input from the corticospinal system reduces the ability to selectively activate sets of muscles needed to perform skilled motor tasks (Lang & Schieber, 2004). Understanding the underlying relationships between damage to the corticospinal system, motor unit activity and muscle activity (illustrated schematically in Figure 20.1) is critical for the interpretation of how stroke affects grasping.

Kinematic and kinetic alterations in grasping post stroke

The corticospinal system is the major neuroanatomical substrate for control and execution of hand and finger movements, such as grasping (for comprehensive reviews see Porter & Lemon, 1993; Wing *et al.*, 1996). While grasping has been studied extensively in healthy, neurologically intact individuals, it is only beginning to be studied in people with stroke. Grasping movements in healthy controls are highly consistent and reproducible, both within and across individuals (see also Chapter 10). This is not the case for grasping movements in people with hemiparesis post stroke. Grasping, like many other movements, is highly variable after stroke, both within subjects and across groups of subjects. Note that increased variability is a hallmark of movement in many diseases affecting the central nervous system, and is not unique to those individuals who have had strokes.

Within any given sample of subjects with hemiparesis, there will be those whose grasping movements appear normal or close to normal, and those whose movements are very far from normal or absent altogether. This means that generalizing group data to specific individuals should be done with caution, since one individual may be quite far from the group mean, or from other individuals within the group. A handful of studies now report on how grasping is altered and how it may recover in the affected, contralesional hand of people with hemiparesis post stroke. As in healthy subjects, hemiparetic grasping typically has been studied

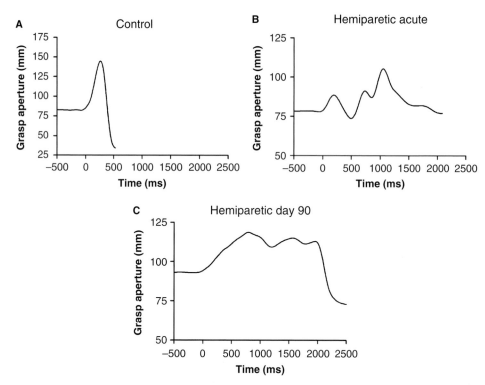

Figure 21.2. Aperture (distance between the thumb and fingertips) traces for a healthy, neurologically intact control subject (A), a subject with hemiparesis 2 weeks post stroke (B), and then again at 3 months post stroke (C). Time 0 indicates start of hand transport toward the target object. The end of the trace indicates the moment of contact with the target object. Note that the y-axis range in panel A is twice as large as the y-axis range in panels B and C. From Lang *et al.* (2006a).

either with kinematic analyses during reach-to-grasp movements or with kinetic analyses during a precision grasp once the fingers are already touching the object. Below, we review the kinematic and kinetic data on hemiparetic grasping with respect to how it is similar to and/or different from grasping in healthy, neurologically intact individuals.

Grasps often occur in the context of a reach-to-grasp movement in everyday life. In healthy individuals, the fingers open and close in a single smooth movement (Figure 21.2A) and the grasp component is temporally coupled with the reach component such that the fingers start opening as the hand starts to move toward the target object (Jeannerod, 1984; see also Chapter 10). In people with hemiparesis post stroke, the ability to open the fingers during a reach-to-grasp movement is altered because stroke impairs the ability to activate finger muscles and/or selectively activate them in the appropriate temporal patterns (see mechanisms outlined in Figure 21.1). The diminished ability to activate the finger extensor muscles (Twitchell, 1951; Trombly & Quintana, 1983; Kamper & Rymer, 2001) and to coordinate activation of flexor with extensor muscles results in highly variable and ineffi-cient patterns of finger opening (Lang *et al.*, 2005). Figure 21.2B shows a typical example of

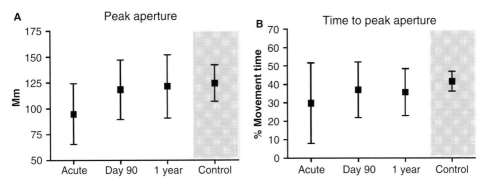

Figure 21.3. Hemiparetic and control group means and SDs for peak aperture (A) and time to peak aperture (B) during a reach-to-grasp movement. Time points are within the first few weeks (acute), 3 months (day 90) and 1 year after stroke. Modified from Lang *et al*. (2006a).

finger opening during a reach-to-grasp movement in a person with hemiparesis 2 weeks after stroke. As the ability to extend the fingers improves over time, finger opening becomes smoother (Figure 21.2C) but does not quite approach normal in this individual (Lang *et al*., 2006a).

The temporal coupling of the reach component and the grasp component during a reach-to-grasp movement is relatively preserved both early (Lang *et al*., 2005) and later (Michaelsen *et al*., 2004; Raghavan *et al*., 2006) after stroke. Those people with hemiparesis who do show alterations in the temporal coupling between the reach and the grasp components are unable to sufficiently activate the finger extensor muscles, resulting in several peaks in the aperture profile, where the largest of these several peaks may occur earlier or later than normal (Lang *et al*., 2005). As the ability to extend the fingers improves, peak aperture during reach-to-grasp movements becomes closer to normal, and time of peak aperture becomes less variable (Figure 21.3) (Lang *et al*., 2006a). In people with hemiparesis, hand shaping occurs gradually as the hand is decelerated during the later part of the reach, and not from the initiation of the reach, as happens in healthy individuals (Raghavan *et al*., 2006). Normally, hand orientation at the end of the reach component is accomplished via rotational control of the upper extremity segments, but in people with hemiparesis, flexion and rotation at the trunk can assist in orienting the hand to grasp the object (Michaelsen *et al*., 2004).

Once the fingers are in contact with the object, people with hemiparesis post stroke have difficulty producing appropriate finger forces for object manipulation (Hermsdörfer *et al*., 2003). The ability to maintain low grasp forces (< 2 N) accurately for short periods of time may be relatively preserved, but the ability to modify fingertip forces, i.e. rapidly grasp and release, is substantially slowed (Mai, 1989; Hermsdörfer *et al*., 2003). This again reflects the underlying mechanisms of hemiparetic movement control (Figure 21.1), a decreased ability to recruit, modulate and turn off motor units and muscle activity. Possibly as a compensatory strategy to prevent an object from slipping, people with hemiparesis often produce higher grip forces relative to load forces in the contralesional hand (Hermsdörfer *et al*., 2003), and in the ipsilesional hand as well (Quaney *et al.*, 2005). Higher force ratios (grip:lift), however, are not a consistent finding across studies (Blennerhassett *et al.*, 2006; McDonnell *et al.*, 2006), but when present, have been attributed to altered somatosensory inputs or altered somatosensory processing (Hermsdörfer *et al.*, 2003; see also Chapters 11 and 19). Interestingly, enhanced somatosensory input can lead to better regulation of these forces (Aruin, 2005).

The ability to proactively and reactively adjust grip forces also is altered post stroke (Grichting *et al.*, 2000; Blennerhassett *et al.*, 2006). This disability can manifest as prolonged times to obtain peak forces, fluctuating force profiles, and/or disordered sequencing of grip and load force production. Such kinetic alterations post stroke may be attributed to altered motor unit activation, motor unit modulation and selective muscle activation post stroke (Figure 21.1). Some evidence, however, suggests that deficits in force production during grasping are not due solely to a motor execution problem. Raghavan and colleagues recently showed that anticipatory grip force control can be improved if the ipsilesional hand experiences the load prior to the contralesional hand (Raghavan *et al.*, 2006b). This result implies that, in addition to an execution problem, some people with hemiparesis post stroke may have a higher-order motor planning deficit.

In neurologically intact subjects, somatosensation is important for grasping (see Chapters 11 and 19). In people with stroke, relationships between somatosensation and grasping are less straightforward. In our own hemiparetic subjects with and without somatosensory loss, the loss of somatosensation has not been shown to be related to the ability to grasp or the ability to use the hand for functional activities (Lang *et al.*, 2005, 2006a; Lang & Beebe, 2007). Other literature that has more specifically examined somatosensory modalities needed for grasping has shown that, after controlling for motor deficits, the degree of somatosensory loss was related to two-fingered grasp function in people with mild hemiparesis post stroke (Blennerhassett *et al.*, 2007). In cases of complete somatosensory loss with normal motor function, visual feedback can partially but not completely compensate for deficits in grasping (Jeannerod *et al.*, 1984; see also Chapter 19). Looking across these studies, we hypothesize that the importance of somatosensation for grasping is dependent upon the presence of adequate motor function. In other words, if motor abilities are diminished enough to substantially disrupt grasping behavior, loss of somatosensation may not worsen the deficit. If motor deficits are relatively mild, however, loss of somatosensation becomes an important factor impeding grasping and hand function.

Recovery of grasping

While much is known about general motor recovery post stroke, far less is known about recovery of grasping post stroke. Motor recovery post stroke is fastest in those individuals who are most mildly affected and slowest in those individuals who are most severely affected (Jorgensen *et al.*, 1995a, 1995b; Duncan *et al.*, 2000). The majority of motor recovery is achieved by the first 3 months post stroke, with only small changes in motor ability occurring after three months, usually in the more severely affected individuals (Jorgensen *et al.*, 1995a, 1995b; Duncan *et al.*, 2000). Functional recovery typically lags motor recovery by approximately 2 weeks (Jorgensen *et al.*, 1995a, 1995b). This may be because it requires practice to discover and then incorporate newly emerging motor abilities into daily functional activities.

For the upper extremity, best possible function was achieved within 3–6 weeks post stroke for those with mild paresis (Nakayama *et al.*, 1994). For those with severe upper extremity paresis, best possible function was achieved within 6–11 weeks (Nakayama *et al.*, 1994). The severity of upper extremity paresis within the first few weeks after stroke is consistently the strongest predictor of eventual upper extremity motor ability and function (Kwakkel *et al.*, 2003). In general, people who are more mildly affected initially are most likely to have full recovery and people who are more severely affected initially are most likely to have little or no recovery. Early observational data from Twitchell showed that about one-third of people with motor deficits after stroke will have complete upper extremity recovery, one-third will have partial upper extremity recovery, and one-third will have little to no recovery of function in the contralesional upper extremity (Twitchell, 1951). These observations are generally consistent with modern clinical experiences.

Very few studies have examined recovery of grasping after stroke. Palmar grasp and release movements have been used as outcome measures during a small trial of specific therapeutic exercises in patients an average of 6 weeks post stroke (Trombly *et al.*, 1986), but grasping was rated by the number of movements made and not examined in a more quantitative or qualitative manner. Grasp movements have also been used in acute imaging studies to see how cortical activation changes after stroke (Staines *et al.*, 2001) (see also Chapters 29 and 30). Unfortunately for the purpose of examining recovery of grasping, the parameters of movements made during scanning had to be experimentally controlled and thus the behavioral data do not permit the evaluation of changes in grasping during recovery after stroke. Kinematic changes in reach-to-grasp movements over the course of recovery were recently examined in a sample of people with mild-to-moderate hemiparesis post stroke (Lang *et al.*, 2006a). Consistent with the time course of recovery of general upper extremity function, most of the recovery in grasping occurred within the first 3 months post stroke, with little change occurring from 3–12 months post stroke. Within the first few months after stroke, recovery of grasping was due to improvements in: the ability to extend the fingers, as measured by larger peak apertures; the ability to open the fingers faster, as measured by higher peak aperture rates; and the ability to efficiently close the fingers in a single smooth movement, as measured by lower aperture path ratios (Lang *et al.*, 2006a). [Aperture path ratio is a measure of how directly the thumb and index fingers close. It is calculated as the ratio of the length of the aperture curve actually traveled to an ideal straight line between the first peak of the aperture trace and the aperture at the end of movement. An aperture path ratio of 1 represents a single, smooth closing of the thumb and index fingers (ideal) and an aperture path ratio of > 1 represents abnormal closing of the fingers, typically seen when subjects make multiple attempts to open and close the fingers as in Figure 21.2B and 21.2C.]

Relationships between the ability to grasp and daily function

Grasping is an important movement for function, where hundreds or perhaps thousands of grasp and release movements are made throughout the day. People automatically adjust the type of grasp used and the amount of force applied based on an object size, weight, surface texture and intended use (see Chapters 10, 11 and 12). Thus, the ability to grasp in people with hemiparesis post stroke may be an important indicator of the functional use of the upper extremity.

A recent paper quantified the importance of grasping for upper extremity function post stroke (McDonnell *et al.*, 2006). Seventeen subjects with subacute hemiparesis post stroke were studied performing a two-finger grip–lift task and then tested on the Action Research Arm Test (ARAT), a well-established standardized clinical measure of upper extremity function. Of the eight kinetic grasp parameters measured, three parameters were correlated to upper extremity function as measured by the ARAT. Preload duration, i.e. the time between the onset of grip force and the onset of the upward lift force, was negatively correlated with ARAT scores (Spearman's rho $= -0.72$), indicating that those subjects who took longer to establish grip forces had poorer upper extremity function. The maximum rate of grip-force production was positively correlated with ARAT scores (rho $= 0.83$), indicating that those subjects who developed grip force rapidly had better upper extremity function. And lastly, the strength of the temporal relationship between the rate of grip-force production and the rate of lift-force production was positively correlated with ARAT scores (rho $= 0.83$), indicating that those subjects who had more synchronous rates of grip- and lift-force production had better upper extremity function.

Another paper has looked at the relationships between grasping and upper extremity function in the early stages of stroke recovery (Lang *et al.*, 2006b). Fifty subjects with hemiparesis were studied performing a reach-to-grasp task and then tested on the ARAT at three time points, an average of 10 days, 26 days and 111 days after stroke. Within a few weeks, within the first month, and around 3 months after stroke, the kinematic parameters of grasping were correlated to upper extremity function as measured by the ARAT. The strength of the correlational relationships at each time point are illustrated in Figure 21.4. In general, the relationships stayed relatively stable across time, despite some fluctuations. Faster (higher peak aperture rates) and more efficient (lower aperture path ratios) grasp performance and a greater ability to open the fingers wider (higher peak apertures) were associated with better function (higher ARAT scores). The strength of these correlations is moderate because the ARAT measures the broad construct of upper extremity (UE) function whereas the kinematic measures capture a specific parameter of one movement essential for UE function. In this same cohort, we found similar relationships between grasping and upper extremity function when function was measured with the Wolf Motor Function Test instead of the ARAT (Edwards *et al.*, in review). Others have reported moderate to strong relationships between kinetic grasp parameters and grip strength (Hermsdörfer *et al.*, 2003), often considered a surrogate measure of upper extremity function in the clinic.

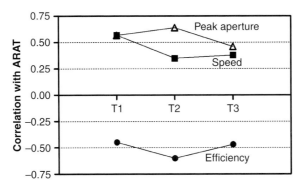

Figure 21.4. Pearson correlation coefficients between kinematic measures of grasping and Action Research Arm Test (ARAT) scores. T1: 10 ± 5 days post stroke; T2: 26 ± 11 days post stroke; T3: 111 ± 21 days post stroke; Speed: peak aperture rate, higher is better; Grasp efficiency: aperture path ratio, lower is better; Peak aperture: higher is better. Modified from Lang *et al.* (2006b).

Thus, available evidence indicates clearly that the ability to grasp is related to the ability to use the upper extremity for functional activities in people with hemiparesis post stroke. In many studies, however, just a few of the many tested grasp parameters are significantly correlated with function (Hermsdörfer *et al.*, 2003; McDonnell *et al.*, 2006). Many measurable parameters of grasping may be uninformative in the post stroke population. Future studies of grasping in this population should focus on those grasp parameters that are consistently correlated with function.

Unique grasping deficits when stroke damages only the posterior parietal lobe

The above discussion has focused on deficits in grasping due to damage to the corticospinal system. In rare cases, a stroke can cause isolated damage to the posterior parietal lobe. The effects of these lesions on grasping are quite distinct, as illustrated by two cases of unilateral posterior parietal lobe damage (Jeannerod, 1986) (see also Chapter 10). When visual feedback is present during the movement, the reach component of reach-to-grasp movements on the contralesional side is relatively normal. When visual feedback is not present, the reach component is characterized by slow and inaccurate movements that often overshoot the target. Unlike the reach component, the shaping of the hand during the grasp component is impaired regardless of whether visual feedback is present or absent. The impairments are most pronounced when visual feedback is absent. Instead of extending the fingers and then quickly closing them on the object, the fingers stay extended and do not shape into a formation that conforms to the shape of the object. Finger closure can be incomplete such that the object is either grasped inefficiently with an unintended portion of the hand or is not grasped at all (Jeannerod, 1986). Thus, grasping deficits after isolated posterior parietal lobe lesions tend to be the opposite of what is seen after corticospinal system lesions. Patients with posterior parietal lobe lesions tend to hold their fingers in extension and have difficulty flexing their fingers to match an object's dimensions, whereas patients with corticospinal system lesions tend to hold their fingers flexed and have difficulty extending them. The observed grasping deficits after isolated lesions to the posterior parietal lobe are consistent with the important role that this region plays in the visuomotor transformations required for grasping (for review see Andersen & Buneo, 2002; Galletti *et al.*, 2003) (see also Chapter 9).

Conclusions

Grasping deficits after stroke are generally a result of a lesion to the corticospinal system. Damage to this system disrupts the descending control of motor units, and consequently the voluntary activation of muscles on the contralesional side of the body. The varied kinematic and kinetic deficits observed after stroke are a consequence of these mechanisms. Future avenues of investigation regarding hemiparetic grasping include the examination of both kinematic and kinetic aspects of grasping in the same subject samples, the examination of different types of grasping (e.g. palmar, precision), and the examinations of different time points post stroke.

References

Andersen, R. A., & Buneo, C. A. (2002). Intentional maps in posterior parietal cortex. *Ann Rev Neurosci*, **25**, 189–220.

Angel, R. W. (1975). Electromyographic patterns during ballistic movement of normal and spastic limbs. *Brain Res*, **99**, 387–392.

Aruin, A. S. (2005). Support-specific modulation of grip force in individuals with hemiparesis. *Arch Phys Med Rehabil*, **86**, 768–775.

Bard, G. & Hirschberg, G. G. (1965). Recovery of voluntary motion in the upper extremity following hemiplegia. *Arch Phys Med Rehabil*, **46**, 567–572.

Blennerhassett, J. M., Carey, L. M. & Matyas, T. A. (2006). Grip force regulation during pinch grip lifts under somatosensory guidance: comparison between people with stroke and healthy controls. *Arch Phys Med Rehabil*, **87**, 418–429.

Blennerhassett, J. M., Matyas, T. A. & Carey, L. M. (2007). Impaired discrimination of surface friction contributes to pinch grip deficit after stroke. *Neurorehabil Neural Repair*, **21**, 263–272.

Boissy, P., Bourbonnais, D., Kaegi, C., Gravel, D. & Arsenault, B. A. (1997). Characterization of global synkineses during hand grip in hemiparetic patients. *Arch Phys Med Rehabil*, **78**, 1117–1124.

Bourbonnais, D. & Vanden Noven, S. (1989). Weakness in patients with hemiparesis. *Am J Occup Ther*, **43**, 313–319.

Bourbonnais, D., Vanden Noven, S., Carey, K. M. & Rymer, W. Z. (1989). Abnormal spatial patterns of elbow muscle activation in hemiparetic human subjects. *Brain*, **112**, 85–102.

Canning, C. G., Ada, L. & O'Dwyer, N. (1999). Slowness to develop force contributes to weakness after stroke. *Arch Phys Med Rehabil*, **80**, 66–70.

Canning, C. G., Ada, L. & O'Dwyer, N. J. (2000). Abnormal muscle activation characteristics associated with loss of dexterity after stroke. *J Neurol Sci*, **176**, 45–56.

Clough, J. F. M., Kernell, D. & Phillips, C. G. (1968). The distributions of monosynaptic excitation from the pyramidal tract and from primary spindle afferents to motoneurons of the baboon's hand and forearm. *J Physiol*, **198**, 145–166.

Colebatch, J. G. & Gandevia, S. C. (1989). The distribution of muscular weakness in upper motor neuron lesions affecting the arm. *Brain*, **112**, 749–763.

Dewald, J. P. & Beer, R. F. (2001). Abnormal joint torque patterns in the paretic upper limb of subjects with hemiparesis. *Muscle Nerve*, **24**, 273–283.

Dewald, J. P., Pope, P. S., Given, J. D., Buchanan, T. S. & Rymer, W. Z. (1995). Abnormal muscle coactivation patterns during isometric torque generation at the elbow and shoulder in hemiparetic subjects. *Brain*, **118**, 495–510.

Dietz, V. & Berger, W. (1984). Interlimb coordination of posture in patients with spastic paresis. Impaired function of spinal reflexes. *Brain*, **107**, 965–978.

Dietz, V., Ketelsen, U. P., Berger, W. & Quintern, J. (1986). Motor unit involvement in spastic paresis. Relationship between leg muscle activation and histochemistry. *J Neurol Sci*, **75**, 89–103.

Dum, R. P. & Strick, P. L. (1996). Spinal cord terminations of the medial wall motor areas in macaque monkeys. *J Neurosci*, **16**, 6513–6525.

Dum, R. P. & Strick, P. L. (2002). Motor areas in the frontal lobe of the primate. *Physiol Behav*, **77**, 677–682.

Duncan, P. W., Lai, S. M. & Keighley, J. (2000). Defining post-stroke recovery: implications for design and interpretation of drug trials. *Neuropharmacology*, **39**, 835–841.

Edwards, D. F., Lang, C. E., Wagner, J. M., Birkenmeier, R. & Dromerick, A. W.. Validation of the Wolf Motor Function Test in the acute stage of stroke recovery. in review.

Farmer, S. F., Swash, M., Ingram, D. A. & Stephens, J. A. (1993). Changes in motor unit synchronization following central nervous lesions in man. *J Physiol*, **463**, 83–105.

Fellows, S. J., Kaus, C. & Thilmann, A. F. (1994). Voluntary movement at the elbow in spastic hemiparesis. *Ann Neurol*, **36**, 397–407.

Fetz, E. E. & Cheney, P. D. (1980). Postspike facilitation of forelimb muscle activity by primate corticomotoneuronal cells. *J Neurophysiol*, **44**, 751–772.

Frontera, W. R., Grimby, L. & Larsson, L. (1997). Firing rate of the lower motoneuron and contractile properties of its muscle fibers after upper motoneuron lesion in man. *Muscle Nerve*, **20**, 938–947.

Galletti, C., Kutz, D. F., Gamberini, M., Breveglieri, R. & Fattori, P. (2003). Role of the medial parieto-occipital cortex in the control of reaching and grasping movements. *Exp Brain Res*, **153**, 158–170.

Gemperline, J. J., Allen, S., Walk, D. & Rymer, W. Z. (1995). Characteristics of motor unit discharge in subjects with hemiparesis. *Muscle Nerve*, **18**, 1101–1114.

Gowland, C., deBruin, H., Basmajian, J. V., Plews, N. & Burcea, I. (1992). Agonist and antagonist activity during voluntary upper-limb movement in patients with stroke. *Phys Ther*, **72**, 624–633.

Granger, C. V., Hamilton, B. B. & Gresham, G. E. (1988). The stroke rehabilitation outcome study – Part I: General description. *Arch Phys Med Rehabil*, **69**, 506–509.

Gray, C. S., French, J. M., Bates, D. *et al.* (1990). Motor recovery following acute stroke. *Age Ageing*, **19**, 179–184.

Grichting, B., Hediger, V., Kaluzny, P. & Wiesendanger, M. (2000). Impaired proactive and reactive grip force control in chronic hemiparetic patients. *Clin Neurophysiol*, **111**, 1661–1671.

Hammond, M. C., Fitts, S. S., Kraft, G. H. *et al.* (1988). Co-contraction in the hemiparetic forearm: quantitative EMG evaluation. *Arch Phys Med Rehabil*, **69**, 348–351.

Han, L., Law-Gibson, D. & Reding, M. (2002). Key neurological impairments influence function-related group outcomes after stroke. *Stroke*, **33**, 1920–1924.

Hermsdörfer, J., Hagl, E., Nowak, D. A. & Marquardt, C. (2003). Grip force control during object manipulation in cerebral stroke. *Clin Neurophysiol*, **114**, 915–929.

Jakobsson, F., Edstrom, L., Grimby, L. & Thornell, L. E. (1991). Disuse of anterior tibial muscle during locomotion and increased proportion of type II fibres in hemiplegia. *J Neurol Sci*, **105**, 49–56.

Jakobsson, F., Grimby, L. & Edstrom, L. (1992). Motoneuron activity and muscle fibre type composition in hemiparesis. *Scand J Rehabil Med*, **24**, 115–119.

Jeannerod, M. (1984). The timing of natural prehension movements. *J Motor Behav*, **16**, 235–254.

Jeannerod, M. (1986). The formation of finger grip during prehension. A cortically mediated visuomotor pattern. *Behav Brain Res*, **19**, 99–116.

Jeannerod, M., Michel, F. & Prablanc, C. (1984). The control of hand movements in a case of hemianaesthesia following a parietal lesion. *Brain*, **107**, 899–920.

Jorgensen, H. S., Nakayama, H., Raaschou, H. O. *et al.* (1995a). Outcome and time course of recovery in stroke. Part I: Outcome. The Copenhagen Stroke Study. *Arch Phys Med Rehabil*, **76**, 399–405.

Jorgensen, H. S., Nakayama, H., Raaschou, H. O. *et al.* (1995b). Outcome and time course of recovery in stroke. Part II: Time course of recovery. The Copenhagen Stroke Study. *Arch Phys Med Rehabil*, **76**, 406–412.

Kamper, D. G. & Rymer, W. Z. (2001). Impairment of voluntary control of finger motion following stroke: role of inappropriate muscle coactivation. *Muscle Nerve*, **24**, 673–681.

Kelly-Hayes, M., Robertson, J. T., Broderick, J. P. *et al.* (1998). The American Heart Association Stroke Outcome Classification. *Stroke*, **29**, 1274–1280.

Kwakkel, G., Kollen, B. J., van der Grond, J. & Prevo, A. J. (2003). Probability of regaining dexterity in the flaccid upper limb: impact of severity of paresis and time since onset in acute stroke. *Stroke*, **34**, 2181–2186.

Lai, S. M., Studenski, S., Duncan, P. W. & Perera, S. (2002). Persisting consequences of stroke measured by the Stroke Impact Scale. *Stroke*, **33**, 1840–1844.

Lang, C. E. & Schieber, M. H. (2004). Reduced muscle selectivity during individuated finger movements in humans after damage to the motor cortex or corticospinal tract. *J Neurophysiol*, **91**, 1722–1733.

Lang, C. E. & Beebe, J. A. (2007). Relating movement control at 9 upper extremity segments to loss of hand function in people with chronic hemiparesis. *Neurorehabil Neural Repair*, **21**, 279–291.

Lang, C. E., Wagner, J. M., Bastian, A. J. *et al.* (2005). Deficits in grasp versus reach during acute hemiparesis. *Exp Brain Res*, **166**, 126–136.

Lang, C. E., Wagner, J. M., Edwards, D. F., Sahrmann, S. A. & Dromerick, A. W. (2006a). Recovery of grasp versus reach in people with hemiparesis poststroke. *Neurorehabil Neural Repair*, **20**, 444–454.

Lang, C. E., Wagner, J. M., Dromerick, A. W. & Edwards, D. F. (2006b). Measurement of upper-extremity function early after stroke: properties of the action research arm test. *Arch Phys Med Rehabil*, **87**, 1605–1610.

Mai, N. (1989). Residual control of isometric finger forces in hemiparetic patients. Evidence for dissociation of performance deficits. *Neurosci Lett*, **101**, 347–351.

McComas, A. J., Sica, R. E., Upton, A. R. & Aguilera, N. (1973). Functional changes in motoneurones of hemiparetic patients. *J Neurol Neurosurg Psychiatry*, **36**, 183–193.

McDonnell, M. N., Hillier, S. L., Ridding, M. C. & Miles, T. S. (2006). Impairments in precision grip correlate with functional measures in adult hemiplegia. *Clin Neurophysiol*, **117**, 1474–1480.

Michaelsen, S. M., Jacobs, S., Roby-Brami, A. & Levin, M. F. (2004). Compensation for distal impairments of grasping in adults with hemiparesis. *Exp Brain Res*, **157**, 162–173.

Nakayama, H., Jorgensen, H. S., Raaschou, H. O. & Olsen, T. S. (1994). Recovery of upper extremity function in stroke patients: the Copenhagen Stroke Study. *Arch Phys Med Rehabil*, **75**, 394–398.

Palmer, E. & Ashby, P. (1992). Corticospinal projections to upper limb motoneurones in humans. *J Physiol*, **448**, 397–412.

Patel, A. T., Duncan, P. W., Lai, S. M. & Studenski, S. (2000). The relation between impairments and functional outcomes poststroke. *Arch Phys Med Rehabil*, **81**, 1357–1363.

Pineiro, R., Pendlebury, S. T., Smith, S. *et al.* (2000). Relating MRI changes to motor deficit after ischemic stroke by segmentation of functional motor pathways. *Stroke*, **31**, 672–679.

Porter, R. & Lemon, R. N. (1993). *Corticospinal Function and Voluntary Movement. Vol. 45.* Oxford, UK: Oxford University Press.

Quaney, B. M., Perera, S., Maletsky, R., Luchies, C. W. & Nudo, R. J. (2005). Impaired grip force modulation in the ipsilesional hand after unilateral middle cerebral artery stroke. *Neurorehabil Neural Repair*, **19**, 338–349.

Raghavan, P., Santello, M., Krakauer, J. W. & Gordon, A. M. (2006). *Shaping the hand to object contours after stroke.* Poster. Atlanta, GA: Society for Neuroscience.

Reding, M. J. & Potes, E. (1988). Rehabilitation outcome following initial unilateral hemispheric stroke. Life table analysis approach. *Stroke*, **19**, 1354–1358.

Rosenfalck, A. & Andreassen, S. (1980). Impaired regulation of force and firing pattern of single motor units in patients with spasticity. *J Neurol Neurosurg Psychiatry*, **43**, 907–916.

Sahrmann, S. A. & Norton, B. J. (1977). The relationship of voluntary movement to spasticity in the upper motor neuron syndrome. *Ann Neurol*, **2**, 460–465.

Staines, W. R., McIlroy, W. E., Graham, S. J. & Black, S. E. (2001). Bilateral movement enhances ipsilesional cortical activity in acute stroke: a pilot functional MRI study. *Neurology*, **56**, 401–404.

Tang, A. & Rymer, W. Z. (1981). Abnormal force – EMG relations in paretic limbs of hemiparetic human subjects. *J Neurol Neurosurg Psychiatry*, **44**, 690–698.

Trombly, C. A. & Quintana, L. A. (1983). The effects of exercise on finger extension of CVA patients. *Am J Occup Ther*, **37**, 195–202.

Trombly, C. A., Thayer-Nason, L., Bliss, G. *et al.* (1986). The effectiveness of therapy in improving finger extension in stroke patients. *Am J Occup Ther*, **40**, 612–617.

Twitchell, T. E. (1951). The restoration of motor function following hemiplegia in man. *Brain*, **74**, 443–480.

Wade, D. T. & Hewer, R. L. (1987). Functional abilities after stroke: measurement, natural history and prognosis. *J Neurol Neurosurg Psychiatry*, **50**, 177–182.

Ward, N. S., Newton, J. M., Swayne, O. B. *et al.* (2006). Motor system activation after subcortical stroke depends on corticospinal system integrity. *Brain*, **129**, 809–819.

Wenzelburger, R., Kopper, F., Frenzel, A. *et al.* (2005). Hand coordination following capsular stroke. *Brain*, **128**, 64–74.

Wing, A. M., Haggard, P. & Flanagan, J. R. (1996). *Hand and Brain. The Neurophysiology and Psychology of Hand Movements*. San Diego, CA: Academic Press.

Young, J. L. & Mayer, R. F. (1982). Physiological alterations of motor units in hemiplegia. *J Neurol Sci*, **54**, 401–412.

22

Prehension characteristics in Parkinson's disease patients

TANIA S. FLINK AND GEORGE E. STELMACH

Summary

Examination of a well-coordinated task such as prehension in patients with Parkinson's disease (PD) provides an opportunity to gain a better understanding of how basic movement control parameters are altered in patients with this disorder, and provides insights into how altered basal ganglia are involved in the control and regulation of movement when compared with healthy control subjects. In this chapter, evidence is presented for prehensile movements that show that patients have reduced amplitudes of maximum grip aperture and are less able to modulate grip aperture to account for changes in object shape and mass. The coordination between the transport and grasp component also shows some dissimilarity between patients and controls, as patients begin opening the fingers later and reach maximum peak aperture later in time. Patients also begin aperture closure closer to the object than controls, and have a reduced ability to regulate grip forces than controls when an object is grasped, as evidenced by delays in grip-force production and variable force profiles. A neural noise hypothesis is discussed as the neural mechanism that leads to the impairments found in Parkinson's disease patients.

Introduction

Fine motor skills are important to tasks of everyday living and include movements such as grasping a door handle, buttoning a shirt, or reaching and holding a beverage. Prehensile actions, more simply referred to as reach-to-grasp movements, are well-practiced movements that require precise control in transporting the hand to a specified object and grasping the object with the grip aperture (see Chapters 2 and 10). What makes studying prehension unique is that it allows for examination of how the transport and grasp components are controlled separately as well as how they are regulated together (see Chapters 9 and 10). Because of its complex multijoint control, it is important to examine the structure of prehensile movements in Parkinson's disease patients to gain a better understanding of basic control characteristics that change with disease onset and progression. Moreover, studies on prehensile movements may provide insight into motor deficits from basal ganglia dysfunction.

Sensorimotor Control of Grasping: Physiology and Pathophysiology, ed. Dennis A. Nowak and Joachim Hermsdörfer. Published by Cambridge University Press. © Cambridge University Press 2009.

Parkinson's disease has been described as a progressive neurological disease that is associated with the depletion of the neurotransmitter dopamine in the substantia nigra pars compacta of the basal ganglia (Mink, 1996; Wichmann & Delong, 2003; Merideth & Kang, 2006). The basal ganglia are part of a complex network of connections in the brain that is formed between the cerebral cortex, the thalamus and other underlying structures. Among these connections is a motor loop, which is comprised of a direct pathway that facilitates activity in the motor cortex, and an indirect pathway that inhibits activity in the motor cortex (Parent & Hazrati, 1995; Kaji, 2001). It is through the dual nature of these pathways that many symptoms and deficits of Parkinson's disease emerge. It is thought that increased signals through the indirect pathway of the basal ganglia and reduced signals in the direct pathway lead to heightened inhibition of the activity of the cortex and thalamus, which leads to parkinsonian symptoms (Pessiglione *et al.*, 2005). It is also postulated that symptoms of PD emerge as a result of overactivity of the globus pallidus externus neurons and disruption of natural oscillations in the globus pallidus–substantia nigra network (Bevan *et al.*, 2002).

Lesions in the substantia nigra pars compacta (SNc) and the internal globus pallidus (GPi) of the basal ganglia in animals with the toxin 1-methyl-4-phenyl-1,2,3,6-tetrahydropyridine elude to possible origins of parkinsonian symptoms. Lesions in the SNc inhibit neural activity in the subthalamic nucleus, substantia nigra pars reticula and the GPi segment, which leads to bradykinesia (slowness of movement), akinesia (difficulty in initiating movements) and cognitive deficits (Boraud *et al.*, 2000; Middleton & Strick, 2000; Wichmann & DeLong, 2003; Merideth & Kang, 2006). Lesions in the GPi of the basal ganglia tend to improve movement control by increasing neural activity in the thalamus. However, GPi lesions also tend to reduce the modulation range of the activity within the GPi and may cause deficits in a variety of movements, particularly movements that require rapid changes (Contreras-Vidal & Stelmach, 1996). While other deficits have been observed in lesion studies, the deficit in the control and regulation of movement is most frequently observed.

Much of the research that has been conducted in our motor control laboratory suggests that excessive neural noise in the basal ganglia pathways is the cause of most PD patient impairments (Guadagnoli *et al.*, 2002; Rand *et al.*, 2002; Van Gemmert & Stelmach, 2002). This is supported by the work of Bergman & Deuschl (2002), Bevan *et al.* (2002), and Gatev *et al.* (2006), who have noted that there are abnormal network properties in the pathways that reduce responsiveness to specific neural signals. Our view is that with the loss of finely differentiated parallel processing, the signal-to-noise ratio in the basal ganglia is reduced in PD patients, causing impairments in movement. It is commonly accepted that PD impairs normal motor performance, but scientists are still searching to understand the nature of the impairments and mechanisms by which they arise.

This chapter will focus on the changes in control of prehensile movements in Parkinson's disease patients. Basic components of prehension are first discussed. A detailed description of the changes in control of transport and grip aperture and the coordination between these two components in patients is then discussed. This is followed by concluding remarks that consider the results and address how prehensile movements reveal changes in movement control in PD.

Components of prehension

While details on the control of prehension can be found throughout this book, the goal of this section is to provide the reader with an overview of the kinematic components of prehension that will be described later in the chapter when the data on PD patients are presented.

Prehensile movements, or reach-to-grasp movements, are commonly studied in motor control literature, as they are highly skilled behaviors. The characteristics of the reach and grasp are well known and the coordination between these components has been established for some time (Gentilucci *et al.*, 1991; Chieffi & Gentilucci, 1993; Paulignan *et al.*, 1997; Haggard & Wing, 1998; Saling *et al.*, 1998; Wang & Stelmach, 2001; Alberts *et al.*, 2002) (see also Chapters 2 and 10). Since the skill has been studied extensively, it is a good candidate to unveil changes in the central nervous system when investigated in patients with neurological disorders. A typical representation of the transport and grasp components in a control subject and a PD patient is found in Figure 22.1. Most prehensile movements start from a stationary position; as the person executes the transport component (solid line, "A"),

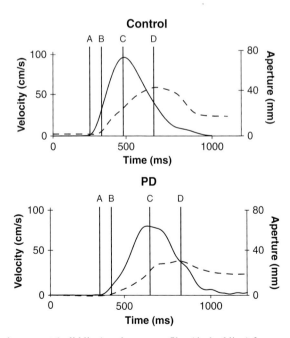

Figure 22.1. Typical transport (solid line) and grasp profiles (dashed line) for a control subject and PD patient. A = transport initiation, B = grasp initiation, C = peak velocity, D = maximum peak aperture.

the arm reaches toward the object and velocity reaches a peak level ("C"). The velocity profiles are generally bell-shaped and symmetrical. The grip aperture (dotted line) opens after the onset of the transport component ("B"), and reaches a maximum (maximum grip aperture, "D") around 78–85% of total reach during deceleration of the limb (Jeannerod, 1984). At that point, the fingers begin to close (decrease in grip aperture amplitude) as the arm delivers that hand to the object. The closure of grip aperture is then adjusted to the movement amplitude, object size and speed of movement. Once the object is grasped, finger forces are modulated so as to successfully lift the object (Gordon *et al.*, 1997; Santello *et al.*, 2004).

When object size is decreased and visual feedback of the arm or object is reduced, the bell-shaped velocity profile of transport becomes less symmetrical due to an increase in the deceleration period (Jeannerod, 1984; Bootsma *et al.*, 1994; Saling *et al.*, 1996; Paulignan *et al.*, 1997; Wang & Stelmach, 2001). Subjects tend to slow down the movement of the arm to compensate for increased accuracy for small objects and increased uncertainty, which is also reflected in reduced amplitude of peak velocity and increased transport time. Once arm transport is initiated, the onset of the opening of the grasp component soon follows (Alberts *et al.*, 2000; Wang & Stelmach, 1998). The amplitude of maximum grip aperture increases to compensate for increasing target size and increasing transport speed (Gentilucci *et al.*, 1991; Bootsma *et al.*, 1994; Rand *et al.*, 2006b).

The coordination between the transport and grasp components can be measured in a temporal or spatial domain. Time to maximum peak aperture measures the *time* the arm is transported from movement onset to maximum peak aperture, and aperture closure time measures the *time* the arm is transported from maximum aperture to object grasp. Similarly, aperture opening distance and aperture closure distance are the *distances* the arm is transported from movement onset to maximum aperture and from maximum aperture to target grasp, respectively. The time to peak aperture is reduced for faster movements in comparison to slower movements and for movements that require high accuracy (e.g. small objects). As a result, closure time increases, and a greater proportion of time is needed to transport the arm and modulate the closing of grip aperture to the object (Paulignan *et al.*, 1997; Alberts *et al.*, 2000; Wang & Stelmach, 2001). Aperture opening distance increases with increases in total transport distance, but despite experimental manipulations in object size, movement amplitude, transport speed, obstacle avoidance and trunk involvement, aperture closure distance remains relatively stable (Saling *et al.*, 1998; Wang & Stelmach, 1998; Alberts *et al.*, 2002). It has been suggested that the central nervous system specifies the distance between the hand and the object to maintain coordination between transport and grasp.

How these prehension characteristics change in individuals with Parkinson's disease are discussed next. The information presented in this section predominantly describes patients in the off-medication state. In order to observe movement deficits in PD patients, movements are commonly studied when patients are off dopaminergic medication after an overnight withdrawal, which constitutes a "quasi"-baseline state for research purposes that allows assessment of assumed basal ganglia dysfunction.

Prehension characteristics for Parkinson's disease patients

Control of transport

Prior to movement execution, movements are planned and prepared. The latency or time of this preparation is 19–20% longer (approximately 100 ms) in PD patients compared with controls, resulting in delays in movement initiation (Stelmach *et al.*, 1986; Jahanshahi *et al.*, 1992; Desmurget *et al.*, 2003). The most distinct feature of arm transport during movement execution in PD patients is movement slowness, or bradykinesia. The general shape of the velocity profile is distorted in PD patients, as evidenced by a smaller magnitude of peak velocity and longer deceleration periods, which contributes to slower movements. As a result, transport time is on average 34% longer in PD patients when reaching for objects compared with age-matched controls (Sheridan *et al.*, 1987; Castiello & Bennett, 1994; Jackson *et al.*, 1995; Gentilucci & Negrotti, 1999; Desmurget *et al.*, 2003; Tunik *et al.*, 2004; Leis *et al.*, 2005; Rand *et al.*, 2006a). An illustration of a 55-cm reach-to-grasp movement to a 2.5 cm diameter object for controls and PD patients can be found in Figure 22.2 (based on data from Rand *et al.*, 2006a). This figure shows the reduced peak velocity and reduced symmetry in the velocity profile in patients compared with controls.

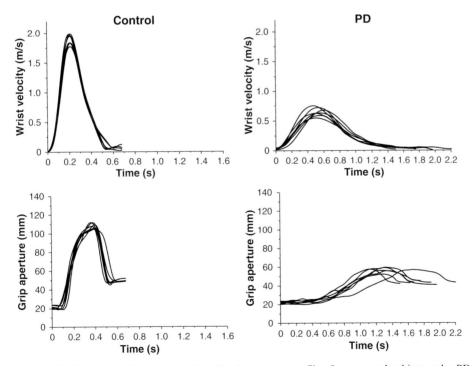

Figure 22.2. Illustration of wrist velocity and grip aperture profiles for a control subject and a PD patient. This figure corresponds to a 55-cm movement to a 2.5-cm diameter object, which was performed as fast as possible for both groups.

Slower transport in patients compared with controls is also evident for small diameter objects versus the large objects and movements performed without visual feedback of the arm (Alberts *et al.*, 1998; Bertram *et al.*, 2005; Gentilucci & Negrotti, 1999; Jackson *et al.*, 1995; Majsak *et al.*, 1998; Poizner *et al.*, 2000; Wang *et al.*, 2006). However, despite movement slowness, patients are able to adjust transport time to account for changes in movement speed, amplitude and object size.

Another typical alteration observed in patients is increased movement variability within the velocity profile during transport. In age-matched controls, the velocity profiles are much smoother with few fluctuations. In PD patients, the velocity profiles are often distorted with more fluctuations in the acceleration and deceleration phases, which suggest a more "noisy" motor system during transport. Fluctuations in the transport and grasp profiles in patients are shown in Figure 22.2. Fluctuations are reflected in the increased value of normalized jerk, which is defined as the change in acceleration of a movement, and is typically used as a measure of movement smoothness. Increased jerk values are found in PD patients compared with controls when they reach to small targets versus large targets (Tresilian *et al.*, 1997; Alberts *et al.*, 2000; Romero *et al.*, 2003).

Parkinson's disease patients also show increased variability of transport time compared with controls across trials, indicating that transport time is less regulated in patients and less reproducible (Sheridan *et al.*, 1987; Tresilian *et al.*, 1997; Desmurget *et al.*, 2003; Rand *et al.*, 2006a). Trial-by-trial variability in patients is illustrated in Figure 22.2 in both the wrist transport component and the grasp component. Moreover, trial-by-trial variability has also been found to increase significantly in patients compared with controls with increased accuracy demands (Tresilian *et al.*, 1997). In order to reduce movement variability, PD patients often choose to decrease movement speed to compensate for more segmented control.

Control of grasp

On average, amplitude of maximum grip aperture is 15% smaller in PD patients compared with controls when utilizing a precision or pinch grip to grasp objects (Jackson *et al.*, 2000; Negrotti *et al.*, 2005; Rand *et al.*, 2006a). Reduced maximum grip aperture in prehensile movements has been observed in patients regardless of object size, movement amplitude or manipulation of visual feedback. Reduced maximum grip aperture in patients is depicted in Figure 22.2, which corresponds to a movement to a 2.5-cm target. As can be seen from the figure, maximum grip aperture amplitude reaches 4 cm in patients compared with 9 cm in controls. Patients open the fingers just wide enough to compensate for the size of the object prior to grasp. Though maximum grip aperture amplitude is reduced, patients are able to modulate maximum grip aperture with changes in object size. The spatial control of grip aperture is therefore not compromised in individuals with PD.

The reasons for the reduced amplitude of maximum grip aperture in Parkinson's disease patients across numerous conditions are unknown. Reduced maximum grip aperture may be due to slow movement execution in patients (e.g. bradykinesia), as slower movements do not necessitate as large an amplitude of grip aperture as faster movements (Bonfiglioli *et al.*,

1998; Wang & Stelmach, 2001; Rand *et al.*, 2006b). Reduced maximum grip aperture amplitude may also be another observable form of hypometria in patients. Hypometric movements in patients are commonly observed in handwriting studies (Longstaff *et al.*, 2003; Van Gemmert *et al.*, 2003), and are evidenced in the undershooting of the primary submovement in reaching studies (Desmurget *et al.*, 2003; Romero *et al.*, 2003). The hypometric results from prehension studies and handwriting studies suggest that the basal ganglia may be involved in the control of movement amplitude, which is affected in both the transport and grasp components during prehension in patients.

Modulation of grip aperture during prehensile movements was further examined by Schettino and colleagues (2003). In this study, PD patients reached and grasped three objects of varying convexity and concavity utilizing all fingers of the hand. Similar to other studies, PD patients showed reduced maximum grip aperture compared with controls. Patients with Parkinson's disease also showed reduced modulation of the fingers and less ability to control for individual finger movements. As patients grasped the object, the index finger position was not modulated to account for the shape of the object until immediately before contact. This is in contrast to controls, who were able to modulate the fingers in a continuous manner prior to object grasp. Furthermore, PD patients showed selective impairments in modulating finger shape without visual feedback (Schettino *et al.*, 2006). Patients with Parkinson's disease produced more incorrect or failed grasps compared with controls and delayed specifying the correct hand shape when visual feedback of the arm and the object was not available. It is suggested that the reduced ability to modulate the grasp component in PD patients may be due to the inability to correctly integrate the grasp component and the transport component. Moreover, patients may be more reliant on visual feedback for movement control, especially when preshaping the hand prior to object grasp.

Coordination between transport and grasp

When patients initiate transport, the temporal onset of the grasp component with respect to transport onset is significantly delayed compared with controls (Castiello *et al.*, 1993; Tresilian *et al.*, 1997; Alberts *et al.*, 2000). Figure 22.3 shows a timeline of reach-to-grasp events as a function of total movement time for PD patients and controls computed using values from multiple studies (Castiello *et al.*, 1993, 1999; Castiello & Bennett, 1994; Alberts *et al.*, 2000; Rand *et al.*, 2006a; Wang *et al.*, 2006). As can be seen from Figure 22.3, onset of grasp occurs, on average, at 5% of movement time in controls, whereas it occurs at 9% of movement time in patients. Evidence for delays in response initiation in PD patients is shown in numerous reaction time paradigms. In a study by Wang *et al.* (1998), a meta-analysis was conducted that examined results from simple and choice reaction time tasks in PD patients and controls. The comparisons between studies demonstrate that PD patients have slower reaction times than age-matched controls, and the difference between groups becomes larger as a function of severity of the disease. Hence, the movement delays that are shown in reaction time studies as well as prior to movement onset and grip aperture

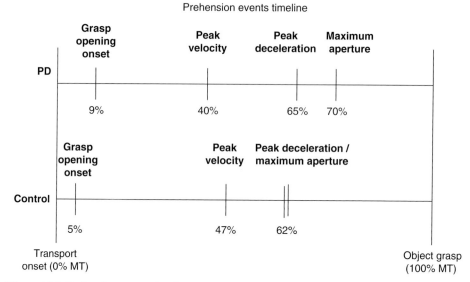

Figure 22.3. Prehension events timeline illustrating the average timing of grasp opening onset, peak velocity, peak deceleration and maximum peak aperture in patients with Parkinson's disease compared with controls. These values were obtained from Alberts *et al.* (2000), Castiello & Bennett (1994), Castiello *et al.* (1993, 1999), Rand *et al.* (2006), Wang *et al.* (2006).

initiation in prehensile movements suggest that basal ganglia may be involved in the initiation of movement.

The time to reach maximum peak aperture is also delayed in PD patients compared with controls (Jackson *et al.*, 1995; Alberts *et al.*, 1998; Schettino *et al.*, 2003; Rand *et al.*, 2006a) (see also Figure 22.3). In controls, maximum peak aperture occurs near peak deceleration time, or at 62% of total movement time. In patients, maximum peak aperture is delayed on average 5% from peak deceleration time to 70% of movement time, which suggests that grip closure is initiated when the arm is slowing down near the final phase of deceleration, and is not well correlated with the timing of peak deceleration. Prolonged initiation of aperture closure in patients indicates a "breakdown" in the temporal coordination between the transport and the grasp components, which alters the way prehensile movements are controlled.

The temporal coordination between transport and grasp components is further disrupted when patients are required to reach for and grasp small targets compared with large targets. Overall, the patients spend a greater proportion of time in aperture closure compared with controls for small objects compared with large objects (Alberts *et al.*, 2000; Negrotti *et al.*, 2005), and spend a greater proportion of time in aperture closure when the vision of the arm or object is also blocked (Jackson *et al.*, 1995; Schettino *et al.*, 2006). It has been theorized that patients have problems implementing precise motor commands for more demanding tasks that require high accuracy constraints or movements with reduced visual feedback, which may result in delays in the modification of the grasp component. Subjects may therefore compensate for the demanding situations by initiating grasp closure early and more slowly.

Changes in the temporal relationship between the transport and grasp components in patients are also demonstrated in experiments that perturb the size of the object to be grasped during movement execution. When the size of the object increased from a small to a large diameter suddenly during the trial, PD patients were delayed by 28% when switching from the use of precision grip for the small object to whole-hand grip compared with controls (Castiello & Bennett, 1994; Castiello *et al.*, 1999). The perturbation also created a plateau in the grip aperture profile, which in turn caused a delay in the occurrence of maximum grip aperture. Control subjects, on the other hand, did not exhibit such a delay in the switching of grip type, and the transition between hand postures was smooth. It is postulated that the delay in switching hand grip in patients may be due to difficulties suppressing one motor plan (precision grip) and initiating a different motor plan (whole-hand grip) in an efficient manner when the context of the movement is manipulated.

Although the temporal coordination of aperture closure is altered in patients, aperture closure distance, or the distance the arm is transported from peak aperture to object grasp, remains relatively stable in patients as well as controls across varying movement amplitudes and object sizes (Alberts *et al.*, 2000; Rand *et al.*, 2006a). The scattergraphs in Figure 22.4

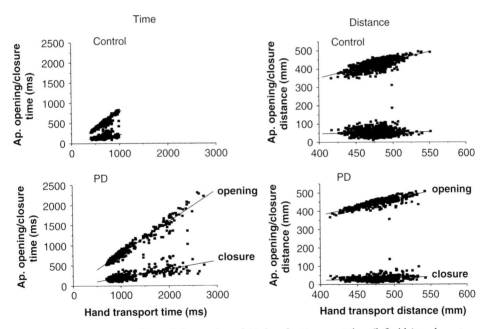

Figure 22.4. Aperture opening and closure time plotted against transport time (left side), and aperture opening and closure distance plotted against transport distance (right side) across all trials for patients with Parkinson's disease and controls. From Rand *et al.* (2006) *Exp Brain Res*, **168**, 131–142. Copyright 2005 by Springer-Verlag. With kind permission from Springer Science and Business Media.

illustrate this concept. Aperture opening and closure distances were computed and pooled across all subjects and trials, and these values were plotted against transport distance. As demonstrated in the figure, both patients and controls modulated aperture opening distance across the varying transport distances. Closure distance, or the distance from the initiation of aperture closure to object grasp, remains relatively stable in both groups, which indicates that the ability to assess the spatial distance between the hand and the object to maintain coordination between the transport and the grasp components is not deficient in Parkinson's disease. However, individuals with PD tend to initiate aperture closure closer to the object in distance compared with controls so there is less distance between the fingers and the object during the final stages of the movement. It has been suggested that the reduced aperture closure distance in PD patients may be simply due to movement speed rather than coordination deficits (Rand *et al.*, 2006a). For fast movements, it is natural to start the initiation of aperture closure further from the object to compensate for the increased speed of the movement performed at the same amplitude. Because movements are commonly performed more slowly in PD patients compared to controls, it may not be necessary to initiate aperture closure as far from the object. This, however, may become problematic, for if the fingers are very close to the object, patients are less able to flexibly position the fingers and modulate the grip aperture amplitude to account for any unexpected changes in the characteristics of the object.

Once the fingers make contact with the object to be grasped, force is exerted by the digits to ensure a successful grasp and lift. The time to stabilize grip and load forces is prolonged in patients (Ingvarsson *et al.*, 1997; Hejdukova *et al.*, 2002; Rearick *et al.*, 2002). The increase in time of force production may be due to more variable and step-like control, which increases with disease severity. Patients with Parkinson's disease also increased finger grip forces when the weight of the object was unpredictable but were able to scale these forces when weight was predictable in a similar manner to controls (Gordon *et al.*, 1997; Ingvarsson *et al.*, 1997). However, PD patients were unable to predict and scale finger forces when the location of the center of mass was unexpectedly changed and used a smaller number of force-sharing patterns between the digits compared with controls (Santello *et al.*, 2004). The reduced ability to predict and flexibly control force-sharing pattern during the grasp in patients suggests impairments in anticipatory force control.

Concluding remarks

The basal ganglia are perhaps one of the least understood brain regions, as they are somewhat insulated from the direct inputs and outputs of the brain. Motor deficiencies in Parkinson's disease result from the degeneration of dopamine receptors in the basal ganglia, which lead to abnormal output discharge from the basal ganglia (Pessiglione *et al.*, 2005). In this chapter, we reviewed the performance in patients during prehension acts. For the most part, the global aspects of reach-to-grasp movements in PD patients are preserved. Patients are able to perform these actions, but it is in the integration of these components that prehension is disrupted. The most prominent impairment reviewed was delays in the initiation and regulation of movement in relation to other components. This loss of integrity was observed in both the transport and grasp components of prehension. Patients display more irregularities in how the hand is transported to the object and how it is grasped. Patients also show reduced modulation in the control of grip aperture and in the production of grip force on the object.

It is suggested that excessive neural noise in the basal ganglia pathways with the loss of finely differentiated parallel processing may be the cause of most PD patient impairments (Guadagnoli *et al.*, 2002; Van Gemmert & Stelmach, 2002), namely the reduced ability to coordinate the initiation and the regulation of the transport and grasp components and movement slowing. Because of excessive neural noise, longer time may be needed to process information prior to and during movement, which can alter the coordination between these two prehensile components. The notion of the excessive neural noise hypothesis is predictive of the types of impairments found in PD. For example, PD patients have a more depressed and variable rate of force production compared with controls (Stelmach *et al.*, 1989; Park & Stelmach, 2007) which may be attributed to irregular bursting of muscle activity, increased co-contraction between agonist and antagonist musculature, and an inability to modulate the duration of the first muscle burst with increasing movement amplitude (Berardelli *et al.*, 1986; Godaux *et al.*, 1992; Pfann *et al.*, 2004). Furthermore, excessive neural noise may also make it difficult for patients to control limb dynamics during multijoint movements as patients show a reduction in the amplitude and timing of the elbow joint relative to the shoulder joint, and exhibit limited control of interactive torques at the elbow produced by the shoulder motion (Alberts *et al.*, 1998; Leiguarda *et al.*, 2000; Seidler *et al.*, 2001; Dounskaia *et al.*, 2005).

The data presented in this chapter fit well with the theory of excessive neural noise within the central nervous system in patients with Parkinson's disease. The largest deficits observed in reach-to-grasp movements in PD patients involve the loss of timing and reduced ability to integrate the transport and grasp components, as evidenced by delays in initiating the transport and grasp component, and delays in modulating grip aperture in response to changing transport and object characteristics. Neural noise therefore reduces the capability of the central nervous system to adequately parameterize movements during prehensile movements. Neuroscientists are still searching for a clearer understanding of the nature of the impairments with PD and neural mechanisms by which they arise. The more precise statements that can be made about the motor deficiencies observed in patients, the better the position the neuroscience community is in to determine the precise role of the intact basal ganglia in prehension control.

Acknowledgments

Preparation of this chapter was supported in part by grants NINDS NS40266 and NIA AG14676.

References

Alberts, J. L., Tresilian, J. R. & Stelmach, G. E. (1998). The co-ordination and phasing of a bilateral prehension task. The influence of Parkinson's disease. *Brain*, **121**, 725–742.

Alberts, J. L., Saling, M., Adler, C. H. & Stelmach, G. E. (2000). Disruptions in the reach-to grasp actions in Parkinson's patients. *Exp Brain Res*, **134**, 353–362.

Alberts, J. L., Saling, M., & Stelmach, G. E. (2002). Alterations in transport path differentially affect temporal and spatial movement parameters. *Exp Brain Res*, **143**, 417–425.

Berardelli, A., Dick, J. P. R., Rothwell, J. C., Day, B. L. & Marsden, C. D. (1986). Scaling of the size of the first agonist EMG burst during rapid wrist movements in patients with Parkinson's disease. *J Neurol Neurosurg Psychiatry*, **49**, 1273–1279.

Bergman, H. & Deuschl, G. (2002). Pathophysiology of Parkinson's disease: from clinical neurology to basic neuroscience and back. *Mov Disord*, **17**, S28–S40.

Bertram, C. P., Lemay, M. & Stelmach, G. E. (2005). The effect of Parkinson's disease on the control of multi-segment coordination. *Brain Cogn*, **57**, 16–20.

Bevan, M. D., Magill, P. J., Terman, D., Bolam, J. P. & Wilson, C. J. (2002). Move to the rhythm: oscillations in the subthalamic nucleus-external globus pallidus network. *Trends Neurosci*, **25**, 525–531.

Bonfiglioli, C., De Berti, G., Nichelli, P., Nicoletti, R. & Castiello, U. (1998). Kinematic analysis of the reach to grasp movement in Parkinson's disease and Huntington's disease subjects. *Neuropsychologia*, **36**, 1203–1208.

Bootsma, R. J., Marteniuk, R. G., MacKenzie, C. L. & Zaal, F. T. J. M. (1994). The speed accuracy trade-off in manual prehension: effects of movement amplitude, object size and object width on kinematic characteristics. *Exp Brain Res*, **98**, 535–541.

Boraud, T., Bezard, E., Bioulac, B. & Gross, C. E. (2000). Ratio of inhibited-to-active pallidal neurons decreases dramatically during passive limb movement in the MPTP-treated monkey. *J Neurophysiol*, **83**, 1760–1763.

Castiello, U. & Bennett, K. M. B. (1994). Parkinson's disease: reorganization of the reach to grasp movement in response to perturbation of the distal motor patterning. *Neuropsychologia*, **32**, 1367–1382.

Castiello, U., Stelmach, G. E. & Lieberman, A. N. (1993). Temporal dissociation of the prehension pattern in Parkinson's disease. *Neuropsychologia*, **31**, 395–402.

Castiello, U., Bennett, K., Bonfiglioli, C., Lim, S. & Peppard, R. F. (1999). The reach-to-grasp movement in Parkinson's disease: response to a simultaneous perturbation of object position and object size. *Exp Brain Res*, **125**, 453–462.

Chieffi, S. & Gentilucci, M. (1993). Coordination between the transport and the grasp components during prehension movements. *Exp Brain Res*, **94**, 471–477.

Contreras-Vidal, J. L. & Stelmach, G. E. (1996). Effect of parkinsonism on motor control. *Life Sci*, **58**, 165–176.

Desmurget, M., Grafton, S. T., Vindras, P., Grea, H. & Turner, R. S. (2003). Basal ganglia network mediates the control of movement amplitude. *Exp Brain Res*, **153**, 197–209.

Dounskaia, N., Ketcham, C. J., Leis, B. C. & Stelmach, G. E. (2005). Disruptions in joint control during drawing arm movements in Parkinson's disease. *Exp Brain Res*, **164**, 311–322.

Gatev, P., Darbin, O. & Wichmann, T. (2006). Oscillations in the basal ganglia under normal conditions and in movement disorders. *Mov Disord*, **21**, 1566–1577.

Gentilucci, M. & Negrotti, A. (1999). The control of an action in Parkinson's disease. *Exp Brain Res*, **129**, 269–277.

Gentilucci, M., Castiello, U., Corradini, M. L. *et al.* (1991). Influence of different types of grasping on the transport component of prehension movements. *Neuropsychologia*, **29**, 361–378.

Godaux, E., Koulischer, D. & Jacquy, J. (1992). Parkinsonian bradykinesia is due to depression in the rate of rise of muscle activity. *Ann Neurol*, **31**, 93–100.

Gordon, A. M., Ingvarsson, P. E. & Forssberg, H. (1997). Anticipatory control of manipulative forces in Parkinson's disease. *Exp Neurol*, **145**, 477–488.

Guadagnoli, M. A., Leis, B., Van Gemmert, A. W. A. & Stelmach, G. E. (2002). The relationship between knowledge of results and motor learning in Parkinsonian patients. *Parkinsonism Related Disord*, **9**, 89–95.

Haggard, P. & Wing, A. (1998). Coordination of hand aperture with the spatial path of hand transport. *Exp Brain Res*, **118**, 286–292.

Hejdukova, B., Hosseini, N., Johnels, B. *et al.* (2002). Manual transport in Parkinson's disease. *Mov Disord*, **18**, 565–572.

Ingvarsson, P. E., Gordon, A. M. & Forssberg, H. (1997). Coordination of manipulative forces in Parkinson's disease. *Exp Neurol*, **145**, 489–501.

Jackson, G. M., Jackson, S. R. & Hindle, J. V. (2000). The control of bimanual reach-to-grasp movements in hemiparkinsonian patients. *Exp Brain Res*, **132**, 390–398.

Jackson, S. R., Jackson, G. M., Harrison, J., Henderson, L. & Kennard, C. (1995). The internal control of action and Parkinson's disease: a kinematic analysis of visually-guided and memory-guided prehension movements. *Exp Brain Res*, **105**, 147–162.

Jahanshahi, M., Brown, R. G. & Marsden, C. D. (1992). Simple and choice reaction time and the use of advance information for motor preparation in Parkinson's disease. *Brain*, **115**, 539–564.

Jeannerod, M. (1984). The timing of natural prehension movements. *J Motor Behav*, **16**, 235–254.

Kaji, R. (2001). Basal ganglia as a sensory gating devise for motor control. *J Med Invest*, **48**, 142–146.

Leiguarda, R., Merello, M., Balej, J. *et al.* (2000). Disruption of spatial organization and interjoint coordination in Parkinson's disease, progressive supranuclear palsy, and multiple system atrophy. *Mov Disord*, **15**, 627–640.

Leis, B. C., Rand, M. K., Van Gemmert, A. W. A. *et al.* (2005). Movement precues in planning and execution of aiming movements in Parkinson's disease. *Exp Neurol*, **194**, 393–409.

Longstaff, M. G., Mahant, P. R., Stacy, M. A. *et al.* (2003). Discrete and dynamic scaling of the size of continuous graphic movements of parkinsonian patients and elderly controls. *J Neurol Neurosurg Psychiatry*, **74**, 299–304.

Majsak, M. J., Kaminski, T., Gentile, A. M. & Flanagan, J. R. (1998). The reaching movements of patients with Parkinson's disease under self-determined maximal speed and visually cued conditions. *Brain*, **121**, 755–766.

Merideth, G. E. & Kang, U. J. (2006). Behavioral models of Parkinson's disease in rodents: a new look at an old problem. *Mov Disord*, **21**, 1595–1606.

Middleton, F. A. & Strick, P. L. (2000). Basal ganglia and cerebellar loops: motor and cognitive circuits. *Brain Res Rev*, **31**, 236–250.

Mink, J. W. (1996). The basal ganglia: focused selection and inhibition of competing motor programs. *Progr Neurobiol*, **50**, 381–425.

Negrotti, A., Secchi, C. & Gentilucci, M. (2005). Effects of disease progression and L-dopa therapy on the control of reaching-grasping in Parkinson's disease. *Neuropsychologia*, **43**, 450–459.

Parent, A. & Hazrati, L. N. (1995). Functional anatomy of the basal ganglia. I. The cortico-basal ganglia-thalamo-cortical loop. *Brain Res Rev*, **20**, 91–127.

Park, J. & Stelmach, G. E. (2007). Force development during object-directed isometric force production in Parkinson's disease. *Neurosci Lett*, **412**, 173–178.

Paulignan, Y., Frak, V. G., Toni, I. & Jeannerod, M. (1997). Influence of object position and size on human prehension movements. *Exp Brain Res*, **114**, 226–234.

Pessiglione, M., Guehl, D., Rolland, A. *et al.* (2005). Thalamic neuronal activity in dopamine-depleted primates: evidence for a loss of functional segregation within basal ganglia circuits. *J Neurosci*, **25**, 1523–1531.

Pfann, K. D., Robichaud, J. A., Gottlieb, G. L. *et al.* (2004). Muscle activation patterns in point-to-point and reversal movements in healthy, older subjects and in subjects with Parkinson's disease. *Exp Brain Res*, **157**, 67–78.

Poizner, H., Feldman, A. G., Levin, M. F. *et al.* (2000). The timing of arm-trunk coordination is deficient and vision-dependent in Parkinson's patients during reaching movements. *Exp Brain Res*, **133**, 279–292.

Rand, M. K., Van Gemmert, A. W. A. & Stelmach, G. E. (2002). Segment difficulty in two stroke movements in patients with Parkinson's disease. *Exp Brain Res*, **143**, 383–393.

Rand, M. K., Smiley-Oyen, A. L., Shimansky, Y. P., Bloedel, J. R. & Stelmach, G. E. (2006a). Control of aperture closure during reach-to-grasp movements in Parkinson's disease. *Exp Brain Res*, **168**, 131–142.

Rand, M. K., Squire, L. M. & Stelmach, G. E. (2006b). Effect of speed manipulation on the control of aperture closure during reach-to-grasp movements. *Exp Brain Res*, **174**, 74–85.

Rearick, M. P., Stelmach, G. E., Leis, B. & Santello, M. (2002). Coordination and control of forces during multifingered grasping in Parkinson's disease. *Exp Neurol*, **177**, 428–442.

Romero, D. H., Van Gemmert, A. W. A., Adler, C. H., Bekkering, H. & Stelmach, G. E. (2003). Altered aiming movements in Parkinson's disease patients and elderly adults as a function of delays in movement onset. *Exp Brain Res*, **151**, 249–261.

Saling, M., Mescheriakov, S., Molokanova, E., Stelmach, G. E. & Berger, M. (1996). Grip reorganization during wrist transport: the influence of an altered aperture. *Exp Brain Res*, **108**, 493–500.

Saling, M., Alberts, J. L., Stelmach, G. E. & Bloedel, J. R. (1998). Reach-to-grasp movements during obstacle avoidance. *Exp Brain Res*, **118**, 251–258.

Santello, M., Muratori, L. & Gordon, A. M. (2004). Control of multidigit grasping in Parkinson's disease: effect of object property predictability. *Exp Neurol*, **187**, 517–528.

Schettino, L. F., Rajaraman, V., Jack, D. *et al.* (2003). Deficits in the evolution of hand preshaping in Parkinson's disease. *Neuropsychologia*, **42**, 82–94.

Schettino, L. F., Adamovich, S. V., Hening, W. *et al.* (2006). Hand preshaping in Parkinson's disease: effects of visual feedback and medication state. *Exp Brain Res*, **168**, 186–202.

Seidler, R. D., Alberts, J. L. & Stelmach, G. E. (2001). Multijoint movement control in Parkinson's disease. *Exp Brain Res*, **140**, 335–344.

Sheridan, M. R., Flowers, K. A. & Hurrell, J. (1987). Programming and execution of movement in Parkinson's disease. *Brain*, **110**, 1247–1271.

Stelmach, G. E., Worringham, C. J. & Strand, E. A. (1986). Movement preparation in Parkinson's disease. *Brain*, **109**, 1179–1194.

Stelmach, G. E., Teasdale, N., Phillips, J. & Worringham, C. J. (1989). Force production characteristics in Parkinson's disease. *Exp Brain Res*, **76**, 165–172.

Tresilian, J. R., Stelmach, G. E. & Adler, C. H. (1997). Stability of reach-to-grasp movement patterns in Parkinson's disease. *Brain*, **120**, 2093–2111.

Tunik, E., Poizner, H., Adamovich, S. V., Levin, M. F. & Feldman, A. G. (2004). Deficits in adaptive upper limb control in response to trunk perturbations in Parkinson's disease. *Exp Brain Res*, **159**, 23–32.

Van Gemmert, A. W. A. & Stelmach, G. E. (2002). Increased motor processing demands reduce stroke size in Parkinsonian handwriting. *J Sport Exercise Psychol*, **24**, S16.

Van Gemmert, A. W., Adler, C. H. & Stelmach, G. E. (2003). Parkinson's disease patients undershoot object size in handwriting and similar tasks. *J Neurol Neurosurg Psychiatry*, **74**, 1502–1508.

Wang, J. & Stelmach, G. E. (1998). Coordination among body segments during reach-to-grasp action involving the trunk. *Exp Brain Res*, **123**, 346–350.

Wang, J. & Stelmach, G. E. (2001). Spatial and temporal control of trunk-assisted prehensile actions. *Exp Brain Res*, **136**, 231–240.

Wang, J., Thomas, J. R. & Stelmach, G. E. (1998). A meta-analysis on cognitive slowing in Parkinson's disease: are simple and choice reaction times differentially impaired? *Parkinsonism Related Disord*, **4**, 17–29.

Wang, J., Bohan, M., Leis, B. C. & Stelmach, G. E. (2006). Altered coordination patterns in parkinsonian patients during trunk-assisted prehension. *Parkinsonism Related Disord*, **12**, 211–222.

Wichmann, T. & DeLong, M. R. (2003). Pathophysiology of Parkinson's disease: the MPTP primate model of the human disorder. *Annals Acad Sci*, **991**, 199–213.

23

Grip-force analysis in Huntington's disease – a biomarker for clinical trials?

RALF REILMANN

Summary

Objective and quantitative measures to assess the severity and progression of Huntington's disease (HD) are desirable. Several studies have demonstrated quantifiable deficits in the coordination of precision grasping in patients with Huntington's disease. Correlation analysis revealed that the amount of grip force variability while holding an object was correlated to the total motor score of the Unified Huntington's Disease Rating Scale (UHDRS) in a cross-sectional study. In addition, grip force variability increased in all HD patients during a 3-year follow-up. The UHDRS total motor score did not change significantly in the same subjects. The results suggest that neurophysiological analysis of isometric grip forces may detect disease progression more sensitively than clinical rating scales. The applicability of the assessment of grip forces in clinical studies is currently tested in large multicenter studies. Possible applications of the technique as a biomarker in clinical studies in HD are discussed.

Introduction

Huntington's disease (HD) is an autosomal dominant neurodegenerative disorder with a prevalence of about 7–10 symptomatic patients per 100 000 individuals and about double the number of pre-symptomatic gene carriers (Harper, 1996). Expansion of a CAG-repeat within exon 1 of the HD gene results in the development of the HD phenotype (The Huntington's Disease Collaborative Research Group, 1993), with longer repeats associated with earlier manifestation and faster progression of disease (Andrew et al., 1993). Gene carriers develop progressive neuronal dysfunction due to a neurodegenerative process, which is particularly pronounced in the caudate nucleus, but extends to other areas of the basal ganglia and cortical structures (Lange et al., 1976; Aylward et al., 1994, 1997; Rosas et al., 2002, 2005). Huntington's disease is clinically characterized by cognitive decline, psychiatric symptoms and a variety of motor symptoms (Harper, 1996; Berardelli et al., 1999). The pathology develops chronically over the course of several years to decades. No objective and quantitative phenotypic measure to assess the severity and progression of HD is available so far.

In the setting of clinical trials, the current "gold standard" for assessment of symptoms in HD, including motor deficits, is a categorical clinical scale, the Unified Huntington's Disease

Sensorimotor Control of Grasping: Physiology and Pathophysiology, ed. Dennis A. Nowak and Joachim Hermsdörfer. Published by Cambridge University Press. © Cambridge University Press 2009.

Rating Scale (UHDRS: Huntington Study Group, 1996). The UHDRS comprises cognitive and functional tests, and a separate section evaluating motor symptoms, the Total Motor Score (TMS). The UHDRS-TMS comprises 31 sub-items, assessing for example oculomotor function, speech, coordination, chorea, dystonia, rigidity, gait and balance. However, categorical rating scales, such as the UHDRS-TMS, have serious limitations: (1) low sensitivity, compromising the detection of possible smaller treatment effects of new drugs; (2) limited statistical power due to the lack of applicability of statistical analysis available for continuous, numerical and objective data, e.g. ANOVA; (3) high inter-rater and intra-rater variability due to subjective error, which is particularly critical in multicenter trials. Thus, more objective and quantitative measures of stage and progression of HD are desirable to serve as biomarkers for clinical trials.

When searching for phenotypic biomarkers of HD, motor symptoms may provide an easier access to objective and quantitative analysis than other symptoms. Besides the characteristic involuntary movements, depicted as "chorea", various other motor impairments, such as bradykinesia, dystonia, sequencing problems, motor impersistency and, particularly in later stages, akinesia, are common (Berardelli *et al.*, 1999; Louis *et al.*, 1999; Gordon *et al.*, 2000). The impairments in motor function are attributed to the widespread and progressive neuro-degeneration in the brains of subjects with HD (Lange *et al.*, 1976; Vonsattel *et al.*, 1985; Aylward *et al.*, 1997; Rosas *et al.*, 2002), which is observed years before clinical symptoms become apparent (Aylward *et al.*, 2000; Thieben *et al.*, 2002; Rosas *et al.*, 2002).

As described in detail in other chapters of this book, the coordination of isometric grip forces in the precision grip between the thumb and index finger is a highly complex motor task (Johansson, 1996) (see also Chapters 1 and 11). It objectively and quantitatively assesses a function that is relevant for successful manipulation of objects in everyday life and has been well studied in healthy subjects and different diseases affecting motor control. Degenerative processes affecting the function of central sensory and motor systems may result in changes of grip-force coordination.

Therefore the coordination of grip forces during precision grasping in patients with HD was investigated with the following objectives: (1) identify impairments in the force coordination during different tasks involving precision grasping in cross-sectional studies; (2) identify those measures that correlate to the severity of HD as assessed clinically in the UHDRS; (3) if such measures are present, test if they progress in the course of the disease within subjects in a small pilot follow-up study; (4) if the measures progress over time in the small pilot study, test them in a multicenter setting in a larger study and in a blinded setting.

The following sections will summarize the most important results of the studies conducted in HD to address the objectives outlined above. At the end the potential of grip-force analysis for use in patients with HD as a motor biomarker will be discussed.

Impairment of grip forces in Huntington's disease shows correlation to clinical measures of disease severity

First evidence for impaired coordination of grip forces in HD was reported after load perturbations (Fellows *et al.*, 1997). Subjects with HD showed increased latencies when

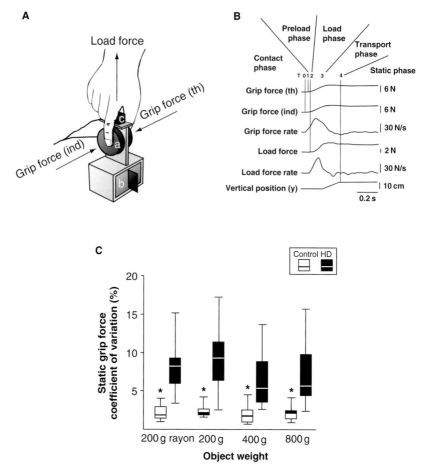

Figure 23.1. A. Grip instrument with force transducers, a; weight of the object adjustable to 200 g and 800 g, b; electromagnetic position sensor, c. B. Sample recording, showing the grip force of each finger, the mean grip force rate, mean load force, mean load force rate, vertical position and the phases evaluated. C. Box–whisker plots indicating the medians and quartiles (boxes) as well as the minimum and maximum values (whiskers) of the coefficient of variation (mean/standard deviation × 100) in grip force during the static phase for each object condition (sandpaper contact surfaces used except where noted). Asterixes above the box–whisker plots indicate a significant difference ($P < 0.05$) from the subjects with Huntington's Disease for that condition. Figures reproduced and modified from Gordon *et al.* (2000). *Exp Neurol*, **163**, 136–148. With kind permission of Elsevier Science Publisher, Orlando, FL, USA.

adapting to sudden increases in loads. However, no correlation with clinical measures, e.g. the UHDRS-TMS, was performed.

A cross-sectional study investigated the coordination of grip forces during grasp initiation and during a static holding phase in HD (n = 12) and normal subjects (n = 12) (Gordon *et al.*, 2000). Using their right, dominant hand, subjects grasped and lifted an object equipped with force transducers and were instructed to hold it stable next to a marker 10 cm high for several seconds (Figure 23.1A). Grip (normal) and load (tangential) forces of both digits and the object's position were recorded (Figure 23.1B). The object's weight (200 g, 400 g, 800 g) and surface texture (sandpaper or rayon) were modified to test the adaptation of grip forces to different sensory stimuli (for details see Gordon *et al.*, 2000). In a second experiment, the same subjects were asked to grasp and transport the object, again the object's weight and surface texture were altered (for details see Quinn *et al.*, 2001).

The results showed that subjects with HD exhibited impaired initiation and delayed transitions between movement sequences during grasp initiation (Gordon *et al.*, 2000; Quinn *et al.*, 2001). In addition, they produced higher and considerably more variable isometric grip forces in the static holding phase compared with controls (Gordon *et al.*, 2000). The increase in grip-force variability, expressed as the coefficient of variation, was one of the most robust findings across all object conditions (see Figure 23.1C). Increased grip-force variability was also seen during object transport (Quinn *et al.*, 2001). In contrast, anticipatory scaling of forces based on the object's expected physical properties (planning) and adjustment of the forces to the object's actual physical properties such as weight and surface texture (sensorimotor integration) was preserved in HD (Gordon *et al.*, 2000; Quinn *et al.*, 2001).

The observed deficits generally were unrelated to the overall disease severity as assessed clinically in the UHDRS (Huntington Study Group, 1996). However, the variability of grip and load forces in the static holding phase was correlated with the deficits seen in the total motor score and the functional capacity score of the UHDRS (Gordon *et al.*, 2000). This observation was independent of the level of chorea observed in the UHDRS. These results implied that patients more severely affected by HD, i.e. patients in more advanced stages, exhibit more grip-force variability.

It was thus hypothesized that grip-force variability might increase during the course of HD within individual subjects, suggesting that it could serve as a biomarker for the stage and progression of motor deficits and the disease process. In order to test this hypothesis, grip-force variability and other measures of grip-force coordination were analysed in a follow-up study in patients with HD.

Assessment of progression in Huntington's disease using grip-force analysis

The impairments in the coordination of isometric grip forces in precision grasping and the severity of symptoms in the UHDRS were assessed in patients with HD (n = 10) with a mean follow-up of 3 years (Reilmann *et al.*, 2001). Patients performed the same paradigms used in the cross-sectional study described in section 2. Grip and load forces were recorded. In addition, the object's 3D position (x-, y-, z-position) and orientation (roll-, pitch-, yaw-orientation) were measured and used to calculate the position-index and orientation-index (both defined as sums of the absolute values of the derivatives of the position or orientation channels, respectively – see Figure 23.2). Mean values of the position- and orientation-index in the static holding phase of the object were thus objective measures of interfering involuntary choreatic movements. The performance of patients was compared between the start and follow-up assessment.

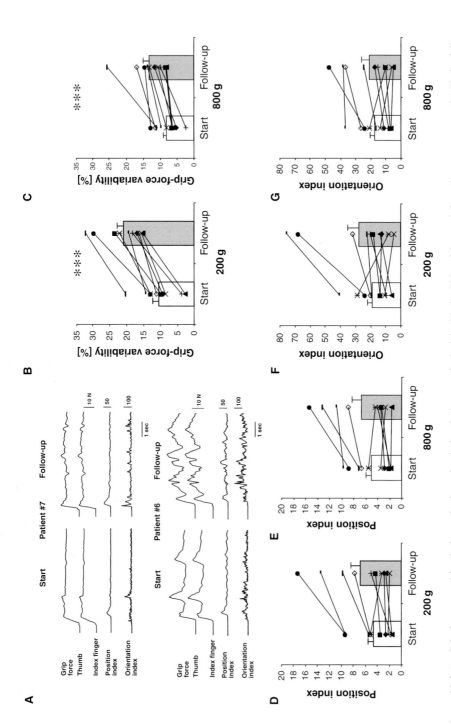

Figure 23.2. A. Grip forces, position index and orientation index during a typical trial with the 800 g object weight at the start compared with the follow-up assessment for two representative patients with Huntington's disease. B, C. Grip-force variability for the two object weights 200 g and 800 g at the start and follow-up assessment. D, E. Position index and (F, G) orientation index of the object at start and follow-up (shown for the 200 g and the 800 g weight). Bars represent means ± SEM. Symbols represent individual patients. (***$P \leq 0.001$) Figures reproduced and modified from Reilmann *et al.* (2001). *Neurology*, 2001; **57**, 920–924, with kind permission of Lippincott Williams & Wilkins Publishers, Baltimore, MD, USA.

Isometric grip-force variability in the static holding phase increased significantly across patients with HD during the follow-up period (Figure 23.2A–C) (Reilmann *et al.*, 2001). The increase was seen in each of the patients and in all object conditions. The amount of chorea as assessed by the UHDRS and using the objective measures of the position- and orientation-index did not increase significantly during the follow-up (Figure 23.2D–G). Accordingly, correlation analysis revealed that the increase in grip-force variability could not be explained by changes in the amount of chorea. Instead the motor variability may represent a distinct motor deficit caused by the neurodegenerative process in HD. It is actually seen in other tasks in patients with HD (Rao *et al.*, 2005; Reilmann *et al.*, pers. comm.).

Conclusion

In conclusion, grip-force variability provided a measure correlated to the UHDRS-TMS in a cross-sectional study and progressed in all subjects participating in a 3-year follow-up study. The increase in grip-force variability was likely caused by the progressive neurodegeneration in HD (Aylward *et al.*, 1997; 2000). Interestingly, the progression of motor deficits in the follow-up study could not be reliably detected using the UHDRS motor scores (Reilmann *et al.*, 2001). Thus, grip-force variability might be a more sensitive measure to assess the progression of HD than the UHDRS.

Further studies to assess the capability of grip-force variability to serve as a phenotypic biomarker of disease progression in HD were initiated. Grip-force analysis is currently investigated in the multicenter 2-year follow-up study TRACK-HD with 360 subjects (120 manifest HD, 120 presymptomatic gene carriers, and 120 controls). The study will also allow us to compare and correlate changes in motor phenotype with changes in neuroimaging and other assessment modalities.

Acknowledgements

RR was supported by grants DFG-RE-1330/1-1, IMF-RE-120-225 (Innovative Medical Research Fund of the University of Münster), the HD-Therapeutics Foundation (New York, USA), the EHDN (European Huntington's Disease Network), the High-Q-Foundation (New York, USA) and the "Cure for Huntington's Disease Initiative" (CHDI). The continuous support of all patients with HD and their relatives is especially appreciated.

References

Andrew, S. E., Goldberg, Y. P., Kremer, B. *et al.* (1993). The relationship between trinucleotide (CAG) repeat length and clinical features of Huntington's disease [see comments]. *Nat Genet*, **4**, 398–403.

Aylward, E. H., Brandt, J., Codori, A. M. *et al.* (1994). Reduced basal ganglia volume associated with the gene for Huntington's disease in asymptomatic at-risk persons. *Neurology*, **44**, 823–828.

Aylward, E. H., Li, Q., Stine, O. C. *et al.* (1997). Longitudinal change in basal ganglia volume in patients with Huntington's disease. *Neurology*, **48**, 394–399.

Aylward, E. H., Codori, A. M., Rosenblatt, A. *et al.* (2000). Rate of caudate atrophy in presymptomatic and symptomatic stages of Huntington's disease [In Process Citation]. *Mov Disord*, **15**, 552–560.

Berardelli, A., Noth, J., Thompson, P. D., *et al.* (1999). Pathophysiology of chorea and bradykinesia in Huntington's disease. *Mov Disord*, **14**, 398–403.

Fellows, S., Schwarz, M., Schaffrath, C., Domges, F. & Noth, J. (1997). Disturbances of precision grip in Huntington's disease. *Neurosci Lett*, **226**, 103–106.

Gordon, A. M., Quinn, L., Reilmann, R. & Marder, K. (2000). Coordination of prehensile forces during precision grip in Huntington's disease. *Exp Neurol*, **163**, 136–148.

Harper, P. S. (1996). *Huntington's Disease*. 2nd edn. London: Saunders.

Huntington Study Group (1996). Unified Huntington's Disease Rating Scale: reliability and consistency. *Mov Disord*, **11**, 136–142.

Johansson, R. S. (1996). Sensory control of dextrous manipulations in humans. In A. M. Wing, P. Haggard & J. R. Flanagan (Eds.), *Hand and Brain: The Neurophysiology and Psychology of Hand Movement* (pp. 381–414). San Diego, CA: Academic Press.

Lange, H., Thorner, G., Hopf, A. & Schroder, K. F. (1976). Morphometric studies of the neuropathological changes in choreatic diseases. *J Neurol Sci*, **28**, 401–425.

Louis, E. D., Lee, P., Quinn, L. & Marder, K. (1999). Dystonia in Huntington's disease: prevalence and clinical characteristics. *Mov Disord*, **14**, 95–101.

Quinn, L., Reilmann, R., Marder, K. & Gordon, A. M. (2001). Altered movement trajectories and force control during object transport in Huntington's disease. *Mov Disord*, **16**, 469–480.

Rao, A. K., Quinn, L. & Marder, K. S. (2005). Reliability of spatiotemporal gait outcome measures in Huntington's disease. *Mov Disord*, **20**, 1033–1037.

Reilmann, R., Kirsten, F., Quinn, L., *et al.* (2001). Objective assessment of progression in Huntington's disease: a 3-year follow-up study. *Neurology*, **57**, 920–924.

Rosas, H. D., Liu, A. K., Hersch, S. *et al.* (2002). Regional and progressive thinning of the cortical ribbon in Huntington's disease. *Neurology*, **58**, 695–701.

Rosas, H. D., Hevelone, N. D., Zaleta, A. K. *et al.* (2005). Regional cortical thinning in preclinical Huntington disease and its relationship to cognition. *Neurology*, **65**, 745–747.

The Huntington's Disease Collaborative Research Group (1993). A novel gene containing a trinucleotide repeat that is expanded and unstable on Huntington's disease chromosomes. [see comments]. *Cell*, **72**, 971–983.

Thieben, M. J., Duggins, A. J., Good, C. D. *et al.* (2002). The distribution of structural neuropathology in pre-clinical Huntington's disease. *Brain*, **125**, 1815–1828.

Vonsattel, J. P., Myers, R. H., Stevens, T. J. *et al.* (1985). Neuropathological classification of Huntington's disease. *J Neuropathol Exp Neurol*, **44**, 559–577.

24

Traumatic brain injury

JOHANN P. KUHTZ-BUSCHBECK

Summary

Upper-limb speed and dexterity are frequently impaired after moderate or severe traumatic brain injury (TBI). The speed of functional hand movements can be assessed with standardized tasks, such as the Developmental Hand Function Test and the Purdue Pegboard test. Kinematic data on reaching and grasping can be obtained by optoelectronic motion analyses. The fingertip forces measured during a precision grip–lift task describe fine motor control. With these methods, a series of studies analyzed recovery of hand function in brain-injured children and adolescents (age 4–15 years) over 5 months of inpatient rehabilitation, starting ~3 months post TBI. Compared with healthy age-matched controls, the patients were slower, their prehension movements exhibited curved and variable movement trajectories, and were delayed especially in the final approach phase. They needed more time to establish a precision grip and showed exaggeratedly high grip forces. Despite substantial recovery, differences in hand function between patients and controls were still present ~8 months after TBI. Young age at injury was not associated with better recovery. Comparable data for adults are lacking so far.

Traumatic brain injury: incidence, severity and imaging

The annual incidence of traumatic brain injury (TBI) in Germany is about 300 per 100 000 inhabitants (Federal Statistical Office, www.destatis.de). Epidemiological studies from other countries report incidences of ~200–500/100 000 per year; these variations reflect different inclusion criteria and study designs (Hillier *et al.*, 1997; Servadei *et al.*, 2002; Andersson *et al.*, 2003). Common causes of TBI are traffic accidents, falls and sport-related accidents. Physical abuse must be considered especially in the pediatric age group (battered child syndrome). The percentage of children sustaining TBI is relatively high. In a recent prospective multicenter survey of ~6800 cases, about 28% of the patients were younger than 16 years, and ~10% of all patients had incurred moderate or severe TBI (Möllmann, 2006). The severity of TBI is often graded with the Glasgow Coma Scale (GCS). Eye opening, the best verbal response and the best motor response, as commonly assessed at the site of the accident, or at the emergency hospital admission, are items of the GCS

Sensorimotor Control of Grasping: Physiology and Pathophysiology, ed. Dennis A. Nowak and Joachim Hermsdörfer. Published by Cambridge University Press. © Cambridge University Press 2009.

(Teasdale & Jennett, 1974). Different levels of severity are defined as corresponding to GCS scores of 13–15 (mild TBI), 8–12 (moderate TBI) and 3–7 (severe TBI).

Traumatic brain injury involves a complex series of pathophysiological events. Primary lesions result as a direct consequence of the mechanical forces that disrupt the skull and brain at the instant occurrence of the trauma (Besenski, 2002). Primary injuries are scalp laceration/hematoma, skull fractures, various extracerebral hemorrhages (epidural, subdural, subarachnoid, intraventricular) and cerebral lesions (e.g. diffuse axonal injury, cortical contusions, intracerebral hemorrhage, brainstem injuries). Secondary lesions can occur within minutes or days after the primary injury and include progression of cerebral edema, increase of intracranial pressure, cerebral herniation, decreased cerebral blood flow that leads to ischemia and infarction, and ongoing cytotoxic neural damage (Gaetz, 2004). Computed tomography (CT), which is easily performed with monitored/ventilated patients, is currently the first imaging technique to be used in the acute phase (Besenski, 2002; Bigler *et al.*, 2006). It can detect skull fractures and complications that require rapid surgical intervention (e.g. epidural hematoma). Magnetic resonance imaging (MRI), which can reveal parenchymal lesions more sensitively than CT, is useful for follow-up and a full assessment of brain damage. It is the method of choice in brainstem injury analyses, which are of prognostic importance (Firsching *et al.*, 2002).

Substantial knowledge has been gained from detailed studies of hand function in stroke patients (see Chapters 21 and 29). However, extrapolation of these findings to TBI is dubious for several reasons. The patterns of brain lesions differ. Impairments of arm and hand function after stroke usually result from damage to the sensorimotor cortex and/or to pathways in the internal capsule or corona radiata, especially after infarctions in the territory of the middle cerebral artery that can cause circumscribed lesions. Such focal damage is uncommon in TBI patients. Instead, diffuse axonal injury often involves widespread microscopic damage to scattered white matter and small vascular structures, so that structures in various motor pathways are affected after severe TBI. Cortical contusions are most likely to occur in the orbitofrontal and anterior temporal brain regions (see Figure 24.1A). Hemorrhagic contusions of the lateral convexity, deep cerebral hemorrhage and transtentorial hemorrhage may affect the sensorimotor cortex and descending motor pathways in TBI (Katz *et al.*, 1998). Most severe TBI will afflict the upper brainstem, where the descending pathways converge (Gennarelli *et al.*, 1982; Katz *et al.*, 1998). Thus, in contrast to stroke, multifocal and diffuse brain injuries are the rule rather than the exception in TBI. Comorbidity differs, since peripheral lesions (e.g. fractures, nerve injuries) are more frequent in TBI than after stroke. The age distribution of TBI patients is bimodal, given that in addition to the elderly (falls as a common cause), many children and young adults are affected (Servadei *et al.*, 2002).

A series of recent studies investigated the recovery of hand function and gross motor proficiency in children and adolescents after moderate and severe TBI with clinical assessments and instrumented measures (Kuhtz-Buschbeck *et al.*, 2003a, 2003b; Gölge *et al.*, 2004; Holzhäuser, 2006). A consecutive series of 23 children (13 boys, 10 girls) with a mean age of 10.6 ± 3.3 (SD) years was included. All were inpatients of a regional rehabilitation

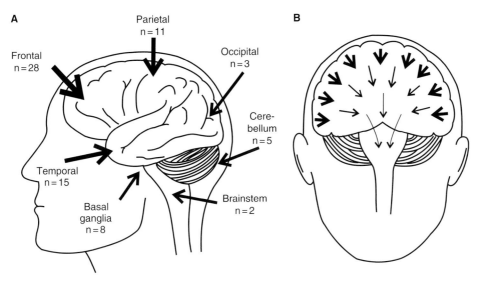

Figure 24.1. Distribution of brain injuries. A. Cumulative numbers of lesions affecting the different lobes, the cerebellum, basal ganglia and brainstem obtained from computed tomography findings of 23 children after TBI. B. According to the Ommaya–Gennarelli model of TBI, acceleration–deceleration forces cause brain injuries on a continuum ranging from the cortical surface inward, with increasing amounts of damage occurring at each level of depth as forces increase (Gaetz, 2004). Hence damage to the cortex and subcortical white matter is more frequent than brainstem damage; see (A).

center, which specializes in the treatment of children with acquired brain injuries, and received physiotherapy (2–5 hours/week), occupational therapy (2–3 hours/week), special schooling and neuropsychological support. Brain injuries had been severe in 17 children and moderate in 6 children according to the initial GCS values. The mean age at injury had been ~10 years (range 4–15 years). The first examination (E0) took place after admission to the rehabilitation center about 3 months after injury; follow-up examinations were scheduled 1 (E1), 2 (E2) and 5 (E5) months later. Control data were collected from 23 healthy children, who were matched to the patients on the basis of age and gender. The CTs (available in all patients) and MRIs (additionally available in seven patients) collectively showed contusions in 21 of the 23 children. Further findings were ventricular or subarachnoid hemorrhage (11 patients), generalized brain edema (8 patients), subdural or epidural hematoma (6 patients) and skull fractures (10 patients). A median number of three affected brain regions per child were counted. The distribution of the lesions (Figure 24.1) is consistent with the Ommaya–Gennarelli model of TBI (Ommaya & Gennarelli, 1974), which predicts that acceleration/deceleration forces cause mechanical strains that operate in a "centripetal sequence," e.g. when the head is propelled through space and suddenly stopped by an object. The cortical surface will be affected first, and deeper structures such as the brainstem will be progressively affected as forces become more severe (Gennarelli *et al.*, 1982; Gaetz, 2004).

Hand-function tests after TBI

Due to the heterogeneity of the brain lesions, the sensorimotor symptoms of TBI patients vary considerably (Haley *et al.*, 1990). Residual spastic or flaccid limb pareses were frequent findings in the aforementioned collective of TBI children (Kuhtz-Buschbeck *et al.*, 2003a). Coordination deficits, impaired balance and reduced movement quality (e.g. dysdiadochokinesia, dysmetria) represent typical neurological soft signs. Several patients exhibited ataxia and tremor. Slowing of psychomotor functions is a frequent clinical finding after TBI (Chaplin *et al.*, 1993; Rossi & Sullivan, 1996; Emanuelson *et al.*, 1998). Its different aspects can be measured with neuropsychological tests of simple processing speed (e.g. reaction time), speed of selective attention performance and manual dexterity speed (Asikainen *et al.*, 1999; Jonsson *et al.*, 2004). Low performance on tasks involving upper-limb speed and dexterity was found 2 years after severe TBI (Wallen *et al.*, 2001). Long-term deficits in fine motor skills, which affected vocational outcome more than 5 years after severe TBI in children and adolescents, were also reported (Emanuelson *et al.*, 1998).

Attention deficits and the short concentration span of the brain-injured children influenced the selection of testing procedures in our patient group (Kuhtz-Buschbeck *et al.*, 2003a). Pauses were often necessary, and the examiners tried to create optimal motivation. Subnormal IQ scores, with a median non-verbal IQ of 72 (range 41–111) and a median verbal IQ of 73 (range 48–98), pointed towards the cognitive deficits of the children after moderate/severe TBI. A simple gait analysis, the Gross Motor Function Measure (GMFM) (Russell *et al.*, 1989; Drouin *et al.*, 1996) and two hand function tests were feasible. Both the dominant hand and the non-dominant hand were examined. Fine motor coordination and speed were assessed with the Purdue Pegboard (Lafayette Co, Lafayette, IN, USA). The children grasped and inserted small pegs (length 25 mm, diameter 2 mm) into board holes as fast as possible in three 30 s test sessions per hand. The number of pegs was averaged from these trials. Functional unilateral hand movements were examined with the Developmental Hand Function Test (DHFT), measuring the time required to complete six tasks: picking up small objects, simulated eating, turning cards, stacking checkers, picking up light and heavy objects (Jebsen *et al.*, 1969). The patients' compliance was better for the DHFT than for the Purdue Pegboard, which is less varied. Normative data and methodological details of both hand-function tests have been published elsewhere (Taylor *et al.*, 1973; Gardner & Broman, 1979; Reddon *et al.*, 1988). Kinematic and force recordings (see below) were scheduled on different days and could not be performed with all brain-injured children.

Data on TBI patients and control children were contrasted with Mann–Whitney U-tests. To detect improvements of hand function during the follow-up interval (5 months), results of the first (E0) and last examination dates (E5) were compared with Wilcoxon tests. Children with brain injuries inserted fewer pegs in the pegboard test and needed more time to complete the DHFT than control children (Figure 24.2). Despite significant improvements during follow-up, considerable deficits in fine motor performance persisted, so that case–control differences were still highly significant at E5, about 8 months after injury

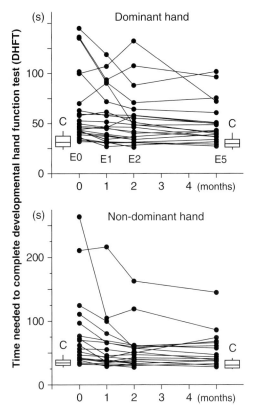

Figure 24.2. Developmental Hand Function Test (DHFT). Individual results of brain-injured children during rehabilitation (black circles). Starting ~3 months post injury, the examinations E0 to E5 covered a follow-up interval of 5 months. Box–whisker plots indicate results (medians, quartiles) of matched control children (C) at E0 and E5. Adapted from Kuhtz-Buschbeck *et al.* (2003a).

(Mann–Whitney U-test, $P < 0.01$). Changes in the DHFT, i.e. faster functional movements of either hand, were correlated with improvements of gross motor proficiency (GMFM) and of an index of daily living activities (Barthel Index; Mahoney & Barthel, 1965; Loewen & Anderson, 1988). Younger age at injury (range 4–15 years) was neither combined with faster recovery nor with smaller case–control differences at any examination date. The control children's fine motor performance changed significantly ($P < 0.05$), but their change rate was slow compared with the patients' recovery rate (Figure 24.2).

Kinematic analysis of prehension and precision grip forces

Reach-to-grasp movements

As a relevant natural hand movement, we analyzed the kinematics of prehension in a subgroup of the young patients after TBI. It consisted of 13 boys and 6 girls with a mean age of ~10 years (range 4–14) at the date of the first examination (E0), which took place about 3 months after injury (Holzhäuser, 2006). The cameras of the motion analysis system (Qualisys, Gothenburg, Sweden) were positioned and calibrated for the examination of the more paretic and/or more affected side, relying on the clinical examination. In controls, the non-dominant hand was examined. Normal developmental kinematic data of reach-to-grasp movements have been published previously (Kuhtz-Buschbeck *et al.*, 1998, 1999).

The children sat in an adjustable chair, the relevant hand resting in a semi-prone posture on a table at a starting point. Thumb and index finger were in a pinched position, lightly touching a small knob. Upon a start signal, the children reached out to grasp and lift cylindrical target objects with the thumb and index finger, often stabilizing with the middle finger (tripod grip). The objects were carried to the starting point and then released. We asked the children to move with their normal comfortable speed, as if "reaching for a building brick on the table." To obtain equivalent conditions for different individuals, the distance of the object from the starting point (60% of arm length) and the object size (diameter: 10% of finger span) were scaled to body proportions. Coordinates of three reflective markers attached to the wrist and to the nails of the thumb and index finger were recorded with a time resolution of 20 ms. Thereby velocity profiles showing the acceleration and deceleration of the reaching hand, and grip aperture curves displaying the simultaneous opening–closing sequence of the grasping fingers were recorded. To illustrate the coordination of both movement components, i.e. of reaching and grasping, hand velocity was plotted against grip aperture (Figure 24.3) (see also Chapter 2).

One session of 12 prehension trials was performed with visual feedback, another session in a no-vision condition (lights off after the start signal). Mean values of the following parameters were calculated for every child and testing session: reaction time from start signal to movement onset, movement duration from movement onset until the object was lifted, peak velocity of the hand, intra-individual variability (standard deviations) of the aforementioned variables, peak height of the wrist, maximum grip aperture, and an index describing the straightness of the path of the reaching hand, i.e. the ratio of the movement trajectory length and the distance between starting point and target object. Furthermore, we measured the timing of peak hand velocity, peak deceleration and maximum aperture.

Irregular kinematic profiles can illustrate a reduced movement quality, which otherwise is described in qualitative clinical terms, such as coordination deficit, dysmetria or ataxia. Examples of such kinematic profiles are shown in Figure 24.3. Patient A was a 12-year-old boy who had incurred lesions of the right frontal and left temporal lobes, with subarachnoid hemorrhage and enlargement of the ventricles as secondary complications. His reach-to-grasp movements were very slow, as the narrow spacing of the data points indicate, but the coordination between reaching (hand velocity) and grip formation was well preserved. Patient B was a 7-year-old girl with lesions of the left thalamus and internal capsule (bilateral, left > right), subarachnoid and intraventricular hemorrhage and multiple cortical contusions of both the frontal lobes. Her prehension movements were highly ataxic, as reflected by the irregular kinematic profiles. Comparison of the collective patient and control group results revealed significant differences ($P < 0.05$): Reaction time and

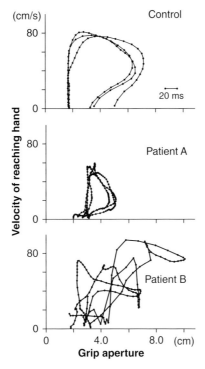

Figure 24.3. Coordination of reaching and grasping. Kinematic profiles of a control subject and of two patients (A and B) after TBI. Hand velocity was plotted against the grip aperture with a sampling rate of 50 Hz (20 ms, dots). Note the disrupted coordination of patient B. Three trials are superimposed for each subject. Adapted from Kuhtz-Buschbeck *et al.* (2003b).

movement duration were prolonged, the reaching trajectories more curved, and the intra-individual variability from trial to trial was clearly higher in TBI patients than controls at the first (E0) examination date (Figure 24.4A,B). The brain-injured children had not yet regained stable hand movement patterns at the onset of follow-up. Deficits were present both in the visual and in the no-vision condition (Holzhäuser, 2006). The grip aperture of the TBI patients was somewhat higher than in controls, but this result did not reach significance. Recovery during the 5-month follow-up period was characterized by significant decreases of the intra-individual variability and reaction time. The temporal structure of the movements was altered in TBI patients. At E0, mainly the last part of the prehension movements (approach phase), between the moments of peak grip aperture and object lifting, was prolonged (Figure 24.4C). This time interval became shorter and approached control values during follow-up. Hence the brain-injured children started to reach out fairly rapidly, but then had difficulties in grasping and lifting the object. These difficulties were alleviated during rehabilitation (from E0 to E5). Congruent results were obtained for the first (preparation) phase of the precision grip–lift task, which is described in the next section.

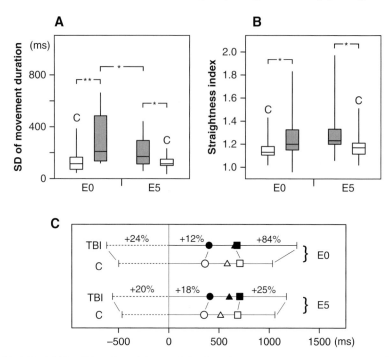

Figure 24.4. Variability (A) and straightness (B) of reach-to-grasp movements. Results of children after TBI (gray-shade) and matched control subjects (C; open) at the onset (E0) and end (E5) of follow-up. Asterisks indicate significant differences (*$P < 0.05$, **$P < 0.01$) between patients and controls (Mann–Whitney U-tests) and, respectively, change during patient rehabilitation (Wilcoxon tests). Box–whisker plots denote medians, quartiles, ranges. Standard deviation of movement duration describes intra-individual variability of repetitive prehension trials. (C) Temporal structure of the movements in children after TBI (black symbols) and in controls (C, open symbols) at examinations E0 and E5. Movement onset is at 0 ms. Median values of reaction time (broken lines), movement duration (solid lines), time to peak hand velocity (circles), peak deceleration (triangles), peak grip aperture (squares) are indicated. Percentages denote the relative prolongations of reaction time, acceleration phase and approach phase in brain-injured children compared with controls.

Precision grip

Analysis of manipulative forces of the precision grip–lift task is an established method for the quantitative assessment of fine motor skills (see Chapter 1). In the pediatric age group, this method has been extensively used, e.g. in individuals with cerebral palsy (Eliasson *et al.*, 1991, 2006; Forssberg *et al.*, 1999) (see also Chapter 31) and to study normal motor development (Forssberg *et al.*, 1991) (see also Chapter 17). To characterize the impairment and recovery of the precision grip in children after moderate and severe TBI (Gölge *et al.*, 2004), we investigated a subsample of the above-mentioned patient group, namely 8 boys and 5 girls (age range 5–14 years). The posture of the children was adjusted so that the forearm was approximately parallel to the table surface during the grip–lift task, which was performed with the more paretic or more affected hand. Control data were obtained from 13 matched healthy children, whose non-dominant hand was examined. After an acoustic signal, a grip instrument of 200 g weight was grasped and lifted with a precision grip of the thumb and index finger. Two parallel grip surfaces spaced 4 cm apart and covered with suede were located at the top of the instrument. The task was carefully explained and demonstrated. Yet, several TBI patients and young control children additionally used the middle finger to stabilize their grip, which was allowed. The instrument was lifted to a height of about 10 cm and held there stationary for about 5 seconds, then replaced on the table and released. This was repeated 15 times per examination. Data from the initial five trials were discarded to exclude early learning effects; data from the following ten trials were averaged to obtain representative mean values for each child and examination date. The grip force and load force (vertical lifting force) were measured with strain-gauge transducers, sampled at 400 Hz with a resolution of 0.05 N, and evaluated interactively with the SC/ZOOM data-acquisition and analysis system (Department of Physiology, University of Umeå, Sweden). The vertical position of the grip instrument was recorded as well.

Temporal variables included the preparation phase (interval between contacts of the thumb and index finger), preload duration (from completion of the grip until onset of a positive load force), and loading duration (from onset of positive load force until the object started to move). Grip-force variables were analyzed separately for the index finger and thumb and included the peak grip force, the mean grip force during static holding, and its standard deviation, which describes oscillations of the isometric force. Load-force parameters comprised the maximal negative and the peak positive load forces. A negative load force indicates that the grip instrument is pressed down onto the table before it is lifted. Further details are given in Gölge *et al.* (2004).

At the onset of follow-up (E0), all grip-force variables differed significantly between patients and controls ($P < 0.05$). The preparation phase of the patients, i.e. the time needed to establish the grip between thumb and index finger, was prolonged (Figure 24.5A). No significant differences were found for the other two temporal parameters (preload and load duration) and for the load-force parameters. However, the TBI children exhibited exaggeratedly high grip forces that fluctuated during static holding, and varied from trial to trial (Figure 24.5B). Only a trend towards increased negative load forces was found in the TBI group. Performance of the precision grip–lift task improved markedly ($P < 0.05$) during rehabilitation. At the end of follow-up (E5), no significant differences between controls and TBI patients were detected but one: the preparation phase of the brain-injured children was still prolonged (Figure 24.5A). Recovery of grasping was paralleled by improvements of functional independence and of gross motor proficiency and gait. In the control group, all grip- and load-force parameters remained stable over the follow-up period of 5 months. Yet the preparation and preload phases shortened significantly, which can be attributed to physiological maturation (Forssberg *et al.*, 1991) (see also Chapter 17). This underscores the importance of parallel repeated examinations of age-matched normal controls. As with the hand function tests (DHFT and Purdue Pegboard), we found no significant correlation between the recovery of grasping (reach-to-grasp, precision grip) and age of the children at

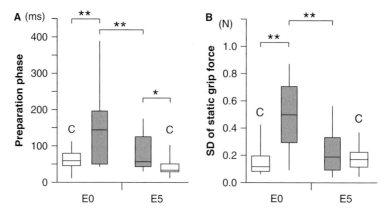

Figure 24.5. Precision grip–lift task. Duration of the preparation phase (A) and variability (SD) of the grip force (B) during static holding. Data from children after TBI (gray-shade) are compared with results of matched control subjects (C; open) at the onset (E0) and end (E5) of follow-up. Box–whisker plots show medians, quartiles and ranges. Significant differences (*$P<0.05$, **$P<0.01$) between patients and controls (Mann–Whitney U-tests) and improvements during patient rehabilitation (Wilcoxon tests) are indicated. Adapted from Gölge *et al.* (2004).

TBI. While age at injury was not relevant, lower initial GCS values and a higher number of affected brain regions were associated with stronger sensorimotor deficits (Kuhtz-Buschbeck *et al.*, 2003a; Gölge *et al.*, 2004; Holzhäuser, 2006).

Discussion

Quantitative analyses of grasping objectify performance and may help to detect subtle characteristics that are not clearly visible (Behbehani *et al.*, 1990). Appropriate measures that are useful for follow-up studies can be identified, as demonstrated for the recovery of prehension movements and of the precision grip in brain-injured children and adolescents. To our knowledge, comparable studies of grasping in adult TBI patients are lacking so far. Despite the advantages of the kinematic and force analyses, some drawbacks have to be considered. The amount of instrumentation, the preparation of the set-up, the time-demanding data evaluation and cost factors impede a more widespread use of such methods. Because of technical and time limitations, it was, for example, not possible to study reach-to-grasp movements of both hands in the TBI patients. The need to focus on selected tasks (e.g. precision grip) restricts the number of eligible patients who can participate and may impede the detection of changes in motor proficiency, since only a limited scope of the motor repertoire can be covered (Kuhtz-Buschbeck *et al.*, 2003b). Comprehensive tests of gross motor proficiency (GMFM) and rather simple hand function tests (Purdue Pegboard, DHFT) were sensitive enough to reveal deficits and recovery of sensorimotor functions after TBI (see Figure 24.2); the additional inclusion of the quantitative analyses did not improve sensitivity.

Although simple hand function tests such as the DHFT can detect a lack of motor speed after TBI, their results neither reflect movement quality nor give insight into pathophysiological mechanisms of this slowness. Similarly, the time that a runner needs for a distance reflects her/his velocity, but does not say anything about the running style. A precise survey of the actual performance is one advantage of the kinematic and force analyses. The reach-to-grasp movements of the brain-injured patients were characterized by a high intra-subject variability and curved movement trajectories, which will increase movement time. In particular the time needed for the final approach, between the time of maximum grip aperture and object lifting, was lengthened, indicating that the TBI children had problems to grasp the objects smoothly and securely, even if visual control was provided (Holzhäuser, 2006). The prolonged time interval between the contacts of the thumb and index finger with the respective grip surfaces in the precision grip–lift task points towards the same difficulty. This time interval was the only precision grip variable which did not reach the range of control values during follow-up (see Figure 24.5A). Moreover, the grip forces were exaggerated, and force profiles were variable and irregular after TBI (Gölge *et al.*, 2004). Since similar deficits in grip-force control have been reported previously in children with various pathologies such as cerebral palsy (Forssberg *et al.*, 1999; Eliasson *et al.*, 2006) (see also Chapter 31), attention-deficit hyperactivity disorder (Pereira *et al.*, 2000), and myelomeningocele associated with hydrocephalus (Gölge *et al.*, 2003), they seem not to be specific for a particular brain lesion. An overshoot of the grip force may be a general strategy to improve grip stability (Eliasson *et al.*, 1991) (see also Chapter 19). All these factors will hamper performance in hand-function tests where objects (e.g. pegs) have to be grasped, transported and released in rapid succession. Deficits of the load force control were not conspicuous in the TBI patients. The load force is primarily produced by contractions of muscles acting on the elbow and shoulder joints, while the grip force of the precision grip is mainly generated by contractions of more distal forearm and intrinsic hand muscles (Westling & Johansson, 1984). Therefore control of distal muscles seemed to be more affected than control of proximal muscles after TBI. Taken together, brain-injured children needed more time to approach and to get hold of an object, and then applied excessive and variable grip forces. It is conceivable also that the release of objects from the grip will be prolonged (Gordon *et al.*, 2003). The preparation phase and the various grip-force parameters were useful to distinguish between patients and controls, and were sensitive enough to document improvements of grasping (Gölge *et al.*, 2004).

By using the Brunnstrom Stages of Recovery to describe the level of paresis (Brunnstrom, 1966), Katz *et al.* (1998) characterized the recovery of arm function in 44 patients after TBI (age range 10–86 years). Stage 1 corresponds to a flaccid paralysis (no active movement possible), stage 7 to normal motor function. Thirty-six patients recovered to stages 5 or 6 by 6 months post TBI. The time needed to recover was best predicted by injury severity (duration of unconsciousness and post-traumatic amnesia) and initial level of paresis, but not by age. Patients with diffuse axonal injury tended towards a more protracted recovery than patients with focal injury. Recent studies of cerebral activation and blood flow have contributed to the understanding of neural mechanisms underlying motor recovery after

TBI. Lotze *et al.* (2006) used functional MRI (fMRI) to examine brain activity in adult patients, who performed a repetitive power grip task about 3 months post injury. Compared with control subjects, the injured patients showed a diminished fMRI signal change in the primary sensorimotor cortex, dorsal premotor cortex and supplementary motor cortex. A decreased signal change of the contralateral primary sensorimotor cortex during movements of the more paretic hand, which implies a reduced local cerebrovascular response, was associated with poor recovery (Lotze *et al.*, 2006). In congruence, Emanuelson *et al.* (1997) detected cerebral hypoperfusion after TBI with single-photon emission computed tomography (SPECT). Slowness after TBI may in part be due to defective motor planning and preparation. Electrophysiological and fMRI studies have disclosed that traumatic lesions of the prefrontal cortex impede the preparation of self-initiated finger movements, and trigger a long-lasting reorganization of the preparatory motor network (Wiese *et al.*, 2004, 2006).

Studies of motor recovery after severe TBI, especially with brain-injured children, bear several problems. Patient compliance can be a limiting factor when tests are attentionally demanding, strenuous or monotonous (e.g. repetitive movements). Cognitive and psychosocial deficits have been considered a more pervasive problem than motor disabilities after TBI (Emanuelson *et al.*, 1998; Fyrberg *et al.*, 2007). The complexity of the brain injuries, which often involve multiple contusions, diffuse axonal injuries and secondary complications (e.g. brain edema), makes it very difficult to assign specific functional losses to focal damage of brain tissue (Kuhtz-Buschbeck *et al.*, 2003a). Circumscribed unifocal lesions of the visuomotor circuit of prehension (see Chapter 21) were not found in our patient collective. Other confounding factors such as additional peripheral injuries, post-traumatic epilepsy and differences in the grade of trauma severity and premorbid development must be considered. Since the recovery rate is high in the first 6–12 months post TBI, but decreases markedly thereafter, a follow-up interval has to be chosen that is long enough to observe improvements (Jaffe *et al.*, 1993, 1995; Massagli *et al.*, 1996). One year after TBI, when recovery approaches a plateau, an interval of 2 months may be too short to detect significant changes of grasping (Kuhtz-Buschbeck *et al.*, 2003b). Moreover, normal maturation of control subjects has to be accounted for. Recovery is not merely characterized by improving performance, but rather by decreasing case–control differences, which indicate that the brain-injured children start to catch up with their non-injured peers.

Based on experimental lesions of the motor cortex in newborn and adult monkeys, the "Kennard principle" postulates that the plasticity of the developing brain allows for good recovery after early lesions (Kennard, 1936). However, young age at injury was not associated with better recovery in the studies of brain-injured children outlined above, which covered the age range of 4–15 years (Kuhtz-Buschbeck *et al.*, 2003a, 2003b; Gölge *et al.*, 2004; Holzhäuser, 2006). A long-term outcome study (23 years after TBI) of 159 children with head injuries, whose age at injury had been 3–16 years, came to the same conclusion (Klonoff *et al.*, 1993). Others even report a tendency towards a worse outcome in younger individuals (Koskiniemi *et al.*, 1995; Benz *et al.*, 1999; Laurent-Vannier *et al.*, 2000). If a brain lesion slows down the ability to learn new skills, the consequences will be particularly deleterious for young children, who can recover fewer previously acquired skills. Detailed studies of sensorimotor functions (such as grasping) in larger, stratified patient collectives may capture the different factors which influence recovery after TBI in more detail.

References

Andersson, E. H., Björklund, R., Emanuelson, I. & Stahlhammer, D. (2003). Epidemiology of traumatic brain injury: a population based study in Western Sweden. *Acta Neurolog Scand*, **107**, 256–259.

Asikainen, I., Nybo, T., Muller, K., Sarna, S. & Kaste, M. (1999). Speed performance and long-term functional and vocational outcome in a group of young patients with moderate or severe traumatic brain injury. *Eur J Neurol*, **6**, 179–185.

Behbehani, K., Kondraske, G. V., Tintner, R., Tindall, R. A. S. & Imrhan, S. N. (1990). Evaluation of quantitative measures of upper extremity speed and coordination in healthy persons and in three patient populations. *Arch Phys Med Rehab*, **71**, 106–111.

Benz, B., Ritz, A. & Kiesow, S. (1999). Influence of age-related factors on long-term outcome after traumatic brain injury (TBI) in children: a review of recent literature and some preliminary findings. *Rest Neurol Neurosci*, **14**, 135–141.

Besenski, N. (2002). Traumatic injuries: imaging of head injuries. *Eur Radiol*, **12**, 1237–1252.

Bigler, E. D., Ryser, D. K., Gandhi, P., Kimball, J. & Wilde, E. A. (2006). Day-of-injury computerized tomography, rehabilitation status, and development of cerebral atrophy in persons with traumatic brain injury. *Am J Phys Med Rehab*, **85**, 793–806.

Brunnstrom, S. (1966). Motor testing procedures in hemiplegia: based on sequential recovery stages. *Phys Ther*, **46**, 357–375.

Chaplin, D., Deitz, J. & Jaffe, K. M. (1993). Motor performance in children after traumatic brain injury. *Arch Phys Med Rehab*, **74**, 161–164.

Drouin, L. M., Malouin, F., Richards, C. L. & Marcoux, S. (1996). Correlation between the gross motor function measure scores and gait spatiotemporal measures in children with neurological impairments. *Dev Med Child Neurol*, **38**, 1007–1019.

Eliasson, A. C., Gordon, A. M. & Forssberg, H. (1991). Basic co-ordination of manipulative forces of children with cerebral palsy. *Dev Med Child Neurol*, **33**, 661–670.

Eliasson, A. C., Forssberg, H., Hung, Y. C. & Gordon, A. M. (2006). Development of hand function and precision grip control in individuals with cerebral palsy: a 13-year follow-up study. *Pediatrics*, **118**, 1226–1236.

Emanuelson, I., von Wendt, L., Bjure, J., Wiklund, L. M. & Uvebrant, P. (1997). Computed tomography and single-photon emission computed tomography as diagnostic tools in acquired brain injury among children and adolescents. *Dev Med Child Neurol*, **39**, 502–507.

Emanuelson, I., von Wendt, L., Beckung, E. & Hagberg, I. (1998). Late outcome after severe traumatic brain injury in children and adolescents. *Ped Rehab*, **2**, 65–70.

Firsching, R., Woischneck, D., Klein, S., Ludwig, K. & Döhring, W. (2002). Brain stem lesions after head injury. *Neurol Res*, **24**, 145–146.

Forssberg, H., Eliasson, A. C., Kinoshita, H., Johansson, R. S. & Westling, G. (1991). Development of human precision grip. I: Basic coordination of force. *Exp Brain Res*, **85**, 451–457.

Forssberg, H., Eliasson, A. C., Redon-Zouitenn, C., Mercuri, E. & Dubowitz, L. (1999). Impaired grip-lift synergy in children with unilateral brain lesions. *Brain*, **122**, 1157–1168.

Fyrberg, A., Marchioni, M. & Emanuelson, I. (2007). Severe acquired brain injury: rehabilitation of communicative skills in children and adolescents. *Int J Rehab Res*, **30**, 153–157

Gaetz, M. (2004). The neurophysiology of brain injury. *Clin Neurophysiol*, **115**, 4–18.

Gardner, R. A. & Broman, M. (1979). The Purdue Pegboard: Normative data on 1334 school children. *J Clin Child Psychol*, **1**, 156–162.

Gennarelli, T. A., Thibault, L. E., Adams, J. H. *et al.* (1982). Diffuse axonal injury and traumatic coma in the primate. *Ann Neurol*, **12**, 564–574.

Gölge, M., Schütz, C., Dreesmann, M. *et al.* (2003). Grip force parameters in precision grip of individuals with myelomeningocele. *Dev Med Child Neurol*, **45**, 249–256.

Gölge, M., Müller, M., Dreesmann, M. *et al.* (2004). Recovery of the precision grip in children after traumatic brain injury. *Arch Phys Med Rehab*, **85**, 1435–1444.

Gordon, A. M., Lewis, S. R., Eliasson, A. C. & Duff, S. V. (2003). Object release under varying task constraints in children with hemiplegic cerebral palsy. *Dev Med Child Neurol*, **45**, 240–248.

Haley, S. M., Cioffi, M. I., Lewin, J. E. & Baryza, M. J. (1990). Motor dysfunction in children and adolescents after traumatic brain injury. *J Head Trauma Rehab*, **5**, 77–90.

Hillier, S. L., Hiller, J. E. & Metzer, J. (1997). Epidemiology of traumatic brain injury in South Australia. *Brain Injury*, **11**, 649–659.

Holzhäuser, M. (2006). Optoelektronische Analyse gezielter Greifbewegungen bei Kindern und Jugendlichen nach Schädel-Hirn-Trauma. Unpublished M.D. thesis (in German). Medizinische Fakultät der Christian-Albrechts-Universität zu Kiel.

Jaffe, K. M., Fay, G. C., Polissar, N. L. *et al.* (1993). Severity of pediatric traumatic brain injury and neurobehavioral recovery at one year – a cohort study. *Arch Phys Med Rehab*, **74**, 587–595.

Jaffe, K. M., Polissar, N. L., Fay, G. C. & Liao, S. (1995). Recovery trends over three years following pediatric traumatic brain injury. *Arch Phys Med Rehab*, **76**, 17–26.

Jebsen, R. H., Taylor, N., Trieschmann, R. B., Trotter, M. J. & Howard, L. A. (1969). An objective and standardized test of hand function. *Arch Phys Med Rehab*, **50**, 311–319.

Jonsson, C. A., Horneman, G. & Emanuelson, I. (2004). Neuropsychological progress during 14 years after severe traumatic brain injury in childhood and adolescence. *Brain Injury*, **18**, 921–934.

Katz, D. I., Alexander, M. P. & Klein, R. B. (1998). Recovery of arm function in patients with paresis after traumatic brain injury. *Arch Phys Med Rehab*, **79**, 488–493.

Kennard, M. A. (1936). Age and other factors in motor recovery from precentral lesions in monkeys. *Am J Physiol*, **115**, 138–146.

Klonoff, H., Clark, C. & Klonoff, P. S. (1993). Long-term outcome of head injuries: a 23 year follow up study of children with head injuries. *J Neurol Neurosurg Psychiatry*, **56**, 410–415.

Koskiniemi, M., Kyykkä, T., Nybo, T. & Jarho L. (1995). Long-term outcome after severe brain injury in preschoolers is worse than expected. *Arch Pediatr Adolesc Med*, **149**, 249–254.

Kuhtz-Buschbeck, J. P., Stolze, H., Jöhnk, K., Boczek-Funcke, A. & Illert, M. (1998). Development of prehension movements in children: a kinematic study. *Exp Brain Res*, **122**, 424–432.

Kuhtz-Buschbeck, J. P., Boczek-Funcke, A., Illert, M., Jöhnk, K. & Stolze, H. (1999). Prehension movements and motor development in children. *Exp Brain Res*, **128**, 65–68.

Kuhtz-Buschbeck, J. P., Hoppe, B., Gölge, M. *et al.* (2003a). Sensorimotor recovery after traumatic brain injury: analyses of gait, gross motor, and fine motor skills. *Dev Med Child Neurol*, **45**, 821–828.

Kuhtz-Buschbeck, J. P., Stolze, H., Gölge, M. & Ritz, A. (2003b). Analyses of gait, reaching, and grasping in children after traumatic brain injury. *Arch Phys Med Rehab*, **84**, 424–430.

Laurent-Vannier, A., Brugel, D. G. & DeAgostini, M. (2000). Rehabilitation of brain-injured children. *Child Nerv System*, **16**, 760–764.

Loewen, S. C. & Anderson, B. A. (1988). Reliability of the Modified Motor Assessment Scale and the Barthel Index. *Phys Ther*, **68**, 1077–1081.

Lotze, M., Grodd, W., Rodden, F. A. *et al.* (2006). Neuroimaging patterns associated with motor control in traumatic brain injury. *Neurorehab Neural Repair*, **20**, 14–23.

Mahoney, F. I. & Barthel, D. W. (1965). Functional evaluation: the Barthel Index. *Maryland State Med J*, **14**, 61–65.

Massagli, T. L., Jaffe, K. M., Fay, G. C. *et al.* (1996). Neurobehavioral sequelae of severe pediatric traumatic brain injury: a cohort study. *Arch Phys Med Rehab*, **77**, 223–231.

Möllmann, F. T. (2006). Epidemiologie, Unfallursachen und akutklinische Initialversorgung beim Schädel-Hirn-Trauma. Unpublished M.D. thesis (in German). Medizinische Fakultät der Westfälischen Wilhelms-Universität Münster.

Ommaya, A. & Gennarelli, T. (1974). Cerebral concussion and traumatic unconsciousness: correlation of experimental and clinical observations on blunt head injuries. *Brain*, **97**, 633–654.

Pereira, H. S., Eliasson, A. C. & Forssberg, H. (2000). Detrimental neural control of precision grip lifts in children with ADHD. *Dev Med Child Neurol*, **42**, 5454–5553.

Reddon, J. R., Gill, D. M., Gauk, S. E. & Maerz, M. D. (1988). Purdue Pegboard: Test-retest estimation. *Percept Motor Skills*, **66**, 503–506.

Rossi, C. & Sullivan, S. J. (1996). Motor fitness in children and adolescents with traumatic brain injury. *Arch Phys Med Rehab*, **77**, 1062–1065.

Russell, D. J., Rosenbaum, P. L., Cadman, D. T. *et al.* (1989). The gross motor function measure: a means to evaluate the effects of physical therapy. *Dev Med Child Neurol*, **31**, 341–352.

Servadei, F., Antonelli, V., Betti, L. *et al.* (2002). Regional brain injury epidemiology as the basis for planning brain injury treatment. The Romagna (Italy) experience. *J Neurosurg Sci*, **46**, 111–119.

Taylor, N., Sand, P. L. & Jebsen, R. H. (1973). Evaluation of hand function in children. *Arch Phys Med Rehab*, **54**, 129–135.

Teasdale, G. M. & Jennett, B. (1974). Assessment of coma and impaired consciousness: practical scale. *Lancet*, **2**, 81–84.

Wallen, M. A., Mackay, S., Duff, S. M., McCartney, L. C. & O'Flaherty, S. J. (2001). Upper-limb function in Australian children with traumatic brain injury: a controlled, prospective study. *Arch Phys Med Rehab*, **82**, 642–649.

Westling, G. & Johansson, R. S. (1984). Factors influencing the force control during precision grip. *Exp Brain Res*, **53**, 277–284.

Wiese, H., Stude, P., Nebel, K. *et al.* (2004). Recovery of movement-related potentials in the temporal course after prefrontal traumatic brain injury: a follow-up study. *Clin Neurophysiol*, **115**, 2677–2692.

Wiese, H., Tönnes, C., de Greiff, A. *et al.* (2006). Self-initiated movements in chronic prefrontal traumatic brain injury: an event-related functional MRI study. *Neuroimage*, **30**, 1292–1301.

25

Focal hand dystonia

SARAH PIRIO RICHARDSON AND MARK HALLETT

Summary

Imprecise and unwanted movements characterize the clinical presentation of focal hand dystonia. It is often task-specific, manifesting as writer's cramp or musician's dystonia (e.g. guitarist's cramp). Repetitive, stereotyped movements play a role in the development of the dystonia, but clearly, a pathophysiological substrate must be present for the disorder to manifest. This substrate is likely due to genetics; however, the exact genetic abnormality is not yet known. Although presenting as a motor disorder, dystonia is also a sensory disorder with subtle abnormalities found in spatial and temporal discrimination and with disordered sensory cortical maps. Abnormal cortical plasticity and a failure of homeostatic mechanisms also are seen in dystonia. Finally, a loss of inhibition from excessive muscle discharge to alterations in cortical circuits has been identified in dystonia. As a result, abnormal sensorimotor integration, abnormal plasticity and a loss of inhibition all are implicated in the pathophysiology of focal hand dystonia. Currently, it is not known which of these pathophysiological abnormalities is primary or secondary to the disorder development. Treatment strategies are aimed at ameliorating these physiological changes by improving the sensory deficit, normalizing plasticity and restoring inhibition.

Focal hand dystonia

Dystonia, a neurological disorder, is characterized by abnormal posturing due to sustained muscle contractions, which interferes with the normal performance of motor tasks (Hallett, 2004). Dystonia can be classified by age at onset, by distribution and by cause (Tarsy & Simon, 2006). When the dystonia is restricted to a limb it is called focal limb dystonia (such as in a foot or hand). The development of focal hand dystonia (FHD) appears to be due to a combination of both environmental and genetic influences (Chen & Hallett, 1998; Defazio et al., 2007). Stereotyped and repetitive behavior play a role; however, as in the case of writer's cramp, clearly, excessive writing cannot be the sole source of the dystonia. A genetic predisposition is likely to play a role in the development of dystonia; however, the exact nature of the abnormality is not known currently but it is unlikely to be a simple Mendelian trait (Defazio et al., 2003, 2007).

Sensorimotor Control of Grasping: Physiology and Pathophysiology, ed. Dennis A. Nowak and Joachim Hermsdörfer. Published by Cambridge University Press. © Cambridge University Press 2009.

Figure 25.1. Dystonic hand posture with pen grasp and handwriting in a patient with writer's cramp.

One of the particularly interesting aspects of focal limb dystonia – especially of the hand – is the task-specificity of the dystonia. For instance, in a task-specific dystonia of the hand known as writer's cramp, handwriting is abnormal due to posturing and muscle spasm, whereas other tasks done with the affected hand are normal. Grasping might be impaired as part of this task specificity. Some patients with writer's cramp, for example, develop dystonia as soon as they grasp the pen or pencil (Figure 25.1). Other grasping movements would be normal. Other examples of a task-specific FHD include pianist's and guitarist's cramp (Frucht, 2004) and even a recently reported task-specific dystonia involving use of a spoon (Song *et al.*, 2007). Although over time the "task-specificity" may be lost, and the dystonia can involve more tasks and become apparent even at rest, it usually remains restricted to the affected limb and does not generalize to other limbs. (Weiss *et al.*, 2006). In some instances, it can involve the contralateral limb if, due to a dystonia in the dominant limb, the task of writing is transferred to the non-dominant limb.

It is not at all apparent how the motor system functions nearly normally in the affected limb for many actions, but is impaired visibly in a particular task. This implies a dysfunctional interaction between a particular effector (i.e. the affected hand) and a particular motor control program (i.e. writing), and a normal interaction between the affected effector and other motor programs (i.e. using utensils and typing). One possible explanation for this dysfunctional interaction is impaired surround inhibition. Surround inhibition is a well-known brain mechanism in visual and sensory systems and likely plays a role in the motor system. The ability to selectively activate particular muscles to perform a specific task likely is generated by suppressing the excitability of a neural network surrounding an activated network (Sohn & Hallett, 2004a). There is evidence that surround inhibition is disturbed in FHD (Sohn & Hallett, 2004b). However, the mechanisms responsible for this phenomenon are not well understood.

Currently, the pathophysiology of focal hand dystonia is characterized primarily by abnormal sensorimotor integration (Abbruzzese & Berardelli, 2003), loss of plasticity (Quartarone *et al.*, 2005; Weise *et al.*, 2006) and loss of inhibition – both in the motor system (Hallett, 2004) and in the somatosensory system (Tinazzi *et al.*, 2003). Which of these pathophysiological hallmarks of dystonia are primary to the disease, secondary or compensatory is not known. There is growing evidence from animal studies that abnormal sensorimotor integration is a key feature that links the various physiological findings together (Evinger, 2005).

Abnormal sensorimotor integration

Abnormal sensorimotor integration has been found in FHD (Rosenkranz *et al.*, 2000; Abbruzzese *et al.*, 2001). It is unclear, however, whether the abnormality manifests due to an aberrant brain response to afferent stimulation or whether the afferent stimulation is a trigger itself (Abbruzzese & Berardelli, 2003).

Clinically, there is no apparent sensory loss in focal dystonia; however, subtle spatial and temporal discrimination abnormalities have been found in patients with FHD, cervical dystonia (CD) and blepharospasm (BSP) (Molloy *et al.*, 2003). In a study with CD and BSP patients, kinesthesia was also found to be abnormal (Putzki *et al.*, 2006). The identification of abnormality of sensation beyond the symptomatic body parts indicates that the sensory abnormality could not be a consequence of abnormal learning, but is more likely a pre-existing physiological state. Studies of asymptomatic carriers of a mutation in the *TOR1A* gene for early-onset primary dystonia (DYT1) show abnormal temporal discrimination, strengthening the argument for a pre-existing physiological susceptibility to dystonia (Fiorio *et al.*, 2007).

Physiological studies, namely somatosensory evoked potentials (SEP), also show sensory dysfunction in FHD. The dipoles of the N20 from stimulation of individual fingers show disordered representation in the primary sensory cortex (Bara-Jimenez *et al.*, 1998). These abnormalities are found on both the affected and unaffected hemispheres of patients with focal hand dystonia, indicating that the sensory abnormality likely represents an endo-phenotypic trait rather than a consequence of repetitive activity (Meunier *et al.*, 2001).

Imaging studies have shown both structural and functional alterations in focal hand dystonia in sensory areas. Echoing the disordered sensory somatotopy seen physiologically, an fMRI study showed dedifferentiation of the cortico-subcortical sensorimotor map in the putamen contralateral to the affected limb (Delmaire *et al.*, 2005). Further, in a position emission tomography (PET) study, the sensory cortex of FHD patients was more activated than normal with writing and when the dystonia symptoms were more severe (Figure 25.2)

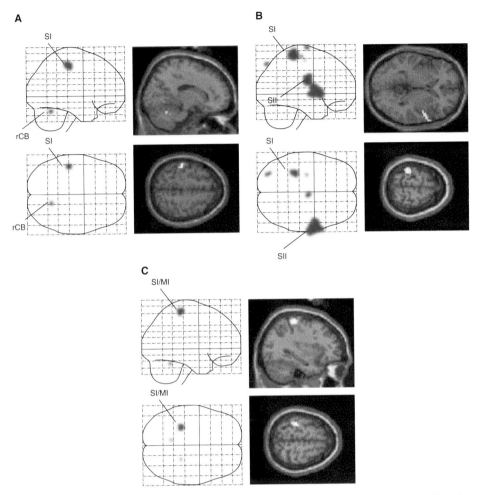

Figure 25.2. Positron emission tomography (PET) findings in FHD patients. A. Significantly increased left SI activity seen during a writing task in patients with FHD compared with controls. Parts (B) and (C) show parametric correlations of the images with (B) symptoms or (C) physiologic measures of the severity of the dystonia during acquisition of the images. B. In seven patients who developed dystonic symptoms during the writing task, increased left SI activity correlated significantly with more severe subjective dystonia scores. C. A trend towards overactivity in the left SI/MI correlated with increased EMG activity in flexor digitorum superficialis during a tapping task in the FHD patients. rCB, right cerebellum; MI, primary motor area; SI, primary sensory area; SII, secondary sensory area. (Reproduced from Lerner *et al.* (2004) with permission.)

(Lerner *et al.*, 2004). Finally, fMRI studies show an increase in basal ganglia activation in patients when doing spatial discrimination assessment – theoretically due to a loss of surround inhibition in sensory processing (Peller *et al.*, 2006). Structurally, voxel-based morphometry studies in patients with focal hand dystonia show an increase in gray matter in the primary sensory cortex (Garraux *et al.*, 2004). The physiology and the imaging findings together suggest that while the clinical manifestations of FHD appear to be solely a motor disorder, clearly dystonia is also a sensory disorder.

Not only is sensory processing abnormal, but the integration of sensory input with motor output is also affected. One piece of evidence comes from evoked potential studies during a reaction time task (Murase *et al.*, 2000). In this experiment, a median nerve stimulus was used to trigger a movement. In patients with FHD, the N30 peak was not gated unlike in healthy volunteers.

Other evidence comes from studies of the influence of a sensory stimulus on the motor evoked potential (MEP) induced by transcranial magnetic stimulation (TMS) (Abbruzzese *et al.*, 2001). Electrical peripheral nerve stimulation in the hand approximately 20 ms (short afferent inhibition, SAI) or approximately 200 ms (long afferent inhibition, LAI) prior to contralateral M1 TMS results in consistent motor inhibition of hand muscles (Classen *et al.*, 2000). Motor cortical inhibition due to electrical stimulation of a digit near to a target muscle (i.e. homotopic) seems to be stronger than the inhibition due to stimulation of a digit distant from a target muscle (i.e. heterotopic) (Classen *et al.*, 2000). Homotopic SAI at rest and during tonic movement was also examined in patients with FHD (Kessler *et al.*, 2005). Homotopic SAI was normal in patients with FHD (specifically writer's cramp) during tonic contraction of abductor pollicis brevis and at rest, but was followed by abnormal facilitation during abductor pollicis brevis relaxation. For LAI at rest, in patients with FHD, the inhibition is converted to facilitation (Abbruzzese *et al.*, 2001). These studies imply a loss of inhibition, which could conceivably be the most fundamental disorder in dystonia.

Abnormal plasticity

Plasticity in the nervous system serves as a mechanism to shape memory, motor plans and sensory representation. While plasticity is essential for normal function, "too much plasticity" may be detrimental (Quartarone *et al.*, 2006). Plasticity of the motor cortex in FHD patients has been shown to be abnormal using the technique of paired associative stimulation (PAS) (Stefan *et al.*, 2000). In PAS a median nerve shock is paired with a TMS pulse to the sensorimotor cortex timed to be immediately after the arrival of the sensory volley. In this protocol, PAS is given for 90 stimuli at 0.05 Hz (30 minutes of stimulation). After PAS, the amplitude of the MEP produced by TMS to the motor cortex increases. The motor learning produced through PAS is similar to long-term potentiation (LTP) (Stefan *et al.*, 2002). In patients with dystonia, PAS produces a larger increase in the MEP than that seen in normal subjects (Quartarone *et al.*, 2003). Paired associative stimulation can also induce inhibition or suppression of the MEP, if appropriately timed, and is thought to be similar to long-term depression (LTD). Patients who have dystonia also have a more marked inhibition compared with normal subjects (Figure 25.3) (Weise *et al.*, 2006).

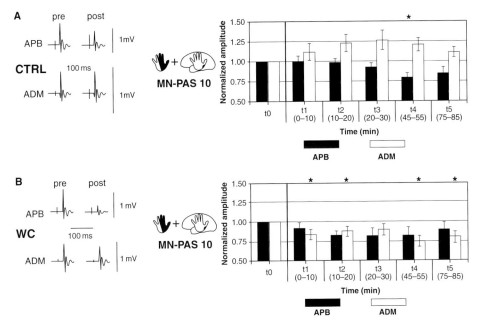

Figure 25.3. Effect of PAS10 (10 ms interval between peripheral stimulation and transcranial magnetic stimulation) in healthy controls and writer's cramp (WC) patients. A. In healthy controls, inhibition of motor evoked potential (MEP) amplitudes was seen only in the muscle (APB) in the territory of the peripheral nerve stimulated (median nerve). B. By contrast, in the WC patients inhibition of MEP amplitudes was seen more widely in both APB and ADM. ADM, abductor digiti minimi; APB, abductor pollicis brevis. Reproduced from Weise *et al.*, (2006) with permission.

Plasticity of the motor system can also be assessed using repetitive low- and high-frequency repetitive TMS (rTMS) and a method called theta burst stimulation (TBS). Using low-frequency rTMS over the primary sensory cortex, SAI was reduced in patients with FHD and was unchanged in healthy volunteers – suggesting abnormal responsiveness of the sensory cortex to rTMS (Baumer *et al.*, 2007). Abnormal plasticity in the motor cortex has also been shown with rTMS (Gilio *et al.*, 2007). After suprathreshold 5 Hz-rTMS, patients with upper limb dystonia had MEP facilitation, which was both more pronounced and more prolonged than in healthy controls.

Increased plasticity playing a role in dystonia had been suspected in the past and an animal model with owl monkeys supported this idea (Byl *et al.*, 1996). Two monkeys were trained to hold a handgrip that opened and closed. After training, their fingers pulled away from the grip. This motor control abnormality was interpreted as a possible hand dystonia. The sensory cortex of these animals was studied pre- and post-training. The sensory receptive fields were 10–20 times larger than normal and often included more than one digit in the representation (i.e. dedifferentiation). The repetitive, synchronous sensory input from the digits was thought to cause the receptive field enlargement and then the sensory changes led to the abnormal motor performance. The same thing might be happening in human focal dystonia: repetitive activity causing sensory receptive field changes and leading to the motor disorder. Changes in sensory receptive fields have been seen in patients with FHD (Meunier *et al.*, 2001; Delmaire *et al.*, 2005). Of course, these studies are all done after the development of FHD leaving the causal role of sensory dediffentiation and plasticity, as yet, unconfirmed.

Another aspect of the abnormal plasticity is that not only is the plasticity increased, but also there is a failure of its homeostatic property (Quartarone *et al.*, 2005). Homeostatic mechanisms ensure that plasticity increases and decreases within bounds. If, for example, the excitability of the motor cortex is high, then it cannot be driven higher, only lower. Taken together, these studies provide evidence that "Deficient homeostatic control might be an important mechanism that triggers maladaptive reorganization and produces symptoms of occupational hand dystonias" (Quartarone *et al.*, 2006).

It is unclear what causes the increased plasticity and disrupted homeostasis. One possibility is decreased inhibition, which may be a more fundamental abnormality in the etiology of dystonia or – at the least – be an important link in demonstrating how environmental influences can trigger dystonia.

Loss of inhibition

Along with abnormal sensorimotor integration and abnormal plasticity, the neurophysiology of focal dystonia is characterized by loss of inhibition (Hallett, 2004). This decreased inhibition is reflected in multiple levels from abnormal patterns of muscle activity, loss of spinal and brain-stem reflexes and impaired inhibition at the motor cortical level. First, the excessive movements consist of prolonged bursts of electromyography (EMG) activity and co-contraction of antagonist muscles (Cohen & Hallett, 1988). Overflow, where activity spreads into other muscles irrelevant for the intended task, also occurs in dystonia. Second, at the spinal and brain-stem level, loss of reciprocal inhibition between flexor and extensor forearm muscles (Nakashima *et al.*, 1989; Panizza *et al.*, 1990) and abnormalities in blink reflex recovery (Berardelli *et al.*, 1985) have been demonstrated in patients with focal dystonia. The loss of reciprocal inhibition may be partially responsible for the phenomenon of co-contraction seen in dystonia.

At the motor cortical level, short- and long-latency intracortical inhibition and the silent period have been shown to be abnormal in dystonia. Intracortical inhibition is studied by applying a conditioning TMS stimulus at a level that causes cortical activity but does not provoke a descending volley to the spinal cord (Ziemann *et al.*, 1996) (see also Chapter 6). A second test stimulus at suprathreshold intensity is delivered at a specified interval. The amplitude of the resultant motor evoked potential (MEP) from the test stimulus alone is compared with the MEP amplitude evoked from the conditioning plus test stimuli condition. The modulation of the MEP amplitude by the conditioning pulse is due predominantly to inhibitory or excitatory interneuron effects on the motor output. When the interval between the conditioning and test stimuli is less than 5 ms, the inhibition seen is called short intra-cortical inhibition (SICI) and is likely due to the GABA^A effect (Di Lazzaro *et al.*, 2000). Deficiencies in SICI have been shown in patients with FHD – interestingly, in the hemi-sphere contralateral to the affected limb as well as the hemisphere ipsilateral to the affected limb (Ridding *et al.*, 1995).

In long intracortical inhibition (LICI), paired suprathreshold TMS pulses are delivered to the motor cortex at intervals from 50–200 ms (Valls-Solé *et al.*, 1992). The degree of LICI and SICI in a given individual do not correlate (Sanger *et al.*, 2001). Further, LICI and SICI differ physiologically as with increasing test pulse strength LICI decreases and SICI increases. In FHD patients, LICI was deficient only in the symptomatic hand (in contrast to the SICI findings) (Chen *et al.*, 1997). Also, the deficiency was only present during background contraction. As this loss of inhibition was demonstrated only in the sympto-matic condition (i.e. during movement) and only on the affected side, it is a possible physiologic correlate for the development of dystonia. Another way to study intracortical inhibition is the silent period (SP), which is a pause in voluntary EMG activity and is evoked by TMS. The SP has been found to be shorter in patients with dystonia (Chen *et al.*, 1997). Cortical networks likely mediate the later part of the silent period through GABA^B receptors (Werhahn *et al.*, 1999). The mechanisms of LICI and the SP may be similar in that both seem to depend on GABA^B receptors.

Using neuroimaging techniques, a loss of inhibition also has been demonstrated in focal dystonia. In a PET study, dopamine D2 receptors have been found to be deficient in focal dystonias (Perlmutter *et al.*, 1997). Using magnetic resonance spectroscopy (MRS), reduced GABA concentration both in basal ganglia and motor cortex was found, adding direct evidence for a loss of inhibition in the pathophysiology of focal dystonia (Levy & Hallett, 2002). Aside from imaging modalities, inhibitory neuron function can be queried through examination of high frequency oscillations after sensory stimulation. These oscillations may be decreased in patients with focal hand dystonia, again signaling a loss of inhibition (Cimatti *et al.*, 2007).

Several experiments in animal models provide evidence that a loss of inhibition not only can produce abnormal, dystonic-like movements directly but also may be the physiological substrate for development of dystonia when combined with the appropriate genetics and/or environmental cues. First, when bicuculline, a GABA antagonist, was applied to the motor cortex in a primate model, disordered movement occurred as well as co-contraction (Matsumura *et al.*, 1991). The motor phenomenon also occurred when bicuculline was applied to the premotor cortex but to a lesser degree. Second, in a rat model for blepharo-spasm, lesioning caused a depletion of dopamine, thereby reducing inhibition (Schicatano *et al.*, 1997). In these animals, the investigators weaken the orbicularis oculi muscle – producing an abnormal reflex drive. Together, but not separately, these two interventions produced spasms of eyelid closure, similar to blepharospasm. Subsequently, several patients with blepharospasm after a Bell's palsy were reported – possible human analogs of the rat model (Chuke *et al.*, 1996; Baker *et al.*, 1997). The idea is that those patients who developed blepharospasm were in some way predisposed. A gold weight implanted into the weak lid of one patient, aiding lid closure, improved the condition, suggesting that when the abnormal increase in reflex drive was removed, the dystonia could be ameliorated (Chuke *et al.*, 1996).

Clinically, focal hand dystonia is both a disorder of imprecise movement and of unwanted movement. Structural abnormalities (i.e. loss of GABA interneurons creating reduced GABA concentration) or inhibitory network dysfunction (i.e. decreased SICI) may explain partially the clinical presentation of FHD; however, a type of functional inhibition – surround inhibition – may be the process that could produce both the imprecision and the extraneous, unwanted movement.

Surround inhibition is a concept well accepted in sensory physiology (Angelucci *et al.*, 2002). Although not as well known in the motor system, it is a logical extension that this process also operates in movement. When making a movement, the brain must activate the motor system. It is possible that the brain just activates the specific movement. On the other hand, it is more likely that the one specific movement is generated, and, simultaneously, other possible movements are suppressed. The suppression of unwanted movements would be surround inhibition, and this should produce a more precise movement, just as surround inhibition in sensory systems produces more precise perceptions. For dystonia, a failure of "surround inhibition" may be particularly important since overflow movement is often seen and is a principal abnormality.

There is now good evidence for surround inhibition in human movement. Sohn *et al.* (2003) have shown that with movement of one finger there is widespread inhibition of muscles in the contralateral limb. Significant suppression of MEP amplitudes was observed when TMS was applied ipsilateral to the right finger movement between 35 and 70 ms after EMG onset. Sohn & Hallett (2004b) have also shown that there is some inhibition of muscles in the ipsilateral limb when those muscles are not involved in any way in the movement. Transcranial magnetic stimulation was delivered to the left motor cortex from 3 ms to 1000 ms after EMG onset in the right flexor digitorum superficialis muscle. Motor evoked potentials from the right abductor digiti minimi were slightly suppressed during the movement of the right index finger in the face of increased F-wave amplitude and persistence, indicating that cortical excitability is reduced.

Surround inhibition was studied similarly in FHD (Sohn & Hallett, 2004b). The MEPs were enhanced similarly in the flexor digitorum superficialis and abductor digiti minimi (a muscle not involved in the task) indicating a failure of surround inhibition. Stinear and Byblow, in another experimental paradigm, have also found a loss of surround inhibition in the hands of patients with FHD (Stinear & Byblow, 2004).

Conclusion

Even though the pathophysiology of focal hand dystonia is not completely known, what we do know has guided therapeutic approaches. Improving sensory deficits by Braille training (Zeuner *et al.*, 2002), reversing abnormal plasticity by motor learning (Candia *et al.*, 2005;

Zeuner *et al.*, 2005) and increasing inhibition using rTMS (Siebner *et al.*, 2003; Murase *et al.*, 2005; Tyvaert *et al.*, 2006) have all been attempted to improve the dystonia. Therapeutic approaches in this disorder will be described in Chapter 33.

References

Abbruzzese, G. & Berardelli, A. (2003). Sensorimotor integration in movement disorders. *Mov Disord*, **18**, 231–240.

Abbruzzese, G., Marchese, R., Buccolieri, A., Gasparetto, B. & Trompetto, C. (2001). Abnormalities of sensorimotor integration in focal dystonia: a transcranial magnetic stimulation study. *Brain*, **124**, 537–545.

Angelucci, A., Levitt, J. B. & Lund, J. S. (2002). Anatomical origins of the classical receptive field and modulatory surround field of single neurons in macaque visual cortical area V1. *Prog Brain Res*, **136**, 373–388.

Baker, R. S., Sun, W. S., Hansan, S. A. *et al.* (1997). Maladaptive neural compensatory mechanisms in Bell's palsy-induced blepharospasm. *Neurology*, **49**, 223–229.

Bara-Jimenez, W., Catalan, M. J., Hallett, M. & Gerloff, C. (1998). Abnormal somatosensory homunculus in dystonia of the hand. *Ann Neurol*, **44**, 828–831.

Baumer, T., Demiralay, C., Hidding, U. *et al.* (2007). Abnormal plasticity of the sensorimotor cortex to slow repetitive transcranial magnetic stimulation in patients with writer's cramp. *Mov Disord*, **22**, 81–90.

Berardelli, A., Rothwell, J. C., Day, B. L. & Marsden, C. D. (1985). Pathophysiology of blepharospasm and oromandibular dystonia. *Brain*, **108**, 593–608.

Byl, N. N., Merzenich, M. M. & Jenkins, W. M. (1996). A primate genesis model of focal dystonia and repetitive strain injury: I. Learning-induced dedifferentiation of the representation of the hand in the primary somatosensory cortex in adult monkeys. *Neurology*, **47**, 508–520.

Candia, V., Rosset-Llobet, J., Elbert, T. & Pascual-Leone, A. (2005). Changing the brain through therapy for musicians' hand dystonia. *Ann NY Acad Sci*, **1060**, 335–342.

Chen, R. & Hallett, M. (1998). Focal dystonia and repetitive motion disorders. *Clin Orthop Relat Res*, **351**, 102–106.

Chen, R., Wassermann, E. M., Canos, M. & Hallett, M. (1997). Impaired inhibition in writer's cramp during voluntary muscle activation. *Neurology*, **49**, 1054–1059.

Chuke, J. C., Baker, R. S., & Porter, J. D. (1996). Bell's palsy-associated blepharospasm relieved by aiding eyelid closure. *Ann Neurol*, **39**, 263–268.

Cimatti, Z., Schwartz, D. P., Bourdain, F. *et al.* (2007). Time-frequency analysis reveals decreased high-frequency oscillations in writer's cramp. *Brain*, **130**, 198–205.

Classen, J., Steinfelder, B., Liepert, J. *et al.* (2000). Cutaneomotor integration in humans is somatotopically organized at various levels of the nervous system and is task dependent. *Exp Brain Res*, **130**, 48–59.

Cohen, L. G. & Hallett, M. (1988). Hand cramps: clinical features and electromyographic patterns in a focal dystonia. *Neurology*, **38**, 1005–1012.

Defazio, G., Aniello, M. S., Masi, G. *et al.* (2003). Frequency of familial aggregation in primary adult-onset cranial cervical dystonia. *Neurol Sci*, **24**, 168–169.

Defazio, G., Berardelli, A. & Hallett, M. (2007). Do primary adult-onset focal dystonias share aetiological factors? *Brain*, **130**, 1183–1193.

Delmaire, C., Krainik, A., Tezenas du Montcel, S. *et al*. (2005). Disorganized somatotopy in the putamen of patients with focal hand dystonia. *Neurology*, **64**, 1391–1396.

Di Lazzaro, V., Oliviero, A., Meglio, M. *et al*. (2000). Direct demonstration of the effect of lorazepam on the excitability of the human motor cortex. *Clin Neurophysiol*, **111**, 794–799.

Evinger, C. (2005). Animal models of focal dystonia. *NeuroRx*, **2**, 513–524.

Fiorio, M., Gambarin, M., Valente, E. M. *et al*. (2007). Defective temporal processing of sensory stimuli in DYT1 mutation carriers: a new endophenotype of dystonia? *Brain*, **130**, 134–142.

Frucht, S. J. (Ed.) (2004). Focal task specific dystonia in musicians. *Advances in Neurology, Dystonia 4*. Philadelphia, PA: Lippincott, Williams & Wilkins.

Garraux, G., Bauer, A., Hanakawa, T. *et al*. (2004). Changes in brain anatomy in focal hand dystonia. *Ann Neurol*, **55**, 736–739.

Gilio, F., Suppa, A., Bologna, M. *et al*. (2007). Short-term cortical plasticity in patients with dystonia: A study with repetitive transcranial magnetic stimulation. *Mov Disord*.

Hallett, M. (2004). Dystonia: abnormal movements result from loss of inhibition. *Adv Neurol*, **94**, 1–9.

Kessler, K. R., Ruge, D., Ilic, T. V. & Ziemann, U. (2005). Short latency afferent inhibition and facilitation in patients with writer's cramp. *Mov Disord*, **20**, 238–242.

Lerner, A., Shill, H., Hanakawa, T. *et al*. (2004). Regional cerebral blood flow correlates of the severity of writer's cramp symptoms. *Neuroimage*, **21**, 904–913.

Levy, L. M. & Hallett, M. (2002). Impaired brain GABA in focal dystonia. *Ann Neurol*, **51**, 93–101.

Matsumura, M., Sawaguchi, T., Oishi, T., Ueki, K. & Kubota, K. (1991). Behavioral deficits induced by local injection of bicuculline and muscimol into the primate motor and premotor cortex. *J Neurophysiol*, **65**, 1542–1553.

Meunier, S., Garnero, L., Ducorps, A. *et al*. (2001). Human brain mapping in dystonia reveals both endophenotypic traits and adaptive reorganization. *Ann Neurol*, **50**, 521–527.

Molloy, F. M., Carr, T. D., Zeuner, K. E., Dambrosia, J. M. & Hallett, M. (2003). Abnormalities of spatial discrimination in focal and generalized dystonia. *Brain*, **126**, 2175–2182.

Murase, N., Kaji, R., Shimazu, H. *et al*. (2000). Abnormal premovement gating of somatosensory input in writer's cramp. *Brain*, **123**, 1813–1829.

Murase, N., Rothwell, J., Kaji, R. *et al*. (2005). Subthreshold low-frequency repetitive transcranial magnetic stimulation over the premotor cortex modulates writer's cramp. *Brain*, **128**, 104–115.

Nakashima, K., Rothwell, J. C., Day, B. L. *et al*. (1989). Reciprocal inhibition in writer's and other occupational cramps and hemiparesis due to stroke. *Brain*, **112**, 681–697.

Panizza, M., Lelli, S., Nilsson, J. & Hallett, M. (1990). H-reflex recovery curve and reciprocal inhibition of H-reflex in different kinds of dystonia. *Neurology*, **40**, 824–828.

Peller, M., Zeuner, K. E., Munchau, A. *et al*. (2006). The basal ganglia are hyperactive during the discrimination of tactile stimuli in writer's cramp. *Brain*, **129**, 2697–2708.

Perlmutter, J. S., Stambuk, M. K., Markham, J. *et al*. (1997). Decreased [18F]spiperone binding in putamen in idiopathic focal dystonia. *J Neurosci*, **17**, 843–850.

Putzki, N., Stude, P., Konczak, J. *et al*. (2006). Kinesthesia is impaired in focal dystonia. *Mov Disord*, **21**, 754–760.

Quartarone, A., Bagnato, S., Rizzo, V. *et al.* (2003). Abnormal associative plasticity of the human motor cortex in writer's cramp. *Brain*, **126**, 2586–2596.

Quartarone, A., Rizzo, V., Bagnato, S. *et al.* (2005). Homeostatic-like plasticity of the primary motor hand area is impaired in focal hand dystonia. *Brain*, **128**, 1943–1950.

Quartarone, A., Siebner, H. R., & Rothwell, J. C. (2006). Task-specific hand dystonia: can too much plasticity be bad for you? *Trends Neurosci*, **29**, 192–199.

Ridding, M. C., Sheean, G., Rothwell, J. C., Inzelberg, R. & Kujirai, T. (1995). Changes in the balance between motor cortical excitation and inhibition in focal, task specific dystonia. *J Neurol Neurosurg Psychiatry*, **59**, 493–498.

Sanger, T. D., Garg, R. R. & Chen, R. (2001). Interactions between two different inhibitory systems in the human motor cortex. *J Physiol*, **530**, 307–317.

Schicatano, E. J., Basso, M. A. & Evinger, C. (1997). Animal model explains the origins of the cranial dystonia benign essential blepharospasm. *J Neurophysiol*, **77**, 2842–2846.

Siebner, H. R., Filipovic, S., Rowe, J. B. *et al.* (2003). Patients with focal arm dystonia have increased sensitivity to slow-frequency repetitive TMS of the dorsal premotor cortex. *Brain*, **126**, 2710–2725.

Sohn, Y. H. & Hallett, M. (2004a). Surround inhibition in human motor system. *Exp Brain Res*, **158**, 397–404.

Sohn, Y. H. & Hallett, M. (2004b). Disturbed surround inhibition in focal hand dystonia. *Ann Neurol*, **56**, 595–599.

Sohn, Y. H., Jung, H. Y., Kaelin-Lang, A. & Hallett, M. (2003). Excitability of the ipsilateral motor cortex during phasic voluntary hand movement. *Exp Brain Res*, **148**, 176–185.

Song, I. U., Kim, J. S., Kim, H. T. & Lee, K. S. (2007). Task-specific focal hand dystonia with usage of a spoon. *Parkinsonism Relat Disord*, **14**, 72–74.

Stefan, K., Kunesch, E., Cohen, L. G., Benecke, R. & Classen, J. (2000). Induction of plasticity in the human motor cortex by paired associative stimulation. *Brain*, **123**, 572–584.

Stefan, K., Kunesch, E., Benecke, R., Cohen, L. G. & Classen, J. (2002). Mechanisms of enhancement of human motor cortex excitability induced by interventional paired associative stimulation. *J Physiol*, **543**, 699–708.

Stinear, C. M. & Byblow, W. D. (2004). Impaired modulation of intracortical inhibition in focal hand dystonia. *Cereb Cortex*, **14**, 555–561.

Tarsy, D. & Simon, D. K. (2006). Dystonia. *N Engl J Med*, **355**, 818–829.

Tinazzi, M., Rosso, T. & Fiaschi, A. (2003). Role of the somatosensory system in primary dystonia. *Mov Disord*, **18**, 605–622.

Tyvaert, L., Houdayer, E., Devanne, H. *et al.* (2006). The effect of repetitive transcranial magnetic stimulation on dystonia: a clinical and pathophysiological approach. *Neurophysiol Clin*, **36**, 135–143.

Valls-Solé, J., Pascual-Leone, A., Wassermann, E. M. & Hallett, M. (1992). Human motor evoked responses to paired transcranial magnetic stimuli. *Electroencephalogr Clin Neurophysiol*, **85**, 355–364.

Weise, D., Schramm, A., Stefan, K. *et al.* (2006). The two sides of associative plasticity in writer's cramp. *Brain*, **129**, 2709–2721.

Weiss, E. M., Hershey, T., Karimi, M. *et al.* (2006). Relative risk of spread of symptoms among the focal onset primary dystonias. *Mov Disord*, **21**, 1175–1181.

Werhahn, K. J., Kunesch, E., Noachtar, S., Benecke, R. & Classen, J. (1999). Differential effects on motorcortical inhibition induced by blockade of GABA uptake in humans. *J Physiol (Lond)*, **517**, 591–597.

Zeuner, K. E., Bara-Jimenez, W., Noguchi, P. S. *et al.* (2002). Sensory training for patients with focal hand dystonia. *Ann Neurol*, **51**, 593–598.

Zeuner, K. E., Shill, H. A., Sohn, Y. H. *et al.* (2005). Motor training as treatment in focal hand dystonia. *Mov Disord*, **20**, 335–341.

Ziemann, U., Rothwell, J. C. & Ridding, M. C. (1996). Interaction between intracortical inhibition and facilitation in human motor cortex. *J Physiol (Lond)*, **496**, 873–881.

26

Cerebellar disorders

MARIO MANTO AND DENNIS A. NOWAK

Summary

Precise control of grasping when manipulating objects depends on intact function of the cerebellum. Given its stereotyped cytoarchitecture, the widespread connections with cortical and subcortical sensorimotor structures and the neural activity of cerebellar Purkinje cells during sensorimotor tasks, the cerebellum is considered to play a major role in the establishment and maintenance of sensorimotor representations related to grasping. Such representations are necessary to predict the consequences of movements. This chapter summarizes anatomical and theoretical aspects, electrophysiological and behavioral data characterizing the cerebellum, a key player in the processing of healthy grasping and in its dysfunction.

The anatomy of the cerebellum and its relation to the control of grasping

The cerebellum has attracted the attention of theorists and modelers for many years. The attraction is that the regular cytoarchitecture of the cerebellar cortex, with only one output cell and four main classes of interneurons, and the functional cerebellar circuitry have been very well documented (Wolpert *et al.*, 1998). The circuitry of the cerebellum is unique by its stereotyped geometric arrangement and its modular organization, highly reminiscent of a machinery designed to process neuronal information in a unique manner (Ito, 2006). The cerebellum appears highly foliated, and this foliation is the reason for subdivision into smaller units (Larouche & Hawkes, 2006). From a structural standpoint, the cerebellum is made of pairs of nuclei embedded in white matter and surrounded by a mantle of cortex (Colin *et al.*, 2002). The fastigial nucleus is medial, the globosus and emboliform nuclei (grouped under the terminology of interpositus nuclei) are intermediate, and the dentate nucleus is lateral. There is a medio-lateral organization in the projections from the cerebellar cortex to nuclei: the vermal cortex projects to the fastigial nuclei, the intermediate cortex projects to the interpositus nuclei and the dentate nuclei receive projections from the lateral cerebellar cortex. Afferents and efferents enter and leave the cerebellum via three pairs of cerebellar peduncles: the inferior, the middle and the superior cerebellar peduncles (Allen & Tsukahara, 1974).

Sensorimotor Control of Grasping: Physiology and Pathophysiology, ed. Dennis A. Nowak and Joachim Hermsdörfer. Published by Cambridge University Press. © Cambridge University Press 2009.

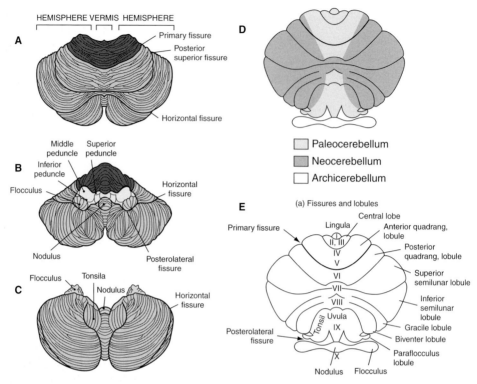

Figure 26.1. (This figure is also reproduced in the color plate section.) Anatomy of the cerebellum. From *The Cerebellum and its Disorders*, Cambridge University Press (2002).

Jansen divided the cerebellum into three major parts: the anterior lobe, the posterior cerebellum and paraflocculus/flocculus area (Jansen, 1969; see Figure 26.1). The primary fissure demarcates the anterior lobe from the posterior lobe. This lobe is separated from the flocculonodular lobe by the posterolateral fissure. The nomenclature of lobules applied in humans is derived from Larsell's works. He identified 10 lobules (I–X) in the vermis and corresponding lobules in the hemispheres (Larsell, 1937). Microzones are the functional units of the cerebellar cortex (Oscarsson, 1976). Microcomplexes refer to the combination of a microzone and the related subcortical structures: small groups of neurons in a cerebellar or vestibular nucleus, the inferior olive and neurons in the red nucleus (Ito, 2006). The human cerebellum might contain about 5000 microcomplexes, each of them playing some specific roles according to its connections in the brain stem, the spinal cord and the cerebral cortex (Ito, 2006).

The cerebellum is generally considered to regulate movement indirectly by adjusting the output of the descending motor system of the brain (Gilman, 1969; Gilman *et al.*, 1981; Bastian, 2006). Lesions of the cerebellum disrupt coordination of limb and eye movements, impair balance and decrease muscle tone (Holmes, 1917, 1939; Glickstein *et al.*, 2005). The most widely accepted idea is that the cerebellum acts as a comparator that compensates for errors in movement by comparing intended movement with actual performance (Gilman *et al.*, 1981; Glickstein & Yeo, 1990). Through comparison of internal and external feedback signals, the cerebellum is able to correct ongoing movements when they deviate from the intended course and to modify central motor commands so that subsequent movements are performed with fewer prediction errors (Miall *et al.*, 1993; Kawato, 1999). These processes depend, at least in part, on the capacities of certain sensory inputs to modify cerebellar circuits for long periods, suggesting that the activity of cerebellar neurons is changed by experience and plays an important role in motor learning (Kitazawa *et al.*, 1998; Imamizu *et al.*, 2000).

The cerebellum receives input from the periphery and from all levels of the central nervous system. Information entering the cerebellum is initially acting on the cerebellar cortex and via collaterals on neurons of the cerebellar nuclei (e.g. the fastigial, interpositus and dentate nuclei) (Colin *et al.*, 2002). Afferent information is processed within the cerebellar cortex. The cerebellar nuclei receive input from the Purkinje cells, the only output cells of the cerebellar cortex. The cerebellar nuclei transmit all output from the cerebellum, primarily to the motor regions of the cerebral cortex and brainstem (Allen & Tsukahara, 1974; Hoover & Strick, 1999). The lateral zone of the cerebellar hemisphere and the basal ganglia participate in the planning of movement by processing information originating primarily in the parietal association cortices, which integrate sensory information for purposeful action (Figure 26.2).

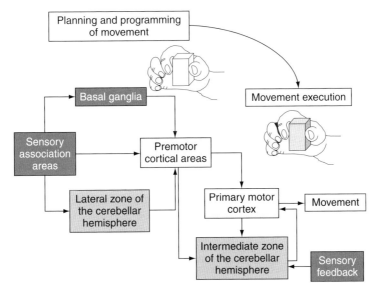

Figure 26.2. The role of the cerebellum in the planning, programming and execution of movement according to the model of Allan & Tsukahara (1974). The lateral zone of the cerebellar hemisphere participates in the planning and programming of movements by integrating sensory information. The intermediate zone of the cerebellar hemisphere contributes to movement execution by monitoring actual sensory feedback and processing error signals that compensate for prediction errors in movement planning.

Such processing is essential for the exact planning and programming of movement issued by the premotor and motor cortical areas and indeed, the lateral zone of the cerebellar hemisphere projects to the dentate nucleus, which connects primarily with these regions of the cerebral cortex via the thalamus (Hoover & Strick, 1999). The intermediate zone of the cerebellar hemisphere is involved in motor execution and monitoring of movement (Gilman *et al.*, 1981; Glickstein & Yeo, 1990). The intermediate zone of the cerebellar hemisphere projects to cortical and brain-stem regions that give rise to the descending motor pathways executing distal movements of the limbs (Hoover & Strick, 1999). The Purkinje cells are the only output cells of the cerebellar cortex and receive input from two distinct sensory afferents: mossy fibers and climbing fibers (Ito, 2006). Mossy fibers constitute the major afferent input to the cerebellar cortex and they project from various brain-stem nuclei and from neurons in the spinal cord giving rise to the spinocerebellar tracts (Gilbert & Thach, 1977; Ito, 1984). Climbing fibers originate in the inferior olivary nucleus in the medulla oblongata and receive input from the cerebral cortex and spinal cord. The activity of climbing fibers is modulated during motor learning, reducing the strength of the mossy fiber input to the Purkinje cells, as demonstrated by changes in the discharge activity within both fiber tracts when monkeys learn to compensate predictive load perturbations applied on a hand-held object (Gilbert & Thach, 1977). Further support to this suggestion comes from single cell recordings in the monkey, which demonstrate that both predictive and reactive mechanisms of grip-force adjustments when lifting and holding objects of different weight and surface friction are well represented in the discharge rates of both cerebellar Purkinje cells and the cerebellar nuclei (Espinoza & Smith, 1990; Dugas & Smith, 1992; Smith *et al.*, 1993; Monzee & Smith, 2004; Mason *et al.*, 2006; see also Chapters 4, 9 and 12).

Nuclear activity as a function of the task

Our knowledge of the roles of the cerebellar nuclei in motor tasks is growing. Studies in monkeys have provided major information regarding the contribution of each nucleus in motor tasks of upper limbs requiring dexterity or inter-segmental coordination. At rest, nuclear neurons fire at high frequencies, in the range of 40–50 Hz (Bastian & Thach, 2002). During movement, the frequencies of discharge are modulated according to the motor activity or goal. While the activities of vestibular nuclei/fastigial nuclei are related to oculomotor control, head orientation, stance and gait, the activities of the interpositus nuclei change mainly during stretch reflexes, contact and placing conditions. Neurons in the interpositus nucleus typically fire when a position is perturbed in reaction tasks aiming at keeping the segments of the limbs in a fixed position (Vilis & Hore, 1977). Recordings in animals suggest that the interpositus nucleus controls or participates in somesthetic reflex behaviors, adapting the activity of the antagonist muscle to damp oscillations in the limbs (Vilis & Hore, 1980), and contributes to stretch reflex excitability by controlling the discharges of gamma motor neurons (Gilman, 1969).

Grasping tasks are associated with increasing firing rates in interpositus nuclei (Monzee & Smith, 2004). The activities of the dentate nuclei are concerned primarily with goal-directed movements of the extremities, such as reaching and grasping (Espinoza & Smith, 1990; Dugas & Smith, 1992; Bastian & Thach, 2002; see also Chapter 4). Dentate lesions induce impaired reaching patterns and difficulties in grasping tasks (Fellows *et al.*, 2001). Changes in firing rates precede the onset of movement, indicating that the cerebellar nuclei probably contribute to movement initiation, presumably by tuning the activities of target structures such as thalamic nuclei or brainstem nuclei. Neuronal activity may start earlier in dentate nuclei as compared with the motor cortex (Lamarre *et al.*, 1983). Thach (1967) has shown that the order of activity was (1) dentate nucleus, (2) motor cortex, (3) interpositus nucleus and finally (4) the muscles in wrist movements in response to light stimulation. Several studies have confirmed that the responses of the dentate nucleus are stronger when movements are triggered by visual or auditory stimuli than somatosthetic stimuli such as a sudden stretch of the muscle. The interpositus nucleus is rather implicated in the numerous tasks based upon peripheral inputs, probably to generate predictive signals.

Theories of cerebellar functions

The feedforward control theory supposes that cerebellar pathways compute the appropriate motor commands in advance of the movement (see also Chapter 8). The second approach is to consider that the cerebellum uses feedback signals related to the actual state (position, velocity) to refine the motor commands. Theories of cerebellar functions, mostly emerging from the engineering field, have attributed to the intermediate cerebellum a role in monitoring of ongoing movement and in error correction. By comparison, anticipation has been ascribed mainly – if not solely – to the lateral cerebellum.

Several other studies underline that a major function of the cerebellum is to monitor sensory signals (Gao *et al.*, 1996; Blakemore *et al.*, 2001; Nowak *et al.*, 2005; Manto, 2006). The cerebellum is compared to a sensory tracking device, allowing the modulation of the efficiency of data acquisition for the sensorimotor system. The cerebellum is indeed very well positioned from the anatomical standpoint to process sensory information, and to participate in optimization of sensory integration and sensorimotor synchronization. A specific role for the cerebellum could be to tune the sensorimotor coupling in a given condition combining reflex and voluntary movement (Manzoni, 2007). This is based on the fact that the relation between sensory signals guiding motion and the movement itself depends on the "context," a notion encompassing the relative position of limb segments, the position of the body in the gravitational field and the external forces interacting with the movement. The idea of an implementation of a correct input–output in sensorimotor transformations fits with the model of Pellionisz & Llinás (1980), according to which the cerebellum transforms an intended movement vector into a motor command.

Coupling between the cerebellum and contralateral thalamic nuclei/primary cortex is well established (Middleton & Strick, 1997). Activity in cerebello-fugal fibers triggers oscillations in thalamic nuclei and motor cortex (Soteropoulos & Baker, 2006). Coherent oscillations between the sensory cortex and the cerebellar cortex have been described. Therefore, the possibility that the sensorimotor cortex and its cerebellar connections behave as pairs of reciprocally coupled oscillators has been raised. The coupling between cerebellar structures and extra-cerebellar targets extends to the contralateral posterior parietal cortex (Butz *et al.*, 2006). Inter-cerebellar coupling might increase bimanual performance (Manto *et al.*, 2006). For example, subjects with cerebellar degeneration lose the ability to transfer information related to the mechanical properties of an object to be grasped and lifted with both hands (Nowak *et al.*, 2005).

One leading theory proposes that the cerebellum houses neural representations, called "internal models," to mimic fundamental natural processes such as a joint movement (Wolpert & Miall, 1996; Nowak *et al.*, 2007; see also Chapter 9). This is based in particular on the inherent time delay of sensory feedback to update motor commands (Manto & Bastian, 2007). The cerebellum could use "forward models" by manipulating efference copies of motor signals to predict sensory effects of movements (Figure 26.3). The predictions need to be very accurate to attenuate the dependence on time-delayed sensory signals. Cerebellar microcircuits would handle error signals to perform predictions. In this scheme, the cortico-ponto-cerebellar tract conveys an efference copy of the motor program to the cerebellar cortex. Cerebellar microcircuits would make a computation of the expected sensory outcome. The results of this calculation would be transferred to the cerebral cortex

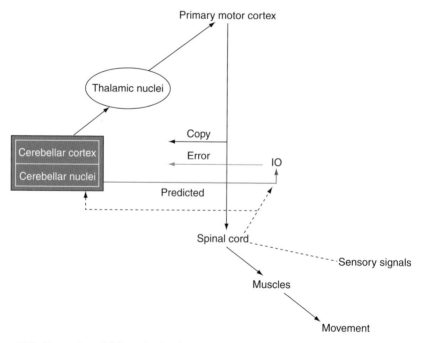

Figure 26.3. Forward model hypothesis. Illustration of the communication flow according to this theory of motor coding. Cerebellar modules receive an efference copy of motor commands, via the cortico-ponto-cerebellar tracts. The expected sensory outcome is sent to the motor cortex and the inferior olivary complex. The inferior olivary complex works as a comparator and generates an error signal to update cerebellar microcircuits.

via thalamic nuclei and to the inferior olivary complex. The theory assumes that this latter works as a comparator, capable of comparing ascending signals from the spinal cord and the signals emerging from cerebellar nuclei, and sending back the signals to the cerebellar cortex with the aim of making more and more accurate predictions. Other studies suggest that the cerebellum addresses an "inverse model" (Bastian, 2006). The input signal is an aimed trajectory. The output is a motor command. Errors are probably best represented in motor coordinates.

Clinical symptoms of cerebellar disorders relevant to grasping

Several deficits are particularly relevant when studies on grasping and goal-directed movements are considered in cerebellar patients. The symptoms of cerebellar disorders are usually influenced more by the location and rate of progression of the disease than by the pathological characteristics (Gilman *et al*., 1981). Patients with slowly progressive lesions may be remarkably asymptomatic for a long time, unlike patients with quickly expanding lesions in whom symptoms are usually severe. Clumsiness in limbs is reported by 53% of patients presenting with a lesion restricted to the cerebellum (Manto, 2002). These patients have difficulties with activities of daily living such as grasping and using tools. Earlier seminal works have delineated the impairment in voluntary limb movements in cases of cerebellar lesion, including in tasks requiring a reaching component or grasping (Holmes, 1917, 1939). The jerky or poorly coordinated character of motion (ataxia) appears in the absence of muscle weakness or sensory deficits, although fatigability is common in cerebellar patients. Dysmetria designates the errors in the metrics of motion. Hypermetria refers to the overshoot of a target, whereas hypometria designates a premature arrest. Table 26.1 lists the clinical signs according to the sagittal zone affected. Ataxia of stance and gait occurring in association with oculomotor deficits indicates a disease affecting the vermis. For lateral lesions, the following signs are often combined: dysmetria, kinetic tremor, dysdiadochokinesia.

Oculomotor deficits can contribute to the difficulties observed in cerebellar patients during grasping tasks. Patients exhibit various combinations of fixation deficits, ocular misalignments, deficits in horizontal and vertical pursuit, dysmetric saccades and nystagmus (Lewis & Zee, 1993). Anatomically, lesions are usually found at the level of the dorsal vermis or fastigial nucleus, the flocculus and paraflocculus, the uvula and nodulus. The dorsal vermis and the fastigial nuclei play key roles in the initiation, accuracy and dynamics of saccades. The flocculus and paraflocculus are primarily involved in stabilization of a visual image on the retina (Dichgans & Fetter, 1993). Disorders of fixation include flutter (brief oscillations of the eyes, usually conjugate and horizontal) and macrosaccadic oscillations (cycles of square waves/jerks) which interfere with head/limb voluntary motion.

Dysmetria of limbs is larger when the movement is performed as fast as possible. It is followed by corrective movements (Figure 26.4). Dysmetria occurs proximally and distally (Hore *et al*., 1991). Kinetic tremor designates oscillations exaggerated at the end of a

Table 26.1. *Clinical signs as a function of the sagittal zone affected.*

Zone	Signs
Vermis	Oculomotor deficits
	Dysarthria
	Head tilt
	Ataxia of stance and gait
	Titubation
Paravermal zone	Dysarthria
Lateral portions	Oculomotor deficits
	Dysarthria
	Head tilt
	Dysmetria
	Tremor: kinetic, action
	Hypotonia
	Dysdiadochokinesia
	Decomposition of movements
	Impaired check and excessive rebound
	Ataxia of stance and gait

Adapted from *The Cerebellum and its Disorders*, Cambridge University Press (2002).

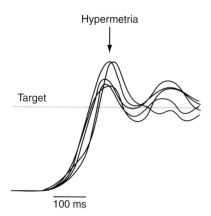

Figure 26.4. Hypermetria followed by attempts to reach the target during a 'ballistic' movement in a cerebellar patient.

voluntary movement, usually assessed clinically with the finger-to-nose test, finger-to-finger test (the patient is asked to touch the index of the examiner, which is moved in various locations in front of the patient), or the knee–tibia test. Tremor is usually perpendicular to the main direction of movement. It may be predominant proximally in

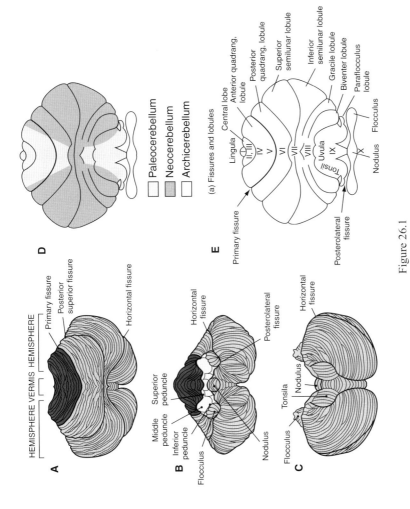

Figure 26.1

A

HEMISPHERE VERMIS HEMISPHERE

Primary fissure

Posterior superior fissure

Horizontal fissure

B

Middle peduncle

Inferior peduncle

Flocculus

Superior peduncle

Horizontal fissure

Posterolateral fissure

Nodulus

C

Flocculus

Tonsila

Nodulus

Horizontal fissure

D

Paleocerebellum

Neocerebellum

Archicerebellum

E

(a) Fissures and lobules

Primary fissure

Lingula

II, III

IV

V

VI

VII

VIII

IX

X

Uvula

Tonsil

Nodulus

Flocculus

Central lobe

Anterior quadrang, lobule

Posterior quadrang, lobule

Superior semilunar lobule

Inferior semilunar lobule

Gracile lobule

Biventer lobule

Paraflocculus lobule

Posterolateral fissure

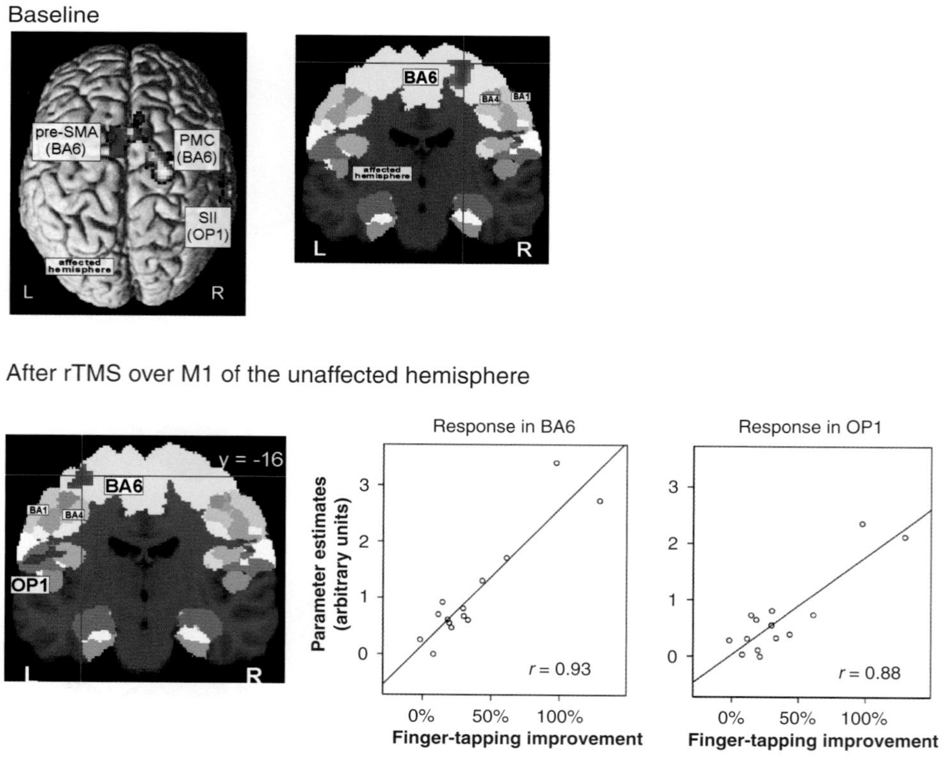
Baseline
pre-SMA (BA6)
PMC (BA6)
SII (OP1)
affected hemisphere
L
R
BA6
BA4
BA1
affected hemisphere
L
R
After rTMS over M1 of the unaffected hemisphere
y = -16
BA6
BA1
BA4
OP1
L
R
Response in BA6
Parameter estimates (arbitrary units)
3
2
1
0
r = 0.93
0%
50%
100%
Finger-tapping improvement
Response in OP1
3
2
1
0
r = 0.88
0%
50%
100%
Finger-tapping improvement

Figure 29.5

A   Movements of the stroke-affected (right) hand

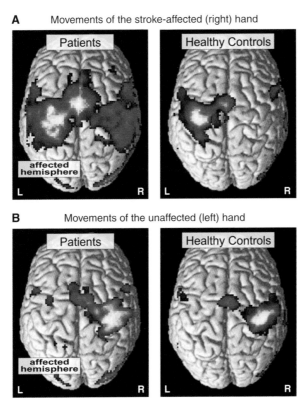

Figure 30.1

Connectivity during right/paretic hand movements

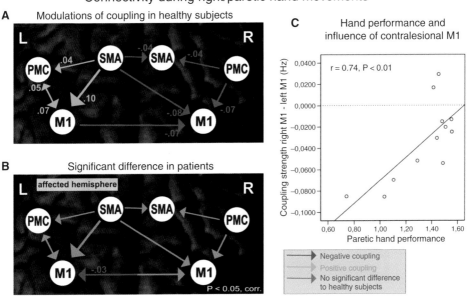

Figure 30.2

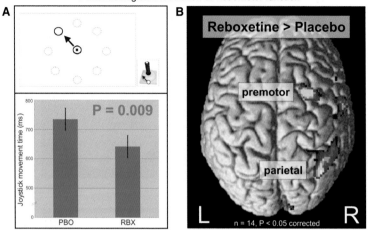

Paramacological modulation of visuomotor function

Figure 30.3

Technical modulation of impaired motor function

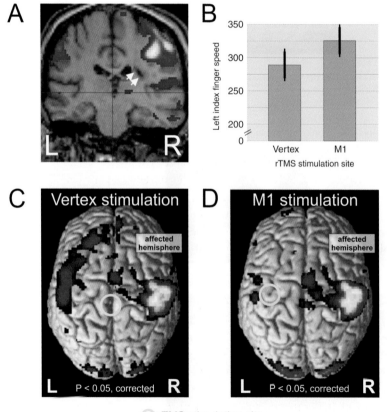

Figure 30.4

some patients (Gilman *et al.*, 1981). Action tremor ("static tremor") is another tremor observed in cerebellar patients. It appears during postural tasks requiring accuracy. Tremor appears usually in the line of gravity, rapidly evolving in lateral movements (Holmes, 1917, 1939). Hypotonia is defined as a reduction in the resistance to the passive manipulation of limbs. In adults, it tends to be more pronounced proximally. Another major feature of cerebellar motion is the decomposition of movement. Compound movements tend to be decomposed into their elemental components. Lack of synergy may be associated with dysdiadochokinesia, characterized by irregular and slow movements during alternate sequential tasks. Isometrataxia refers to the inability of cerebellar patients to maintain constant forces during skilled tasks with hands. It may be masked by an action tremor.

Pathophysiology of grasping in cerebellar disorders

Humans and higher primates are particularly skilled at finger use (Glickstein *et al.*, 2005). The cerebellum controls the activities of many antagonist and synergist muscles used in an automatic manner in most natural movements (Bastian & Thach, 2002). In general, goal-directed voluntary motion in chronic cerebellar patients tends to be slower, with a reduction in the ratio of acceleration/deceleration. Hand paths are larger than required, being characterized by curved trajectories. Holmes observed a delay in the initiation of movement during grasping in his patients (Holmes, 1917, 1939). He also found a reduction in phasic muscle strength and noted that the time to reach a maximal muscle force was prolonged.

   Deficits in the control of the digits observed in monkeys or in severely affected patients are strikingly reminiscent of pyramidal tract lesions (Glickstein *et al.*, 2005). These lesions induce a loss of independence of movements of digits (Lawrence & Kuypers, 1968). It is possible that deficits in finger use are partly due to the deprivation of the cerebellum from collaterals to the cerebellar hemispheres, conveying the efferent copy (Glickstein *et al.*, 2005). Corollary discharges from pyramidal tract fibers to the cerebellum appear required for skilled sequential movements. Studies addressing the lack of coordination of movement in cerebellar patients have pointed out that the cerebellum may indeed play a specific role in coordination of joints, which is more than just a summation of actions at the level of individual joints. The relationship between the change in behavior of the movement over time (kinematics) and the forces driving movements (dynamics) may be described by sets of non-linear equations of motion (Schneider *et al.*, 1990). Net torques can be computed from muscle torques, external forces and dynamic interaction forces. In multi-segmental movements, interaction forces need to be monitored closely and muscle forces need to be optimized accordingly to avoid inaccuracy of motion. It has been demonstrated that lack of coordination in movement may result from an inappropriate compensation of dynamic interaction forces. This theory of deficient compensation of dynamic joint interactions in cerebellar patients is compatible with

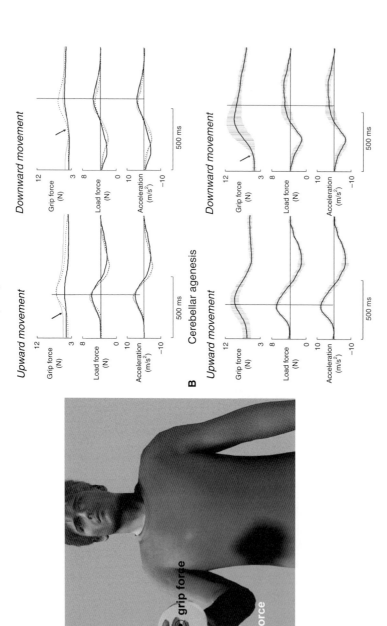

Figure 26.5. Coupling between grip and load forces during vertical point-to-point arm movements in cerebellar agenesis. A. The upper panels refer to the performance of the healthy control subjects. Illustrated are average force traces. Grip force is modulated in parallel with movement-induced load-force profile, regardless of movement direction. There is an early increase in grip force during the upward movement and a late increase in grip force during the downward movement. A peak in grip force coincides with a peak in load force. B. The lower panels refer to the performance of a 63-year-old female (HK) with cerebellar agenesis. Illustrated are average force traces (±1 SD). In contrast to the performance of healthy subjects, HK started to increase grip force right from the movement onset, regardless of whether the movement was directed upward or downward. This finding suggests inability to predict the load force arising from the intended movement with a failure to plan the grip force profile differentially to the loading requirements of the upcoming movement. (Modified according to Nowak et al., 2007.)

the hypothesis of internal models, according to which the cerebellum holds an internal representation of the biomechanical properties of each joint, which is updated on a regular basis by sensory inputs (Sainburg *et al.*, 1999).

Inertial loads arise when we transport grasped objects during arm movements (see also Chapter 1, 11 and 12). The size and direction of inertial loads change in dependence of the acceleration vector of movement. When a grasped object is moved around in space, grip force is not maintained at a constant level. Rather inertial load fluctuations evoke simultaneous fluctuations in grip force (see Chapters 1 and 11). The absence of a temporal delay between the grip and load force profiles implies that our motor system is able to precisely predict the load variations induced by voluntary transport movements and regulates grip force in anticipation (Nowak *et al.*, 2007). Cerebellar deficits cause significant impairments of the control of isometric grip forces when grasping and handling objects in the environment (Serrien & Wiesendanger, 1999; Fellows *et al.*, 2001; Nowak *et al.*, 2002, 2004, 2005; Rost *et al.*, 2005). In particular, the parallel modulation of grip force with the movement-induced load variations to be found in healthy people during transport movements of hand-held objects is decalibrated in cerebellar degeneration (Nowak *et al.*, 2002) (Figure 26.5). This observation has been taken as indicative that the cerebellum is involved in predictive control of movement based on internal models (Nowak *et al.*, 2007). In addition, subjects with cerebellar lesions usually generate excessive grip forces in relation to the loads (Serrien & Wiesendanger, 1999; Fellows *et al.*, 2001; Nowak *et al.*, 2002, 2004, 2005; Rost *et al.*, 2005).

Also when grasping and lifting, healthy people adjust grip forces very precisely to the mechanical constraints of an object to be lifted and there is a parallel processing of grip force and lift force (see Chapters 1 and 11). Fellows *et al.* (2001) studied grip-force control in patients with a variety of topographically different cerebellar lesions lifting and holding objects of different weight. It appears as if damage to the main output nucleus of the cerebellum, the dentate nucleus (territory of the superior cerebellar artery) and more markedly to its afferent relay area, the Purkinje cells of the cerebellar cortex (in subjects with cerebellar degeneration) cause an overshoot in grip force and at the same time a disruption of the predictive coupling between grip- and load-force profiles. A lesion within the territory of the posterior inferior cerebellar artery, however, does not cause a lasting overshoot in grip force nor a dyscoordination between grip- and load-force profiles.

Cerebellar tremors are better explained by a generation within central loops. Several experimental models have shown that terminal tremor may appear following a lesion in the dentate nucleus or at the level of the superior cerebellar peduncle (Flament & Hore, 1986). Cooling of dentate nuclei is typically associated with a kinetic tremor. The anticipatory character of muscle bursts normally generated to stabilize the limb and posture would be lost, giving rise to successive attempts to correct the instability. Therefore, non-anticipatory spinal and transcortical reflexes would become predominant. Cerebellar tremor could be attributable to a faulty predictive signal (error in the feedforward signal) to the antagonist muscle (Bastian, 2006).

References

Allen, G. I., Tsukahara, N. (1974). Cerebrocerebellar communication systems. *Physiol Rev*, **54**, 957–1006.

Bastian, A. J. (2006). Learning to predict the future: the cerebellum adapts feedforward movement control. *Curr Opin Neurobiol*, **16**, 645–649.

Bastian, A. J. & Thach, W. T. (2002). Structure and function of the cerebellum. In M. Manto & M. Pandolfo (Eds.), *The Cerebellum and its Disorders* (pp. 49–66). Cambridge, UK: Cambridge University Press.

Blakemore, S. J., Frith, C. D. & Wolpert, D. M. (2001). The cerebellum is involved in predicting the sensory consequences of action. *Neuroreport*, **12**, 1879–1884.

Butz, M., Timmermann, L., Gross, J. *et al.* (2006). Oscillatory coupling in writing and writer's cramp. *J Physiol Paris*, **99**, 14–20.

Colin, F., Ris, L. & Godaux, E. (2002). Neuroanatomy of the cerebellum. In M. Manto & M. Pandolfo (Eds.), *The Cerebellum and its Disorders* (pp. 6–29). Cambridge, UK: Cambridge University Press.

Dichgans, J. & Fetter, M. (1993). Compartmentalized cerebellar functions upon the stabilization of body posture. *Rev Neurol* (Paris), **149**, 654–664.

Dugas, C. & Smith, A. M. (1992). Responses of cerebellar Purkinje cells to slip of a hand-held object. *J Neurophysiol*, **67**, 483–495.

Espinoza, E., Smith, A. M. (1990). Purkinje cell simple spike activity during grasping and lifting objects of different textures and weights. *J Neurophysiol*, **64**, 698–714.

Fellows, S. J., Ernst, J., Schwarz, M., Töpper, R. & Noth, J. (2001). Precision grip in cerebellar disorders in man. *Clin Neurophysiol*, **112**, 1793–1802.

Flament, D. & Hore, J. (1986). Movement and electromyographic disorders associated with cerebellar dysmetria. *J Neurophysiol*, **55**, 1221–1233.

Gao, J. H., Parsons, L. M., Bower, J. M. *et al.* (1996). Cerebellum implicated in sensory acquisition and discrimination rather than motor control. *Science*, **272**, 545–547.

Gilbert, P. F. C. & Thach, W. T. (1977). Purkinje cell activity during motor learning. *Brain Res*, **128**, 309–328.

Gilman, S. (1969). The mechanism of cerebellar hypotonia. *Brain*, **92**, 621–638.

Gilman, S., Bloedel, J. R. & Lechtenberg, R. (1981). *Disorders of the Cerebellum. Contemporary Neurology Series*. Philadelphia, PA: Davis.

Glickstein, M. & Yeo, C. (1990). The cerebellum and motor learning. *J Cogn Neurosci*, **2**, 69–80.

Glickstein, M., Waller, J., Baizer, J. S., Brown, B. & Timmann, D. (2005). Cerebellum lesions and finger use. *Cerebellum*, **4**, 189–197.

Holmes, G. (1917). The symptoms of acute cerebellar injuries from gunshot wounds. *Brain*, **40**, 461–535.

Holmes, G. (1939). The cerebellum of man. The Hughlings Jackson memorial lecture. *Brain*, **62**, 1–30.

Hore, J., Wild, B. & Diener, H. C. (1991). Cerebellar dysmetria at the elbow, wrist, and fingers. *J Neurophysiol*, **65**, 563–571.

Hoover, J. & Strick, P. (1999). The organization of cerebellar and basal ganglia outputs to primary motor cortex as revealed by retrograde transneural transport of herpes simplex virus type I. *J Neurosci*, **19**, 1446–1463.

Imamizu, H., Miyauchi, S., Tamada, T. *et al.* (2000). Human cerebellar activity reflecting an acquired internal model of a new tool. *Nature*, **403**, 192–195.

Ito, M. (1984). *The Cerebellum and Neural Control*. New York, NY: Raven Press.

Ito, M. (2006). Cerebellar circuitry as a neuronal machine. *Prog Neurobiol*, **78**, 272–303.

Jansen, J. (1969). On cerebellar evolution and organization, from the point of view of a morphologist. In R. Llinas (Ed.), *Neurobiology of Cerebellar Evolution and Development* (pp. 881–893). Chicago, IL: AMA-ERF Inst Biomed Res.

Kawato, M. (1999). Internal models for motor control and trajectory planning. *Curr Opin Neurobiol*, **9**, 718–727.

Kitazawa, S., Kimura, T. & Yin, P. B. (1998). Cerebellar complex spikes encode both destinations and errors in arm movements. *Nature*, **392**, 494–497.

Lamarre, Y., Spidalieri, G. & Chapman, C. E. (1983). A comparison of neuronal discharge recorded in the sensori-motor cortex, parietal cortex, and dentate nucleus of the monkey during arm movements triggered by light, sound or somesthetic stimuli. *Exp Brain Res*, **7**, 140–156.

Larouche, M. & Hawkes, R. (2006). From clusters to stripes: the developmental origins of adult cerebellar compartmentation. *Cerebellum*, **5**, 77–88.

Larsell, O. (1937). The cerebellum. A review and interpretation. *Arch Neurol Psychiatric (Chicago)*, **38**, 580–607.

Lawrence, D. G. & Kuypers, H. G. J. M. (1968). The functional organization of the motor system in the monkey. I. The effects of bilateral pyramidal lesions. *Brain*, **91**, 1–14.

Lewis, R. F. & Zee, D. S. (1993). Ocular motor disorders associated with cerebellar lesions: pathophysiology and topical localization. *Rev Neurol (Paris)*, **149**, 665–677.

Manto, M. (2002). Clinical signs of cerebellar disorders. In M. Manto & M. Pandolfo (Eds.), *The Cerebellum and its Disorders* (pp. 97–120). Cambridge, UK: Cambridge University Press.

Manto, M. (2006). On the cerebello-cerebral interactions. *Cerebellum*, **5**, 286–288.

Manto, M. & Bastian, A. (2007). Cerebellum and the deciphering of motor coding. *Cerebellum*, **6**, 3–6.

Manto, M., Nowak, D. A. & Schutter, D. J. L. G. (2006). Coupling between cerebellar hemispheres and sensory processing. *Cerebellum*, **5**, 187–188.

Manzoni, D. (2007). The cerebellum and sensorimotor coupling: looking at the problem from the perspective of vestibular reflexes. *Cerebellum*, **6**, 24–37.

Mason, C. R., Hendrix, C. M. & Ebner, T. J. (2006). Purkinje cells signal hand shape and grasp force during reach-to-grasp in the monkey. *J Neurophysiol*, **95**, 144–158.

Miall, R. C., Weir, D. J., Wolpert, D. M. & Stein, J. F. (1993). Is the cerebellum a Smith predictor? *J Mot Behav*, **25**, 203–216.

Middleton, F. A. & Strick, P. L. (1997). Cerebellar output channels. In J. D. Schmahmann (Ed.), *The Cerebellum and Cognition* (pp. 61–82). San Diego, CA: Academic Press.

Monzee, J. & Smith, A. M. (2004). Responses of cerebellar interpositus neurons to predictable perturbations applied to an object held in a precision grip. *J Neurophysiol*, **911**, 230–239.

Nowak, D. A., Hermsdörfer, J., Marquardt, C. & Fuchs, H. H. (2002). Grip and load force coupling during discrete vertical movements in cerebellar atrophy. *Exp Brain Res*, **145**, 28–39.

Nowak, D. A., Hermsdörfer, J., Rost, K., Timmann, D. & Topka, H. (2004). Predictive and reactive finger force control during catching in cerebellar degeneration. *Cerebellum*, **3**, 227–235.

Nowak, D. A., Hermsdörfer, J., Timmann, D., Rost, K. & Topka, H. (2005). Impaired generalization of weight-related information in cerebellar degeneration. *Neuropsychologia*, **43**, 20–27.

Nowak, D. A., Timmann, D. & Hermsdörfer, J. (2007). Dexterity in cerebellar agenesis. *Neuropsychologia*, **45**, 696–703.

Nowak, D. A., Topka, H., Timmann, D., Boecker, H. & Hermsdörfer, J. (2007). The role of the cerebellum for predictive control of grasping. *Cerebellum*, **6**, 7–17.

Oscarsson, O. (1976). Functional organization of spinocerebellar paths. In A. Iggo (Ed.), *Handbook of Sensory Physiology, Vol II. Somatosensory System* (pp. 339–380). Berlin: Springer-Verlag.

Pellionisz, A. & Llinás, R. (1980). Tensorial approach to the geometry of brain function: cerebellar coordination via a metric tensor. *Neuroscience*, **5**, 1125–1138.

Rost, K., Nowak, D. A., Timmann, D. & Hermsdörfer, J. (2005). Preserved and impaired aspects of predictive grip force control in cerebellar patients. *Clin Neurophysiol*, **116**, 1405–1414.

Sainburg, R. L., Ghez, C. & Kalakanis, D. (1999). Intersegmental dynamics are controlled by sequential anticipatory, error correction, and postural mechanisms. *J Neurophysiol*, **81**, 1045–1056.

Schneider, K., Zernicke, R. F., Ulrich, B. D., Jensen, J. L. & Thelen, E. (1990). Understanding movement control in infants through the analysis of limb intersegmental dynamics. *J Motor Behav*, **22**, 493–520.

Serrien, J. D. & Wiesendanger, M. (1999). Grip-load coordination in cerebellar patients. *Exp Brain Res*, **128**, 76–80.

Smith, A. M., Dugas, C., Fortier, P., Kalaska, J. & Picard, N. (1993). Comparing cerebellar and motor cortical activity in reaching and grasping. *Can J Neurol Sci*, **3**, S53–S61.

Soteropoulos, D. S. & Baker, S. N. (2006). Cortico-cerebellar coherence during a precision grip task in the monkey. *J Neurophysiol*, **95**, 1194–1206.

Thach, W. T. (1967). Discharges of Purkinje and cerebellar nuclear neurons during rapidly alternating arm movements in the monkey. *J Neurophysiol*, **31**, 785–796.

Vilis, T. & Hore, J. (1977). Effects of changes in mechanical state of limb on cerebellar intention tremor. *J Neurophysiol*, **40**, 1214–1224.

Vilis, T. & Hore, J. (1980). Central neuronal mechanisms contributing to cerebellar tremor produced by limb perturbations. *J Neurophysiol*, **43**, 279–291.

Wolpert, D. M. & Miall, R. C. (1996). Forward models for physiological motor control. *Neural Networks*, **9**, 1265–1279.

Wolpert, D. M., Miall, R. C. & Kawato, M. (1998). Internal models in the cerebellum. *Trends Cogn Sci*, **2**, 338–347.

27

Tremor

LARS TIMMERMANN, JAN RAETHJEN AND GÜNTHER DEUSCHL

Summary

Tremor is defined as a "… rhythmical, involuntary oscillatory movement of a body part …" (Deuschl *et al.*, 1998). These involuntary movements can easily affect the voluntary movements of reaching and grasping up to the total loss of control in patients with severe tremor disorders. The following chapter will review the clinical characteristics and pathophysiological concepts of the most frequent and pathophysiologically important tremor disorders and link the findings to the control of grasping and other hand functions.

Physiological tremor

Clinical characteristics

Any movement or isometric contraction is accompanied by the mostly invisible normal physiological tremor. The limits between normal and pathological tremors can be difficult to define. A pragmatic clinical approach is to define abnormal tremor whenever it is visible to the naked eye. The frequency of physiological tremor is usually greater than 7–8 Hz. It has recently been proposed that any tremor at lower frequencies is likely to be pathological (Elbe *et al.*, 2005), but in cases of gradual transitions this clinical criterion can be problematic.

Pathophysiology

Theoretically tremor oscillations can emerge from two basic mechanisms. Any movable limb can be regarded as a pendulum with the capability to swing rhythmically (oscillate). These oscillations will automatically assume the resonant frequency of this limb which is dependent on its mechanical properties; the greater its weight the lower its resonant frequency, the greater the joint stiffness the higher its frequency (Elbe & Randall, 1978; Lakie *et al.*, 1986). Any mechanical perturbation can activate such an oscillation. In the case of the hands the main and most direct mechanical influence comes from the forearm muscles. Indeed, it has been shown that the physiological tremor measured in normal subjects during muscle activation mainly emerges from an amplification of the muscles' effect on the hand at its resonant frequency (Elbe & Randall, 1978; Timmer *et al.*, 1998).

Sensorimotor Control of Grasping: Physiology and Pathophysiology, ed. Dennis A. Nowak and Joachim Hermsdörfer. Published by Cambridge University Press. © Cambridge University Press 2009.

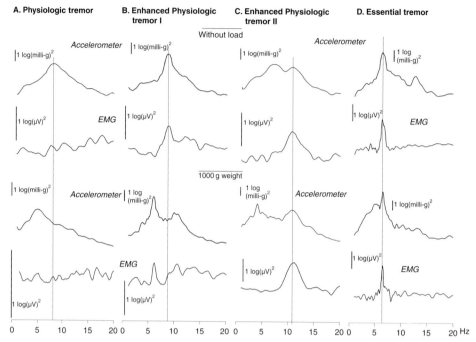

Figure 27.1. Power spectra of accelerometric recordings from the dorsum of the hand and surface EMG recordings from the forearm extensor while the hand is held against gravity (without load, top traces) and while the hand holds an additional load of 1000 g (1000 g weight, bottom traces). A. The typical situation encountered in the majority of normal subjects. There is a broad peak in the acclerometric recordings around 8 Hz which drops significantly by more than 2 Hz under additional weight. The EMG spectra are flat indicating a lack of rhythmic burst activity. Thus this physiologic tremor is a mere resonant phenomenon without any obvious neural involvement. In (B) there is a peak at the accelerometric tremor frequency and also in the EMG indicating a rhythmic burst of activity at this frequency. However the accelerometric peak and the EMG peak drop in parallel under weight load. Thus the rhythmic EMG activity reflects a rhythmic reflex entrainment by the mechanical resonant tremor ("mechanical-reflex-component"). In (C) there are again peaks in EMG and accelerometer recordings at the tremor frequency. But these peaks' frequencies remain constant under weight load indicating a centrally driven rhythm that is largely independent from the peripheral mechanics. Nevertheless there is a second, lower frequency peak which clearly drops under weight load. Thus this is an example of a subject exhibiting a central and peripheral mechanical component of physiological tremor. (D) An example of a pathological essential tremor which clearly is a centrally driven tremor with constant EMG and accelerometer frequencies under weight load.

Spectral analysis of the accelerometrically measured tremor will show a peak at the tremor frequency and the activity of the muscles typically is flat without a distinct peak at this frequency (Figure 27.1A). Such a pure resonant phenomenon does not produce pathologic tremors as its amplitude is typically quite low. However, as this low-amplitude oscillation leads to rhythmic activation of muscle receptors it activates segmental (spinal) or long (e.g. transcortical) reflex loops which can greatly enhance this oscillation. In case of a relevant reflex enhancement of the oscillation the muscle spectrum will not remain flat as in pure resonance but it will show a peak at the tremor frequency driven by the oscillating reflex loop (Figure 27.1B). As the limb mechanics and possibly reflex loops play a role in these oscillations they are termed "mechanical-reflex oscillations". Such reflex mechanisms are one basis of enhanced physiologic tremor.

The second basic mechanism of tremors is a transmission of oscillatory activity within the CNS to the peripheral muscles. The rhythmic activity of the muscles then leads to tremor. Spectral analysis will again show a peak at the tremor frequency in the accelerometer spectrum as well as the muscle spectrum (Figure 27.1C). These oscillations are called "central oscillations." In contrast to the mechanical-reflex oscillations central oscillations occur at the centrally determined frequency and are independent of the limbs' mechanics. The limb mechanics can easily be influenced by putting additional weight on the limb under study. In the case of mechanical-reflex oscillations the tremor frequency becomes lower while it remains unchanged in the case of central oscillations (Figure 27.1A and B). In about 30% of normal subjects such a central tremor component of the physiological tremor can be found (Raethjen *et al.*, 2000b). This central component can also be enhanced under certain circumstances (e.g. drugs) and can lead to an increase in physiologic tremor amplitudes. The vast majority of pathological tremors are centrally driven oscillations, and in their early stage (e.g. low amplitude pathological postural/action tremors in early essential tremor (ET)) they can be difficult to distinguish from enhanced physiological tremor (Figure 27.1D).

The fact that any slow voluntary movement is composed of short rhythmic pulses at the frequency of this tremor has led to the hypothesis that physiological tremor is only an expression of the pulsatile nature of voluntary movements. As these oscillations have been shown to originate from cortical and subcortical motor centers they may represent a common carrier frequency in central motor control.

Interference with grasping movements

Physiological tremor may be necessary for grasping movements rather than interfere with them. Nevertheless, when its amplitude increases, for example in enhanced physiological tremor, it may well be a relevant handicap for such movements. Although there are currently no data available on the effect of enhanced physiological tremor on grasping movements it has been argued that the higher frequency tremor interfering with reach-to-grasp movements in PD may be related to physiologic tremor rather than the dopaminergic deficit (see below).

Metabolic tremor: "Mini-asterixis" in hepatic encephalopathy

Clinical characteristics

Metabolic tremor disorders can be seen in a variety of diseases affecting the liver, the endocrine system or the renal system and manifest clinically often in a postural tremor (Deuschl *et al.*, 1998). One special clinical entitity is the sudden lapse of posture called asterixis (Leavitt & Tyler, 1964). "Asterixis" often occurs with high-grade hepatic encephalopathy (HE) (Butterworth, 2000) and is often accompanied by a mild to moderate irregular postural tremor syndrome which has been termed "mini-asterixis" (Young & Shahani, 1986). "Mini-asterixis" usually occurs with manifest HE in the range of 6–12 Hz and starts after a latent period of about 2–30 s (Leavitt & Tyler, 1964).

Pathophysiology

The pathophysiological mechanisms leading to the generation of "mini-asterixis" in HE have been partly unraveled in the last few years (Timmermann *et al.*, 2002, 2003a, 2005, 2008). Physiologically, corticomuscular coherence, indicating coupling between the primary motor cortex (M1) and the surface EMG, can be found during weak isometric contraction in normal subjects in the 15–35 Hz range (Conway *et al.*, 1995; Salenius *et al.*, 1997; Brown, 2000; Gross *et al.*, 2000; Schnitzler *et al.*, 2000). In patients with hepatic encephalopathy pathological alterations in the sensorimotor system could already be deduced from changes in the excitability of the motor cortex (Nolano *et al.*, 1997) and prolongation of late cortical responses to somatosensory stimulation (Yang *et al.*, 1985, 1998).

To analyze the origin of "mini-asterixis" Timmermann *et al.* compared corticomuscular coherence in patients with "mini-asterixis" at higher stages of HE, in patients with liver cirrhosis without any clinical or subclinical HE and in controls (Timmermann *et al.*, 2002). The authors used magnetoencephalography (MEG) and calculated the coupling between muscle activity in the tremulous arm and cerebral activity. With this approach they could demonstrate that "mini-asterixis" results from a pathologically increased and slowed drive of the primary motor cortex (M1) (Timmermann *et al.*, 2002). This means that "mini-asterixis" as a form of metabolic tremor results from a pathological drive from M1 to the spinal motor neuron pool. In a following study the same group investigated whether this motor deficit is due solely to an alteration in primary motor cortical activity or whether cerebro-cerebral coupling in the motor system of patients with HE and "mini-asterixis" is pathologically altered (Timmermann *et al.*, 2003a). Interestingly, thalamo-cortical coupling was significantly altered in cirrhotics with manifest HE towards pathologically low frequencies. In the cirrhotics with manifest HE the frequency of thalamo-cortical coupling matched the individual frequency of the "mini-asterixis." One could therefore speculate that the pathological motor cortical drive in these patients arises from an altered diencephalic activity that is coupled to M1 (Timmermann *et al.*, 2003a).

These findings gave rise to the hypothesis that motor deficits in encephalopathies in general, and hepatic encephalopathy in particular, are due to a pathological slowing and increased synchronization of oscillatory activity in cortico-subcortical neuronal networks (Timmermann *et al.*, 2005). As a proof of principle, a recent study investigated whether increasing stages of HE lead to a progressive alteration of oscillatory activity in the sensorimotor system (Timmermann *et al.*, 2008). Interestingly, higher stages of HE led not only to a slowing of the cortico-muscular drive on the spinal motor neuron pool as assessed by a M1–EMG coherence analysis, but also to a progressive slowing in the detection of the "critical flicker frequency." The slowing within the motor system, due to a progressive encephalopathy, is probably one of the possible pathophysiological mechanisms in the generation of the clinical deficit of tremor. However, the simultaneous slowing in the visual system of HE patients, as assessed by the critical flicker frequency in HE (Kircheis *et al.*, 2002), indicated that these alterations in cerebral oscillatory networks are not limited to the motor system but can also interfere with other neuronal structures like the visual system. The etiology of the changes in the oscillatory cortico-subcortical loops is probably due to a large number of different factors like changes in the dopaminergic system (Bergeron *et al.*, 1989; Mousseau *et al.*, 1993, 1997), a focal cerebral edema (Haussinger *et al.*, 1994) as well as changes in the cerebral and systemic ammonia levels (Conn, 1993; Butterworth, 2000).

Interference with (grasping) movements

Clinically, "mini-asterixis" certainly interferes with the postural function. However, the delayed onset of "mini-asterixis" and the occurrence during maintained posture prevents a severe effect on voluntary movements like reaching. However, higher grades of hepatic encephalopathy have been shown to affect kinesia mainly by reducing movement initiation (Joebges *et al.*, 2003).

Essential tremor and cerebellar tremor

Clinical characteristics

Classical essential tremor (ET) is a monosymptomatic, bilateral, predominantly postural and action tremor which is usually slowly progressive over years. It may be somewhat asymmetric at onset and hardly ever manifests as a purely unilateral tremor. In 60% of the patients the condition is autosomal dominant. Twin studies suggest a strong heritability above 90% and an almost complete penetrance above 60 years of age. Some chromosomal loci have been linked to essential tremor but the genes causing ET have not yet been identified. About 50–90% of the patients improve with ingestion of alcohol. The topographic distribution shows hand tremor in 94%, head tremor in 33%, voice tremor in 16%, jaw tremor in 8%, facial tremor in 3%, leg tremor in 12% and tremor of the trunk in 3% of the patients. Essential tremor is usually a mainly postural tremor, but rarely even resting tremors do occur which are not related to an abnormality of dopaminergic function or a specific dopaminergic degeneration.

Pathophysiology

Essential tremor is a centrally driven tremor that hardly changes its frequency under differing mechanical conditions (Figure 27.1). Nevertheless, strong peripheral perturbations are capable of resetting the tremor as well as central cortical stimuli applied by magnetic stimulation. These resetting studies indicate that peripheral mechanical as well as central influences can affect even the relatively stable pattern of a centrally driven pathological tremor. The most likely explanation is a reset of the central oscillation that can be achieved by both stimuli, as a strong peripheral stimulus will inevitably also lead to a central response.

The oscillations in ET do not originate from a single central oscillator but rather from a network involving different cortical and subcortical motor structures. This network is dynamically organized with a shifting pattern of cooperation between cortical and subcortical centers. Within this large-scale oscillating network there are several independent loops for each extremity that is involved in the tremor (Raethjen *et al.*, 2000a). Such independent oscillators for different extremities is characteristic of most pathological organic tremors and is contrasted by the strong correlation between different extremities in voluntary and a proportion of psychogenic tremors.

Whereas the cortical oscillations in ET can be measured directly by EEG, and the thalamus has been shown to be involved in deep brain recordings during electrode implantation, there is no direct access to the more subcortical parts of the oscillating network. Only functional imaging studies have looked at the activity in these regions. All of these studies in ET have found an intense bilateral cerebellar activation which is likely related to the tremor.

Interference with (grasping) movements

On the basis of the imaging studies showing cerebellar activation and in view of the findings in animal models and in patients with symptomatic palatal tremor and olivar hypertrophy the hypothesis emerged that the olivo-cerebellar system is the most likely subcortical constituent of the oscillating network in ET. In line with this hypothesis there are a number of abnormalities during movement execution in ET patients resembling a cerebellar deficit. About half of the patients show at least some intention tremor during goal-directed movements that can even be seen clinically, and the trajectories of reach-to-grasp movements in ET and cerebellar patients show a striking similarity when looking at ET patients who present with intention tremor (Deuschl *et al.*, 2000). The rest of the ET patients show movement trajectories that do not differ from those of normal controls (Figure 27.2). The pathophysiological basis of such oscillations in the terminal phase of a ballistic movement is a disturbance of the muscle activation pattern which in single-joint movements typically consists of one agonist burst, followed by an antagonist burst which is then followed by a weaker second agonist burst to stabilize the final position. This final burst is delayed in cerebellar disease and leads to a destabilization and oscillation (intention tremor) at the end of the movement. This same delay in the second agonist burst can be found in ET patients with intention tremor. Whereas normal gait is usually relatively unaffected in ET the more difficult tandem gait also shows severe abnormalities in ET patients presenting with intention tremor. Again the foot trajectories in these patients are very similar to those of cerebellar patients with gait ataxia. Oculomotor function is clinically normal in the vast majority of ET patients even when they show intention tremor. This is one means to distinguish them clinically from patients with cerebellar disease. However, when looking at the eye movements quantitatively one can find a mild abnormality consisting of a reduced

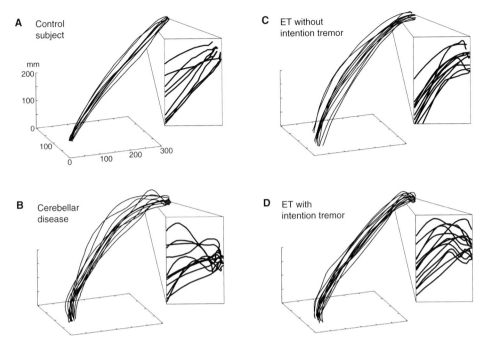

Figure 27.2. Examples of hand trajectories during reach-to-grasp movements in essential tremor patients (ET), patients with cerebellar disease and normal controls. In ET without intention tremor (C) the trajectories look straight up to the final approach and do not differ from those in the control subjects (A). In ET with intention tremor (D) there is overshoot and irregular oscillations in the final phase of the approach of the target, and this trajectory shape resembles the final phase in patients with cerebellar disease (B). (Modified from Deuschl *et al.*, 2000.)

initial eye acceleration in smooth pursuit movements and a pathological suppression of the vestibulo-ocular reflex time constants by head tilts.

As arm movements, gait and oculomotor function are all affected there seems to be a pancerebellar deficit in ET. The mechanism of this loss of cerebellar function remains unclear, however. It has recently been postulated that ET is a neurodegenerative disease. Such a degeneration could then lead to the cerebellar malfunction making it more prone to pathological oscillations.

Tremors in Parkinson's disease

Clinical characteristics

Most patients with Parkinson's disease present with tremor. Resting tremor with pill-rolling is typical. However, around 40% of patients have different forms of postural and action tremor (Findley *et al.*, 1981; Raethjen *et al.*, 2005) which can occur isolated or together with resting tremor. Isolated rest tremor around 4–6 Hz which may or may not re-emerge under steady postural conditions ("re-emergent tremor") is termed Type I or Classical Parkinsonian tremor. A combination of the classical rest tremor with a higher-frequency non-harmonically related postural/action tremor is called Type II Parkinsonian tremor. A mild higher-frequency action tremor is common in patients with PD and has often been attributed to a combination of ET and PD. Type III Parkinsonian tremor denotes an isolated higher frequency action tremor which can rarely occur without a concomitant rest tremor in PD and may be very disabling for the patients.

The severity of the different tremors in PD is not related to the severity of the other cardinal symptoms (akinesia, rigidity and postural instability) in PD. The classical rest tremor can occur many years before the patients develop the other Parkinsonian features. Such monosymptomatic rest tremors can be difficult to distinguish from ET. However, imaging of the basal ganglia dopamine system can help, as all of the monosymptomatic rest tremor patients show the typical dopaminergic deficit of PD.

Pathophysiology

Rest tremor

Resting tremor in Parkinson's disease has been extensively studied in the last decade. Parkinson's disease resting tremor is often a tricky clinical symptom because of its insufficient response to dopaminergic therapy (Lang & Lozano, 1998). However, the effect of levo dopa on resting tremor in PD is clinically mainly a reduction in amplitude and increase in tremor frequency (Sturman et al., 2004).

The pathological substrate of the peripheral oscillatory activity in the EMG, leading to the clinical symptom of PD tremor, is obviously a pathological synchronization of oscillatory activity within a cortico-subcortical network (Timmermann et al., 2003b). One of the key players involved in cerebral tremor generation is the primary motor cortex (M1), which shows strong coupling with the peripheral tremor EMG (Tass et al., 1998; Volkmann, 1998; Timmermann et al., 2003b). Surprisingly, M1 activity does not occur primarily at tremor frequency around 3–6 Hz, but is mainly established at double tremor frequency around 8–12 Hz (Tass et al., 1998; Timmermann et al., 2003b). The transformation of the cerebral 8–12 Hz M1 activity to the peripheral tremor in the range of 3–6 Hz is probably realized by a 2:1 coupling (Tass et al., 1998; Timmermann et al., 2003b; Wang et al., 2005). It has been hypothesized that one of the mechanisms leading to this 2:1 transformation is a spinal "flip-flop effect" leading to an alternating drive of the 8–12 Hz M1 activity to the agonistic and antagonistic muscle (Timmermann et al., 2003b; Timmermann et al., 2007). This could easily explain the alternating clinical pattern of the PD resting tremor.

The primary motor cortex is not an isolated brain area involved in PD tremor generation but integrated in an extensive network of cortical and subcortical sensorimotor brain areas, comprising higher motor areas like the premotor cortex and the supplementary motor area as well as sensory areas like the primary and secondary somatosensory cortex and diencephalic and cerebellar brain areas (Timmermann et al., 2003b). These brain areas show strong coupling among each other at tremor frequency and, even stronger, at double tremor frequency around 8–12 Hz. Additionally, coupling is realized in the range of 20 Hz. One might speculate about the physiological role of this pathological oscillatory network. Interestingly, a further MEG study revealed that the same brain areas are involved in the voluntary simulation of PD resting tremor in healthy volunteers as in PD patients showing involuntary resting tremor (Pollok *et al.*, 2004). The major difference between patients and healthy controls was that controls showed strong coupling between premotor areas and M1 whereas patients showed stronger coupling between M1 and thalamic areas. This indicates that the generation of PD resting tremor is established in a preformed physiological network of sensorimotor areas with pathological driving from deep, subcortical structures.

One of the central questions remains open: what is the cerebral generator of PD resting tremor? The involvement of the cerebellum in central PD tremor generation was clearly demonstrated in MEG studies (Timmermann *et al.*, 2003b). However, the fact that PD patients without a cerebellum can still develop tremor, even though altered, indicates that the cerebral generator of the PD resting tremor is not located in the cerebellum, but that PD tremor patterns are obviously modulated within the cerebello-thalamo-cortical loops (Deuschl *et al.*, 1999). One likely answer could be that there is no single generator, but a number of generators are giving rise to tremor activity in different limbs (Hurtado *et al.*, 2000; Raethjen *et al.*, 2000a). Functionally, PD resting tremor seems to be mainly generated within the contralateral hemisphere; however, there is good experimental evidence that interhemispheric cross-talk exists (Liu *et al.*, 2002).

One explanation why resting tremor is not completely abolished with dopaminergic therapy is that the pathological synchronization of oscillatory activity in the primate model of Parkinson's disease is not completely abolished by dopamine replacement therapy (Heimer *et al.*, 2006). Interestingly, the cerebral oscillatory activity giving rise to the peripheral resting tremor is dynamically organized. Wang and colleagues showed that in local field potentials in the subthalamic nucleus of PD patients oscillatory activity in the range of 10–30 Hz was suppressed and displaced by activity in the tremor range (3.0–4.5 Hz) before an episode with tremor was detectable in the EMG (Wang *et al.*, 2005). Moreover, in a single case report a PD patient showed transient synchronization between different local field potentials within the internal globus pallidus (GPi) with otherwise separated tremor "subloops" (Hurtado *et al.*, 1999). That indicates that the described pathological subcortical-cortical oscillatory activity is (i) organized in distinct functional tremor "subloops" and (ii) dynamically modulated with alternating frequency patterns giving rise to subsequent episodes of different clinical deficits like tremor.

Postural/action tremor

Postural tremor in PD is common. However, in the majority of patients this is only a re-emergence of the rest tremor under steady isometric muscle contractions (Raethjen *et al.*, 2005). As soon as the patients initiate an active movement this rest and re-emergent tremor is inhibited. An analysis of the hand velocity profile in a reach-to-grasp paradigm shows typical lower frequency (4–6 Hz) rest tremor oscillations while the hand rests before the movement starts. These oscillations stop as soon as the reaching movement starts, and during the fast

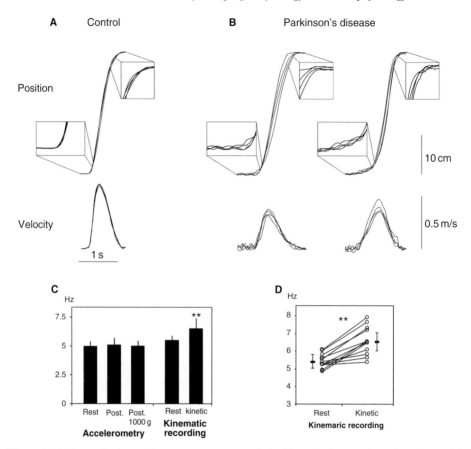

Figure 27.3. Tremor during reach-to-grasp movements in Parkinson's disease. The trajectories and the velocity profiles (A/B) show oscillations in the resting state before the start of the movement and in the final phase of the target approach. The mean frequency (n = 13) of the resting oscillations is significantly lower than the kinetic oscillations in the final phase (C/D, asterisks indicate statistical significance). The frequency of the resting oscillations before the movements start does not differ from the accelerometrically measured rest and re-emergent tremor frequency (C). Modified from Wenzelburger *et al.* (2000).

ballistic movement components hardly any tremor can be seen. In the final phase of the movement when the target is approached and the grasp is initiated another tremor occurs which has a lower amplitude in the majority of patients and has a higher frequency (5.5–7.5 Hz) than the rest tremor before movement onset (Wenzelburger *et al.*, 2000). In a quantitative movement analysis of a reach-to-grasp paradigm such a tremor can be detected in virtually all PD patients (Wenzelburger *et al.*, 2000). Two examples of reach-to-grasp trajectories and velocity profiles from two PD patients are demonstrated in Figure 27.3 showing these oscillations. When counting the frequencies of these oscillations there is a significant difference between the higher frequencies of the action tremor in the final phase and the rest tremor before movement onset, which is in the same range as the accelerometrically measured tremor frequencies of the rest and re-emergent postural tremor in the same patients (Figure 27.3). These results underline that the higher frequency kinetic tremor is a common phenomenon in PD and clearly is an entity separate from re-emergent rest tremor (Wenzelburger *et al.*, 2000). In a recent study of the finger forces during pinch-grip object manipulation these findings were extended. Patients were asked to grasp an object, lift it and hold it in a steady position for some time (Raethjen *et al.*, 2005). Again in the final phase of the lift and right after the steady holding position was reached there were fast force oscillations between 6 and 10 Hz corresponding to the higher frequency action tremor. After some seconds of holding the object in a steady position a higher amplitude and lower frequency (4–6 Hz) tremor appears matching the re-emerging resting tremor (Figure 27.4). Under levodopa treatment this re-emergent rest tremor improves and even disappears in the majority of subjects, whereas the higher frequency action tremor remains almost unchanged (Figure 27.4). Thus the dopaminergic deficit seems to be less important in the pathogenesis of the higher-frequency action tremor than in the classical rest tremor (Raethjen *et al.*, 2005). However, in contrast to the emerging knowledge on the central tremor-generating network of the classical rest tremor in PD (see above) little is known on the origin of the action tremor. As similar oscillations also occur in healthy subjects it has been hypothesized that it might be an enhancement of physiological tremor. It is not clear if the higher-frequency action tremor is related to the oscillations at the first harmonic of the rest tremor frequency (see above).

Interference with grasping movements

The classical resting tremor (type I tremor) hardly influences the movement itself because it is inhibited as soon as the movement is initiated. Nevertheless, there is some evidence that it paces the movement onset in PD patients. It has been shown that the movement onset typically coincides with the corresponding part of the tremor cycle (Wenzelburger *et al.*, 2000).

As the action tremor is only present during active, e.g. grasping, movements it has the potential to interfere with them. However, it is usually of low amplitude and only few PD patients exhibit obviously disabling action tremor. Nevertheless, even smaller amplitude action tremor seems to have some influence on the timing of grasping movements. The stronger the action tremor during grasping the slower the grip formation. This effect was seen particularly in those patients who were severely affected by akinesia. This in conjunction with the akinesia action tremor in PD interferes with the patients' manual dexterity by further altering the fast sequence of fine finger movements during grasping (Raethjen *et al.*, 2005).

Summary and conclusions

Although clinically well described for decades, our knowledge on the cerebral pathophysiological mechanisms of the different tremor disorders is sparse. However, new electrophysiological and functional imaging techniques are promising new methodological steps

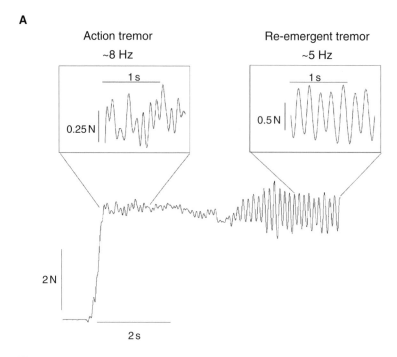

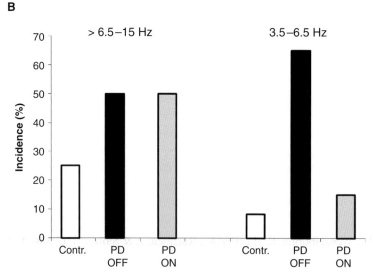

Figure 27.4. Action and re-emergent tremor in a lift-to-hold paradigm. A. An example of a lift force trace exhibiting force-oscillations in the upward direction of the index finger lifting a mobile object of 220 g while holding it with the thumb in a pinch grip. Force transducers at the touch points measure the force in all directions. The magnified insets show that there are fast oscillations around 8 Hz right in the final phase of the lift movement and during the initial holding phase. After a short time these higher frequency oscillations diminish and are replaced by the typical higher amplitude re-emergent tremor around 5 Hz. B. The proportion of patients (n = 20) exhibiting these tremors with (ON) and without (OFF) levodopa treatment is displayed and compared with age-matched controls (n = 18). It can be seen that the low frequency re-emergent tremor in 3.5–6.5 band regularly disappears under levodopa whereas the proportion of patients with the higher frequency (> 6.5–15 Hz) action tremor remains unchanged. Modified from Raethjen *et al.* (2005).

in the understanding of tremor disorders. Our knowledge regarding how different tremor disorders affect hand function and grasping remain incomplete even though these aspects are of great importance in everyday life and disability of our patients. The new possibilities of therapeutic intervention with deep brain stimulation in pathological cerebral tremor activity resulting in often dramatic clinical improvement have encouraged a new generation of research in the field of tremor. However, with respect to a number of unresolved questions in the field of tremor: we are just at the beginning.

Acknowledgments

Support from the German Research Council (Deutsche Forschungsgemeinschaft, DFG) is gratefully acknowledged (Grant RA 1005, 1-1). LT is supported by grants of the "Manfred and Ursula Müller Foundation," the "Klüh-Foundation," the Cologne Fortune program, University of Cologne, Faculty of Medicine and the "Bundesministerium für Bildung und Forschung" (BMBF).

References

Bergeron, M., Reader, T. A., Layrargues, G. P. & Butterworth, R. F. (1989). Monoamines and metabolites in autopsied brain tissue from cirrhotic patients with hepatic encephalopathy. *Neurochem Res*, **14**, 853–859.

Brown, P. (2000). Cortical drives to human muscle: the Piper and related rhythms. *Prog Neurobiol*, **60**, 97–108.

Butterworth, R. F. (2000). Complications of cirrhosis. III. Hepatic encephalopathy. *J Hepatol*, **32**, 171–180.

Conn, H. O. (1993). Hepatic encephalopathy. In B. Schiff & E. R. Schiff (Eds.), *Diseases of the Liver* (pp. 1036–1060). Philadelphia, PA: Lippincott.

Conway, B. A., Halliday, D. M., Farmer, S. F. *et al.* (1995). Synchronization between motor cortex and spinal motoneuronal pool during the performance of a maintained motor task in man. *J Physiol (Lond)*, **489**, 917–924.

Deuschl, G., Bain, P. & Brin, M. (1998). Consensus statement of the Movement Disorder Society on Tremor. Ad Hoc Scientific Committee. *Mov Disord*, **13** (Suppl. 3), 2–23.

Deuschl, G., Wilms, H., Krack, P., Wurker, M. & Heiss, W. D. (1999). Function of the cerebellum in Parkinsonian rest tremor and Holmes' tremor. *Ann Neurol*, **46**, 126–128.

Deuschl, G., Wenzelburger, R., Loffler, K., Raethjen, J. & Stolze, H. (2000). Essential tremor and cerebellar dysfunction clinical and kinematic analysis of intention tremor. *Brain*, **123 (Pt. 8)**, 1568–1580.

Elble, R. J. & Randall, J. E. (1978). Mechanistic components of normal hand tremor. *Electroencephalogr Clin Neurophysiol*, **44**, 72–82.

Elble, R. J., Higgins, C. & Elble, S. (2005). Electrophysiologic transition from physiologic tremor to essential tremor. *Mov Disord*, **20**, 1038–1042.

Findley, L. J., Gresty, M. A. & Halmagyi, G. M. (1981). Tremor, the cogwheel phenomenon and clonus in Parkinson's disease. *J Neurol Neurosurg Psychiatry*, **44**, 534–546.

Gross, J., Tass, P. A., Salenius, S. *et al.* (2000). Cortico-muscular synchronization during isometric muscle contraction in humans as revealed by magnetoencephalography. *J Physiol*, **527**, 623–631.

Haussinger, D., Laubenberger, J., vom Dahl, S. *et al*. (1994). Proton magnetic resonance spectroscopy studies on human brain myo-inositol in hypo-osmolarity and hepatic encephalopathy. *Gastroenterology*, **107**, 1475–1480.

Heimer, G., Rivlin-Etzion, M., Bar-Gad, I. *et al*. (2006). Dopamine replacement therapy does not restore the full spectrum of normal pallidal activity in the 1-methy-4-phenyl-1,2,3,6-tetra-hydropyridine primate model of Parkinsonism. *J Neurosci*, **26**, 8101–8114.

Hurtado, J. M., Gray, C. M., Tamas, L. B. & Sigvardt, K. A. (1999). Dynamics of tremor-related oscillations in the human globus pallidus: a single case study. *Proc Natl Acad Sci USA*, **96**, 1674–1679.

Hurtado, J. M., Lachaux, J. P., Beckley, D. J., Gray, C. M. & Sigvardt, K. A. (2000). Inter- and intralimb oscillator coupling in parkinsonian tremor. *Mov Disord*, **15**, 683–691.

Joebges, E. M., Heidemann, M., Schimke, N. *et al*. (2003). Bradykinesia in minimal hepatic encephalopathy is due to disturbances in movement initiation. *J Hepatol*, **38**, 273–280.

Kircheis, G., Wettstein, M., Timmermann, L., Schnitzler, A. & Haussinger, D. (2002). Critical flicker frequency for quantification of low-grade hepatic encephalopathy. *Hepatology*, **35**, 357–366.

Lakie, M., Walsh, E. G. & Wright, G. W. (1986). Passive mechanical properties of the wrist and physiological tremor. *J Neurol Neurosurg Psychiatry*, **49**, 669–676.

Lang, A. E. & Lozano, A. M. (1998). Parkinson's disease. Second of two parts. *N Engl J Med*, **339**, 1130–1143.

Leavitt, S. & Tyler, H. R. (1964). Studies in asterixis. *Arch Neurol*, **10**, 360–368.

Liu, X., Ford-Dunn, H. L., Hayward, G. N. *et al*. (2002). The oscillatory activity in the Parkinsonian subthalamic nucleus investigated using the macro-electrodes for deep brain stimulation. *Clin Neurophysiol*, **113**, 1667–1672.

Mousseau, D. D., Perney, P., Layrargues, G. P. & Butterworth, R. F. (1993). Selective loss of pallidal dopamine D2 receptor density in hepatic encephalopathy. *Neurosci Lett*, **162**, 192–196.

Mousseau, D. D., Baker, G. B. & Butterworth, R. F. (1997). Increased density of catalytic sites and expression of brain monoamine oxidase A in humans with hepatic encephalopathy. *J Neurochem*, **68**, 1200–1208.

Nolano, M., Guardascione, M. A., Amitrano, L. *et al*. (1997). Cortico-spinal pathways and inhibitory mechanisms in hepatic encephalopathy. *Electroencephalogr Clin Neurophysiol*, **105**, 72–78.

Pollok, B., Gross, J., Dirks, M., Timmermann, L. & Schnitzler, A. (2004). The cerebral oscillatory network of voluntary tremor. *J Physiol*, **554**, 871–878.

Raethjen, J., Lindemann, M., Schmaljohann, H. *et al*., (2000a). Multiple oscillators are causing parkinsonian and essential tremor. *Mov Disord*, **15**, 84–94.

Raethjen, J., Pawlas, F., Lindemann, M., Wenzelburger, R. & Deuschl, G. (2000b). Determinants of physiologic tremor in a large normal population. *Clin Neurophysiol*, **111**, 1825–1837.

Raethjen, J., Pohle, S., Govindan, R. B. *et al*. (2005). Parkinsonian action tremor: interference with object manipulation and lacking levodopa response. *Exp Neurol*, **194**, 151–160.

Salenius, S., Portin, K., Kajola, M., Salmelin, R. & Hari, R. (1997). Cortical control of human motoneuron firing during isometric contraction. *J Neurophysiol*, **77**, 3401–3405.

Schnitzler, A., Gross, J. & Timmermann, L. (2000). Synchronised oscillations of the human sensorimotor cortex. *Acta Neurobiol Exp*, **60**, 271–287.

Sturman, M. M., Vaillancourt, D. E., Metman, L. V., Bakay, R. A. & Corcos, D. M. (2004). Effects of subthalamic nucleus stimulation and medication on resting and postural tremor in Parkinson's disease. *Brain*, **127**, 2131–2143.

Tass, P., Rosenblum, M. G., Weule, J. *et al.* (1998). Detection of n:m phase locking from noisy data: Application to magnetoencephalography. *Phys Rev Lett*, **81**, 3291–3294.

Timmer, J., Lauk, M., Pfleger, W. & Deuschl, G. (1998). Cross-spectral analysis of physiological tremor and muscle activity. I. Theory and application to unsynchronized electromyogram. *Biol Cybern* **78**, 349–357.

Timmermann, L., Gross, J., Kircheis, G., Haussinger, D. & Schnitzler, A. (2002). Cortical origin of mini-asterixis in hepatic encephalopathy. *Neurology*, **58**, 295–298.

Timmermann, L., Gross, J., Butz, M. *et al.* (2003a). Mini-asterixis in hepatic encephalopathy induced by pathologic thalamo-motor-cortical coupling. *Neurology*, **61**, 689–692.

Timmermann, L., Gross, J., Dirks, M. *et al.* (2003b). The cerebral oscillatory network of parkinsonian resting tremor. *Brain*, **126**, 199–212.

Timmermann, L., Butz, M., Gross, J. *et al.* (2005). Neural synchronization in hepatic encephalopathy. *Metab Brain Dis*, **20**, 337–346.

Timmermann, L., Florin, E. & Reck, C. (2007). Pathological cerebral oscillatory activity in Parkinson's disease: a critical review on methods, data and hypotheses. *Expert Rev Med Devices*, **4**, 651–661.

Timmermann, L., Butz, M., Gross, J. *et al.* (2008). Impaired cerebral oscillatory processing in hepatic encephalopathy. *Clin Neurophysiol*, **119**, 265–272.

Volkmann, J. (1998). Oscillations of the human sensorimotor system as revealed by magnetoencephalography. *Mov Disord*, **13**, 73–76.

Wang, S. Y., Aziz, T. Z., Stein, J. F. & Liu, X. (2005). Time-frequency analysis of transient neuromuscular events: dynamic changes in activity of the subthalamic nucleus and forearm muscles related to the intermittent resting tremor. *J Neurosci Meth*, **145**, 151–158.

Wenzelburger, R., Raethjen, J., Loffler, K. *et al.* (2000). Kinetic tremor in a reach-to-grasp movement in Parkinson's disease. *Mov Disord*, **15**, 1084–1094.

Yang, S. S., Chu, N. S. & Liaw, Y. F. (1985). Somatosensory evoked potentials in hepatic encephalopathy. *Gastroenterology*, **89**, 625–630.

Yang, S. S., Wu, C. H., Chiang, T. R. & Chen, D. S. (1998). Somatosensory evoked potentials in subclinical portosystemic encephalopathy: a comparison with psychometric tests. *Hepatology*, **27**, 357–361.

Young, R. R. & Shahani, B. T. (1986). Asterixis: one type of negative myoclonus. *Adv Neurol*, **43**, 137–156.

28

Schizophrenia

DENNIS A. NOWAK

Summary

It is widely held that schizophrenia is associated with a variety of subtle sensory and motor impairments – so called *neurological soft signs* – that may impact on manual dexterity. Neurological soft signs (NSS) in schizophrenia appear to be part of the underlying disorder. The motor deficit of the hand, however, may also worsen as a side effect of antipsychotic treatment. Within the theoretical framework of internal models schizophrenia has been associated with a deficit of self-monitoring and awareness of action. Deficient monitoring of the sensory consequences of voluntary movement may be directly related to the motor deficit to be found in schizophrenia. This chapter summarizes kinetic and kinematic aspects of impaired manual dexterity in schizophrenia and discusses the motor disability within the context of internal models for the sensorimotor processing of voluntary actions.

Introduction

Early in the 20th century, Bleuler (1908) and Kraepelin (1919) described several motor abnormalities in schizophrenia, such as problems in the sequencing and spacing of steps when walking and dyscoordination of hand and arm movements when performing handiwork and crafts. In this era antipsychotic drugs did not exist and, consequently, these early clinical observations cannot simply be considered a side effect of antipsychotic treatment. Today, deficits of fine motor performance, also referred to as neurological soft signs (NSS),[1] are still observed in a substantial proportion of schizophrenic subjects, but their nature is still not completely understood and their semiology is not easily distinguishable from side effects of antipsychotic treatment. Importantly, the presence of NSS in schizophrenia has been found to be significantly correlated with poor premorbid social functioning, early onset of the disease and poor prognosis (Quitkin *et al.*, 1976; Johnstone *et al.*, 1990; Gupta *et al.*, 1995; Tosato & Dazzan, 2005; Jahn *et al.*, 2006). Therefore, a clear-cut distinction between discrete motor signs related to the underlying pathology on the one hand and Parkinson-like side effects of antipsychotic medication on the other hand is very desirable for both clinical and research purposes.

Sensorimotor Control of Grasping: Physiology and Pathophysiology, ed. Dennis A. Nowak and Joachim Hermsdörfer. Published by Cambridge University Press. © Cambridge University Press 2009.

Neurological soft signs in schizophrenia

Pertinent clinical reports describe a general dyscoordination and clumsiness of voluntary upper limb movement in a great proportion (about 60%) of schizophrenic subjects (Heinrichs & Buchanan, 1988; Jahn, 1999; Boks *et al.*, 2000; Chen *et al.*, 2000). Despite the significance of NSS in schizophrenia several essential aspects are still controversial. Some earlier data imply that NSS may be used as a subclinical marker for subjects free of psychopathological symptoms, but at risk for schizophrenia (McNeil *et al.*, 1993; Chen *et al.*, 2000; Yazici *et al.*, 2002; Gourion *et al.*, 2004; Varambally *et al.*, 2006). Others suggest a correlation between NSS and the psychopathological symptoms of the disease (Browne *et al.*, 2000; Varambally *et al.*, 2006). However, attempts to correlate the occurrence of NSS with clinical indices of psychopathology generated inconsistent results (for a recent review see Tosato & Dazzan, 2005). Whereas some investigators observed significant correlations with negative symptoms (Merriam *et al.*, 1990; Henkel *et al.*, 2004; Jahn *et al.*, 2006; Varambally *et al.*, 2006), positive symptoms (Green & Walker, 1985; Browne *et al.*, 2000) or both negative *and* positive symptoms (Mohr *et al.*, 1996; Mittal *et al.*, 2007), others did not observe any significant correlations between psychopathologic symptom severity and the presence of NSS (Bartko *et al.*, 1988).

Also of debate are the time of occurrence of NSS, their development over the course of the disease and their interaction with antipsychotic treatment. Neurological soft signs may occur prior to the psychopathologic manifestation of schizophrenia in affected individuals (Walker 1994; Compton *et al.*, 2007). Interestingly, several studies have shown a reduction in the severity of NSS after a single psychotic episode and over the course of the disease (Schröder *et al.*, 1992, 1998; Jahn *et al.*, 2006). Recent data suggest that the severity of NSS to be found in untreated subjects with schizophrenia is negatively correlated with the degree of improvement during antipsychotic treatment (Scheffer, 2004; Mittal *et al.*, 2007). That is, the greater the amount of NSS in untreated schizophrenia, the smaller the beneficial effect of antipsychotic therapy regarding psychopathological symptom relief (Mittal *et al.*, 2007). To judge the potential prognostic value of NSS in schizophrenia (Quitkin *et al.*, 1976; Johnstone *et al.*, 1990; Gupta *et al.*, 1995), it appears essential to clearly distinguish them from the side effects of antipsychotic treatment.

Parkinson-like side effects of antipsychotic treatment in schizophrenia

Antipsychotics are widely used in the treatment of schizophrenic psychosis and frequently associated with basal ganglia dysfunction causing Parkinson-like symptoms, such as tremor, increased muscle stiffness (rigidity), slowness of movement (bradykinesia) and postural imbalance (Vaughan *et al.*, 1991; Farde *et al.*, 1992). First-generation antipsychotic drugs are high-affinity antagonists of dopamine-D2 receptors that are most effective against psychotic symptoms, but have high rates of Parkinson-like side effects (Quitkin *et al.*, 1976; Merriam *et al.*, 1990; Henkel *et al.*, 2004). Importantly, it appears as if Parkinson-like side effects may also occur with modern antipsychotic treatment (Nowak *et al.*, 2006a).

There is some evidence that the severity of NSS correlate with the doses of antipsychotics administered and/or the clinical severity of Parkinson-like side effects associated with such treatment (Quitkin *et al.*, 1976; Merriam *et al.*, 1990; Henkel *et al.*, 2004). However, other investigators did not observe such associations (Heinrichs & Buchanan, 1988; Buchanan *et al.*, 1994; Gupta *et al.*, 1995; Jahn *et al.*, 2006).

Within this context it is important to consider that both NSS and Parkinson-like side effects are more frequent in the chronic stages of schizophrenia with substantial clinical overlap between both entities in late-stage schizophrenic subjects (Schröder *et al.*, 1992). Longitudinal studies found a dissociation between the severity of NSS and the occurrence of Parkinson-like side effects in drug-naïve schizophrenic patients after the initiation of antipsychotic therapy (Schröder *et al.*, 1998). Interestingly, an upregulation of striatal dopamine-D2 receptors as indexed by single photon emission computed tomography (SPECT) was significantly correlated with scores of Parkinson-like side effects, but not with scores of NSS (Schröder *et al.*, 1998). This is an essential observation given the fact that striatal dopamine-D2 receptor blockade is accepted to be the cause of Parkinson-like side effects in antipsychotic therapy (Farde *et al.*, 1989, 1992). In contrast, the pathophysiology of NSS has been associated with functional changes in various sites of the central nervous system, such as the cerebellum (Blakemore *et al.*, 2000, 2001) or sensorimotor cortex (Schröder *et al.*, 1995).

Kinematic motion analysis as an objective measure of NSS and Parkinson-like side effects of antipsychotic treatment in schizophrenia

Kinematic motion analysis allows for an objective evaluation and documentation of both NSS and Parkinson-like side effects associated with antipsychotic therapy in schizophrenia (Carnahan *et al.*, 1996; Saoud *et al.*, 2000; Tigges *et al.*, 2000; Delevoye-Turrell *et al.*, 2003; Henkel *et al.*, 2004; Jahn *et al.*, 2006; Nowak *et al.*, 2006a). Diadochokinesia, the coordination of alternating activity in antagonistic muscles, is easily accessible to clinical testing and frequently found to be hampered in schizophrenia (Jahn *et al.*, 1995; Putzhammer *et al.*, 2005). Impaired diadochokinesia, also referred to as dysdiadochokinesia, is considered a NSS in schizophrenic subjects. Several lines of behavioral evidence imply that there is a primary deficit of motor performance in schizophrenia, probably causing NSS, and that this sensorimotor deficit may be further enhanced by antipsychotic treatment.

Using an ultrasound-based 3D motion analyzer, Putzhammer *et al.* (2005) comparatively investigated the kinematics of alternating pronation/supination movements of the forearm in 20 healthy subjects, 20 drug-naïve schizophrenic subjects and schizophrenic subjects treated with olanzapine (n = 20; average dosage: 16.8 ± 6 mg per day), haloperidol (n = 13; average dosage: 7.6 ± 4 mg per day) or fluphenazine (n = 7; average dosage: 6.7 ± 3 mg per day). The schizophrenic subjects' psychopathology was rated using the Positive and Negative Syndrome Scale (PANNS) (Kay *et al.*, 1987). There were no significant differences in the PANNS scores assessed for overall, negative or positive psychopathology among the groups of schizophrenic subjects. These authors found a significant reduction in the amplitude and velocity of diadochokinetic forearm movements in drug-naïve schizophrenic subjects, suggesting the presence of NSS. Importantly, administration of either first-generation (haloperidol, fluphenazine) or second-generation (olanzapine) antipsychotics further worsened motor performance. Slowing of movement and deficient automation was most pronounced within the group of patients receiving first-generation antipsychotics and this is probably related to the greater affinity of first-generation antipsychotics to striatal dopamine-D2 receptors.

In an attempt to disentangle NSS from the side effects of antipsychotic treatment, Henkel *et al.* (2004) investigated writing movements of 16 schizophrenic subjects (12 paranoid type, 1 disorganized type, 3 undifferentiated type) in comparison with healthy control subjects matched for age and gender. Schizophrenic subjects were tested at baseline (after a pause from antipsychotic drugs of at least 4 weeks) and after a treatment period with haloperidol over 14 days (average dosage: 10 ± 4 mg per day). Haloperidol is a first-generation antipsychotic (butyrophenon) with high affinity to striatal dopamine-D2 receptors and great potential to produce Parkinson-like side effects. At baseline, all patients suffered from positive symptoms, such as delusions, hallucinations and disorganized speech. Symptom severity was rated using the PANNS (Kay *et al.*, 1987). Kinematic analysis of handwriting of the dominant hand comprised drawing of superimposed concentric circles with a diameter of 12 mm as fast as possible and writing of a test sentence. There were three important results: (i) in comparison with healthy controls, schizophrenic subjects free of antipsychotic drugs exhibited subtle, but clearly measurable motor impairments at baseline, such as slower velocities and reduced automation of drawing and writing movements; (ii) in schizophrenic subjects a significant correlation between the severity of negative schizophrenic symptoms (as indexed by the PANNS negative score) and the velocities and automation of drawing and writing were found both at baseline and after 14 days of haloperidol treatment; and (iii) Parkinson-like side effects of haloperidol treatment, such as bradykinesia and rigidity, were positively correlated with an increased variability of writing peak velocity. Again, these data strongly support the idea that subtle deficits of sensorimotor performance in schizophrenia are part of the underlying pathology and that Parkinson-like side effects associated with antipsychotic therapy may enhance this deficit.

Delevoye-Turrell *et al.* (2003) addressed the question of whether schizophrenia affects the predictive control mechanisms necessary for an accurate timing, scaling and sequencing of grip forces. Sixteen subjects with schizophrenia and 16 age- and gender-matched healthy subjects performed three different manipulative tasks: (i) grasping and lifting objects of different mass (450 g, 710 g, 880 g) and texture (sandpaper, cotton, silk); (ii) using a hand-held manipulandum to hit a pendulum; and (iii) using a hand-held manipulandum to resist impacts produced by a pendulum. Severity of positive (mean: 32 ± 26 points) and negative symptoms (mean: 42 ± 20 points) of schizophrenia was rated using the PANNS (Kay *et al.*, 1987). Fourteen of the 16 subjects with schizophrenia were receiving long-term antipsychotic treatment (average dosage of chlorpromazine or chlorpromazine equivalents: 244 ± 171 mg per day). Two patients were not receiving any treatment. Compared to healthy subjects (see also Chapters 1 and 12), schizophrenic subjects showed several abnormalities when grasping and lifting objects of different weight and texture: they took more time to establish a firm grip as indexed by longer time lags from the onset of grip-force increase to the onset of load-force increase and they applied greater and more variable grip force to lift and hold the object. However, schizophrenic subjects, like healthy controls, scaled grip forces differentially to object texture and mass: peak rates of grip-force increase and peak grip force were scaled higher for objects that were more slippery and heavier. When hitting the pendulum schizophrenic subjects exhibited an abnormal timing and scaling of grip force compared with healthy subjects: patients started to increase grip force too late with respect to the onset of the hitting movement, peak grip force lagged behind the time of impact and the rate of grip force increase was generally slower for schizophrenic subjects. Interestingly, for the resist task the timing and scaling of grip forces were not significantly different between schizophrenic subjects and healthy controls. The authors interpret their data to reflect a specific problem in the fluid sequencing of voluntary motor actions, namely the programming and execution of motor commands to the arm and fingers when handling objects in schizophrenia. However, a main shortcoming of the study was that the authors did not consider possible side effects due to antipsychotic treatment in their schizophrenic subjects.

Another recent study addressed the investigation of Parkinson-like side effects associated with antipsychotic therapy in schizophrenia (Nowak *et al.*, 2006a). When performing arm movements with a hand-held object, healthy subjects modulate grip force in parallel with load force without an obvious time delay, suggesting that the central nervous system can predict the load variations induced by voluntary movements before their occurrence (Flanagan & Wing, 1993) (see also Chapters 1 and 12). If, however, the force acting on the object is unexpectedly changed, for example by dropping a mass into a hand-held receptacle, then the adjustment of grip force lags behind the increase in load force, showing a switch from predictive to reactive control (Johansson & Westling, 1988). In contrast, if the subject can estimate the time of impact, for example by dropping the weight from one hand into a receptacle held by the opposite hand, then grip force is adjusted in a predictive manner before impact. Eighteen subjects with schizophrenia and 18 age- and gender-matched healthy subjects performed (i) vertical point-to-point arm movements with a hand-held instrumented object and (ii) caught a weight that was dropped into a hand-held receptacle either expectedly from the opposite hand or unexpectedly from the experimenter's hand (Nowak *et al.*, 2006a). Five schizophrenic subjects were completely drug naïve, 13 schizophrenic subjects were under regular and stable antipsychotic therapy. The average scores on the scale for the assessment of positive and negative symptoms (Kay *et al.*, 1987) were 10.4 ± 6 points for positive symptoms and 15 ± 6 points for negative symptoms. The average motor subscore (subitems 18–31) of the unified Parkinson's disease rating scale (UPDRS; Fahn & Elton, 1987), used to rate the patients' motor disability associated with antipsychotic treatment, was 15 ± 13 points. Figure 28.1 illustrates representative data obtained from vertical point-to-point arm movements of two subjects with schizophrenia. Subject 7 did receive antipsychotic treatment, while subject 5 had never received antipsychotic drugs. Subject 5 performed similarly to healthy subjects when moving the hand-held object. Compared to the performance of subject 5, subject 7 produced greater and more variable

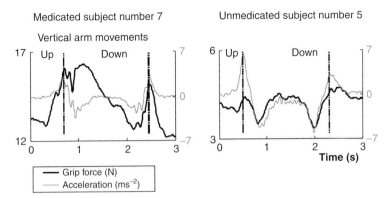

Figure 28.1. Grip force and acceleration traces obtained from single vertical arm movements with the hand-held object performed by subjects with schizophrenia. Subject 7 was receiving antipsychotic drugs, while subject 5 had never received antipsychotic treatment. The dotted vertical lines indicate peak acceleration. (Modified from Nowak *et al.*, 2006b.)

grip forces when moving the object. In addition, subject 7 generated smaller arm accelerations when transporting the object. Obviously, antipsychotic drugs may cause an overflow of grip force and bradykinesia of voluntary movements. The clinical scores of motor disability confirmed these findings: the motor subscore of the UPDRS was 52 points for subject 7, but 0 points for subject 5.

Quantitative data analysis revealed that schizophrenic subjects under antipsychotic therapy exhibited significantly lower peak accelerations and greater grip forces than healthy subjects or drug-naïve schizophrenic subjects when performing vertical point-to-point arm movements. Also, the temporo-spatial coupling between grip- and load-force profiles (as indexed by the *r* correlation coefficients obtained from correlation analysis) was hampered during vertical arm movements with the hand-held object in schizophrenic subjects receiving antipsychotic drugs, compared with healthy subjects or drug-naïve patients. For the weight-catching tasks, schizophrenic subjects showed a predictive mode of grip-force processing in the self-release condition and a reactive mode of grip-force processing in the experimenter-release condition, irrespective of whether they were on regular antipsychotic medication or not. Thus the ability to program grip forces in either a predictive or reactive mode, depending on the task requirements, is unaffected by schizophrenia or antipsychotic therapy (Delevoye-Turrell *et al.*, 2002, 2003, 2006). However, subjects with schizophrenia treated with antipsychotic drugs produced greater grip forces than healthy subjects or drug-naïve schizophrenic subjects. In accordance with earlier data, these observations indicate that (i) there is a subtle impairment of fine motor performance in schizophrenia and (ii) antipsychotic drugs may enhance that impairment causing an overflow of grip force and bradykinesia of voluntary movement (Delevoye-Turrell *et al.*, 2003; Henkel *et al.*, 2004; Putzhammer *et al.*, 2005). Indeed, in medicated schizophrenic subjects the overshoot of grip-force scaling (as measured by the ratio between grip and load forces) and the deficit of coordination between grip and load forces (as indexed by the *r* correlation

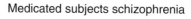

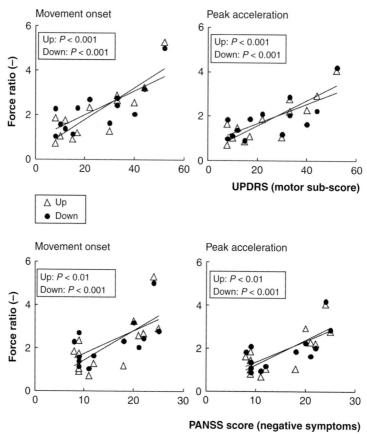

Figure 28.2. Plots of average force ratios at the times of movement onset and peak acceleration for vertical point-to-point arm movements performed by schizophrenic subjects under regular antipsychotic therapy versus the average values of the UPDRS motor subscore and the negative symptom score of the PANNS. Significant correlations (Spearman rank) between force ratios and clinical scores are illustrated for both movement directions.

coefficients) were significantly correlated with the severity of negative schizophrenic symptoms (as assessed with the PANNS; Kay *et al.*, 1987) and the severity of Parkinson-like side effects (as measured with the UPDRS motor subscore; Fahn & Elton, 1987). Figure 28.2 illustrates this observation.

Impaired awareness of action in schizophrenia

Daprati *et al.* (1997) asked 30 subjects with schizophrenia (18 paranoid type, 7 undifferentiated type, 5 residual type) under regular antipsychotic therapy (average dosage:

427 ± 320 mg chlorpromazine equivalents) and healthy control subjects to perform a movement with their gloved dominant right hand (extend thumb, extend index, extend index and middle finger, open hand wide). During the voluntary movement, a video screen displayed either the subject's own movement or the gloved experimenter's hand performing either the same type of movement or a different movement. Subjects were asked if the image of a moving hand was their own hand performing the requested movement or not. Subjects with schizophrenia and healthy subjects made virtually no errors when they saw their own hand or the experimenter's hand performing another type of movement. However, subjects with schizophrenia made more errors when they saw the experimenter's hand performing a similar movement they had performed before. In this condition the median error rate was 5% for healthy subjects, 17% in schizophrenic subjects without delusions and 23% in schizophrenic subjects suffering from delusions.

Franck *et al.* (2001) conducted a similar experiment using a realistic virtual hand displayed on a screen that could reproduce joystick movements of the subject's own hand towards a target on the screen, but with systematic distortions in the temporal and spatial domain. Twenty-nine subjects with schizophrenia (12 paranoid type, 3 disorganized type, 11 undifferentiated type, 3 residual type) under regular antipsychotic therapy (principally risperidone, olanzapine, clozapine and levomepromazine) participated. The task consisted of executing simple movements with the joystick during which a spot was displayed for 1 second on the left, on the right, or on the top of the screen. The image of the virtual hand then appeared for 2 seconds, during which the subject had to execute a movement of the joystick in the direction indicated by the position of the spot. During performance subjects saw a movie of the virtual hand performing a hand movement. Subjects were then asked if the movement they saw on the screen corresponded to that they performed with their own hand. Three categories of trials were used: (1) neutral trials, in which the movements of the virtual hand exactly replicated those made by the joystick, (2) trials with angular biases, in which the movements of the virtual hand deviated by a given angular value with respect to those made by the joystick and (3) trials with temporal biases, in which the movements of the virtual hand were delayed by a given time with respect to those made by the joystick. All subjects (healthy and schizophrenic) showed errors when the distorted movement of the virtual hand varied little from their own hand movement. The error rate of healthy subjects and schizophrenic subjects without delusions improved significantly when the spatial bias displayed in angles reached between 15–20°. Schizophrenic subjects with delusions reached a similar reduction in error rates only when the spatial bias increased to 30–40° distortion. For the temporal bias healthy subjects detected very small changes (100–150 ms), whereas the error rate of schizophrenic subjects improved only when temporal distortions around 300 ms were reached, regardless of whether they suffered from delusions or not.

Taken together these data provide evidence that subjects with schizophrenia have problems in correctly monitoring the execution of their own hand movements (Daprati *et al.*, 1997; Franck *et al.*, 2001; Sukhwinder *et al.*, 2005).

Theoretical considerations

Several authors explained the deficit of monitoring voluntary actions in schizophrenia within the framework of internal models (Arbib & Mundhenk, 2005; Frith *et al.*, 2000; Lindner *et al.*, 2005). The following should be considered a theoretical approach to explain impaired monitoring of action during grasping in schizophrenia based on these earlier models. Skillful performance of grasping involves different modes of control which rely on prediction and sensory feedback to different extents. Motor prediction refers to the estimation of the future states of a system. For example, motor prediction could estimate how our hand and fingers move in response to a motor command. To predict the consequences of our motor commands requires a system that can simulate the kinematics and dynamics of our body and the external environment. Such a system has been termed an internal forward model as it models the causal relationship between voluntary actions and their consequences (Wolpert *et al.*, 1998; Blakemore *et al.*, 2001) (see also Chapter 9). As the dynamics and kinematics of our body change during ontogenetic development and we experience objects that have their own intrinsic dynamics, we need to acquire new models and update existing ones. Therefore forward models are not fixed entities but learned and updated with motor experience. Forward models compute prediction errors resulting from comparison of the actual sensory outcome and the predicted sensory outcome of a motor command (Wolpert *et al.*, 1998; Blakemore *et al.*, 2001). The predicted sensory outcome, also termed *corollary discharge*, represents an internal sensory signal that may be produced by an internal forward model in conjunction with a copy of the descending motor command, referred to as *efference copy*. A mismatch between the predicted and actual sensory outcomes (*prediction error*) may then trigger force corrections along with updating the relevant internal models (Figure 28.3).

   The idea is that schizophrenia causes a specific impairment to generating the corollary discharge signal, which in consequence makes a comparison between the intended and real sensory outcome of a voluntary action impossible. This may cause a deficit in the self-monitoring of voluntary action and is illustrated in Figure 28.3 within the context of object manipulation. The possibility to predict the sensory outcome of a voluntary movement enables a very fast correction of errors, which relies on the realization that the predicted sensory outcome does not match the desired sensory outcome. In the latter circumstance the movement could be corrected online before its actual sensory consequence is known (Wolpert & Flanagan, 2001). In schizophrenia such rapid error correction has been found to be hampered (Malenka *et al.*, 1982; Delevoye-Turrell *et al.*, 2003). If schizophrenic subjects are unable to establish an internal representation of the predicted sensory outcome of their own movement then they have problems in coping with unexpected sensory feedback and are more likely to attribute sensory effects of their own actions to external sources. This may be the pathophysiologic origin of delusions of motor control, such as the feeling that one's own hand moves not in response to one's own will, but as a consequence of some outside force (Blakemore *et al.*, 2000; Frith *et al.*, 2000; Lindner *et al.*, 2005).

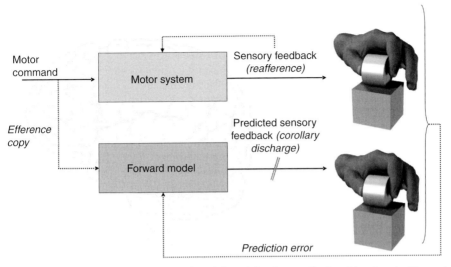

Figure 28.3. The theory of internal forward models and the abnormality in schizophrenia. The motor system generates a descending motor command that results in sensory feedback (*reafference*). A forward model of this system uses a copy of the descending motor command (*efference copy*) and generates an estimate of the sensory feedback likely to result from the movement (*corollary discharge*). A mismatch between the predicted and actual sensory outcomes (*prediction error*) triggers force corrections along with an updating of the relevant internal models. In schizophrenia a deficit in generating the corollary discharge signal is proposed to cause a deficit in the self-monitoring of voluntary action. Modified from Nowak *et al.* (2006b).

Note

1 Neurological soft signs in schizophrenia comprise a variety of deficits, such as subtle impairments of sensory processing and integration, discrete motor dyscoordination and clumsiness of movement, or even occurrence of primitive reflexes. They may also be found in non-affected first-degree relatives (McNeil *et al.*, 1993; Chen *et al.*, 2000; Yazici *et al.*, 2002; Compton *et al.*, 2007) and have therefore been considered as indicators of an underlying, possibly genetically transmitted enhanced susceptibility for the disease (Fish *et al.*, 1992; Gourion *et al.*, 2004).

References

Arbib, M. A. & Mundhenk, T. N. (2005). Schizophrenia and the mirror system: an essay. *Neuropsychologia*, **43**, 268–280.

Bartko, G., Zador, G., Horvath, S. & Herczeg, I. (1988). Neurological soft signs in chronic schizophrenic patients: clinical correlates. *Biol Psychiatry*, **24**, 458–460.

Blakemore, S. J., Smith, J., Steel, R., Johnstone, C. E. & Frith, C. D. (2000). The perception of self-produced sensory stimuli in patients with auditory hallucinations and passivity experiences: evidence for a breakdown in self-monitoring. *Psychol Med*, **30**, 1131–1139.

Blakemore, S. J., Frith, C. D. & Wolpert, D. M. (2001). The cerebellum is involved in predicting the sensory consequences of action. *Neuroreport*, **12**, 1879–1884.

Bleuler, E. (1908). *Dementia praecox or the Group of Schizophrenias*. Translated by J. Zinkin (1950). New York, NY: International Universities Press.

Boks, N. P. M., Russo, S., Knegtering, R. & van den Bosch, R. J. (2000). The specificity of neurological signs in schizophrenia: a review. *Schizophr Res*, **43**, 109–116.

Browne, S., Clarke, M., Gervin, M. *et al.* (2000). Determinants of neurological dysfunction in first episode schizophrenia. *Psychol Med*, **30**, 1433–1441.

Buchanan, R. W., Kirkpatrick, B., Heinrichs, D. W. & Carpenter, W. T. Jr. (1990). Clinical correlates of the deficit syndrome of schizophrenia. *Am J Psychiatry*, **147**, 290–294.

Buchanan, R. W., Koeppl, P. & Breier, A. (1994). Stability of neurological signs with clozapine treatment. *Biol Psychiatry*, **36**, 198–200.

Carnahan, H., Elliott, D. & Velomoor, V. R. (1996). Influence of object size on prehension in leukotomized and unleukotomized individuals with schizophrenia. *J Clin Exp Neuropsychol*, **18**, 136–147.

Chen, E. Y., Kwok, C. L., Au, J. W., Chen, R. Y. & Lau, B. S. (2000). Progressive deterioration of soft neurological signs in chronic schizophrenic patients. *Acta Psychiat Scand*, **102**, 342–349.

Compton, M. T., Bollini, A. M., McKenzie, M. L. *et al.* (2007). Neurological soft signs and minor physical anomalies in patients with schizophrenia and related disorders, their first-degree biological relatives, and non-psychiatric controls. *Schizophr Res* May 16 [Epub ahead of print].

Daprati, E., Franck, N., Georgieff, N. *et al.* (1997). Looking for the agent: an investigation into consciousness of action and self-consciousness in schizophrenic patients. *Cognition*, **65**, 71–86.

Delevoye-Turrell, Y., Giersch, A. & Danion, J. M. (2002). A deficit in the adjustment of grip force responses in schizophrenia. *Neuroreport*, **13**, 1537–1539.

Delevoye-Turrell, Y., Giersch, A. & Danion, J. M. (2003). Abnormal sequencing of motor actions in patients with schizophrenia: evidence from grip force adjustments during object manipulation. *Am J Psychiatry*, **160**, 134–141.

Delevoye-Turrell, Y., Thomas, P. & Giersch, A. (2006). Attention for movement production: abnormal profiles in schizophrenia. *Schizophr Res*, **84**, 430–432.

Fahn, S. & Elton, R. L. (1987). The unified Parkinson's disease rating scale. In S. Fahn, C. D. Marsden, C. B. Calne, *et al.* (Eds.), *Recent Developments in Parkinson's Disease* (pp. 153–163). Florham Park, NJ: MacMillan Healthcare Information.

Farde, L., Wiesel, F. A. & Nordström, A. L. (1989). D1- and D2-dopamine receptor occupancy during treatment with conventional and atypical neuroleptics. *Psychopharmacology*, **99**, S28–S31.

Farde, L., Nordström, A. L. & Wiesel, F. A. (1992). Positron emission tomographic analysis of central D1 and D2 dopamine receptor occupancy in patients treated with classical neuroleptics and clozapine, relation to extrapyramidal side effects. *Arch Gen Psychiatry*, **49**, 538–544.

Fish, B., Marcus, J., Hans, S. L., Auerbach, J. G. & Perdue, S. (1992). Infants at risk for schizophrenia: sequelae of a genetic neurointegrative defect. A review and replication analysis of pandysmaturation in the Jerusalem Infant Development Study. *Arch Gen Psychiatry*, **49**, 221–235.

Flanagan, J. R. & Wing, A. M. (1993). Modulation of grip force with load force during point-to-point arm movements. *Exp Brain Res*, **95**, 131–143.

Franck, N., Farrer, C., Georgieff, N. *et al.* (2001). Defective recognition of one's own actions in patients with schizophrenia. *Am J Psychiatry*, **158**, 454–459.

Frith, C. D., Blakemore, S. & Wolpert, D. M. (2000). Explaining the symptoms of schizophrenia: abnormalities in the awareness of action. *Brain Res Rev*, **31**, 357–363.

Gourion, D., Goldberger, C., Olie, J., Loo, H. & Krebs, M. (2004). Neurological and morphological anomalies and the genetic liability to schizophrenia: a composite phenotype. *Schizophr Res*, **67**, 23–31.

Green, M. & Walker, E. (1985). Neuropsychological performance and positive and negative symptoms in schizophrenia. *J Abnorm Psychol*, **94**, 460–469.

Gupta, S., Rajaprabhakaran, R., Arndt, S., Flaum, M. & Andreasen, N. C. (1995). Premorbid adjustment as a predictor of phenomenological and neurobiological indices in schizophrenia. *Schizophr Res*, **16**, 189–197.

Heinrichs, D. W. & Buchanan, R. W. (1988). Significance and meaning of neurological signs in schizophrenia. *Am J Psychiatry*, **145**, 11–18.

Henkel, V., Mergl, R., Schäfer, M. *et al.* (2004). Kinematic analysis of motor function in schizophrenic patients: a possibility to separate negative symptoms from extrapyramidal dysfunction induced by neuroleptics? *Pharmacopsychiatry*, **37**, 110–118.

Jahn, T. (1999). *Diskrete motorische Störungen bei Schizophrenie*. Weinheim, Germany: Beltz/Psychologie Verlags Union.

Jahn, T., Mai, N., Ehrensperger, M. *et al.* (1995). Untersuchung der fein- und grobmotorischen Dysdiadochokinese schizophrener Patienten. *Z Klin Psychol*, **24**, 300–315.

Jahn, T., Hubmann, W., Karr, M. *et al.* (2006). Motoric neurological soft signs and psychopathological symptoms in schizophrenic psychoses. *Psychiatry Res*, **142**, 191–199.

Johansson, R. S. & Westling, G. (1988). Programmed and triggered actions to rapid load changes during precision grip. *Exp Brain Res*, **71**, 72–86.

Johnstone, E. C., MacMillan, J. F., Frith, C. D., Benn, D. K. & Crow, T. J. (1990). Further investigation of the predictors of outcome following first schizophrenic episodes. *Br J Psychiatry*, **157**, 182–189.

Kay, S. R., Fiszbein, A. & Opler, L. A. (1987). The positive and negative syndrome scale (PANSS) for schizophrenia. *Schizophr Bull*, **13**, 261–276.

Kraepelin, E. (1919). *Dementia Praecox and Paraphrenia*. Translated by R. M. Barclay, edited by G. M. Robertson (1971). New York, NY: Robert E. Krieger.

Lehoux, C., Everett, J., Laplante, L. *et al.* (2003). Fine motor dexterity is correlated to social functioning in schizophrenia. *Schizophr Res*, **62**, 269–273.

Lindner, A., Thier, P., Kircher, T. T., Haarmeier, T. & Leube, D. T. (2005). Disorders of agency in schizophrenia correlate with an inability to compensate for the sensory consequences of actions. *Curr Biol*, **15**, 1119–1124.

Malenka, R. C., Angel, R. W., Hampton, B. & Berger, P. A. (1982). Impaired central error correcting behaviour in schizophrenia. *Arch Gen Psychiatry*, **39**, 101–107.

McNeil, T. F., Harty, B., Blennow, G. & Cantor-Graae, E. (1993). Neuromotor deviation in offspring of psychotic mothers: a selective developmental deficiency in two groups of children at heightened psychiatric risk? *J Psychiatr Res*, **27**, 39–54.

Merriam, A. E., Kay, S. R., Opler, L. A., Kushner, S. F. & van Praag, H. M. (1990). Neurological signs and the positive–negative dimension in schizophrenia. *Biol Psychiatry*, **28**, 181–192.

Mittal, V. A., Hasenkamp, W., Sanfilipo, M. *et al.* (2007). Relation of neurological soft signs to psychiatric symptoms in schizophrenia. *Schizophr Res*, May 30 [Epub ahead of print].

Mohr, F., Hubmann, W., Cohen, R. *et al.* (1996). Neurological soft signs in schizophrenia: assessment and correlates. *Eur Arch Psychiatry Clin Neurosci*, **246**, 240–248.

Nowak, D. A., Connemann, B. J., Alan, M. & Spitzer, M. (2006a). Sensorimotor dysfunction of grasping in schizophrenia: a side effect of antipsychotic treatment? *J Neurol Neurosurg Psychiatry*, **77**, 650–657.

Nowak, D. A., Topka, H., Timmann, D., Boecker, H. & Hermsdörfer, J. (2006b). The role of the cerebellum for predictive control of grasping. *Cerebellum*, **6**, 7–17.

Putzhammer, A., Perfahl, M., Pfeiff, L. *et al.* (2005). Performance of diadochokinetic movements in schizophrenic patients. *Schizophr Res*, **79**, 271–280.

Quitkin, F., Rifkin, A. & Klein, D. F. (1976). Neurologic soft signs in schizophrenia and character disorders. Organicity in schizophrenia with premorbid asociality and emotionally unstable character disorders. *Arch Gen Psychiatry*, **33**, 845–853.

Saoud, M., Coello, Y., Dumas, P. *et al.* (2000). Visual pointing and speed/accuracy trade-off in schizophrenia. *Cogn Neuropsychiatry*, **5**, 123–134.

Scheffer, R. E. (2004). Abnormal neurological signs at the onset of psychosis. *Schizophr Res*, **70**, 19–26.

Schröder, J., Niethammer, R., Geider, F. J. *et al.* (1992). Neurological soft signs in schizophrenia. *Schizophr Res*, **6**, 25–30.

Schröder, J., Wenz, F., Baudendistel, K., Schad, L. R. & Knopp, M. V. (1995). Sensorimotor cortex and supplementary motor area activation changes in schizophrenia: a study with functional magnetic resonance imaging. *Br J Psychiatry*, **167**, 197–201.

Schröder, J., Silvestri, S., Bubeck, B. *et al.* (1998). D2 dopamine receptor upregulation, treatment response, neurological soft signs, and extrapyramidal side effects in schizophrenia: a follow-up study with 123I-IBZM SPECT in the drug-naive state and after neuroleptic treatment. *Biol Psychiatry*, **42**, 660–665.

Sukhwinder, S. S., Samson, G., Bays, P. M., Frith, C. D. & Wolpert, D. M. (2005). Evidence for sensory prediction deficits in schizophrenia. *Am J Psychiatry*, **162**, 2384–2386.

Tigges, P., Mergl, R., Frodl, T. *et al.* (2000). Digitized analysis of abnormal hand-motor performance in schizophrenic patients. *Schizophr Res*, **45**, 133–143.

Tosato, S. & Dazzan, P. (2005). The psychopathology of schizophrenia and the presence of neurological soft signs: a review. *Curr Opin Psychiatry*, **18**, 285–288.

Varambally, S., Venkatasubramanian, G., Thirthalli, J., Janakiramaiah, N. & Gangadhar, B. N. (2006). Cerebellar and other neurological soft signs in antipsychotic-naive schizophrenia. *Acta Psychiatr Scand*, **114**, 352–356.

Vaughan, S., Oquendo, M. & Horwath, E. (1991). A patient's psychotic interpretation of a drug side effect. *Am J Psychiatry*, **148**, 393–394.

Walker, E. F. (1994). Neurodevelopmental precursors of schizophrenia. In A. S. David & J. C. Cutting (Eds.), *The Neuropsychology of Schizophrenia* (pp. 119–129). Hillsdale, NJ: Lawrence Erlbaum Associates.

Wolpert, D. M. & Flanagan, J. R. (2001). Motor prediction. *Curr Biol*, **11**, R729–R732.

Wolpert, D. M., Miall, R. C. & Kawato, M. (1998). Internal models in the cerebellum. *Trends Cogn Sci*, **2**, 338–347.

Yazici, A. H., Demir, B., Yazici, K. M. & Gogus, A. (2002). Neurological soft signs in schizophrenic patients and their nonpsychotic siblings. *Schizophr Res*, **58**, 241–246.

Part IV

Therapy of impaired grasping

29

Stroke therapy

DENNIS A. NOWAK AND JOACHIM HERMSDÖRFER

Summary

Stroke is the leading cause of disability in the adult worldwide. The most common neurological impairment following stroke is weakness or loss of sensibility of the extremities contralateral to the side of the brain lesion. Only about 40% of affected individuals regain full recovery; the remaining 60% have persistent neurological deficits that impact on their social functioning in private and community life. By now, much of our clinical and scientific interest is focused on stroke prevention and acute stroke therapy. In contrast, there is less effort in developing novel strategies for hand motor rehabilitation after stroke. This is surprising since about two-thirds of stroke survivors are left with permanent sensory or motor impairment. This chapter discusses the intrinsic capacity of the cortical motor system for reorganization and gives an overview of established and novel concepts for sensorimotor rehabilitation of the hand after stroke.

Introduction

Stroke is the leading cause of disability in the adult worldwide (Kolominsky-Rabas et al., 2001). The annual incidence of stroke is 100–300 per 100 000 (Broderick et al., 1998). The most common impairment following stroke is weakness of the limbs contralateral to the side of the brain lesion (Kelly-Hayes et al., 1998). Only about 40% of stroke survivors recover completely (Hankey et al., 2002) and among the remaining 60% permanent sensory and/or motor disability of the hand constitutes a major problem (Stein, 1998). Chronic motor problems cause difficulties in using the hand in functional activities of daily life, such as grasping and holding an object, picking up a glass, tying shoe laces or buttoning a shirt. The severity of these motor impairments and their negative impact on social functioning have encouraged the development of therapeutic strategies to improve hand function following stroke. Within recent years, progress in technology has provided several useful objective measures to quantify the impairments of both the kinetics and kinematics of grasping following stroke (see Chapter 21). We have only started to use these novel technologies in the clinical setting to quantify the degree of disability, monitor recovery and evaluate treatment strategies targeting the improvement of impaired hand function after stroke

Sensorimotor Control of Grasping: Physiology and Pathophysiology, ed. Dennis A. Nowak and Joachim Hermsdörfer. Published by Cambridge University Press. © Cambridge University Press 2009.

(Mai, 1989; Hermsdörfer *et al.*, 1999, 2003; Nowak *et al.*, 2003, 2007; Nowak & Hermsdörfer, 2005; Lang *et al.*, 2005, 2006b; Blennerhassett *et al.*, 2006; McDonnell *et al.*, 2006; Nowak, 2006).

Organization of the motor system promoting hand function and its intrinsic capacity for reorganization after stroke

A first step towards the development of novel concepts for hand motor rehabilitation after stroke is to understand how the healthy brain works and how the motor system promoting hand function reorganizes after brain damage. The primary motor system for grasping movements comprises *upper motor neurons* in the primary motor cortex (M1) that connect via the pyramid tract to the *lower motor neurons* within the contralateral anterior gray matter of the cervical spinal cord (Figure 29.1). The spinal motor neurons innervate hand and arm muscles. In the middle of the last century Penfield and co-workers studied the representation

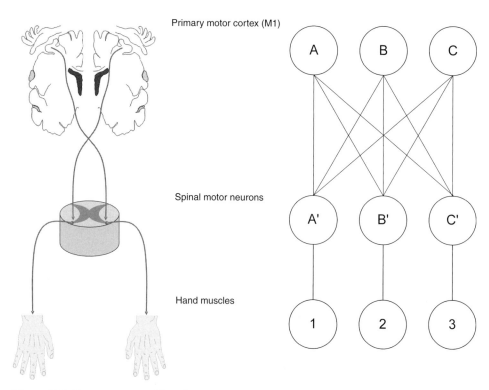

Figure 29.1. Current concept of the primary motor cortex (M1) organization with divergence of motor output from M1 towards multiple spinal motor neuron pools and convergence of motor output from various M1 regions towards individual motor neuron pools of the cervical spinal cord (Schieber, 2001). Within this concept different overlapping regions of M1 promote movements of individual hand and arm muscles.

of sensory and motor functions of the body within the cerebral cortex by electrical cortex stimulation in humans (Penfield & Boldrey, 1937; Penfield & Rasmussen, 1950). Their work established the concept of somatotopy within the primary motor cortex meaning that different regions of the body, such as single fingers of the hand, are controlled by spatially separated regions within M1. Our current understanding of M1 is very different from the simplified concept of somatotopy. Clinical experience with people recovering from cortical lesions who regained motor abilities that have been lost early after brain damage and experimental data imply that different spinal motor neuron pools receive input from broad and overlapping cortical territories within M1 and that M1 neurons send out projections to several different spinal motor neuron pools (see Schieber, 2001 for a recent review). This current concept of motor output from M1 is illustrated schematically in Figure 29.1 and implies that upper motor neurons within a wide cortical region are active during hand and finger movements.

Such an organization of M1 has essential implications for recovery of function after brain damage. In a case where a particular region of M1 is destroyed by a stroke causing hand paresis, spared regions of M1 may take over its function and this is not because these latter regions acquire new duties, but rather because they have been involved in controlling the hand before the stroke (Nudo *et al.*, 1996). Thus, the parallel distributed organization of M1 with its inherent plasticity may provide an intrinsic capacity for recovery of function after stroke when parts of M1 are spared by the lesion. Reorganization within M1 after stroke could be enhanced by practice and novel treatment approaches. For example, deafferentation of the upper arm by regional anesthesia to the upper brachial plexus in combination with hand motor practice significantly improved hand motor function in chronic stroke subjects, suggesting that somatosensory deprivation of M1 neurons primarily involved in proximal muscle control activated new resources for the control of hand and fingers (Muellbacher *et al.*, 2002). The improvement in hand function was associated with an enhancement in motor output from M1 of the affected hemisphere as indexed by an increase in the size of motor-evoked potentials in intrinsic hand muscles probed by transcranial magnetic stimulation.

Within the hierarchical model of motor planning and execution, the non-primary motor areas of the frontal and parietal lobes have been associated with motor planning and preparation, while M1 has been associated with motor execution. The idea was that the non-primary motor areas send a motor plan to M1 which sends out a descending motor command to the spinal motor neurons. Within this concept, dysfunction of each single area causes impaired motor function. The concept of such a sequential control of voluntary movement, however, has been replaced by a parallel, distributed model. In a parallel, distributed model the primary motor areas and M1 work in concert when planning and executing a voluntary motor command. Again, clinical observations in people with focal brain lesions make the existence of a parallel, distributed model of motor control very likely. Studies in the primate brain have detected several distinct non-primary motor areas in the frontal lobe (Rizzolatti *et al.*, 1998), such as the ventral (PMv) and dorsal premotor (PMd) areas on the lateral surface of the hemisphere, the supplementary motor area (SMA) and the

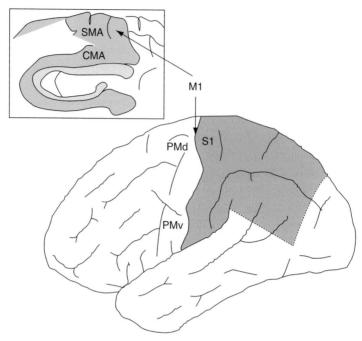

Figure 29.2. Current understanding of the anatomical locations of the primary motor (M1), primary somatosensory (S1) and non-primary motor areas in the frontal lobe of the human brain. The dorsal (PMd) and ventral premotor areas (PMv) are located on the lateral surface of the brain. The supplementary motor area (SMA) is located on the medial surface as are the cingulate motor areas (CMA).

rostral, dorsal and caudal cingulate motor areas on the medial surface (Figure 29.2). Essentially, there are also projections from PMv, PMd, SMA and the cingulate motor areas (CMA) to the intermediate zone and the ventral horn of the gray matter of the spinal cord (Dum & Strick, 2002). It is important to remember in this context that our current understanding of the anatomy and function of human non-primary motor areas stems from work with the monkey brain. For the sake of simplicity we focus on the frontal lobe, well aware also that subcortical structures, such as the cerebellum (Chapter 24) and the basal ganglia (Chapter 18), and also the parietal cortex, play essential roles for hand motor control.

The parallel, distributed control of hand movements with primary and non-primary motor areas working in concert provides an intrinsic capacity for recovery after damage to one part of the network. The idea is that non-primary motor areas may compensate for loss of function when M1 is inactivated by stroke. Neurons within non-primary motor areas are more flexibly integrated into neural circuits than M1 neurons, providing a greater capacity for plastic changes in function. In addition, non-primary motor areas also project to the spinal motor neuron pools (Dum & Strick, 2002), and it has been shown that the activation patterns of neural discharge in non-primary motor areas during grasping movements are similar to those of M1 (Cadoret & Smith, 1997). Recently, functional imaging data have shown that the premotor areas and the supplementary motor area are commonly more active during hand and finger movements with the paretic hand after stroke, suggesting that non-primary motor areas take over when M1 is out of order (see Chapter 30).

Physiotherapeutic concepts

There are a huge variety of therapeutic concepts aiming to enhance function of the affected hand after stroke. Many of these approaches follow the various schools of physiotherapy, such as the *Bobath concept* (Bobath, 1987; Lennon *et al.*, 2001; Luke *et al.*, 2004). Based on neurophysiological principles, Bobath therapy aims to modulate the somatosensory input from the affected limb to the brain to facilitate motor output from the brain to the affected limb. Other physiotherapeutic concepts focusing on the interaction between somatosensory perception and motor output that are frequently used in rehabilitation of impaired hand function following stroke are *proprioceptive neuromuscular facilitation* (Myers, 1989), the *Klein–Vogelbach functional kinetics* (FBL) concept (Suppé & Spirgi-Gantert, 2007) and the *spiral dynamics* concept (Larsen & Schneider, 2007). Modern approaches include cognitive processes in the rehabilitative concept, known as *neurocognitive rehabilitation* (Perfetti, 2006). Within this theoretical framework rehabilitation is thought to reflect learning under pathological conditions. The concept works with different sensory inputs, such as tactile or kinesthetic stimuli, that should be cognitively evaluated in order to adjust pathological components of movements and enable a fragmented re-adjusted motor performance. Only a few studies, however, evaluated the effectiveness of physiotherapeutic concepts and recent systematic reviews of the available data found no or insignificant evidence in terms of functional outcome for traditional rehabilitation approaches (Van Peppen *et al.*, 2004). Meta-analysis of the effects of exercise therapy on activities of daily living, walking and dexterity after stroke revealed that the intensity of exercise therapy in terms of training hours is critical (Kwakkel *et al.*, 2004).

Repetitive motor training

Several therapeutic concepts focus on repetitive practice of impaired movements. These concepts differ in the complexity of the motor task to be trained, the time after stroke the training commences and the specific training environments.

Constraint-induced movement therapy

The theory of *learned non-use* claims that people learn to avoid using the stroke-affected hand in activities of daily life given their frequent experience of problems in performance (Taub *et al.*, 1993). To modify this maladaptive behavior the concept of constraint-induced movement (CIM) therapy was developed that inactivates the unaffected limb over most of the waking day to force people to use their affected arm and hand (Taub *et al.*, 1999). The inactivation of the unaffected upper limb is maintained without interruption over a 2-week period and is typically accompanied by intensive training sessions over 6 hours daily on each working day. During the training sessions simple movements relevant for daily life, such as prehensile movements, are practiced under supervision by therapists. Importantly, several studies have shown that CIM therapy improves the function of the affected hand after stroke and there is retention of improvement over several months following the treatment period (Miltner *et al.*, 1999; Taub *et al.*, 1999). A recent multi-center study including 222 chronic stroke subjects demonstrated a significant improvement of motor function due to CIM therapy that was evident up to one year after therapy (Wolf *et al.*, 2006). Functional imaging studies revealed plastic changes within the cortical motor system after CIM therapy that correlated with recovery of function (Hamzei *et al.*, 2006). However, the contribution of inactivation of the unaffected upper limb to the effectiveness of CIM therapy remains unclear. Therapies implementing practice with both the affected and unaffected hands appear to be of comparable effectiveness as CIM therapy in cases where the number of training hours is comparable (Van der Lee, 2001). A limitation of the approach is definitely that CIM therapy is tailored to stroke subjects with marked residual hand function, for example a capacity of at least 10°–20° extension at the wrist is a common inclusion criterion.

Movement training

Daily training of very simple single-joint fast-as-possible wrist movements of the paretic hand has been found to improve the velocity of contraction and peak force of the trained muscle groups after stroke (Bütefisch *et al.*, 1995). Interestingly, improvement was not limited to the trained muscle groups, but also appeared to generalize to other muscle segments of the arm as indexed by an improvement of clinical scales of arm function. In contrast, repetitive training of more complex movements that required a greater amount of accuracy, such as grasping and transporting objects, did not result in similar improvements (Woldag *et al.*, 2003). Another approach including repetitive training of basal motor abilities, such as rapid arm and finger movements, is the *arm ability training* that is applied to people with mild to moderate paresis of the arm and hand (Platz *et al.*, 2001; Platz, 2004).

Strength training

It is a currently controversial whether strength training enhances the motor deficit of the paretic arm and hand following stroke or not. The main argument against strength training is that it may have negative consequences by enhancing pathological co-contraction and spasticity. However, several studies did not find a relevant increase in spasticity after strength training of the paretic arm after stroke (Bourbonnais *et al.*, 1997; Miller & Light, 1997). In a comparison of strength training, task-specific training and standard training, the strength and task-specific training strategies caused significant improvements in several performance measures as compared with standard training; however, and most importantly, retainment of the effects of task-specific training was more effective (Winstein *et al.*, 2004).

Bilateral training

Bilateral training includes the unaffected arm and hand in training motor function of the affected arm and hand (Mudie & Matyas, 2000; Whitall *et al.*, 2000). Within this concept stroke subjects perform simultaneous bilateral movements. Some pertinent data suggest that bilateral training may have some advantages in comparison to unilateral training of the affected hand (Mudie & Matyas, 2000); however, there are also reports suggesting that bilateral training is similarly effective to other treatment strategies (Tijs & Matyas, 2006). More data are needed to provide definite conclusions regarding the usefulness of bilateral training.

Feedback and virtual reality-based training

Feedback-based training refers to the measurement of intrinsic body signals and their display to the person in training. This strategy allows the bypassing of an impaired sensory channel, e.g. hampered somatosensory input from the grasping fingers, via alternative sensory cues, such as visual or acoustic signals. Classic movement feedback uses electromyographic (EMG) recording of muscle activity. This allows acoustic display of muscle activity and is commonly used to enhance activity in agonistic muscle groups while at the same time reducing activity in antagonistic muscle groups. An earlier meta-analysis of EMG-based feedback training implied positive effects on motor performance of the affected upper limb after stroke (Glanz *et al.*, 1995), whereas a more recent meta-analysis did not find significant evidence for its effectiveness (Van Peppen *et al.*, 2004).

People with sensorimotor dysfunction of the hand following stroke frequently exhibit excessive and irregular grip forces when grasping objects (Hermsdörfer *et al.*, 2003; Nowak *et al.*, 2003, 2007; McDonnell *et al.*, 2006; see also Chapter 19). Within the context of grasping, a study analyzed the effectiveness of visual feedback training to reduce force overshoot and irregularities at the grasping digits when handling objects in a precision grip between the index finger and thumb (Kriz *et al.*, 1995). The grip force that stroke subjects used to track a slowly decreasing and increasing target force was displayed on a computer screen. After 10 weekly training sessions, nine out of ten chronic stroke subjects were able to reduce grip-force overshoot and irregularities to or close to the normal range.

Visual feedback is also the essential sensory cue used in *virtual reality training* strategies (Merians *et al.*, 2006; Broeren *et al.*, 2007). People with hand paresis following stroke used instrumented gloves enabling the measurement of thumb and finger movements (Merians *et al.*, 2006). The kinematic measures were transferred into movements of a virtual hand within a virtual reality environment displayed on a computer screen. Chronic stroke subjects were asked to control the virtual hand and to perform repetitive movement sequences, such as playing a piano, in order to train individual finger movements. Task difficulty was individually adjusted to the patients' current performance. Kinematic performance and functional abilities of the affected hand, as indexed by a functional hand assessment, improved significantly over a 2-week training program. The interpretation of these data is made difficult by the lack of a real control condition without virtual reality, including training sessions of similar intensity. As control data are lacking, there is not yet evidence that functional improvement of hand function by virtual reality training is due to the therapeutic concept, and not just related to the intensity of training.

In conclusion, the evaluation of feedback and virtual reality-based treatment approaches is still limited by the lack of sufficient data with larger patient collectives and relevant control groups. Novel techniques, such as *telerehabilitation*, which allows supervision of rehabilitation in a home-based setting, may be future applications for these therapy approaches (Broeren *et al.*, 2007; Carey *et al.*, 2007).

Mirror reflection training

A recent concept combines *bilateral training* and *visual feedback* in a particular way. The subject is sitting at a table with a mirror fixed in a sagittal plane at the level of the shoulder of the paretic arm. When the subject is looking at the mirror a mirror image of the unaffected arm is seen that appears to be in a similar position to the affected arm. During bilateral training the subject receives visual feedback of the movements of the unaffected arm. A pilot study including nine chronic stroke subjects revealed that mirror reflection training was more effective than a comparable bilateral training without a mirror (Altschuler *et al.*, 1999). Studies on larger patient populations have not been done (Stevens & Stoykov, 2004).

Motor imagery

From sports kinesiology we know that mental training may help to optimize movement performance. In addition, recent neuroimaging studies have shown that real and mental movement execution are associated with an increase in activity within the same cortical motor network (Buccino *et al.*, 2006). The idea is that motor imagery of movement may improve the performance of movement with the affected arm and hand (Mulder, 2007) and, indeed, recent pilot studies provide provisional evidence for the correctness of this assumption (Page *et al.*, 2005, 2007).

Movement observation

The observation of movement, similar to motor imagery, causes activation of the cortical motor network that resembles that to be found during movement execution. Within this context, observation of healthy unimpaired motor performance may assist the recovery of hampered hand movements following stroke, simply by activating the affected cortical motor network in a similar way to that during movement execution and thereby enhancing plastic changes within this network (Sharma *et al.*, 2006; De Vries & Mulder, 2007). Preliminary data imply that movement observation may reduce the deficit of motor execution of the upper limb following stroke (Ertelt *et al.*, 2007). Eight chronic stroke subjects with moderate paresis of the upper limb as a consequence of medial artery infarction performed movement observation training over a period of 4 weeks. Significant improvement of motor functions of the arm and hand, as compared with the stable pre-treatment baseline and with a control group, was found. The improvement lasted for at least 8 weeks after the end of the intervention. Functional magnetic resonance imaging (fMRI) during object manipulation showed that in comparison to the neural activations before training, movement observation caused a significant rise in activity in the bilateral ventral premotor cortex, bilateral superior temporal gyrus, the supplementary motor area and the contralateral supramarginal gyrus.

Robotic-assisted training

Mechanical devices were developed to guide active movement or to provide passive movement of the paretic arm following stroke. The application of these methods within repetitive movement training resulted in significant improvements of specific performance aspects after stroke (Masiero *et al.*, 2007). The transfer of these changes in performance to upper limb movements relevant for activities in daily life is still limited (Kahn *et al.*, 2006). An interesting approach is a bilateral training modus during which movements at the elbow and shoulder joint of the unaffected arm are recorded and used to passively impose movements to the affected arm (Lum *et al.*, 2002). An evaluation of this approach, however, did not yield clear advantages as compared with conventional therapy after a 6-month follow-up period and showed little carry-over effect (Lum *et al.*, 2006). Bilateral training with a mechanical device assisting wrist and forearm movements had significant positive effects on strength and on a motor performance scale that were stable across a retention period of 6 months (Hesse *et al.*, 2005).

Sensorimotor discrimination training

Manual dexterity is the result of a subtle interplay between action and perception and strongly depends on intact processing of both somatosensory feedback and motor output. Somatosensory deficits are frequent after stroke and hamper dexterity of the affected hand. The amount of functional impairment of grasping after stroke appears to be correlated with the severity of somatosensory impairment at the grasping fingers (Hermsdörfer *et al.*, 2003; Nowak *et al.*, 2007). Training of somatosensory perception and discrimination can alter the somatotopy of cortical representation of the hand (Jenkins *et al.*, 1990). Training of individual aspects of somatosensory perception, such as discrimination of different textures or weights, is possible after stroke, but typically the improvement of performance is not generalized within a wider context of manipulative tasks. A combined training of both somatosensory and motor aspects of dexterity can facilitate the transfer of learning to manual activities relevant for daily life (Carey & Matyas, 2005).

Functional neuromodulation

Within recent years neurophysiological methods, such as repetitive transcranial magnetic stimulation (rTMS), electrical muscle or peripheral nerve stimulation have been used to either facilitate peripheral muscle contraction (electrical muscle stimulation) or to interfere with cortical excitability (rTMS, peripheral nerve stimulation) in order to enhance recovery of function of the affected arm and hand after stroke. The detailed processes underlying each of these approaches are still not well understood. In the following sections current theoretical concepts will be presented that are based on data from the pertinent literature. The reader should be aware that these concepts may be modified in the future as additional knowledge is gathered (see also Chapter 30).

Functional electrical muscle stimulation

Direct electrical stimulation of the muscles of the affected hand via electrode pads attached to the skin of the affected arm can induce muscle contraction and thereby result in movement. This kind of therapy appears to be most attractive for subjects with profound paresis of the upper limb. *Functional electrical stimulation* aims to establish a lasting improvement of functional movement synergies and is frequently combined with some kind of motor training (Cauraugh *et al.*, 2000; Cauraugh & Kim, 2002). In contrast, a transient electrical stimulation aims to modulate cortical excitability and thereby the processing of somatosensory input and motor output. A recent overview of the current literature concludes that electrical muscle stimulation has the potential to facilitate motor recovery of the affected upper limb following stroke (De Kroon *et al.*, 2002). In a comparative study, an active muscle stimulation triggered internally by a predefined level of voluntary muscle activity as obtained from surface electromyography appears to be more effective than a passive muscle stimulation triggered externally (Hummelsheim *et al.*, 1996, 1997). The idea is that EMG-triggered muscle stimulation provides a precise temporal coupling between motor output and somatosensory feedback associated with voluntary movement (re-afference) and thereby an internal relearning of physiological movement patterns. Combination of active muscle stimulation and motor training is more effective than motor training alone (Cauraugh & Kim, 2002; Bhatt *et al.*, 2007). Also, an externally triggered functional muscle stimulation of paretic hand and finger muscles caused significant functional improvements of grasping movements (Popovic *et al.*, 2003; Santos *et al.*, 2006). There is, however, still a lack of proof regarding transfer of relearned motor ability achieved by these promising techniques to manual activities relevant for daily life (De Kroon *et al.*, 2002; Bolton *et al.*, 2004).

Repetitive peripheral nerve stimulation, repetitive transcranial magnetic stimulation (rTMS) and transcranial direct current stimulation (tDCS)

Within the concept of interhemispheric competition, cortical excitability of M1 of the affected hemisphere is reduced after stroke, based on enhanced transcallosal inhibition from M1 of the unaffected hemisphere (see Floel & Cohen, 2006; Hummel & Cohen, 2006, for recent reviews). This concept has been established based on both electrophysiological (Liepert *et al.*, 2000; Murase *et al.*, 2004) and functional imaging data (Grefkes *et al.*, 2008; see also Chapter 30). Reduction in cortical excitability of M1 of the affected hemisphere may contribute to the motor deficit of the affected hand after stroke (Figure 29.3). Consequently, a possible therapeutic strategy to normalize the imbalance of cortical excitability in between both hemispheres after stroke may be to increase excitability of M1 of the

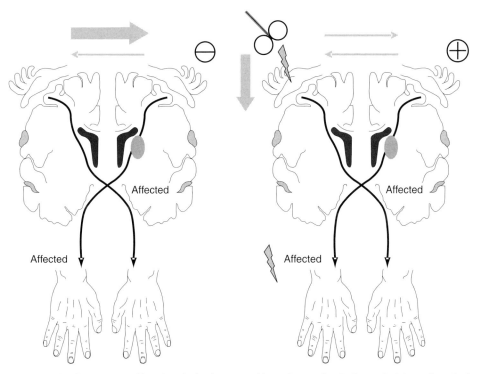

Figure 29.3. The concept of interhemispheric competition after stroke. Left panel: After subcortical stroke the primary motor cortex (M1) of the unaffected hemisphere is released from transcallosal inhibition and thereby causes enhanced inhibition of M1 of the affected hemisphere. Cortical excitability of M1 of the affected hemisphere is reduced. This may worsen the sensorimotor deficit of the affected hand. Right panel: Cortical excitability of M1 of the affected hemisphere could be increased, for example, by repetitive electrical stimulation of somatosensory (and motor) fibers of a peripheral nerve of the affected hand or, alternatively, by inhibition of cortical excitability of M1 of the affected hemisphere by inhibitory rTMS reducing transcallosal inhibition towards M1 of the affected hemisphere. Also, facilitatory rTMS of M1 of the affected hemisphere may normalize the unbalance of cortical excitability in between both hemispheres (not shown).

affected hemisphere (see Figure 29.3). Cortical excitability of M1 of the affected hemi-sphere could be increased by repetitive electrical stimulation of a peripheral nerve of the affected hand, inhibition of cortical excitability of M1 of the unaffected hemisphere by inhibitory rTMS or by facilitatory rTMS of M1 of the affected hemisphere. Repetitive electrical stimulation of the median nerve of the affected hand with a stimulation protocol exciting primarily proprioceptive fibers causes significant increase of maximum grip force of the affected hand as well as dexterity measures (Conforto *et al.*, 2002, 2007; Wu *et al.*, 2006). Combined motor and somatosensory stimulation of the median nerve of the affected hand by repetitive magnetic stimulation (Struppler *et al.*, 2003) and repetitive electrical whole-hand stimulation by means of a mesh glove (Dimitrijevic & Soroker, 1994) resulted in functional improvements of motor function of the affected hand after stroke.

Facilitation of cortical excitability of M1 of the affected hemisphere by anodal tDCS (Hummel *et al.*, 2005) or high-frequency (3 Hz) rTMS (Khedr *et al.*, 2005) has the potential to moderately improve simple motor tasks with the affected hand after stroke. Similarly, inhibition of cortical excitability of M1 of the unaffected hemisphere by cathodal tDCS (Fregni *et al.*, 2005) or low-frequency (1 Hz) rTMS (Mansur *et al.*, 2005) can improve dexterity at the affected hand after cortical or subcortical stroke. An example of how inhibitory (1 Hz) rTMS over M1 of the unaffected hemisphere improves kinetic aspects of grasping with the affected hand in subcortical stroke is illustrated in Figure 29.4. Twelve right-handed stroke subjects suffering from a first-ever subcortical ischemic stroke in the territory of the middle cerebral artery (four females, age range 24–56 years, mean age 45 ± 9 years) participated in the experiments. Eight patients had a left hemispheric stroke, four patients had a right hemispheric stroke. Subjects grasped, lifted and held an instrumented object in a precision grip between the index finger and thumb with both the affected and unaffected hand prior to (baseline) and following 10 minutes of 1 Hz rTMS over (i) the vertex (control stimulation) and (ii) M1 of the unaffected hemisphere. Compared with baseline, 1 Hz rTMS over M1 of the unaffected hemisphere, but not over the vertex, improved the efficiency of grip-force scaling as indexed by the ratio between peak grip and lift forces and the timing between peak grip and lift forces when grasping and lifting with the affected hand.

Also, the kinematic aspects of grasping movements with the affected hand after sub-cortical stroke may improve with downregulation of excitability within M1 of the unaffected hemisphere. In 15 right-handed subjects (four females, age range 37–54 years, mean age 46 ± 8 years) with impaired dexterity due to subcortical middle cerebral artery stroke, the effects of 1 Hz rTMS over M1 of the unaffected hemisphere on movement kinematics and neural activation within the motor system were studied. Subjects received 1 Hz rTMS for 10 minutes applied over (i) vertex (control stimulation) and (ii) M1 of the unaffected hemi-sphere. For behavioral testing, subjects performed index finger-tapping and reach-to-grasp movements with both hands at two baseline conditions, separated by one week, and following each rTMS application. For functional magnetic resonance imaging (fMRI), subjects performed hand-grip movements with their affected or unaffected hand before and after each rTMS application. One Hz rTMS applied over M1 of the unaffected

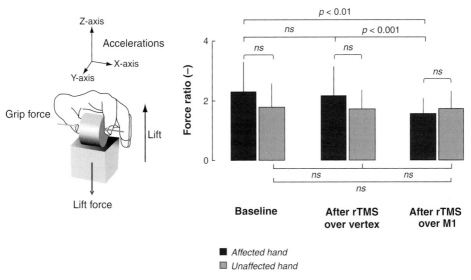

Figure 29.4. Low-frequency (1 Hz) repetitive transcranial magnetic stimulation (rTMS) to improve the efficiency of grip-force scaling at the affected hand in subcortical stroke. Twelve stroke subjects were asked to grasp and lift an instrumented test object in a precision grip between the index finger and thumb. At baseline, subjects applied excessive grip forces to lift the object as indexed by the ratio between peak grip and lift forces. In healthy subjects, the ratio between peak grip and lift forces varies between 1.0 to 1.3 (Nowak & Hermsdörfer, 2005). One Hz rTMS over M1 of the unaffected hemisphere, but not over the vertex, significantly reduces the force ratio when grasping and lifting the object with the affected hand. (Modified from Dafotakis *et al.*, 2008.)

hemisphere, but not over the vertex, improved the kinematics of finger-tapping and reach-to-grasp movements with the affected hand. At the neural level, rTMS applied over M1 of the unaffected hemisphere reduced overactivity of contralesional primary and non-primary motor areas. Importantly, overactivity of contralesional dorsal premotor cortex (BA6), contralesional parietal operculum (SII) and ipsilesional mesial frontal cortex at baseline and an increase in activity within ipsilesional BA6 and SII after rTMS treatment were significantly correlated with improvement of finger-tapping with the affected hand after rTMS over M1 of the unaffected hemisphere (Figure 29.5). This observation suggests that fMRI may be a useful tool to establish surrogate markers that suggest the individual's response to rTMS treatment. More research is necessary to implement this technique in individualized rehabilitation strategies developed to improve dexterity of the affected hand after stroke.

Conclusion

In summary, a rich variety of very heterogeneous concepts for therapy of hand function deficits following stroke is currently under investigation. Although the data on some

Baseline

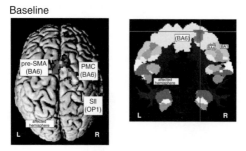

After rTMS over M1 of the unaffected hemisphere

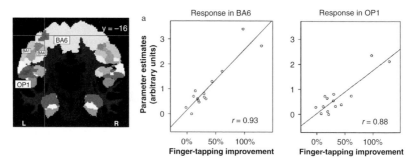

Figure 29.5. This figure is also reproduced in the color plate section. Neural activity in motor areas during movements with the affected hand in subcortical stroke. Right corresponds to the unaffected hemisphere, left corresponds to the affected hemisphere. Activity strengths at baseline (prior to repetitive transcranial magnetic stimulation (rTMS) intervention) that indexed a positive correlation with a subsequent rTMS-induced increase in finger-tapping frequency with the affected hand were found in the contralesional precentral gyrus (PMd, BA6, $r = 0.89$), ipsilesional mesial frontal cortex (pre-SMA, $r = 0.90$) and contralesional parietal operculum (area OP1, SII, $r = 0.90$) ($P < 0.05$, corrected for multiple comparisons). That means that a higher BOLD (blood-oxygenation level dependent signal) activity in these areas at baseline (compared with movements of the unaffected hand) was associated with a relevant improvement in index finger tapping performance after rTMS treatment over M1 of the unaffected hemisphere.

In addition, neural activity within two clusters – one situated on the ipsilesional precentral gyrus (PMd, BA6) and the other on the ipsilesional parietal operculum (SII, area OP1) – linearly increased with the improvement of index finger-tapping frequency of the affected hand after rTMS over M1 of the unaffected hemisphere ($P < 0.05$, corrected for multiple comparisons). Strong correlation coefficients were found for both activity clusters (BA6: $r = 0.93$; OP1: $r = 0.88$), i.e. the greater the increase in tapping performance due to rTMS the higher the neural signal in these two areas.

concepts are quite promising, there is still a need for thorough evaluation. Studies involving different therapeutic approaches differ greatly in the number of patients involved, methods of data acquisition and randomization, the severity of disability of patients included, time since onset of disease, the amount and location of damaged brain tissue, the length and intensity of training, selection of the control sample, type of outcome measures and many other factors. Based on the heterogeneity of signs and symptoms after stroke, the variability of stroke location and distribution and the at times loose association between the clinical syndrome and the location of tissue damage, it seems very unlikely that one particular therapy will suit each patient effectively. Rather, the efficiency of the therapeutic concepts will depend on individual patient characteristics. Our vision for the future is to develop strategies that match the individual deficit of a patient to the most effective therapy for that deficit.

References

Altschuler, E. L., Wisdom, S. B., Stone, L. *et al.* (1999). Rehabilitation of hemiparesis after stroke with a mirror. *Lancet*, **353**, 2035–2036.

Bhatt, E., Nagpal, A., Greer, K. H. *et al.* (2007). Effect of finger tracking combined with electrical stimulation on brain reorganization and hand function in subjects with stroke. *Exp Brain Res*, **182**, 435–447.

Blennerhassett, J. M., Carey, L. M. & Matyas, T. A. (2006). Grip force regulation during pinch grip lifts under somatosensory guidance: comparison between people with stroke and healthy controls. *Arch Phys Med Rehabil*, **87**, 418–429.

Bobath, B. (1987). *Adult Hemiplegia. Evaluation and Treatment*. London: Butterworth-Heinemann.

Bolton, D. A., Cauraugh, J. H. & Hausenblas, H. A. (2004). Electromyogram-triggered neuromuscular stimulation and stroke motor recovery of arm/hand functions: a meta-analysis. *J Neurol Sci*, **223**, 121–127.

Bourbonnais, D., Bilodeau, S., Cross, P. *et al.* (1997). A motor reeducation program aimed to improve strength and coordination of the upper-limb of a hemiparetic subject. *Neurorehabilitation*, **9**, 3–15.

Broderick, J., Brott, T., Kothari, R. *et al.* (1998). The Greater Cincinnati/Northern Kentucky Stroke Study: preliminary first-ever and total incidence rates of stroke among blacks. *Stroke*, **29**, 415–421.

Broeren, J., Rydmark, M., Bjorkdahl, A. & Sunnerhagen, K. S. (2007). Assessment and training in a 3-dimensional virtual environment with haptics: a report on 5 cases of motor rehabilitation in the chronic stage after stroke. *Neurorehabil Neural Repair*, **21**, 180–189.

Buccino, G., Solodkin, A. & Small, S. L. (2006). Functions of the mirror neuron system: implications for neurorehabilitation. *Cogn Behav Neurol*, **19**, 55–63.

Bütefisch, C., Hummelsheim, H., Denzler, P. & Mauritz, K. H. (1995). Repetitive training of isolated movements improves the outcome of motor rehabilitation of the centrally paretic hand. *J Neurol Sci*, **130**, 59–68.

Cadoret, G. & Smith, A. M. (1997). Comparison of the neuronal activity in the SMA and the ventral cingulate cortex during prehension in the monkey. *J Neurophysiol*, **77**, 153–166.

Carey, J. R., Durfee, W. K., Bhatt, E. *et al.* (2007). Comparison of finger tracking versus simple movement training via telerehabilitation to alter hand function and cortical reorganization after stroke. *Neurorehabil Neural Repair*, **21**, 216–232.

Carey, L. M. & Matyas, T. A. (2005). Training of somatosensory discrimination after stroke: facilitation of stimulus generalization. *Am J Phys Med Rehabil*, **84**, 428–442.

Cauraugh, J. H. & Kim, S. (2002). Two coupled motor recovery protocols are better than one: electromyogram-triggered neuromuscular stimulation and bilateral movements. *Stroke*, **33**, 1589–1594.

Cauraugh, J. H., Light, K., Kim, S., Thigpen, M. & Behrman, A. (2000). Chronic motor dysfunction after stroke: recovering wrist and finger extension by electromyography-triggered neuromuscular stimulation. *Stroke*, **31**, 1360–1364.

Conforto, A. B., Kaelin-Lang, A. & Cohen, L. G. (2002). Increase in hand muscle strength of stroke patients after somatosensory stimulation. *Ann Neurol*, **51**, 122–125.

Conforto, A. B., Cohen, L. G., dos Santos, R. L., Scaff, M. & Marie, S. K. (2007). Effects of somatosensory stimulation on motor function in chronic cortico-subcortical strokes. *J Neurol*, **254**, 333–339.

Dafotakis, M., Grefkes, C., Eickhoff, S. B. *et al.* (2008). Effects of rTMS on grip force control following subcortical stroke. *Exp Neurol*, **211**, 407–412.

De Kroon, J. R., van der Lee, J. H., IJzerman, M. J. & Lankhorst, G. J. (2002). Therapeutic electrical stimulation to improve motor control and functional abilities of the upper extremity after stroke: a systematic review. *Clin Rehabil*, **16**, 350–360.

De Vries, S. & Mulder, T. (2007). Motor imagery and stroke rehabilitation: a critical discussion. *J Rehabil Med*, **39**, 5–13.

Dimitrijevic, M. M. & Soroker, N. (1994). Mesh-glove. 2. Modulation of residual upper limb motor control after stroke with whole-hand electric stimulation. *Scand J Rehabil Med*, **26**, 187–190.

Dum, R. P. & Strick P. L. (2002). Motor areas in the frontal lobe of the primate. *Physiol Behav*, **77**, 677–682.

Ertelt, D., Small, S., Solodkin, A. *et al.* (2007). Action observation has a positive impact on rehabilitation of motor deficits after stroke. *Neuroimage*, **36**, 164–173.

Floel, A. & Cohen, L. G. (2006). Translational studies in neurorehabilitation: from bench to bedside. *Cogn Behav Neurol*, **19**, 1–10.

Fregni, F., Boggio, P. S., Mansur, C. G. *et al.* (2005). Transcranial direct current stimulation of the unaffected hemisphere in stroke patients. *Neuroreport*, **16**, 1551–1555.

Glanz, M., Klawansky, S., Stason, W. *et al.* (1995). Biofeedback therapy in poststroke rehabilitation: a meta-analysis of the randomized controlled trials. *Arch Phys Med Rehabil*, **76**, 508–515.

Grefkes, C., Nowak, D. A., Eickhoff, S. *et al.* (2008). Cortical connectivity after subcortical stroke assessed with fMRI. *Arch Neurol*, **65**, 741–747.

Hamzei, F., Liepert, J., Dettmers, C., Weiller, C. & Rijntjes, M. (2006). Two different reorganization patterns after rehabilitative therapy: an exploratory study with fMRI and TMS. *Neuroimage*, **31**, 710–720.

Hankey, G. J., Jamrozik, K., Broadhurst, R. J., Forbes, S. & Anderson, C. S. (2002). Long-term disability after first-ever stroke and related prognostic factors in the Perth Community Stroke Study, 1989–1990. *Stroke*, **33**, 1034–1040.

Hermsdörfer, J., Marquardt, C., Wack, S. & Mai, N. (1999). Comparative analysis of diadochokinetic movements. *J Electromyogr Kinesiol*, **9**, 283–295.

Hermsdörfer, J., Hagl, E., Nowak, D. A. & Marquardt, C. (2003). Grip force control during object manipulation in cerebral stroke. *Clin Neurophysiol*, **114**, 915–929.

Hesse, S., Werner, C., Pohl, M. *et al.* (2005). Computerized arm training improves the motor control of the severely affected arm after stroke: a single-blinded randomized trial in two centers. *Stroke*, **36**, 1960–1966.

Hummel, F., Celnik, P., Giraux, P. *et al.* (2005). Effects of non-invasive cortical stimulation on skilled motor function in chronic stroke. *Brain*, **128**, 490–499.

Hummel, F. C. & Cohen, L. G. (2006). Non-invasive brain stimulation: a new strategy to improve neurorehabilitation after stroke? *Lancet Neurol*, **5**, 708–712.

Hummelsheim, H., Amberger, S. & Mauritz, K. H. (1996). The influence of EMG initiated electrical muscle stimulation on motor recovery of the centrally paretic hand. *Eur J Neurosci*, **3**, 245–254.

Hummelsheim, H., Maierloth, M. L. & Eickhof, C. (1997). The functional value of electrical muscle stimulation for the rehabilitation of the hand in stroke patients. *Scand J Rehabil Med*, **29**, 3–10.

Jenkins, W. M., Merzenich, M. M., Ochs, M. T., Allard, T. & Guíc-Robles, E. (1990). Functional reorganization of primary somatosensory cortex in adult owl monkeys after behaviorally controlled tactile stimulation. *J Neurophysiol*, **63**, 82–104.

Kahn, L. E., Lum, P. S., Rymer, W. Z. & Reinkensmeyer, D. J. (2006). Robot-assisted movement training for the stroke-impaired arm: does it matter what the robot does? *J Rehabil Res Dev*, **43**, 619–630.

Kelly-Hayes, M., Robertson, J. T., Broderick, J. P. *et al.* (1998). The American Heart Association Stroke Outcome Classification: executive summary. *Circulation*, **97**, 2474–2478.

Khedr, E. M., Ahmed, M. A., Fathy, N. & Rothwell, J. C. (2005). Therapeutic trial of repetitive transcranial magnetic stimulation after acute ischemic stroke. *Neurology*, **65**, 466–468.

Kolominsky-Rabas, P. L., Weber, M., Gefeller, O., Neundörfer, B. & Heuschmann, P. U. (2001). Epidemiology of ischemic stroke subtypes according to the TOAST criteria: incidence, recurrence, and long-term survival in ischemic stroke subtypes: a population-based study. *Stroke*, **32**, 2735–2740.

Kowalczewski, J., Gritsenko, V., Ashworth, N., Ellaway, P. & Prochazka, A. (2007). Upper-extremity functional electric stimulation-assisted exercises on a workstation in the subacute phase of stroke recovery. *Arch Phys Med Rehabil*, **88**, 833–839.

Kriz, G., Hermsdörfer, J., Marquardt, C. & Mai, N. (1995). Feedback-based training of grip force control in patients with brain damage. *Arch Phys Med Rehabil*, **76**, 653–659.

Kwakkel, G., Kollen, B. J. & Wagenaar, R. C. (2002). Long-term effects of intensity of upper and lower limb training following stroke: a randomised trial. *J Neurol Neurosurg Psychiatry*, **72**, 473–479.

Kwakkel, G., van Peppen, R., Wagenaar, R. C. *et al.* (2004). Effects of augmented exercise therapy time after stroke: a meta-analysis. *Stroke*, **35**, 2529–2539.

Lang, C. E., Wagner, J. M., Bastian, A. J. *et al.* (2005). Deficits in grasp versus reach during acute hemiparesis. *Exp Brain Res*, **166**, 126–136.

Lang, C. E., Reilly, K. T. & Schieber, M. H. (2006a). Human voluntary motor control and dysfunction. In M. Selzer, S. Clarke, L. Cohen, P. Duncan & F. Gage (Eds.), *Neural Repair and Rehabilitation* (Vol. II, pp. 24–36). New York, NY: Cambridge University Press.

Lang, C. E., Wagner, J. M., Edwards, D. F., Sahrmann, S. A. & Dromerick, A. W. (2006b). Recovery of grasp versus reach in people with hemiparesis poststroke. *Neurorehabil Neural Repair*, **20**, 444–454.

Larsen, C. & Schneider, W. (2007). *Spiraldynamische Körperarbeit. Hands-on-Techniken der 3D-Massage*. Stuttgart, Germany: Thieme.

Lennon, S., Baxter, D. & Ashburn, A. (2001). Physiotherapy based on the Bobath concept in stroke rehabilitation: a survey within the UK. *Disabil Rehabil*, **23**, 254–262.

Liepert, J., Hamzei, F. & Weiller, C. (2000). Motor cortex disinhibition of the unaffected hemisphere after acute stroke. *Muscle Nerve*, **23**, 1761–1763.

Luke, C., Dodd, K. J. & Brock, K. (2004). Outcomes of the Bobath concept on upper limb recovery following stroke. *Clin Rehabil*, **18**, 888–898.

Lum, P. S., Burgar, C. G., Shor, P. C., Majmundar, M. & Van der Loos, M. (2002). Robot-assisted movement training compared with conventional therapy techniques for the rehabilitation of upper-limb motor function after stroke. *Arch Phys Med Rehabil*, **83**, 952–959.

Lum, P. S., Burgar, C. G., Van der Loos, M. *et al.* (2006). MIME robotic device for upper-limb neurorehabilitation in subacute stroke subjects: a follow-up study. *J Rehabil Res Dev*, **43**, 631–642.

Mai, N. (1989). Residual control of isometric finger forces in hemiparetic patients. Evidence for dissociation of performance deficits. *Neurosci Lett*, **101**, 347–351.

Mansur, C. G., Fregni, F., Boggio, P. S. *et al.* (2005). A sham stimulation-controlled trial of rTMS of the unaffected hemisphere in stroke patients. *Neurology*, **64**, 1802–1804.

Masiero, S., Celia, A., Rosati, G. & Armani, M. (2007). Robotic-assisted rehabilitation of the upper limb after acute stroke. *Arch Phys Med Rehabil*, **88**, 142–149.

McDonnell, M. N., Hillier, S. L., Ridding, M. C. & Miles, T. S. (2006). Impairments in precision grip correlate with functional measures in adult hemiplegia. *Clin Neurophysiol*, **117**, 1474–1480.

Merians, A. S., Poizner, H., Boian, R., Burdea, G. & Adamovich, S. (2006). Sensorimotor training in a virtual reality environment: does it improve functional recovery poststroke? *Neurorehabil Neural Repair*, **20**, 252–267.

Miller, G. J. T. & Light, K. E. (1997). Strength training in spastic hemiparesis – should it be avoided? *Neurorehabilitation*, **9**, 17–28.

Miltner, W. H., Bauder, H., Sommer, M., Dettmers, C. & Taub, E. (1999). Effects of constraint-induced movement therapy on patients with chronic motor deficits after stroke: a replication. *Stroke*, **30**, 586–592.

Mudie, M. H. & Matyas, T. A. (2000). Can simultaneous bilateral movement involve the undamaged hemisphere in reconstruction of neural networks damaged by stroke? *Disabil Rehabil*, **22**, 23–37.

Muellbacher, W., Richards, C., Ziemann, U. *et al.* (2002). Improving hand function in chronic stroke. *Arch Neurol*, **59**, 1278–1282.

Mulder, T. (2007). Motor imagery and action observation: cognitive tools for rehabilitation. *J Neural Transmiss*, **114**, 1265–1278.

Murase, N., Duque, J., Mazzocchio, R. & Cohen, L. G. (2004). Influence of interhemispheric interactions on motor function in chronic stroke. *Ann Neurol*, **55**, 400–409.

Myers, B. J. (1989). Proprioceptive neuromuscular facilitation (PNF) approach. In C. A. Thromby (Ed.), *Occupational Therapy for Physical Dysfunction* (pp. 135–155). Baltimore, MD: Williams & Wilkins.

Naito, E., Roland, P. E. & Ehrsson, H. H. (2002). I feel my hand moving: a new role of the primary motor cortex in somatic perception of limb movement. *Neuron*, **36**, 979–988.

Nakayama, H., Jorgensen, H. S., Raaschou, H. O. & Olsen, T. S. (1994). Recovery of upper extremity function in stroke subjects: the Copenhagen Stroke Study. *Arch Phys Med Rehabil*, **75**, 394–398.

Nowak, D. A. (2006). Toward an objective quantification of impaired manual dexterity following stroke: the usefulness of precision grip measures. *Clin Neurophysiol*, **117**, 1409–1411.

Nowak, D. A. & Hermsdörfer, J. (2005). Grip force behaviour during object manipulation in neurological disorders: toward an objective evaluation of manual performance deficits. *Mov Dis*, **20**, 11–25.

Nowak, D. A., Hermsdörfer, J. & Topka, H. (2003). Deficits of predictive grip force control during object manipulation in acute stroke. *J Neurol*, **250**, 850–860.

Nowak, D. A., Grefkes, C., Dafotakis, M. *et al.* (2007). Dexterity is impaired at both hands following unilateral subcortical middle cerebral artery stroke. *Eur J Neurosci*, **25**, 3173–3184.

Nudo, R. J., Milliken, G. W., Jenkins, W. M. & Merzenich, M. M. (1996). Use-dependent alterations of movement representation in primary motor cortex of adult squirrel monkeys. *J Neurosci*, **16**, 785–807.

Page, S. J., Levine, P. & Leonard, A. C. (2005). Effects of mental practice on affected limb use and function in chronic stroke. *Arch Phys Med Rehabil*, **86**, 399–402.

Page, S. J. Levine, P. & Leonard, A. (2007). Mental practice in chronic stroke: results of a randomized, placebo-controlled trial. *Stroke*, **38**, 1293–1297.

Penfield, W. & Boldrey, E. (1937). Somatic motor and sensory representation in the cerebral cortex of man as studied by electrical stimulation. *Brain*, **37**, 389–443.

Penfield, W. & Rasmussen, T. (1950). *The Cerebral Cortex of Man*. New York, NY: MacMillan.

Perfetti, C. (2006). *Rehabilitieren mit Gehirn. Kognitiv – Therapeutische Übungen in Neurologie und Orthopädie*. Munich, Germany: Pflaum.

Platz, T. (2004). Impairment-oriented training (IOT) – scientific concept and evidence-based treatment strategies. *Restor Neurol Neurosci*, **22**, 301–315.

Platz, T., Winter, T., Muller, N. *et al.* (2001). Arm ability training for stroke and traumatic brain injury patients with mild arm paresis: a single-blind, randomized, controlled trial. *Arch Phys Med Rehabil*, **82**, 961–968.

Popovic, M. B., Popovic, D. B., Sinkjaer, T., Stefanovic, A. & Schwirtlich, L. (2003). Clinical evaluation of Functional Electrical Therapy in acute hemiplegic subjects. *J Rehabil Res Dev*, **40**, 443–453.

Rizzolatti, G., Luppino, G. & Matelli, M. (1998). The organization of the cortical motor system: new concepts. *Electroencephalogr Clin Neurophysiol*, **106**, 283–296.

Santos, M., Zahner, L. H., Mckiernan, B. J., Mahnken, J. D. & Quaney, B. (2006). Neuromuscular electrical stimulation improves severe hand dysfunction for individuals with chronic stroke: a pilot study. *J Neurol Phys Ther*, **30**, 175–183.

Schieber, M. H. (2001). Constraints on somatotopic organization in the primary motor cortex. *J Neurophysiol*, **86**, 2125–2143.

Sharma, N., Pomeroy, V. M. & Baron, J. C. (2006). Motor imagery: a backdoor to the motor system after stroke? *Stroke*, **37**, 1941–1952.

Stein, D. G. (1998). Brain injury and theories of recovery. In L. B. Goldstein (Ed.), *Restorative Neurology: Advances in Pharmacotherapy for Recovery After Stroke* (pp. 1–34). Armonk, NY: Futura Publishing.

Stevens, J. A. & Stoykov, M. E. (2004). Simulation of bilateral movement training through mirror reflection: a case report demonstrating an occupational therapy technique for hemiparesis. *Top Stroke Rehabil*, **11**, 59–66.

Struppler, A., Havel, P. & Müller-Barna, P. (2003). Facilitation of skilled finger movements by repetitive peripheral magnetic stimulation (RPMS) – a new approach in central paresis. *Neurorehabilitation*, **18**, 69–82.

Suppé, B. & Spirgi-Gantert, I. (2007). *FBL Klein-Vogelbach Functional Kinetics: Die Grundlagen. Bewegungsanalyse, Untersuchung, Behandlung*. Berlin, Germany: Springer.

Taub, E., Miller, N. E., Novack, T. A. *et al.* (1993). Technique to improve chronic motor deficit after stroke. *Arch Phys Med Rehabil*, **74**, 347–354.

Taub, E., Uswatte, G. & Pidikiti, R. (1999). Constraint-induced movement therapy – a new family of techniques with broad application to physical rehabilitation – a clinical review. *J Rehabil Res Develop*, **36**, 237–251.

Tijs, E. & Matyas, T. A. (2006). Bilateral training does not facilitate performance of copying tasks in poststroke hemiplegia. *Neurorehabil Neural Repair*, **20**, 473–483.

Van der Lee, J. H. (2001). Constraint-induced therapy for stroke: more of the same or something completely different? *Curr Opin Neurol*, **14**, 741–744.

Van Peppen, R. P., Kwakkel, G., Wood-Dauphinee, S. *et al.* (2004). The impact of physical therapy on functional outcomes after stroke: what's the evidence? *Clin Rehabil*, **18**, 833–862.

Whitall, J., Waller, S. M., Silver, K. H. C. & Macko, R. F. (2000). Repetitive bilateral arm training with rhythmic auditory cueing improves motor function in chronic hemiparetic stroke. *Stroke*, **31**, 2390–2395.

Winstein, C. J., Rose, D. K., Tan, S. M. *et al.* (2004). A randomized controlled comparison of upper-extremity rehabilitation strategies in acute stroke: a pilot study of immediate and long-term outcomes. *Arch Phys Med Rehabil*, **85**, 620–628.

Woldag, H., Waldmann, G., Heuschkel, G. & Hummelsheim, H. (2003). Is the repetitive training of complex hand and arm movements beneficial for motor recovery in stroke patients? *Clin Rehabil*, **17**, 723–730.

Wolf, S. L., Winstein, C. J., Miller, J. P. *et al.* for the EXCITE Investigators (2006). Effect of constraint-induced movement therapy on upper extremity function 3 to 9 months after stroke: the EXCITE randomized clinical trial. *J Am Med Assoc*, **296**, 2095–2104.

Wu, C. W., Seo, H. J. & Cohen, L. G. (2006). Influence of electric somatosensory stimulation on paretic-hand function in chronic stroke. *Arch Phys Med Rehabil*, **87**, 351–357.

30

Functional reorganization and neuromodulation

CHRISTIAN GREFKES AND GEREON R. FINK

Summary

The human brain has a great potential for reorganizing itself after lesions to regain lost function. In recent years, functional imaging techniques such as functional magnetic resonance imaging (fMRI) and positron emission tomography (PET) have revealed complex changes in cortical networks that are functionally relevant for recovery of function following stroke. In this chapter, we demonstrate how stroke may influence cortical activity over time depending on structural damage and functional outcome. We furthermore discuss different techniques to modulate human brain function, e.g. via pharmacological interventions and transcranial magnetic stimulation (TMS), and their effects on cortical activity in both healthy subjects and patients. Assessing the changes in cortical network architecture following neuromodulation in individual patients will help to design novel treatment strategies based on neurobiological principles to minimize functional impairment resulting from brain lesions.

Introduction

As discussed in the preceding chapters, stroke is the leading cause of permanent disability in Europe and the USA (Gresham *et al.*, 1975; Whisnant, 1984; Taylor *et al.*, 1996) (see Chapters 21 and 29). Treatment of stroke patients in specialized facilities such as neurological intensive care and stroke units has led to a significant reduction of mortality rates in the acute stage of cerebral ischemia or hemorrhage in the past decades (Howard *et al.*, 2001). This positive development is, however, associated with an increasing number of people living with residual neurological symptoms such as hemiparesis, aphasia or other neuropsychological deficits. At present only a small fraction of stroke patients can be subjected to thrombolysis therapy in order to provide a causal treatment of cerebral ischemia (NINDS rt-PA Stroke Study Group, 1995); for most patients treatment relies on rehabilitation. While it is well documented in the literature that patients having suffered a stroke may significantly benefit from rehabilitative therapies beyond the natural recovery occurring without any targeted therapy (Gresham, 1986; Ottenbacher & Jannell, 1993; Maulden *et al.*, 2005) (see Chapter 29), there is a great variance in individual responses to rehabilitative treatment with

Sensorimotor Control of Grasping: Physiology and Pathophysiology, ed. Dennis A. Nowak and Joachim Hermsdörfer. Published by Cambridge University Press. © Cambridge University Press 2009.

some patients recovering better than others. Population-based studies have revealed a number of predictors for a better functional outcome such as age, gender, size and side of the lesion, initial severity of symptoms and the onset/initiation of rehabilitation after stroke (Maulden *et al.*, 2005). By contrast, the neural mechanisms underlying recovery of function following stroke remain poorly understood. Thus, advancing our understanding of how the brain reorganizes itself subserving recovery of function has the great potential to enable the development of individual treatment regimes designed to interact with the pathophysiological consequences resulting from cerebral ischemia (Ward, 2007). Neuroimaging techniques such as positron emission tomography (PET) or functional magnetic resonance imaging (fMRI) may detect changes in cerebral metabolism and blood flow, thereby allowing in-vivo measurement of task-related neural activity in humans with excellent spatial resolution. In this chapter, we discuss the impact of stroke on the neural networks subserving motor activity, and ways of modulating these networks by means of pharmacological and technical interventions.

Stroke and plasticity

Focal ischemia of the brain due to the obstruction of a supplying artery, e.g. via arterio-arterial embolism, releases a complex cascade of metabolic and cytotoxic reactions in the acute phase of stroke leading to a loss of neuronal integrity. Work in animal models demonstrates that even at very early stages of cerebral ischemia, developmental proteins and other substrates of molecular plasticity are expressed in both perilesional and remote brain regions (Schallert *et al.*, 2000). These mechanisms may also promote structural changes such as dendritic branching and synaptogenesis, and finally alter cortical excitability due to changes in neurotransmitter systems, e.g. reduced GABAergic inhibition in perilesional and distant brain areas (Witte & Stoll, 1997). Thus, a focal lesion in the brain triggers mechanisms of increased neuronal plasticity, thereby enabling changes in structural and functional organization of neural networks.

Changes in the motor system following stroke

The first PET and fMRI studies examining the motor system in stroke patients described task-dependent over-activations in a number of motor-related areas such as the contralesional primary motor cortex (M1) and premotor cortex (PMC), ipsilesional cerebellum, bilateral supplementary motor area (SMA) and parietal cortex when compared with healthy control subjects (Chollet *et al.*, 1991; Weiller *et al.*, 1992). A typical finding is a bilateralization of neural activity during movements of the paretic hand (Figure 30.1) which is not present in healthy subjects or when the patients move their unaffected hand (Feydy *et al.*, 2002; Bütefisch *et al.*, 2005; Grefkes *et al.*, 2008; Nowak *et al.*, 2008).

The observation of increased neural activity during movements of the stroke-affected hand has stimulated the discussion regarding the role of non-primary motor areas for

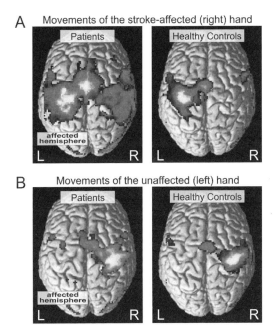

Figure 30.1. This figure is also reproduced in the color plate section. Neural activity during movements of the left or right hand in healthy subjects and in stroke patients with left-sided subcortical lesions ($P<0.05$, corrected on the cluster level). Activation clusters were surface rendered onto a canonical brain shown from above. In stroke patients, movements of the impaired hand were associated with significant activations in ipsilateral (= contralesional) motor areas, which were absent in the healthy controls (A) or when moving the unaffected hand (B).

recovery of motor function. For example, therapy-induced improvements in upper limb function in chronic stroke patients have been associated with increased neural activation in ipsilesional dorsal premotor cortex (Johansen-Berg *et al.*, 2002b). A cross-sectional fMRI study with a group of chronic stroke patients with infarcts sparing M1 demonstrated that those patients with a poorer outcome recruited more of the primary and secondary motor areas in both the affected and unaffected hemisphere than those subjects with good functional recovery (Ward *et al.*, 2003b). These observations are supported by other studies demonstrating that those patients with better functional outcome exhibit more lateralized activation patterns towards the ipsilesional hemisphere, and reduced activity in contralesional areas (Marshall *et al.*, 2000; Carey *et al.*, 2002; Nelles, 2004). A similar result was obtained from a longitudinal study examining patients from 10 days to up to 12 months post stroke (Ward *et al.*, 2003a). After observing initially a widespread activation of motor areas in both hemispheres, neural activity in bilateral sensorimotor cortex, PMC, SMA, cingulate motor areas, cerebellum, basal ganglia and thalamus progressively diminished over weeks and months, and reduction of overactivity in these areas was strongly correlated with the individual recovery scores (Ward *et al.*, 2003a). In other words, the motor network activated during movements of the stroke-affected hand resembled more a physiological pattern (as would be observed in healthy subjects) the better patients recovered from their initial motor deficit. Such reductions in motor system activation with improved performance resemble observations in healthy subjects demonstrating a decrease of neural activity in motor areas during the acquisition of new skills (Hikosaka *et al.*, 2002). This implies similar principles underlying learning new movements and motor rehabilitation. Furthermore, stroke patients with greater motor deficits also engage attentional networks comprising contralesional, intraparietal and ipsilesional rostro-dorsal premotor cortex, and the recruitment of these networks occurs more in the early than in a later post-stroke phase (Ward *et al.*, 2004). This observation has led to the hypothesis that different neural compensatory mechanisms might be active depending on the time since stroke, which may also impact on the therapeutic approaches to be used at different stages post stroke.

Stroke and connectivity in the motor system

Anatomical studies in non-human primates revealed that not only M1 but also PMC and SMA are directly connected to spinal motor neurons (Strick, 1988; see also Chapters 6 and 29). Corticospinal projections from these secondary motor areas seem to terminate at spinal motor neurons in a distribution that is quantitatively less, but qualitatively similar to axonal projections from M1 (Dum & Strick, 2002). For the recovery of function following brain injury, such a parallel organization of cortical motor output for voluntary movements raises the hypothesis that disruption of one motor area may be compensated by other areas within this network. Evidence from clinical studies and neuroimaging experiments suggests that the integrity of the corticospinal tract is critical for cortical reorganization and the restitution of motor function after stroke (Strick, 1988; Ward & Cohen, 2004; Stinear *et al.*, 2007). For example, analyzing the degree of disruption of corticofugal fibers from M1, SMA and PMC in a sample of subcortical stroke patients by means of fMRI and diffusion tensor imaging (DTI) demonstrated that successful recovery was related to intact descending fibers originating from M1 (Newton *et al.*, 2006). Another study showed that when stroke patients performed isometric hand grips (compressing two bars) with their impaired hand, a shift in average motor system recruitment from primary to secondary motor networks was observed in those patients with a greater disruption of the corticospinal system (as assessed with TMS) (Ward *et al.*, 2006). This shift from M1 to premotor regions was also associated with changes in functional properties of these two motor regions: while in both those patients with less corticospinal damage and healthy subjects neural activity in ipsilesional (contralateral) M1 co-varied with the amount of grip force, patients with greater damage to the corticospinal tract showed this force-related modulation of activity mainly in PMC, but less in M1. This observation suggests that by taking over some of the executive functions of M1, PMC may contribute to the recovery of function of the affected hand (Ward, 2007).

While most imaging studies suggest a compensatory role of secondary motor areas (especially for patients with subcortical lesions), the significance of contralesional M1 is still under debate. Whereas some studies using rTMS to disrupt cortical function of contralesional M1 have failed to find any functional significance of this area upon recovery of hand function

(Johansen-Berg *et al.*, 2002a; Werhahn *et al.*, 2003), others have shown significant improvement of dexterous movements with the stroke-affected hand when contralesional M1 function is inhibited by means of low-frequency rTMS (Mansur *et al.*, 2005;Takeuchi *et al.*, 2005). The latter data are consistent with interhemispheric competition models of sensory and motor processing implying that the function of M1 within the lesioned hemisphere is hampered by enhanced transcallosal inhibition from contralesional M1 (Ferbert *et al.*, 1992; Hummel & Cohen, 2006). Evidence supporting this view comes from a recent fMRI study analyzing effective connectivity among motor areas within and across hemispheres in patients with subcortical stroke lesions (Grefkes *et al.*, 2008): compared with healthy controls, movements of the paretic hand were associated with an additional inhibitory influence originating from M1 of the unaffected hemisphere (Figure 30.2), that was most pronounced in patients with stronger motor deficits (see correlation analysis in Figure 30.2C). Lesions in this subgroup of patients often affected the medial putamen, globus pallidus and the adjacent internal capsule. This suggests that the observed pathological adaptation processes among both M1 may result from lesions to these subcortical structures which are known to be densely connected with SMA and M1 (Grefkes *et al.*, 2008). Furthermore, the analysis of the *intrinsic* coupling rates revealed a disturbance of the interaction between SMA and ipsilesional M1, which was independent of whether the patients moved their affected or healthy hand, again especially in those patients with stronger motor impairments. These results indicate that pathological intra- and interhemispheric interactions among key motor regions may constitute an important pathophysiological aspect of disability following subcortical stroke. In this context, purposeful modulation of cortical excitability may render a useful tool to promote functional recovery of the affected hand following stroke (Ziemann, 2005; Hummel & Cohen, 2006).

Neuromodulation and recovery of function

Imaging of the structural and functional changes following stroke not only provides better insights into the mechanism of adaptation and reorganization, but has frequently been used to assess the effects of modulating the motor function by means of behavioral (e.g. physical therapy), pharmacological (e.g. neurotransmitter-enhancing drugs) or technical interventions (e.g. TMS; see also Chapter 29).

Pharmacological stimulation

Studies in both animals and humans demonstrated that stimulation of various neurotransmitter systems may impact on stroke recovery (Pariente *et al.*, 2001; Scheidtmann *et al.*, 2001; Feeney *et al.*, 2004). In particular, those drugs enhancing the concentration of monoaminergic neurotransmitters such as (nor-)adrenaline, dopamine or serotonin may improve motor function in both healthy subjects and stroke patients. For example, administration of reboxetine, a selective noradrenaline reuptake inhibitor, may improve hand coordination abilities in a visuomotor joystick task in healthy subjects (Figure 30.3A). At the neural level, reboxetine

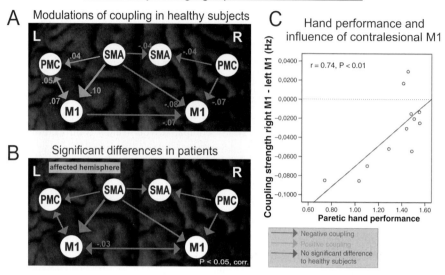

Figure 30.2. This figure is also reproduced in the color plate section. Modulation of neural coupling during unimanual hand movements. A. Coupling parameters for right-hand movements in healthy subjects. B. Coupling parameters for movements of the paretic (right) hand in stroke patients which were significantly different from those of the healthy control subjects. C. Correlation between paretic hand performance and interhemispheric inhibition exerted from contralesional M1 on ipsilesional M1 (Grefkes *et al.*, 2008).

selectively increased activity (compared with placebo) in parietofrontal areas known to be involved in visuospatial processing and hand–eye coordination (Figure 30.3B). In stroke patients, stimulating the noradrenergic and the dopaminergic system by means of amphetamines significantly improved motor performance in stroke patients (Walker-Batson *et al.*, 1995). Similar effects can be achieved by the amphetamine-like drug methylphenidate, which increases synaptic noradrenaline and dopamine levels by inhibition of reuptake transporters. Transcranial magnetic stimulation studies demonstrated that methylphenidate enhances cortical excitability and intracortical facilitation in the primary motor cortex (Ilic *et al.*, 2003). At the neural level, a single dose of methylphenidate increased activity in ipsilesional primary motor cortex and contralesional premotor cortex in an auditory-paced finger flexion task (Tardy *et al.*, 2006). We know from TMS studies that these two regions play a crucial role in promoting recovery of function after stroke (Johansen-Berg *et al.*, 2002b). Stimulating the serotonergic system by means of serotonin reuptake inhibitors (SSRI) like fluoxetine, paroxetine or fluvoxamine demonstrated improvements of motor performance and activation strength of the sensorimotor cortex in healthy subjects (Loubinoux *et al.*, 1999). Likewise, chronic application of paroxetine over 30 days has been shown to improve motor performance in healthy subjects (Loubinoux *et al.*, 2005). However, at the neural level, chronic paroxetine administration yielded a decrease of activation in ipsilesional sensorimotor cortex. This

Pharmacological modulation of visuomotor function

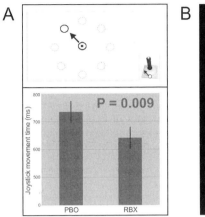
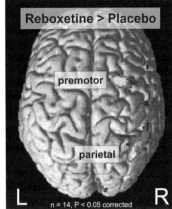

Figure 30.3. This figure is also reproduced in the color plate section. Effects of adrenergic stimulation by means of reboxetine. A. Healthy subjects moved a joystick with the right hand to guide a cursor from the center to a peripheral circle while being scanned with fMRI. Compared with placebo, subjects were significantly faster when their adrenergic system was stimulated by a single dose of reboxetine. B. The imaging data show that these behavioral improvements were associated with increased neural activity in right fronto-parietal circuits known to be involved in hand–eye coordination and visuomotor transformation.

opposite action of single versus chronic paroxetine administration on neural activity was attributed to a downregulation of cortical serotonin receptors (Loubinoux *et al.*, 2005). The data suggest that modulating cortical function by means of a pharmacological approach may over time induce differential mechanisms of plasticity.

Transcranial magnetic stimulation (TMS)

Another way of modulating human brain function is to apply TMS to focally stimulate the cerebral cortex. Simultaneous TMS pulse application and acquisition of MR images is technically possible (Bestmann *et al.*, 2003); it requires, however, considerable modifications of the TMS unit (e.g. non-magnetic TMS coil, shielding of the stimulator) and imaging parameters (e.g. interleaved TMS–EPI protocols). For example, applying short trains of high-intensity (110% of the resting motor threshold) repetitive TMS (3 Hz for blocks of 10 s) over the left dorsal PMC while subjects lay in the MR scanner in a relaxed position with their eyes closed (3T Siemens Trio scanner, TR = 3.3 s, TE = 36 ms) demonstrated increased neural activity in the left precentral sulcus, but also in remote areas such as the homotopic right dorsal PMC, bilateral ventral PMC, SMA, left thalamus, bilateral caudate nucleus and in the cerebellar hemispheres (Bestmann *et al.*, 2005). There was a close spatial correspondence with the regions showing increased activity during rTMS of left dorsal PMC at rest and during voluntary finger movements of the right hand, except for left M1 (which was only active *during*

Technical modulation of impaired motor function

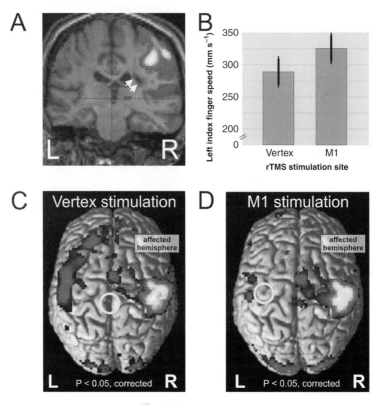

Figure 30.4. This figure is also reproduced in the color plate section. Technical modulation of neural function using repetitive TMS in a 54-year-old stroke patient 8 weeks post stroke. A. Neural activity during whole hand closing movements of the left paretic hand reveals bilateral activity in the motor cortex. Arrows mark the subcortical stroke lesion near the right lateral ventricle. B. Effects of inhibitory 1 Hz stimulation over contralesional M1 and over the vertex (control stimulation site) on finger-tapping speed of the paretic hand. C. Neural activity after vertex stimulation. D. Neural activity after stimulation of contralesional M1 demonstrates reductions in over-activity in the left hemisphere. $P < 0.05$, family-wise error corrected on the voxel level. Yellow circle: rTMS stimulation site.

voluntary finger movements). Accordingly, focal rTMS in absence of overt motor responses may induce an activation of the whole motor network including subcortical regions.

An alternative approach is to stimulate the volunteer outside the MR room using TMS protocols with a sufficiently long behavioral after-effect, for example inhibitory 1-Hz repetitive TMS for 10 min (see also Chapter 29). Applying 600 rTMS pulses at 1 Hz and 100% resting motor threshold over contralesional M1 can improve motor performance of the impaired hand in subcortical stroke patients (Nowak *et al.*, 2008) (see Chapter 29). Figure 30.4 demonstrates neural activity in a patient with a small periventricular ischemic lesion in the right hemisphere causing slowness of movements of the left hand. During control stimulation (vertex), movements of the stroke-affected hand evoked increased neural activity also in the contralesional motor cortex (Figure 30.4C, $P < 0.05$, family-wise error corrected at the voxel level). However, inhibiting the contralesional hand area by means of 1-Hz rTMS reduced over-activity not only in the vicinity of the precentral gyrus but also in contralesional premotor and prefrontal areas (Figure 30.4D). These data are consistent with online rTMS–fMRI experiments showing not only local but also remote effects in the motor network during focal stimulation of the cerebral cortex as reported above (Bestmann *et al.*, 2005).

Brain–computer interfaces

Brain–computer interfaces (BCI) or brain–machine interfaces (BMI) use neurophysiological signals originating in the brain to activate or deactivate external devices or computers (Birbaumer & Cohen, 2007). A possible application of this approach is to use neural signals in paralyzed patients to activate muscles and prosthetic devices. These signals can be obtained from local field potentials, electrocortiograms (ECoG), EEG oscillations, event-related brain potentials and metabolic responses such as the BOLD signal in fMRI. For example, a pilot MEG study with subcortical stroke patients showed that activation of the sensorimotor cortex of the lesioned hemisphere by movement imagination could be used to control a hand prosthesis for grasping (Birbaumer & Cohen, 2007). Weiskopf and colleagues demonstrated that modifying acquisition and processing techniques in fMRI may allow for the generation of online activation maps of the brain with high spatial resolution ("real time fMRI") (Weiskopf *et al.*, 2003). For example, BOLD activity can be extracted from a specific area (e.g. M1 of the affected hemisphere), and directly presented to the patient in the scanner by means of a visual feedback. Such maps showing the ability of a patient to activate a specific target region may, therefore, be used for neurofeedback training. Accordingly, real-time fMRI is a novel approach to endogenously modulate neural function in order to support recovery of function in patients. In principle, this approach allows training of the different neural components of the motor network, e.g. for grasping movements, which may result in an improved network structure enabling a better behavioral control of the paretic hand.

Neuroimaging and the prediction of functional outcome

Finally, as neuroimaging allows direct insights into the structural and functional consequences of stroke, several prognostic markers have been probed to predict functional outcome and responsiveness to rehabilitation. So far, surprisingly few imaging parameters have been identified to significantly correlate with the potential of recovery. While structural markers such as lesion size as defined with diffusion-weighted imaging (DWI) do not correlate with functional outcome beyond clinical measures (Hand *et al.*, 2006), hypo-activation in ipsilesional M1 prior to physical therapy has been shown to correlate with a poorer functional outcome (Cramer *et al.*, 2007). Likewise, the strengths of activation in ipsilesional M1, somatosensory cortex (S1) and insula in the first 3 weeks after stroke were indicative of the quality of recovery one year later (Loubinoux *et al.*, 2007).

The fact that only a few factors have thus far been identified as prognostic factors for recovery of function can partly be attributed to the fact that most neuroimaging studies investigated relatively small sample sizes (usually between 10 and 20 patients) with heterogeneous lesion age and location. Thus, combining different neuroimaging and neurophysiological techniques may further increase the sensitivity to predict functional outcomes. For example, DTI can be used to assess the structural damage to the internal capsule by measuring the fractional anisotropy (FA) – a DTI parameter depending on the diffusibility of water in tissue which is reduced in stroke lesions (Le Bihan *et al.*, 2001). As mentioned above, several studies have thus far been able to demonstrate that the integrity of the corticospinal tract is crucial for cortical reorganization (Newton *et al.*, 2006; Ward *et al.*, 2006; Ward, 2007). Furthermore, patients in whom motor-evoked potentials (MEPs) can be elicited by means of TMS within 30 days post stroke are more likely to make a better motor recovery than those patients with absent MEPs (Pennisi *et al.*, 1999). Stinear *et al.* (2007) combined DTI, fMRI, TMS and clinical scales to investigate the effect of a 30-day physical therapy in chronic stroke patients. They found that in patients with MEPs, meaningful gains in motor performance were still possible even 3 years after stroke, albeit with a decline of the functional potential with increasing time since stroke. In patients without MEPs, the current clinical score was strongly predicted by the FA asymmetry between the affected and unaffected corticospinal tract as assessed with DTI. In this group of patients, fMRI of the motor system demonstrated that ipsilesional activity was associated with better upper limb function, whereas cortical activity was lateralized to the contralesional hemisphere in those patients with the poorest clinical outcomes. However, for patients both with and without MEPs the lateralization of cortical activity assessed by fMRI did not predict functional improvement following motor practice. Therefore, Stinear and colleagues demonstrated that by combining different techniques patients may be stratified in subgroups with different functional potential (even if the vascular accident has occurred some years in the past) which may influence the type of individual rehabilitation therapy in the light of the emerging pharmacological or technical approaches.

Conclusion

In summary, functional neuroimaging in combination with structural analyses and other measures reflecting the integrity of neural networks has the great potential to provide a better understanding of the mechanisms underlying recovery of function from stroke and other neurological conditions. For a more individualized analysis of the neural consequences following stroke, greater sample sizes than usually employed in fMRI studies seem to be necessary. That is, however, difficult to achieve for various technical and practical reasons. Therefore, future fMRI studies should focus on more homogeneous samples of patients (with respect to, for example, lesion location or time from stroke onset) paired with a comprehensive analysis of the individual lesion anatomy. In the same vein, more longitudinal fMRI studies are needed to unravel the impact of factors such as lesion location, pre-stroke morbidity or (novel) treatment protocols on recovery of function following stroke. In this context, a translational approach making use of animal models and neuroimaging techniques (e.g. ultra-highfield fMRI or microPET) may further our understanding of the biological determinants of spontaneous recovery (and technical/pharmacological/practice-induced modulation thereof). Such research might finally enable the tailoring of treatment regimes to the individual needs of the patient based on the individual network pathologies following stroke.

References

Bestmann, S., Baudewig, J. & Frahm, J. (2003). On the synchronization of transcranial magnetic stimulation and functional echo-planar imaging. *J Magn Reson Imaging*, **17**, 309–316.

Bestmann, S., Baudewig, J., Siebner, H. R., Rothwell, J. C. & Frahm, J. (2005). BOLD MRI responses to repetitive TMS over human dorsal premotor cortex. *Neuroimage*, **28**, 22–29.

Birbaumer, N. & Cohen, L. G. (2007). Brain-computer interfaces: communication and restoration of movement in paralysis. *J Physiol*, **579**, 621–636.

Bütefisch, C. M., Kleiser, R., Müller, K. *et al.* (2005). Recruitment of contralesional motor cortex in stroke patients with recovery of hand function. *Neurology*, **64**, 1067–1069.

Carey, J. R., Kimberley, T. J., Lewis, S. M. *et al.* (2002). Analysis of fMRI and finger tracking training in subjects with chronic stroke. *Brain*, **125**, 773–788.

Chollet, F., DiPiero, V., Wise, R. J. *et al.* (1991). The functional anatomy of motor recovery after stroke in humans: a study with positron emission tomography. *Ann Neurol*, **29**, 63–71.

Cramer, S. C., Parrish, T. B., Levy, R. M. *et al.* (2007). Predicting functional gains in a stroke trial. *Stroke*, **38**, 2108–2114.

Dum, R. P. & Strick, P. L. (2002). Motor areas in the frontal lobe of the primate. *Physiol Behav*, **77**, 677–682.

Feeney, D. M., De Smet, A. M. & Rai, S. (2004). Noradrenergic modulation of hemiplegia: facilitation and maintenance of recovery. *Restor Neurol Neurosci*, **22**, 175–190.

Ferbert, A., Priori, A., Rothwell, J. C. *et al.* (1992). Interhemispheric inhibition of the human motor cortex. *J Physiol*, **453**, 525–546.

Feydy, A., Carlier, R., Roby-Brami, A. *et al.* (2002). Longitudinal study of motor recovery after stroke: recruitment and focusing of brain activation. *Stroke*, **33**, 1610–1617.

Grefkes, C., Nowak, D. A., Eickhoff, S. B. *et al.* (2008). Cortical connectivity after subcortical stroke assessed with functional magnetic resonance imaging. *Ann Neurol*, **63**, 236–246.

Gresham, G. E. (1986). Stroke outcome research. *Stroke*, **17**, 358–360.

Gresham, G. E., Fitzpatrick, T. E., Wolf, P. A. *et al.* (1975). Residual disability in survivors of stroke – the Framingham study. *N Engl J Med*, **293**, 954–956.

Hand, P. J., Wardlaw, J. M., Rivers, C. S. *et al.* (2006). MR diffusion-weighted imaging and outcome prediction after ischemic stroke. *Neurology*, **66**, 1159–1163.

Hikosaka, O., Nakamura, K., Sakai, K. & Nakahara, H. (2002). Central mechanisms of motor skill learning. *Curr Opin Neurobiol*, **12**, 217–222.

Howard, G., Howard, V. J., Katholi, C., Oli, M. K. & Huston, S. (2001). Decline in US stroke mortality: an analysis of temporal patterns by sex, race, and geographic region. *Stroke*, **32**, 2213–2220.

Hummel, F. C. & Cohen, L. G. (2006). Non-invasive brain stimulation: a new strategy to improve neurorehabilitation after stroke? *Lancet Neurol*, **5**, 708–712.

Ilic, T. V., Korchounov, A. & Ziemann, U. (2003). Methylphenidate facilitates and disinhibits the motor cortex in intact humans. *Neuroreport*, **14**, 773–776.

Johansen-Berg, H., Dawes, H., Guy, C. *et al.* (2002a). Correlation between motor improvements and altered fMRI activity after rehabilitative therapy. *Brain*, **125**, 2731–2742.

Johansen-Berg, H., Rushworth, M. F., Bogdanovic, M. D. *et al.* (2002b). The role of ipsilateral premotor cortex in hand movement after stroke. *Proc Natl Acad Sci USA*, **99**, 14518–14523.

Le Bihan, D., Mangin, J. F., Poupon, C. *et al.* (2001). Diffusion tensor imaging: concepts and application. *J Magn Reson Imaging*, **13**, 534–546.

Loubinoux, I., Boulanouar, K., Ranjeva, J. P. *et al.* (1999). Cerebral functional magnetic resonance imaging activation modulated by a single dose of the monoamine neurotransmission enhancers fluoxetine and fenozolone during hand sensorimotor tasks. *J Cereb Blood Flow Metab*, **19**, 1365–1375.

Loubinoux, I., Tombari, D., Pariente, J. *et al.* (2005). Modulation of behavior and cortical motor activity in healthy subjects by a chronic administration of a serotonin enhancer. *Neuroimage*, **27**, 299–313.

Loubinoux, I., Dechaumont, S., Castel-Lacanal, E. *et al.* (2007). Prognostic value of fMRI in recovery of hand function in subcortical stroke patients. *Cereb Cortex* [epub ahead of print].

Mansur, C. G., Fregni, F., Boggio, P. S. *et al.* (2005). A sham stimulation-controlled trial of rTMS of the unaffected hemisphere in stroke patients. *Neurology*, **64**, 1802–1804.

Marshall, R. S., Perera, G. M., Lazar, R. M. *et al.* (2000). Evolution of cortical activation during recovery from corticospinal tract infarction. *Stroke*, **31**, 656–661.

Maulden, S. A., Gassaway, J., Horn, S. D., Smout, R. J. & DeJong, G. (2005). Timing of initiation of rehabilitation after stroke. *Arch Phys Med Rehabil*, **86**, S34–S40.

Nelles, G. (2004). Cortical reorganization – effects of intensive therapy. *Restor Neurol Neurosci*, **22**, 239–244.

Newton, J. M., Ward, N. S., Parker, G. J. *et al.* (2006). Non-invasive mapping of corticofugal fibres from multiple motor areas – relevance to stroke recovery. *Brain*, **129**, 1844–1858.

NINDS rt-PA Stroke Study Group (1995). Tissue plasminogen activator for acute ischemic stroke. The National Institute of Neurological Disorders and Stroke rt-PA Stroke Study Group. *N Engl J Med*, **333**, 1581–1587.

Nowak, D. A., Grefkes, C., Dafotakis, M. *et al.* (2008). Improving dexterity following stroke: Effects of low-frequency rTMS over contralesional M1 on movement kinematics and neural activity in subcortical stroke. *Arch Neurol*, **65**, 741–747.

Ottenbacher, K. J. & Jannell, S. (1993). The results of clinical trials in stroke rehabilitation research. *Arch Neurol*, **50**, 37–44.

Pariente, J., Loubinoux, I., Carel, C. *et al.* (2001). Fluoxetine modulates motor performance and cerebral activation of patients recovering from stroke. *Ann Neurol*, **50**, 718–729.

Pennisi, G., Rapisarda, G., Bella, R. *et al.* (1999). Absence of response to early transcranial magnetic stimulation in ischemic stroke patients: prognostic value for hand motor recovery. *Stroke*, **30**, 2666–2670.

Schallert, T., Leasure, J. L. & Kolb, B. (2000). Experience-associated structural events, subependymal cellular proliferative activity, and functional recovery after injury to the central nervous system. *J Cereb Blood Flow Metab*, **20**, 1513–1528.

Scheidtmann, K., Fries, W., Müller, F. & Koenig, E. (2001). Effect of levodopa in combination with physiotherapy on motor recovery after stroke: a prospective, randomised, double-blind study. *Lancet*, **358**, 787–790.

Stinear, C. M., Barber, P. A., Smale, P. R. *et al.* (2007). Functional potential in chronic stroke patients depends on corticospinal tract integrity. *Brain*, **130**, 170–180.

Strick, P. L. (1988). Anatomical organization of multiple motor areas in the frontal lobe: implications for recovery of function. *Adv Neurol*, **47**, 293–312.

Takeuchi, N., Chuma, T., Matsuo, Y., Watanabe, I. & Ikoma, K. (2005). Repetitive transcranial magnetic stimulation of contralesional primary motor cortex improves hand function after stroke. *Stroke*, **36**, 2681–2686.

Tardy, J., Pariente, J., Leger, A. *et al.* (2006). Methylphenidate modulates cerebral post-stroke reorganization. *Neuroimage*, **33**, 913–922.

Taylor, T. N., Davis, P. H., Torner, J. C. *et al.* (1996). Lifetime cost of stroke in the United States. *Stroke*, **27**, 1459–1466.

Walker-Batson, D., Smith, P., Curtis, S., Unwin, H. & Greenlee, R. (1995). Amphetamine paired with physical therapy accelerates motor recovery after stroke. Further evidence. *Stroke*, **26**, 2254–2259.

Ward, N. S. (2007). Future perspectives in functional neuroimaging in stroke recovery. *Eura Medicophys*, **43**, 285–294.

Ward, N. S. & Cohen, L. G. (2004). Mechanisms underlying recovery of motor function after stroke. *Arch Neurol*, **61**, 1844–1848.

Ward, N. S., Brown, M. M., Thompson, A. J. & Frackowiak, R. S. (2003a). Neural correlates of motor recovery after stroke: a longitudinal fMRI study. *Brain*, **126**, 2476–2496.

Ward, N. S., Brown, M. M., Thompson, A. J. & Frackowiak, R. S. (2003b). Neural correlates of outcome after stroke: a cross-sectional fMRI study. *Brain*, **126**, 1430–1448.

Ward, N. S., Brown, M. M., Thompson, A. J. & Frackowiak, R. S. (2004). The influence of time after stroke on brain activations during a motor task. *Ann Neurol*, **55**, 829–834.

Ward, N. S., Newton, J. M., Swayne, O. B. *et al.* (2006). Motor system activation after subcortical stroke depends on corticospinal system integrity. *Brain*, **129**, 809–819.

Weiller, C., Chollet, F., Friston, K. J., Wise, R. J. & Frackowiak, R. S. J. (1992). Functional reorganization of the brain in recovery from striatocapsular infarction in man. *Ann Neurol*, **31**, 463–472.

Weiskopf, N., Veit, R., Erb, M. *et al.* (2003). Physiological self-regulation of regional brain activity using real-time functional magnetic resonance imaging (fMRI): methodology and exemplary data. *Neuroimage*, **19**, 577–586.

Werhahn, K. J., Conforto, A. B., Kadom, N., Hallett, M. & Cohen, L. G. (2003). Contribution of the ipsilateral motor cortex to recovery after chronic stroke. *Ann Neurol*, **54**, 464–472.

Whisnant, J. P. (1984). The decline of stroke. *Stroke*, **15**, 160–168.

Witte, O. W. & Stoll, G. (1997). Delayed and remote effects of focal cortical infarctions: secondary damage and reactive plasticity. *Adv Neurol*, **73**, 207–272.

Ziemann, U. (2005). Improving disability in stroke with rTMS. *Lancet Neurol*, **4**, 454–455.

31

Intensive training of upper extremity function in children with cerebral palsy

ANDREW M. GORDON AND KATHLEEN M. FRIEL

Summary

Cerebral palsy (CP) is the most common cause of severe physical disability in childhood. Spastic hemiplegia, characterized by motor impairments largely affecting one side of the body, is the most common form of CP. The resulting impaired hand function is one of the most disabling symptoms of hemiplegia, affecting self-care activities such as feeding, dressing and grooming. Consequently, children with hemiplegic CP tend not to use the more affected extremity. This "developmental non-use" can lead to further deficits, most notably affecting bimanual coordination. To date, there is unfortunately little evidence of efficacy of any specific treatment approach. Nevertheless, several lines of evidence suggest the impairments are not static. Upper extremity performance in children with CP may improve with practice and development, indicating that hand function may well be amenable to treatment. In this chapter we review this evidence along with studies involving intensive unilateral practice; i.e. constraint-induced movement therapy (CIMT). We then discuss important limitations of CIMT (most importantly, bimanual impairments underlie functional limitations) and introduce a new form of intensive training to address these limitations: Hand–Arm Bimanual Intensive Training (HABIT). The clinical implications of these findings and future directions for pediatric rehabilitation research are discussed.

Introduction

Cerebral palsy (CP) is a development disorder of movement and posture causing limitations in activity and deficits in motor skill (Bax *et al.*, 2005) and is attributed to non-progressive disturbances in the developing fetal or infant brain. Cerebral palsy is the most common cause of severe physical disability in childhood, with an incidence of about 2–2.5/1000 live births (Lin, 2003; Koman *et al.*, 2004). Spastic hemiplegia, characterized by motor impairments largely affecting one side of the body, is the most common type, accounting for 30–40% of new cases (Stanely *et al.*, 2000; Himmelmann *et al.*, 2005). Congenital hemiplegia is generally the result of middle cerebral artery infarct, hemi-brain atrophy, periventricular white matter damage, brain malformation or posthemorrhagic porencephaly (Uvcbrant, 1988; Bouza *et al.*, 1994; Okumura *et al.*, 1997; Cioni *et al.*, 1999). The integrity

Sensorimotor Control of Grasping: Physiology and Pathophysiology, ed. Dennis A. Nowak and Joachim Hermsdörfer. Published by Cambridge University Press. © Cambridge University Press 2009.

of the motor cortex and corticospinal tract (CST) underlying dexterity is often compromised (Duque *et al.*, 2003; Krageloh-Mann, 2004, 2005; Staudt *et al.*, 2004, 2005).

Unilateral damage to motor areas results in a failure of the affected CST to secure and maintain normal terminations in the spinal cord (Eyre *et al.*, 2007; see Martin, 2005; Martin *et al.*, 2007, for reviews) (see Chapter 24). Termination requires activity-dependent competition between the two sides of the developing motor system. In normal development, the CST projects bilaterally from both motor cortices at birth and is pruned into the mature contralateral projection pattern during the first few years of life (Eyre *et al.*, 2001). Damage to one side of the motor cortex, as occurs often in hemiplegic CP, results in aberrant organization of the motor system, with the damaged side failing to establish normal connections. Concurrently, the undamaged/less-damaged CST maintains exuberant bilateral projections, invading the normal termination zone of the contralateral side. Responsiveness of the CST in children with hemiplegia has been examined using transcranial magnetic stimulation (TMS); TMS weakly activates the damaged side while the undamaged side maintains bilateral terminations (Eyre *et al.*, 2007).

Abnormalities in the connectivity of the motor system result in abnormal development of skilled independent finger movements and hand dexterity (Brown *et al.*, 1987; Himmelmann *et al.*, 2006). The involved upper extremity may exhibit abnormal muscle tone with posturing into wrist flexion, ulnar deviation, elbow flexion and shoulder internal or external rotation (Brown *et al.*, 1987). There may be weakness and tactile and proprioceptive disturbances, which further impact fine motor skills (Brown *et al.*, 1987; Gordon & Duff, 1999b). The resulting impaired hand function is one of the most disabling symptoms of hemiplegia (Skold *et al.*, 2004). Consequently, children with hemiplegic CP tend not to use the more affected extremity. This "developmental non-use" can lead to further deficits (Eyre *et al.*, 2007), most notably affecting activities involving bimanual coordination. To date, strong evidence-based treatment approaches are unfortunately lacking (see Boyd *et al.*, 2001; Koman *et al.*, 2004).

Mechanisms underlying impaired grasping

Our approach to understanding the mechanisms underlying impaired hand function in children with CP began more than 15 years ago with a series of studies on prehensile force control (Eliasson *et al.*, 1991, 1992, 1995; Gordon & Duff, 1999a, 1999b; Eliasson & Gordon, 2000; Duff & Gordon, 2003; Gordon *et al.*, 2003, 2006a). These studies focused on the sensorimotor control of grasping by quantifying fingertip force control during object manipulation (see Chapters 1, 11 and 12). Our initial studies demonstrated that compared with typically developing children (see Chapter 17), the performance of children with CP consistently differed in several important ways. First, their application of force is much slower, with prolonged transitions between task phases (Eliasson *et al.*, 1991; Eliasson & Gordon, 2000; Gordon *et al.*, 2003). Second, they have sequential force generation. While typically developing children at that age (Forssberg *et al.*, 1991; Gordon, 2000) and adults (Johansson & Westling, 1984) simultaneously apply grip and load force in a functional

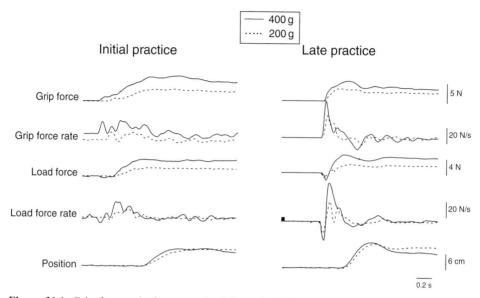

Figure 31.1. Grip force, grip force rate, load force, load force rate and vertical position from a representative child with hemiplegic CP during one lift with 200 g (dashed lines) and one lift with 400 g (solid lines) during initial lifts and after 25 lifts. The traces are aligned at the onset of vertical force increase and the force rates are shown using a ±20 point numerical differentiation. Modified from Gordon *et al.* (1999).

synergy, children with hemiplegia initially display a negative load force (first pushing the object down against its support) while developing a large grasp force (Eliasson *et al.*, 1991; Forssberg *et al.*, 1999). Also, children with CP are less able to adjust their grasping forces to the weight and texture of the object (Eliasson *et al.*, 1992, 1995). Finally, children with CP were unable to utilize sensory information signaling the object's properties from prior performance to develop their initial force increase prior to availability of sensory feedback (i.e. they had impaired motor planning) (Eliasson *et al.*, 1992).

While the above studies provided some insight into some of the causes of impaired grasp control in children with CP, subsequent studies provided three lines of evidence that these impairments are not static. First, we noted anecdotally that hand function improved markedly during the course of a one-hour testing session. This led us to systematically study the effects of extended practice by asking children with CP to repeatedly lift an object of a given weight 25 times. We found that the impairments in fine manipulative capabilities and force regulation during grasp are partially ameliorated with this extended practice (Gordon & Duff, 1999a; Duff & Gordon, 2003). This can be seen in Figure 31.1, which shows fingertip forces, force rates and position as children grasp and lift an object weighing either 200 g or 400 g. The rates of force increase were initially not scaled according to the object's weight, and occurred slowly in a stepwise fashion. After 25 lifts, the rates of force increase were smoother and faster and were scaled to the object's weight. This suggests that the initial impaired performance, at least in part, may be due to the lack of use. Importantly, practice may provide a window of opportunity for improvement.

Second, initial use of the non-involved extremity facilitates subsequent performance with the involved extremity (Gordon *et al.*, 1999, 2006a). This suggests a potential role of the non-involved extremity in the learning of motor skills by the involved extremity. Finally, during a 13-year follow-up study, we tested most of the participants in our original studies (Eliasson *et al.*, 1991) and found that hand function in children with CP improves with age (Eliasson *et al.*, 2006). The application of grip and load force characteristic of this population improved over the period of development, as indicated by a decreased mean grip-force/load-force path ratio and less sequential development of the two forces. These changes were accompanied by decreases in the time to perform the lifts and the time to perform standardized tests of motor performance. Together, these lines of evidence contradict traditional clinical assumptions that motor impairments in CP are static. Upper extremity performance in children with CP may improve with practice and development. More importantly, this implies that hand function may well be amenable to treatment.

Constraint-induced therapy: historical perspectives in animals and adult humans

There is a rich history of theoretical constructs derived from basic science underlying the application of intensive practice-based models to rehabilitation. Tower (1940) noted that after unilateral pyramidal tract lesions, monkeys failed to use their affected upper limb spontaneously, but limb use could improve if the unaffected limb was restrained, forcing use of the affected limb. Subsequently, Taub (see Taub & Shee, 1980) explored the effects of the unilateral surgical deafferentation of the monkey forelimb (see also Chapter 29). It was noted that monkeys often ceased using the deafferented limb. The monkeys learned compensatory strategies with the non-deafferented limb. This phenomenon was termed "learned non-use" and suggested that the monkeys never realized the functional potential of the deafferented limb because it was masked by their compensation with the unaffected limb (Taub *et al.*, 1975; Taub & Shee, 1980). Importantly, they subsequently found that learned non-use could actually be ameliorated when the uninvolved limb was restrained.

The above theoretical construct concerning "residual (masked) capability" that could potentially be tapped into as the result of "forced use" of the deafferented or impaired limb drove the development of intensive practice-based therapies in humans. This line of research began with studies of "forced use" in adults with hemiparetic stroke by Wolf and colleagues more than 25 years ago (Ostendorf & Wolf, 1981; Wolf *et al.*, 1989). The unaffected upper extremity was restrained during waking hours for 2 weeks. Improvements in timed or force-generating activities were observed during a variety of functional tasks. Improvements were maintained up to 12 months after the intervention. Subsequently, Taub and colleagues added 6 hours of structured activities incorporating principles of behavioral psychology (shaping). Active intervention involving restraint plus structured practice evolved to become known as "constraint-induced movement therapy" (CIMT) (Taub & Wolf, 1997). Since these pioneering studies, there have been many studies of CIMT in adults with hemiparesis. In one of the first multi-site randomized control trials in physical rehabilitation, Wolf *et al.* (2006) reported positive outcomes of CIMT across a large number of adults with hemiparetic stroke (see Wolf *et al.*, 2002; Dromerick *et al.*, 2006; Taub *et al.*, 2006; Bonaiuti *et al.*, 2007) (see Chapter 29).

Constraint-induced movement therapy in children with hemiparesis

Constraint-induced movement therapy has not been studied in CP nearly to the extent it has been in adults with stroke, nor is the level of evidence of efficacy (Sackett level) nearly as strong. Based on our own findings concerning practice effects in children with hemiplegia (see above), we began to modify CIMT for use in children with hemiplegic CP in the late 1990s. In a preliminary study (Charles *et al.*, 2001), we examined the efficacy of modified CIMT in three children (age 8, 11 and 13 years) with hemiplegic CP in the home environment. Children donned a cotton sling with the distal end sewn closed on their non-involved upper extremity for 6 hours per day for 14 days. During these 6 hours, they were engaged with a trained interventionist in a variety of child-friendly functional and play activity requiring unilateral use of their involved upper extremity. The children showed improvements in unilateral motor performance as measured by a standardized test of timed motor activities (Jebsen–Taylor Test of Hand Function), as well as both temporal and force coordination during grasping (Charles *et al.*, 2001).

These initial findings were promising, and drove our subsequent development (Gordon *et al.*, 2005) and testing (Charles *et al.*, 2006; Gordon *et al.*, 2006b; Charles & Gordon, 2007) of a modified form of CIMT. These latter studies were provided in a day camp setting with two to four children in each session. In one study (Charles *et al.*, 2006) children (ages 4–8 years) were randomized to receive 2 weeks (10 consecutive weekdays) of CIMT or no additional treatment (both groups continued to receive their usual and customary care). Significant improvements were noted in movement efficiency and functional limitations (standardized and criterion referenced tests of upper extremity and hand function) and environmental functional limitations (caregiver report).

An example of these improvements can be seen in Figure 31.2, which shows that children in the intervention group improved more than children in the no-treatment control group on the Jebsen–Taylor Test of Hand Function 1 week after the intervention. The decrease in time for the intervention group was maintained over the 6-month follow-up period. Interestingly, in this small randomized control trial we found that severity of hand function and attention significantly predicted improvement, with children who were more mildly to moderately affected and had better attention improving the most. In a subsequent study (Gordon *et al.*, 2006b) we showed similar gains in 9–13-year-old children.

There is a growing number of studies of forced use (Willis *et al.*, 2002) and CIMT (Eliasson *et al.*, 2003, 2005; Taub *et al.*, 2004; Sung *et al.*, 2005; Naylor & Bower, 2005; Bonnier *et al.*, 2006; Charles *et al.*, 2006) in children with hemiplegia. These studies (reviewed in Charles & Gordon, 2005; Hoare *et al.*, 2007; Eliasson & Gordon, 2008) differ in age of participants (ranging from 9 months to 18 years), inclusion criteria, duration

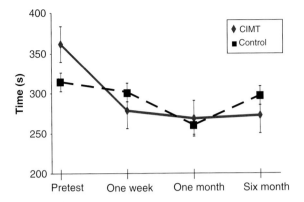

Figure 31.2. Mean ± SEM time to complete the six timed items (writing excluded) of the Jebsen–Taylor Test of Hand Function for the constraint-induced movement therapy (CIMT) (solid line) (n = 11) and control (dashed line) (n = 11) groups at each testing session. Faster times correspond to better performance. The maximum allowable time to complete each item was capped at 120 s, resulting in a maximum score of 720 s. Modified from Charles *et al.* (2006).

and intensity of treatment (ranging from usual and customary care schedules to adult models of CIMT), restraint (gloves, mitts, slings and casts) and outcome measures. While the evidence is not yet conclusive, all of these studies have reported positive outcomes.

While it has been claimed that greater improvements are seen using the more intense models with a cast worn continuously for 3 weeks (Taub *et al.*, 2007), there are two lines of evidence against these claims. First, these authors are unable to compare their outcomes to the other studies, as their outcome measures are all self-designed (with no validity or reliability studies) and largely rely on perceptions of caregivers. Second, positive outcomes using standardized measures have been reported using far less restrictive restraints (mitts) during just 2 hours per day (Eliasson *et al.*, 2005). In a recent study we administered 2 weeks of CIMT to 8 children with hemiplegia (ages 5–13 years) for 6 hours per day, and then re-administered the intervention 12 months later (Charles & Gordon, 2007). As seen in Figure 31.3, time to complete the Jebsen–Taylor test of Hand Function significantly decreased after the initial intervention, with the improvement maintained 12 months later. Following the second dose of CIMT, a similar decrease in time occurred. Given that children benefit from repeated bouts of CIMT (Charles & Gordon, 2007), there is no advantage to maintaining the potentially invasive schedule and restraints (see below), and instead, one could administer CIMT through repeated, and less intensive, bouts.

Limitations of CIMT

Despite the promise of CIMT, there are several conceptual problems in applying it to children. First, CIMT was developed to overcome *learned* non-use in adults with hemiplegia. These adults lost upper extremity function as a result of their condition, and may be motivated to regain previously learned functional behaviors. In contrast, children with

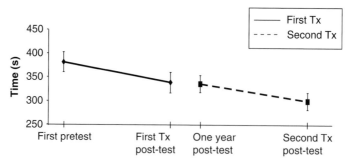

Figure 31.3. Mean ± SEM time to complete the six timed items (writing excluded) of the Jebsen–Taylor Test of Hand Function before and after the first CIMT treatment (Tx), 12 months later and following a second Tx. Modified from Charles & Gordon (2007).

hemiplegia must overcome "developmental disuse," whereby they have never learned how to use their involved extremity during many tasks and may need to learn how to use it for the first time. Thus, treatment must be developmentally focused, and take into account principles of motor learning.

Second, restraining a child's non-involved extremity (especially with casts) is potentially invasive, and thus should not be performed on young or severely impaired children with the same intensity as in adults. Evidence suggests the key element for improving motor function is the elicited practice rather than the restraint (Sunderland & Tuke, 2005). In fact, neuro-anatomical data in the developing kitten indicate that refinement and maintenance of corticospinal terminations in the spinal cord is activity dependent. Restriction of limb use on one side during a critical period in early development reduces the topographic distribution, branch density and density of presynaptic boutons on the side of restricted use (Martin *et al.*, 2004). It is unknown whether such a critical period exists in human infants. Nevertheless, this suggests that there may be a risk in restraining the non-involved limb for long periods at too early an age.

Finally, CIMT focuses on unimanual impairments, which do not greatly impact functional independence and quality of life (see Skold *et al.*, 2004). Children with hemiplegia have impairments in spatial and temporal coordination of the two hands (Steenbergen *et al.*, 1996; Utley & Sugden, 1998; Hung *et al.*, 2004; Utley & Steenbergen, 2006; Gordon & Steenbergen, 2008) as well as global impairments in motor planning (Steenbergen *et al.*, 2007). Thus, we propose that the goal of upper extremity rehabilitation should be to increase functional independence by improving use of both hands in cooperation. CIMT does not directly target this goal.

Rationale for intensive bimanual training in children with CP

Neurophysiological studies of children with hemiplegia posit a biological rationale for the use of bimanual training in rehabilitation, particularly in children older than 2 years.

Transcranial magnetic stimulation has shown that hemiplegia causes an age-dependent change in the excitability of the involved and non-involved corticospinal tracts (Carr *et al.*, 1993; Eyre *et al.*, 2001; Staudt *et al.*, 2004). Immediately after neurological insult (at or near the time of birth, in most cases), the CST from the damaged hemisphere is readily excitable by TMS. The excitability of the damaged CST decreases in a time-dependent manner, even years after the initial damage. Concurrently, the hemisphere contralateral to the damage increases in excitability; particularly, the ipsilateral response is heightened, suggesting a strengthening of these projections. This is believed to be a consequence of activity-dependent competition between the two sides – the more active side "wins out" over the less active (damaged) side (Eyre *et al.*, 2007; Martin *et al.*, 2007).

Consistent with the human TMS data, neuroanatomical and behavioral studies in a cat model of hemiplegia indicate that behavioral deficits that emerge with hemiplegia are caused by aberrant organization of the CST that appears to worsen with age (Martin *et al.*, 2007; see also Chapters 29 and 30). Balancing activity of the two sides immediately after unilateral brain trauma, by pharmacologically decreasing activity of the uninvolved side, restores motor function, normal anatomical organization of the CST and the motor representational map in the primary motor cortex (Friel & Martin, 2007).

The model points to the importance of increasing activity of the involved extremity, a principle incorporated into both CIMT and bimanual training (see below). On the one hand, the model encourages very early intensive training of the affected side exclusively, to balance activity between the two sides before the less-affected CST "outcompetes" the affected CST for synaptic space in the spinal cord. The time window for restoring CST connectivity by balancing activity may be quite narrow, occurring at a very young age. However, performing intensive physical training on very young children, when their motor repertoire is limited (< 1–2 years of age), may not be appropriate. This encourages the development of therapies that do not require intensive attention or movement (e.g. pharmacotherapy, TMS).

On the other hand, the data from children with CP and from the cat model show that unilateral motor cortex damage/activity loss results in the maintenance and age-dependent strengthening of ipsilateral connections from the less-affected side that control movement of the affected extremity. Essentially, the less-affected motor cortex provides control signals to both upper extremities. In fact, preliminary evidence suggests that CIMT is less effective and actually results in a *decrease* in cortical hand representation in individuals with hemiplegic CP who demonstrate ipsilateral CST innervation of the paretic hand (Juenger *et al.*, 2007). Bimanual training is likely to train both cortices to control both limbs in a skilled, functionally relevant way. Additionally, as described above, most activities of daily living are bimanual in nature; training the motor system to coordinate both hands in a skilled manner will likely provide gains in long-term, spontaneous skilled performance of bimanual activities.

There is increasing evidence of efficacy of functional training for improving functional skills (Ketelaar *et al.*, 2001; Ahl *et al.*, 2005). Principles of motor learning (practice specificity) would suggest that the best way to achieve improved bimanual control would be to practice bimanual control directly. There is emerging evidence that other forms of bimanual training may be efficacious in adults with hemiparesis (see Rose & Winstein, 2004; Cauraugh & Summers, 2005), although these largely employ repetitive non-functional tasks (i.e. are not child friendly). Thus there is a need for an intensive bimanual training protocol that is child friendly.

Hand–arm bimanual intensive therapy (HABIT)

The above limitations of CIMT and rationale for bimanual training compelled us to develop a form of intensive functional training, hand–arm bimanual intensive therapy (HABIT), which aims to improve the amount and quality of involved hand use during *bimanual* tasks (see Charles & Gordon, 2006). Like CIMT, HABIT is performed in a day-camp setting and is based on (1) our basic work delineating the mechanisms of hand impairments in CP, (2) our discovery of the key ingredient in CIMT (intensive practice) and (3) principles of motor learning and neuroplasticity (e.g. specificity of training).

Hand–arm bimanual intensive therapy retains the two major elements of pediatric CIMT (intensive structured practice and child friendliness) and engages the child in bimanual activities 6 hours/day for 10 days (60 hours). However, it differs from CIMT in that there is an absence of a physical restraint. Activities that necessitate bimanual upper extremity coordination are employed.

The HABIT approach also differs from conventional rehabilitation approaches in several ways. The intensity of HABIT is much greater than conventional therapies, providing ample opportunity for practice using principles of motor learning. Furthermore it is far more structured than conventional therapy, as described below. Finally, rather than encouraging compensatory use of the involved hand, we ask children to use the hand as a typically developing child uses their non-dominant hand. Specifically, we ask them to focus on how the hand and arm are performing at the end-point of the movement.

To conduct HABIT, specific bimanual activities involving performance of play or functional activities are chosen (see Table 1 in Charles & Gordon, 2006). Careful selection of these activities, along with the training environment, determine how the involved hand and arm are used. Task demands are graded to ensure success. During these activities, children receive instructions from the interventionist and also engage in their own active problem solving. Positive reinforcements and communication of results are used to motivate performance and reinforce target movements.

Like CIMT, HABIT is conducted in groups of up to four children to provide social interaction, modeling and encouragement. The choice of activities is not as important as the movements they elicit. Movement deficits of the involved extremity and bimanual coordination deficits are determined during the pre-intervention evaluation and early during the intervention. Bimanual activities are then selected that will focus on these deficits and engage the child in activities of increasingly complex coordination. Directions specifying how each hand will be used during the activity are provided before the start of each task in order to prevent use of compensatory strategies (performing the task unimanually with the non-involved extremity).

As with CIMT, two types of practice are employed in HABIT. During performance of *whole task practice*, activities are performed continuously for at least 15–20 minutes in the context of playing a game or performing a functional task. Targeted movements and spatial and temporal coordination are practiced within the context of completing the task or activity. *Part task practice* involves practicing a targeted movement in isolation. Specifically, we employ symmetric bimanual movements, such as putting game pieces away with both hands simultaneously, along with asymmetric tasks such as transferring items between hands. The task difficulty is changed by requiring greater speed or accuracy, or by providing tasks that require more skilled use of the involved hand and arm (for example, moving from activities in which the involved limb is used as a passive stabilizer to activities where it is used as an active manipulator (see Krumlinde-Sundholm & Eliasson, 2003; Krumlinde-Sundholme *et al.*, 2007).

We recently conducted a preliminary randomized control trial of HABIT (Gordon *et al.*, 2007) on 20 children with hemiplegic CP between the ages of 3.5 and 14 years. The Assisting Hand Assessment (AHA), which measures and describes the effectiveness with which a child with unilateral disability makes use of their affected (assisting) hand during bimanual activities, was used as a primary outcome measure (Krumlinde-Sundholm & Eliasson, 2003; Eliasson *et al.*, 2005; Holmefur *et al.*, 2007; Krumlinde-Sundholme *et al.*, 2007). We also used accelerometers on the wrists during performance of the AHA to measure frequency of use of each extremity during the AHA (see Uswatte *et al.*, 2006). Finally, we examined the kinematics of a drawer-opening task in which children were asked to open a drawer with one (drawer) hand and to insert their contralateral (task) hand inside manipulate a switch (see Hung *et al.*, 2004).

Figure 31.4A shows the results of the Assisting Hand Assessment. Scores for children in the treatment group significantly improved initially after the intervention, while scores for the no-treatment control group did not change. Scores for the treatment group decreased by the 1-month post-test, but remained significantly higher than at pretest (Gordon *et al.*, 2007).

Figure 31.4B shows the percentage of time the involved extremity was used during performance of the AHA, as determined by accelerometry. Use of the involved extremity increased for children who received HABIT, but not for the control group. The change in use of the involved extremity did not correlate with the change in AHA scores.

As seen in Figure 31.5, the time between fully opening the drawer with one hand and pressing the light switch with the other (goal synchronization time) decreased nearly three-fold for the children in the treatment group. This suggests that children learned to coordinate the timing of the movement of each extremity.

Recently, we began to examine specificity of training by providing 1 week of CIMT followed by 1 week of HABIT in four children with hemiplegic CP between the ages of 3.5 and 13 years (Gordon, unpublished data). Figure 31.6 shows the scores of six bimanual items across the bilateral coordination, upper limb coordination, and upper limb speed and dexterity subtests of the Bruininks–Oseretsky Test of Motor Proficiency. Performance does

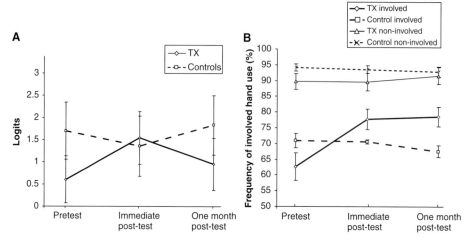

Figure 31.4. A. Mean ± SEM scores (in logits) on the Assisting Hand assessment for the HABIT treatment and no-treatment control groups at each testing session. B. Frequency of upper extremity movement as a percentage of task times for the involved (bold lines) and non-involved (thin lines) hands of the HABIT treatment (TX, solid lines) and control (dashed lines) groups. Higher scores represent better bimanual performance in both plots. Modified from Gordon *et al.* (2007).

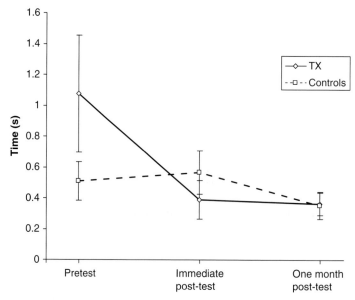

Figure 31.5. Mean ± SEM goal synchronization duration (time difference between the two hands completing the task) during the drawer-opening task for the HABIT treatment (TX, solid line) and no-treatment control (dashed line) groups at each testing session. Lower times correspond to improved performance. Modified from Gordon *et al.* (2007).

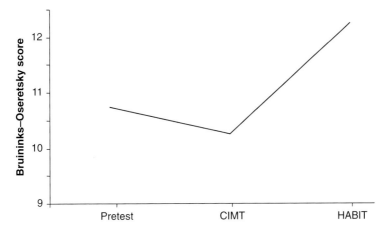

Figure 31.6. Mean score on the six bimanual items of the Bruininks–Oseretsky Test of Motor Proficiency during the pretest, after 1 week of constraint-induced therapy, followed by 1 week of HABIT. Higher scores correspond to better performance. Gordon (unpublished data).

not change following 1 week of CIMT, but increases following the crossover to HABIT. This emphasizes the importance of training specificity, and the need to identify end-goals before selecting training paradigms.

Discussion

The results of this preliminary randomized controled trial suggest that providing 10 consecutive days of intensive bimanual training results in improved bimanual use of the upper extremities. These findings are consistent with the CIMT studies in terms of the benefits of intensive practice, although we show that intensive *bimanual training* can improve the quality and quantity of bimanual hand use. This suggests specificity of training, and is consistent with motor learning theories (e.g. Welford, 1968). It also suggests that restrictive and potentially invasive restraints such as casts are not needed to elicit changes.

Surprisingly, improvements in the quality of bimanual hand use and frequency of use did not correlate, suggesting a need to measure both of these components separately. It should be noted, however, that the extent to which quality and quantity are related may be task-dependent.

The HABIT approach was designed specifically for use in children with unilateral upper-extremity impairments based on our experience with CIMT (Charles *et al.*, 2001, 2006; Gordon *et al.*, 2006b; Charles & Gordon, 2007) to target deficits in spatial and temporal control (Steenbergen *et al.*, 1996; Hung *et al.*, 2004; Skold *et al.*, 2004b; Utley & Steenbergen, 2006; Gordon *et al.*, 2007) and developmental disuse, especially during bimanual activities (Charles & Gordon, 2006). It utilizes principles of practice (specificity of training, e.g. Welford, 1968), as well as principles of plasticity, whereby neuroplastic changes are induced by increasing task complexity and reward (Kleim *et al.*, 2002; Nudo, 2003) (see Charles & Gordon, 2006).

The corticospinal system underlying human dexterity is capable of considerable reorganization after damage, and this reorganization likely underlies recovery of function (see Eyre, 2003). As described above, unilateral damage to the motor system results in a reorganization of the CST on both sides (Eyre *et al.*, 2007; see Martin, 2005; Martin *et al.*, 2007 for reviews). The CST from the damaged side shows an age-dependent reduction in size and efficacy, while the CST from the less-affected cortex maintains bilateral projections to the spinal cord that strengthen during development. The contralateral projection controls the less-affected limb while the ipsilateral pathway from the less-affected cortex plays a role in the control of the involved extremity (Staudt *et al.*, 2002). Involvement of the ipsilateral pathway appears to correlate with the extent of the damage in the affected hemisphere, indicating that the ipsilateral pathway may mediate recovery when the contralateral pathway controlling the affected limb is significantly damaged and unable to control movements (Staudt *et al.*, 2002). Emergence of the ipsilateral pathway does not occur until approximately 12 months of age, even when the brain trauma occurs at or before birth (Eyre *et al.*, 2007), suggesting that activity-dependent competition between the two sides during postnatal development progressively drives the organization and refinement of the CST. By the time children reach an age appropriate for CIMT (> 3 years), the ipsilateral pathways have likely been cemented. Recruitment of these pathways may promote long-term functional recovery. Ipsilateral pathways are also implicated in the recovery of function after stroke, particularly during the early stages of recovery and in cases involving large lesions that do not spare enough motor areas to support movement (Marshall *et al.*, 2000).

Thus, task recruitment of ipsilateral pathways, such as symmetric bilateral movements (Hallett, 2001; Stinear & Byblow, 2004), may be beneficial. Bilateral practice may result in changes in cortical representations (Cauraugh & Summers, 2005) and excitability in the undamaged hemisphere (Stinear & Byblow, 2004), which in turn may provide a neural basis for bimanual therapy.

To motivate children, participation must be fun. Thus, HABIT is consistent with the recent emphasis on functional training and practicing predefined goals in therapeutic environments (Law *et al.*, 1998; Ketelaar *et al.*, 2001; Bower *et al.*, 2001; Ahl *et al.*, 2005). The emphasis of HABIT on functional activity performance also directly addresses the recent modification of the definition of CP, whereby it is considered a "disorder of movement and posture *causing activity limitation*" (Bax *et al.*, 2005; Steenbergen & Gordon, 2006).

While HABIT is potentially less invasive than CIMT since there no restraints, in our experience, administering HABIT is often more difficult. Children with hemiplegia often have developmental disuse. They are strikingly adept at compensating by using only their non-involved extremity to perform tasks for which their typically developing peers require both hands, even if it is at the cost of efficiency (e.g. performing tasks sequentially or using body parts as a brace). During CIMT, the restraint forces the participant to use the involved extremity to accomplish the task, with the drawback that the tasks must be unimanual.

During HABIT, tasks must be bimanual to train specific coordination skills since children with hemiplegia naturally would choose to use their non-involved extremity for unimanual tasks. In many instances, we observed spatial and temporal dyscoordination associated with using the two extremities together sometimes for the very first time (see Steenbergen *et al.*, 2007; Gordon & Steenbergen, 2008). Often, their natural tendency would be to over-compensate with their non-involved extremity (e.g. reach into the involved extremity's hemispace). Interventionists rarely use verbal prompting to use the involved extremity as many caregivers and therapists constantly prod their children to use their involved extremity which children quickly attenuate and it is potentially psychologically invasive. Thus, far more attention must be provided to select activities minimizing the possibility of unimanual compensation and intentionally plan the optimal placement of objects in the workspace to maximize the likelihood of bimanual use. We find that rule-setting prior to an activity is far more effective since the child is asked to verbally agree prior to participation. In this way the interventionist uses these rules and the environment as a new type of restraint.

Although HABIT is more difficult to administer than CIMT, bimanual activities are generally more motivating since children do not have to continually focus on their impaired hand in isolation. In fact, compared with our prior CIMT work (Charles *et al.*, 2006), children spent *c.* 25% more time on tasks during the HABIT intervention. Activities in general are more salient and may be selected to maximize interest (e.g. video games). Thus, motivational and social aspects of types of practice need to be considered in intervention design (Ochsner & Lieberman, 2001).

Conclusions

The results of a variety of studies now suggest that children with CP benefit from intensive training. While all protocols focus on functional training, the specific ingredients that prove to be beneficial are not known. Although it is not compatible with the frequency of therapeutic services offered through most health-care models, it is conceivable that the key ingredient is merely intensity of treatment. This prospect needs to be tested with control groups that receive the same intensity of "conventional therapy." The results thus far suggest that it is important to put the training goal first, and then choose the appropriate training protocol to meet those goals. Here we provide evidence that training bimanual skills can improve bimanual function. While substantial improvements can result from intensive training, the underlying impairments still persist, and these approaches should be integrated with, rather than replace, other long-term pediatric care practices.

Despite the improvement, larger studies of both unimanual and bimanual training across a more diverse subject population with a long-term follow-up are required. While provision of these interventions in a group setting may be advantageous in providing a supportive and competitive environment, it is not known whether our initial strategy of providing CIMT in the home environment is equally effective. It is not known whether this intervention is advisable for all children with hemiplegia. The child's age and severity of hand function need to be considered. The appropriate age and impairment levels need to be identified.

Factors such as side and location of the lesion, attention span, balance of whole versus part practice, optimal dosage and means of documenting efficacy all need to be considered to ultimately define the most efficacious rehabilitation strategy. Finally, the neural mechanisms of plasticity need to be studied to understand the mechanisms of recovery.

Acknowledgments

This work was supported by grants from the National Institutes of Health, United Cerebral Palsy Research and Education Foundation and the Thrasher Research Foundation.

References

Ahl, L. E., Johansson, E., Granat, T. & Carlberg, E. B. (2005). Functional therapy for children with cerebral palsy: an ecological approach. *Dev Med Child Neurol*, **47**, 613–619.

Bax, M., Goldstein, M., Rosenbaum, P. *et al.* (2005). Proposed definition and classification of cerebral palsy, April 2005. *Dev Med Child Neurol*, **47**, 571–576.

Bonaiuti, D., Rebasti, L. & Sioli, P. (2007). The constraint induced movement therapy: a systematic review of randomised controlled trials on the adult stroke patients. *Eura Medicophys*, **43**, 139–146.

Bonnier, B., Eliasson, A. C. & Krumlinde-Sundholm, L. (2006). Effects of constraint-induced movement therapy in adolescents with hemiplegic cerebral palsy: a day camp model. *Scand J Occup Ther*, **13**, 13–22.

Bouza, H., Dubowitz, L., Rutherford, M. & Pennock, J. M. (1994). Prediction of outcome in children with congenital hemiplegia: a magnetic resonance imaging study. *Neuropediatrics (Stuttgart)*, **25**, 60–66.

Bower, E., Michell, D., Burnett, M., Campbell, M. J. & Mclellan, D. L. (2001). Randomized controlled trial of physiotherapy in 56 children with cerebral palsy followed for 18 months. *Dev Med Child Neurol*, **43**, 4–15.

Boyd, R. N., Morris, M. E. & Graham, H. K. (2001). Management of upper limb dysfunction in children with cerebral palsy: a systematic review. *Eur J Neurol*, **8** Suppl 5, 150–166.

Brown, J. K., Rensburg Van, E., Walsh, G., Lakie, M. & Wright, G. W. (1987). A neurological study of hand function of hemiplegic children. *Dev Med Child Neurol (Lond)*, **29**, 287–304.

Carr, L. J., Harrison, L. M., Evans, A. L. & Stephens, J. A. (1993). Patterns of central motor reorganization in hemiplegic cerebral palsy. *Brain*, **116**, 1223–1247.

Cauraugh, J. H. & Summers, J. J. (2005). Neural plasticity and bilateral movements: a rehabilitation approach for chronic stroke. *Prog Neurobiol*, **75**, 309–320.

Charles, J. & Gordon, A. M. (2005). A critical review of constraint-induced movement therapy and forced-use in children with hemiplegia. *Neural Plasticity*, **12**, 245–262.

Charles, J. & Gordon, A. M. (2006). Development of hand–arm bimanual intensive therapy (HABIT) for improving bimanual coordination in children with hemiplegic cerebral palsy. *Dev Med Child Neurol*, **48**, 931–936.

Charles, J. & Gordon, A. M. (2007). A repeated course of constraint-induced movement therapy results in further improvement. *Dev Med Child Neurol*, **49**, 770–773.

Charles, J., Lavinder, G. & Gordon, A. M. (2001). The effects of constraint induced therapy on hand function in children with hemiplegic cerebral palsy. *Ped Phys Ther*, **13**, 68–76.

Charles, J. R., Wolf, S. L., Schneider, J. A. & Gordon, A. M. (2006). Efficacy of a child-friendly form of constraint-induced movement therapy in hemiplegic cerebral palsy: a randomized control trial. *Dev Med Child Neurol*, **48**, 635–642.

Cioni, G., Sales, B., Paolicelli, P. B. *et al.* (1999). MRI and clinical characteristics of children with hemiplegic cerebral palsy. *Neuropediatrics*, **30**, 249–255.

Dromerick, A. W., Lum, P. S. & Hidler, J. (2006). Activity-based therapies. *NeuroRx*, **3**, 428–438.

Duff, S. V. & Gordon, A. M. (2003). Learning of grasp control in children with hemiplegic cerebral palsy. *Dev Med Child Neurol*, **45**, 746–757.

Duque, J., Thonnard, J. L., Vandermeeren, Y. *et al.* (2003). Correlation between impaired dexterity and corticospinal tract dysgenesis in congenital hemiplegia. *Brain (London)* 732–747.

Eliasson, A. C. & Gordon, A. M. (2000). Impaired force coordination during object release in children with hemiplegic cerebral palsy. *Dev Med Child Neurol*, **42**, 228–234.

Eliasson, A. C. & Gordon, A. M. (2008). Constraint-induced movement therapy for children with hemiplegia. In A. C. Eliasson & P. Burtner (Eds.), *Child with Cerebral Palsy: Management of the Upper Extremity. Clinics in Developmental Medicine* (pp. 308–319). London: MacKeith Press.

Eliasson, A. C., Gordon, A. M. & Forssberg, H. (1991). Basic coordination of manipulative forces in children with cerebral palsy. *Dev Med Child Neurol (Lond)*, **33**, 659–658.

Eliasson, A. C., Gordon, A. M. & Forssberg, H. (1992). Impaired anticipatory control of isometric forces during grasping by children with cerebral palsy. *Dev Med Child Neurol (Lond)*, **34**, 216–225.

Eliasson, A. C., Gordon, A. M. & Forssberg, H. (1995). Tactile control of isometric fingertip forces during grasping in children with cerebral palsy. *Dev Med Child Neurol (Lond)*, **37**, 72–84.

Eliasson, A. C., Bonnier, B. & Krumlinde-Sundholm, L. (2003). Clinical experience of constraint induced movement therapy in adolescents with hemiplegic cerebral palsy – a day camp model. *Dev Med Child Neurol*, **45**, 357–359.

Eliasson, A. C., Krumlinde-Sundholm, L., Shaw, K. & Wang, C. (2005). Effects of constraint-induced movement therapy in young children with hemiplegic cerebral palsy: an adapted model. *Dev Med Child Neurol*, **47**, 266–275.

Eliasson, A. C., Forssberg, H., Hung, Y. C. & Gordon, A. M. (2006). Development of hand function and precision grip control in individuals with cerebral palsy: a 13-year follow-up study. *Pediatrics*, **118**, e1226–e1236.

Eyre, J. A. (2003). Development and plasticity of the corticospinal system in man. *Neural Plasticity*, **10**, 93–106.

Eyre, J. A., Taylor, J. P., Villagra, F., Smith, M. & Miller, S. (2001). Evidence of activity-dependent withdrawal of corticospinal projections during human development. *Neurology*, **57**, 1543–1554.

Eyre, J. A., Smith, M., Dabydeen, L. *et al.* (2007). Is hemiplegic cerebral palsy equivalent to amblyopia of the corticospinal system? *Ann Neurol*, 493–503.

Forssberg, H., Eliasson, A. C., Kinoshita, H., Johansson, R. S. & Westling, G. (1991). Development of human precision grip. I: Basic coordination of force. *Exp Brain Res*, **85**, 451–457.

Forssberg, H., Eliasson, A. C., Redon-Zouitenn, C., Mercuri, E. & Dubowitz, L. (1999). Impaired grip-lift synergy in children with unilateral brain lesions. *Brain (Lond)*, **122**, 1157–1168.

Friel, K. M. & Martin, J. H. (2007). Bilateral activity-dependent interactions in the developing corticospinal system. *J Neurosci*, **27**, 11083–11090.

Gordon, A. M. (2000). The development of hand motor control. In A. F. Kalverboer & A. Gramsbergen (Eds.), *Brain and Behavior in Human Development* (pp. 513–537). Dordrecht, Netherlands: Kluwer Academic Publishers.

Gordon, A. M. & Duff, S. V. (1999a). Fingertip forces during object manipulation in children with hemiplegic cerebral palsy. I: Anticipatory scaling. *Dev Med Child Neurol*, **41**, 166–175.

Gordon, A. M. & Duff, S. V. (1999b). Relation between clinical measures and fine manipulative control in children with hemiplegic cerebral palsy. *Dev Med Child Neurol*, **41**, 586–591.

Gordon, A. M. & Steenbergen, B. (2008). Bimanual coordination in children with cerebral palsy. In A. C. Eliasson & P. Burtner (Eds.), *Child with Cerebral Palsy: Management of the Upper Extremity. Clinics in Developmental Medicine* (pp. 160–175). London: MacKeith Press.

Gordon, A. M., Charles, J. & Duff, S. V. (1999). Fingertip forces during object manipulation in children with hemiplegic cerebral palsy. II: Bilateral coordination. *Dev Med Child Neurol*, **41**, 176–185.

Gordon, A. M., Lewis, S. R., Eliasson, A. C. & Duff, S. V. (2003). Object release under varying task constraints in children with hemiplegic cerebral palsy. *Dev Med Child Neurol*, **45**, 240–248.

Gordon, A. M., Charles, J. & Wolf, S. L. (2005). Methods of constraint-induced movement therapy for children with hemiplegic cerebral palsy: development of a child-friendly intervention for improving upper-extremity function. *Arch Phys Med Rehabil*, **86**, 837–844.

Gordon, A. M., Charles, J. & Steenbergen, B. (2006a). Fingertip force planning during grasp is disrupted by impaired sensorimotor integration in children with hemiplegic cerebral palsy. *Pediatr Res*, **60**, 587–591.

Gordon, A. M., Charles, J. & Wolf, S. L. (2006b). Efficacy of constraint-induced movement therapy on involved upper extremity use in children with hemiplegic cerebral palsy is not age dependent. *Pediatrics*, **117**, 363–373.

Gordon, A. M., Schneider, J. A., Chinnan, A. & Charles, J. (2007). Efficacy of hand-arm bimanual intensive therapy (HABIT) in children with hemiplegic cerebral palsy: a randomized control trial. *Dev Med Child Neurol*, **49**, 730–739.

Hallett, M. (2001). Plasticity of the human motor cortex and recovery from stroke. *Brain Res Brain Res Rev*, **36**, 169–174.

Himmelmann, K., Hagberg, G., Beckung, E., Hagberg, B. & Uvebrant, P. (2005). The changing panorama of cerebral palsy in Sweden. IX. Prevalence and origin in the birth-year period 1995–1998. *Acta Paediatr*, **94**, 287–294.

Himmelmann, K., Beckung, E., Hagberg, G. & Uvebrant, P. (2006). Gross and fine motor function and accompanying impairments in cerebral palsy. *Dev Med Child Neurol*, **48**, 417–423.

Hoare, B. J., Wasiak, J., Imms, C. & Carey, L. (2007). Constraint-induced movement therapy in the treatment of the upper limb in children with hemiplegic cerebral palsy. *Cochrane Database Syst Rev*, CD004149.

Holmefur, M., Krumlinde-Sundholme, L. & Eliasson, A. C. (2007). Interrater and intrarater reliability of the Assisting Hand Assessment. *Am J Occup Ther*, **61**, 80–85.

Hung, Y. C., Charles, J. & Gordon, A. M. (2004). Bimanual coordination during a goal-directed task in children with hemiplegic cerebral palsy. *Dev Med Child Neurol*, **46**, 746–753.

Johansson, R. S. & Westling, G. (1984). Roles of glabrous skin receptors and sensorimotor memory in automatic control of precision grip when lifting rougher or more slippery objects. *Exp Brain Res (Berlin)*, **56**, 550–564.

Juenger, H., Linder-Lucht, M., Walther, M. *et al.* (2007). Cortical neuromodulation by constraint-induced movement therapy in congenital hemiparesis: an FMRI study. *Neuropediatrics*, **38**(3): 130–136.

Ketelaar, M., Vermeer, A., Hart, H., Van Petegem-Van Beek, E. & Helders, P. J. (2001). Effects of a functional therapy program on motor abilities of children with cerebral palsy. *Phys Ther*, **81**, 1534–1545.

Kleim, J. A., Barbay, S., Cooper, N. R. *et al.* (2002). Motor learning-dependent synaptogenesis is localized to functionally reorganized motor cortex. **77**, 63–77.

Koman, L. A., Smith, B. P. & Shilt, J. S. (2004). Cerebral palsy. *Lancet*, **363**, 1619–1631.

Krageloh-Mann, I. (2004). Imaging of early brain injury and cortical plasticity. *Exp Neurol*, **190** Suppl 1, S84–S90.

Krageloh-Mann, I. (2005). Cerebral palsy: towards developmental neuroscience. *Dev Med Child Neurol*, **47**, 435.

Krumlinde-Sundholm, L. & Eliasson, A. C. (2003). Development of the assisting hand assessment: a Rasch-built measure intended for children with unilateral upper limb impairments. *Scan J Occup Ther*, **10**, 16–26.

Krumlinde-Sundholme, L., Holmefur, M., Kottorp, A. & Eliasson, A. C. (2007). The Assisting Hand Assessment: current evidence of validity, reliability and responsiveness to change. *Dev Med Child Neurol*, **49**, 259–264.

Law, M., Darrah, J., Pollock, N. *et al.* (1998). Family-centered functional therapy for children with cerebral palsy: an emerging practical model. *Phys Occup Ther Pediatr*, **18**, 83–102.

Lin, J. P. (2003). The cerebral palsies: a physiological approach. *J Neurol Neurosurg Psychiatry*, **74** Suppl. 1, i23–i29.

Marshall, R. S., Perera, G. M., Lazar, R. M. *et al.* (2000). Evolution of cortical activation during recovery from corticospinal tract infarction. *Stroke*, **31**, 656–661.

Martin, J. H. (2005). The corticospinal system: from development to motor control. *Neuroscientist*, **11**, 161–173.

Martin, J. H., Choy, M., Pullman, S. & Meng, Z. (2004). Corticospinal system development depends on motor experience. *J Neurosci*, **24**, 2122–2132.

Martin, J. H., Friel, K. M., Salimi, I. & Chakrabarty, S. (2007) Activity- and use-dependent plasticity of the developing corticospinal system. *Neurosci Biobehav Rev*, **31**, 1125–1135.

Naylor, C. E. & Bower, E. (2005). Modified constraint-induced movement therapy for young children with hemiplegic cerebral palsy: a pilot study. *Dev Med Child Neurol*, **47**, 365–369.

Nudo, R. J. (2003). Adaptive plasticity in motor cortex: implications for rehabilitation after brain injury. *J Rehabil Med*, **41**, 7–10.

Ochsner, K. & Lieberman, M. (2001). The emergence of social cognitive neuroscience. *Am Psychol*, **56**, 717–734.

Okumura, A., Kato, T., Kuno, K., Hayakawa, F. & Watanabe, K. (1997). MRI findings in patients with spastic cerebral palsy. II: correlation with type of cerebral palsy. *Dev Med Child Neurol (Lond)*, **39**, 369–372.

Ostendorf, C. G. & Wolf, S. L. (1981). Effect of forced use of the upper extremity of a hemiplegic patient on changes in function. *Phys Ther*, **61**, 1022–1028.

Rose, D. K. & Winstein, C. J. (2004). Bimanual training after stroke: are two hands better than one? *Top Stroke Rehabil*, **11**, 20–30.

Skold, A., Josephsson, S. & Eliasson, A. C. (2004). Performing bimanual activities: the experiences of young persons with hemiplegic cerebral palsy. *Am J Occup Ther*, **58**, 416–425.

Stanely, F., Blair, E. & Alberman, E. (2000). *Cerebral Palsies: Epidemiology and Causal Pathways. Clinics in Developmental Medicine No. 151.* London: MacKeith Press.

Staudt, M., Grodd, W., Gerloff, C. *et al.* (2002). Two types of ipsilateral reorganization in congenital hemiparesis: a TMS and fMRI study. *Brain*, **125**, 2222–2237.

Staudt, M., Gerloff, C., Grodd, W. *et al.* (2004). Reorganization in congenital hemiparesis acquired at different gestational ages. *Ann Neurol*, **56**, 854–863.

Staudt, M., Krageloh-Mann, I. & Grodd, W. (2005). Ipsilateral corticospinal pathways in congenital hemiparesis on routine magnetic resonance imaging. *Pediatr Neurol*, **32**, 37–39.

Steenbergen, B. & Gordon, A. M. (2006). Activity limitation in hemiplegic cerebral palsy: evidence for disorders in motor planning. *Dev Med Child Neurol*, **48**, 780–783.

Steenbergen, B., Hulstijn, W., De Vries, A. & Berger, M. (1996). Bimanual movement coordination in spastic hemiparesis. *Exp Brain Res (Berlin)*, **110**, 91–98.

Steenbergen, B., Verrel, J. & Gordon, A. M. (2007). Motor planning in congenital hemiplegia. *Disabil Rehabil*, **29**, 13–23.

Stinear, J. W. & Byblow, W. D. (2004). Rhythmic bilateral movement training modulates corticomotor excitability and enhances upper limb motoricity poststroke: a pilot study. *J Clin Neurophysiol*, **21**, 124–131.

Sunderland, A. & Tuke, A. (2005). Neuroplasticity, learning and recovery after stroke: a critical evaluation of constraint-induced therapy. *Neuropsychol Rehab*, **15**, 81–96.

Sung, I. Y., Ryu, J. S., Pyun, S. B. *et al.* (2005). Efficacy of forced-use therapy in hemiplegic cerebral palsy. *Arch Phys Med Rehabil*, **86**, 2195–2198.

Taub, E. & Shee, L. P. (1980). *Somatosensory Deafferentation Research with Monkeys: Implications for Rehabilitation Medicine.* Baltimore, MD: Williams Wilkins.

Taub, E. & Wolf, S. L. (1997). Constraint induction techniques to facilitate upper extremity use in stroke patients. *Topics Stroke Rehab*, **3**, 38–61.

Taub, E., Goldberg, I. A. & Taub, P. B. (1975). Deafferentation in monkeys: pointing at a target without visual feedback. *Exp Neurol*, **46**, 178–186.

Taub, E., Ramey, S. L., Deluca, S. & Echols, K. (2004). Efficacy of constraint-induced movement therapy for children with cerebral palsy with asymmetric motor impairment. *Pediatrics*, **113**, 305–312.

Taub, E., Uswatte, G., Mark, V. W. & Morris, D. M. (2006). The learned nonuse phenomenon: implications for rehabilitation. *Eura Medicophys*, **42**, 241–256.

Taub, E., Griffin, A., Nick, J. *et al.* (2007). Pediatric CI therapy for stroke-induced hemiparesis in young children. *Dev Neurorehabil*, **10**, 3–18.

Tower, S. S. (1940). Pyramidal lesion in the monkey. *Brain (London)*, **63**, 36–90.

Uswatte, G., Giuliani, C., Winstein, C. *et al.* (2006). Validity of accelerometry for monitoring real-world arm activity in patients with subacute stroke: evidence from the extremity constraint-induced therapy evaluation trial. *Arch Phys Med Rehabil*, **87**, 1340–1345.

Utley, A. & Steenbergen, B. (2006). Discrete bimanual co-ordination in children and young adolescents with hemiparetic cerebral palsy: recent findings, implications and future research directions. *Pediatr Rehabil*, **9**, 127–136.

Utley, A. & Sugden, D. (1998). Interlimb coupling in children with hemiplegic cerebral palsy during reaching and grasping at speed. *Dev Med Child Neurol (Lond)*, **40**, 396–404.

Uvebrant, P. (1988). Hemiplegic cerebral palsy aetiology and outcome. *Acta Paediatr Scand Suppl (Stockholm)*, 1–100.

Welford, A. T. (1968). *Fundamentals of Skill*. London: Methuen & Co.

Willis, J. K., Morello, A., Davie, A., Rice, J. C. & Bennett, J. T. (2002). Forced use treatment of childhood hemiparesis. *Pediatrics*, **110**, 94–96.

Wolf, S. L., Lecraw, D. E., Barton, L. A. & Jann, B. B. (1989). Forced use of hemiplegic upper extremities to reverse the effect of learned nonuse among chronic stroke and head-injured patients. *Exp Neurol (New York)*, **104**, 125–132.

Wolf, S. L., Blanton, S., Baer, H., Breshears, J. & Butler, A. J. (2002). Repetitive task practice: a critical review of constraint-induced movement therapy in stroke. *Neurologist*, **8**, 325–338.

Wolf, S. L., Winstein, C. J., Miller, J. P. *et al.* (2006). Effect of constraint-induced movement therapy on upper extremity function 3 to 9 months after stroke: the EXCITE randomized clinical trial. *J Am Med Assoc*, **296**, 2095–2104.

32

Therapy of sensorimotor dysfunction of the hand in Parkinson's disease

ROLAND WENZELBURGER

Summary

Dopaminergic medication or deep brain stimulation of the subthalamic nucleus (STN DBS) impact on many aspects of the grip–lift task in idiopathic Parkinson's disease (PD). The rate of both grip- and load-force generation were normalized by the levodopa test, whereas the maximum vertical acceleration was not improved in all studies. Other dopa-responsive factors included load preparation time, which was shortened, and maximal grip force that showed an extra increase in the test. The overflow of grip force and maximum negative load force was correlated with the intensity of levodopa-induced dyskinesias (LID) in patients affected by this symptom. Maximal negative load force and tremor were not dopa-sensitive. Subthalamic nucleus DBS exerted a dopa-like effect on most parameters of the grip–lift task except for grip force in the long-term comparison. In patients with LID the preoperative overflow of force in on-state and the severity of LID were both ameliorated by STN-DBS, although the force level did not return to normal values in all studies. A dopa-resistant action tremor of higher frequency can be seen during the grip–lift task, while the rest tremor of PD is suppressed at onset of the movement. Further therapies involve facilitation of movements with a training augmented by external cues like auditory or visual signals to overcome akinesia. The grip–lift task offers a valuable instrument to study therapeutic effects in PD.

Introduction

Dopaminergic medication or deep brain stimulation of the subthalamic nucleus impact on almost any aspect of motor deficits in idiopathic Parkinson's disease (PD). An impressive awakening effect can result from the levodopa test or when switching on the stimulator. Going into more detail, however, this abrupt change of clinical symptoms is the net result of a multitude of different effects and some of them are not always beneficial. The temporal and topographical distribution of these effects can be very complex. After intake of medication many late-stage patients pass through rapid cycles from severest akinesia with painful off-dystonia, to peak-dose dyskinesias combined with optimal mobility and vice versa. Virtually any measure can be obtained when the patients are tested within these

Sensorimotor Control of Grasping: Physiology and Pathophysiology, ed. Dennis A. Nowak and Joachim Hermsdörfer. Published by Cambridge University Press. © Cambridge University Press 2009.

cycles, even if the most sophisticated instruments are used for measurement. Studies of therapeutic effects should therefore consider carefully the state of activation at the time of testing. There is also a striking difference of symptoms between different patients. While tremor is the major symptom in many parkinsonian subjects, others suffer from severe akinesia, followed by dyskinesias at peak on-state. Many studies on the effects of therapy on hand function have stratified patients according to their main symptoms that impact on the upper extremity, i.e. akinesia and potentially dyskinesias or tremor. Therefore, the discussion of the impact of antiparkinsonian therapy on sensorimotor dysfunction of the hand will also follow the predominance of those cardinal symptoms.

The effect of levodopa on akinesia and on levodopa-induced dyskinesias

Akinesia (or bradykinesia) and dyskinesias appear contradictory but in late-stage idio-pathic Parkinson's disease these symptoms are often two sides of the same medal. Many patients suffer from off-states for parts of the day or night and it is a challenge to promote a stable on-state by oral therapy. If the titration of dopaminergic therapy in such patients goes "low and slow" until the threshold for satisfactory mobility is obtained they will often encounter no dyskinesias, but their mobility will stay below an optimal on-state. An implicit under-dosage can be found in many cases as the patients and their doctor are often cautious with high-dose therapy. However, if the same patients are challenged with a suprathreshold dose of levodopa (in a levodopa test usually with $1.5–2 \times$ the morning dose) many of them will have dyskinesias and a better mobility (Wenzelburger, 2005). Some studies have used such well-defined conditions for analysis of a grip–ift task (see Chapters 1 and 12) in PD (Ingvarsson *et al.*, 1997; Fellows *et al.*, 1998; Wenzelburger *et al.*, 2002b; Wenzelburger *et al.*, 2003; Benice *et al.*, 2007) while healthy controls were tested under comparable conditions only in one recent study (Benice *et al.*, 2007). These analyses demonstrated that levodopa impacts on many parameters of the grip–lift task, whereas the drug has no measurable impact on controls. The slower generation of grip and lift force, as measured by the grip–lift parameters maximal grip velocity and maximal vertical acceleration, reflect slowness of muscle recruitment and bradykinesia. Parkinsonian patients in off-state have longer load preparation time, lower maximal vertical acceleration, lower maximal grip velocity and higher maximal negative force than normal controls. Although many parameters of the task were altered by the disease, only some of them were improved by levodopa. Parkinsonian subjects exhibited a lower rate of both grip- and load-force generation during the task in most of the studies (Gordon *et al.*, 1997; Ingvarsson *et al.*, 1997; Fellows *et al.*, 1998; Guo *et al.*, 2004; Benice *et al.*, 2007), and this abnormality was completely normalized by levodopa. While Nowak *et al.* (2005) reported that levodopa did not improve the maximum vertical acceleration in late-stage PD subjects, this parameter was improved by the levodopa test in another study on patients at earlier stages (Benice *et al.*, 2007). A factor analysis performed by Benice *et al.* extracted two factors, a dopa-responsive factor and a dopa-resistant factor, that together explained about 75% of the variance in grip–lift performance (Benice *et al.*, 2007). The dopa-responsive factor was loaded by load preparation time, maximal vertical acceleration, maximal grip velocity and maximal grip force. The dopa-resistant factor was loaded with maximal negative load force and tremor during the lift. In summary the grip–lift task is well suited for quantitative analysis of the anti-akinetic effects of dopaminergic therapy.

The published results on the peak or static forces exerted on an object held in the precision grip by parkinsonian patients depended strongly on the group of patients and other settings. Patients generated a wide range of grip forces in different studies, making it difficult to clearly attribute this finding to the disease, its stage or the effects of therapy. The parameter maximum grip force in parkinsonian off-state was equal to that produced by healthy controls in some studies (Ingvarsson *et al.*, 1997; Wenzelburger *et al.*, 2002b; Benice *et al.*, 2007), but was described as elevated in others (Wenzelburger *et al.*, 2002a; Fellows & Noth, 2004; Guo *et al.*, 2004). Possible contributing factors to the discrepancy include the difference in disease severity and the use of lifted objects of different size, weight and degrees of freedom of the apparatus used. An overshooting of grip force was seen also in de novo or early-stage parkinsonian patients (Benice *et al.*, 2007; Fellows & Noth, 2004), and in a multitude of other diseases (see Chapters 19, 21 and 26). This led to the conclusion that an increased safety margin (see Chapters 1 and 11) could be the reason for this non-specific effect. Off-dystonia which occurs in some patients at later stages may also have a role for elevated grip force in this condition, because a large force overflow can be seen in such patients (Wenzelburger *et al.*, 2002a). All the published studies agreed that levodopa can increase the maximum or static grip forces in late-stage PD subjects to a level higher than in unmedicated state. Levodopa-induced dyskinesias (LID) are a serious problem in late-stage Parkinson's disease and impact on the measures of the grip–lift task as they are evoked by and interfere with motor activities. In two studies the levodopa-induced overflow in grip force during peak on-state has been correlated with the severity of levodopa-induced dyskinesias (LID) occurring in this condition (Wenzelburger *et al.*, 2002b; Nowak *et al.*, 2005). One study focused on PD patients at different stages of the disease compared 10 PD patients without levodopa-induced dyskinesias (PD-LID, mean disease duration 5.3 years, mean motor score 22.2 out of 108) with 23 PD patients with levodopa-induced dyskinesias (PD+LID, mean disease duration 16.3 years, mean motor score 47.9 out of 108, mean LID score 9.1 on a 28-point scale), and with age-matched controls. At off-state, neither PD-LID group nor PD+LID group had higher maximal grip force than normal controls. After levodopa, the maximal grip force increased only in the PD+LID group, not in the PD-LID group. At on-state, PD+LID subjects, but not PD-LID subjects, had higher maximal grip force than controls (Figure 32.1). Interestingly, the maximal grip force while "on" correlated strongly with the peak dose dyskinesia-score in the PD+LID group (Wenzelburger *et al.*, 2002b) and this was in accordance with the findings of Nowak *et al.* who found a moderate correlation in a group of six patients with LID scores ranging from 0–4 on a 28-point scale (Nowak *et al.*, 2005). Such coherence can only be demonstrated in patients selected for the presence of LID which usually indicate a late-stage disease. Patients at early or intermediate stages may also exhibit increased grip forces in the levodopa test without clinical signs of

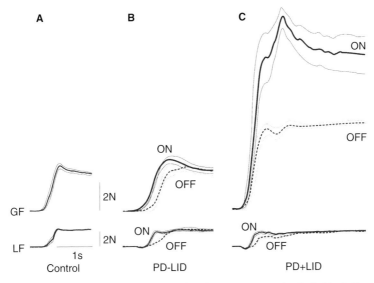

Figure 32.1. Force profiles in a control subject (A) and two representative individuals from the group without (B: PD-LID) and with levodopa-induced dyskinesias (C: PD+LID), in off-drug (dashed lines) and ON-drug (solid lines) conditions. The thick lines depict mean grip force of thumb (GF) and mean load force of thumb and index (LF). The thin lines show standard error of mean of 10 repetitions.The figure illustrates the overshooting of grip force in PD patients with LID both in OFF-drug and ON-drug conditions, whereas there was no general force excess in the patients without LID. From Wenzelburger *et al.* (2002b).

LID (Fellows & Noth, 2004; Benice *et al.*, 2007). It is an ongoing discussion whether the grip–lift task may be sensitive for detection of early signs of dyskinesias keeping in mind that the clinical syndrome of LID is often transient and could be evoked in some patients only using high dosage in the levodopa test. Another aspect of force overflow in Parkinson's disease is the pressing down of the object before lifting it that has been depicted as negative load force. This is an action particularly involving proximal muscles (Lemon *et al.*, 1995). Parkinsonian subjects showed increased negative load force during the action of grasping the object in several studies (Ingvarsson *et al.*, 1997; Wenzelburger *et al.*, 2002b; Benice *et al.*, 2007) and this parameter did not return to healthy control levels after levodopa administration (Benice *et al.*, 2007; Ingvarsson *et al.*, 1997). In contrast, in PD subjects with verified LID, levodopa administration actually increased the maximum negative load force (Wenzelburger *et al.*, 2002b). Therefore the increase in the maximum negative force may be related to the force overflow associated with dyskinesias or another parkinsonian condition. In summary, force overflow in the grip–lift task can arise from the parkinsonian condition and is promoted by dopaminergic medication or LID.

Oscillations in gain between the on- and off-conditions may lead to disturbances in calibrating the correct force and results in LID. However, future longitudinal studies are required to clarify this issue.

The effect of deep brain stimulation on akinesia
and levodopa-induced dyskinesias

Subthalamic nucleus DBS is a highly effective treatment for motor symptoms in levodopa-responsive PD. Its overall benefits on akinesia resemble those of levodopa (Krack *et al.*, 1998; Brown *et al.*, 1999). The substantial anti-akinetic effects of DBS had been evaluated mainly in movements where axial and proximal muscles are the prime movers like walking or arm movements. The grip–lift task offers the opportunity for detailed studies of the effects of stimulation on a natural movement that involves the parallel control of both grip force, exerted by distal muscles, and load force, exerted mainly by axial appendicular muscles. Subthalamic nucleus DBS resolves bradykinesia of the grip–lift synergy as indicated by elevation of peak rates of grip force and of vertical acceleration of the instrumented object (Figure 32.2), whereas the minimum grip force necessary to prevent the object slipping from the fingertips was not affected by any of the treatment conditions (Nowak *et al.*, 2005). Subthalamic nucleus DBS shortened all phases of the grip–lift task in a study on 18 patients, with a predominant impact on the latest phase (load phase duration) (Wenzelburger *et al.*, 2003), which is governed by proximal arm muscles. This may be related to the observation that STN DBS exerts the most outstanding effects on axial and proximal appendicular akinesia, whereas grip formation and other hand functions show gradually less improvement. In total the effects of STN DBS on akinesia are very close to those of levodopa whereas DBS may exert an additional influence on brain-stem nuclei like the pedunculopontine nucleus which are involved in the control of axial movements.

The grip–lift task may be suited for analysis of a potential reversal of LID in late-stage parkinsonian patients. Once a patient develops LID, the sensitization to dopaminergic treatment cannot be reversed by currently available drugs (Nutt, 2000) and whereas LID are promoted by acute STN stimulation, they improve with chronic stimulation (Krack *et al.*, 1999). This reversal of the susceptibility to LID in the on-drug state is called desensitization. An inhibition of the excess of command sent to the muscle could underlie the anti-dyskinetic effects of chronic STN DBS, and may therefore impact on the calibration of forces of the precision grip. Indeed a change of grip force towards physiological values induced by chronic STN DBS was found in a study on 10 patients (Wenzelburger *et al.*, 2002a). In preoperative off-state these patients applied much higher peak grip forces than healthy controls, but grip forces in off-state were almost normal following 3 months of STN DBS (Figure 32.3).

Similarly, the pre-operative levodopa challenge caused an extra overshooting of grip force in line with previous findings, but this excess was ameliorated in the post-operative challenge with the same dosage. Although forces remained at elevated levels compared with healthy controls, grip-force overshoot was reduced significantly if the pre-operative condition was compared to the evaluation 3 months post surgery (Wenzelburger *et al.*, 2002a). On the other hand Nowak *et al.* described a clear elevation of grip force by levodopa in a group of six parkinsonian patients who had received STN DBS for at least 3 months (Figure 32.2) (Nowak *et al.*, 2005). If the stimulator was switched on, however, the effect

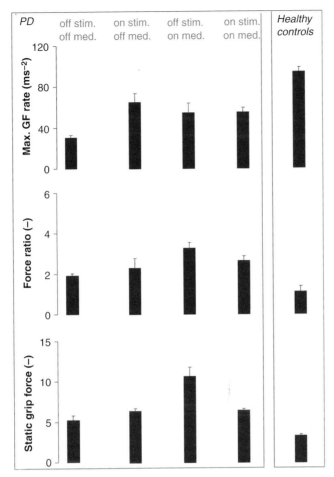

Figure 32.2. Lifting the object. Group means and standard deviations of maximum rate of grip force development, the ratio between maximum grip and load forces and the static grip force established to hold the object stationary for patients under each treatment condition and healthy controls. From Nowak *et al.* (2005).

of medication to increase the force ameliorated significantly. These observations both show desensitizing effects of STN DBS for the overshooting of grip force by levodopa, similarly for the pre- vs. post-operative comparison (Wenzelburger *et al.*, 2002a) and for the post-operative off- vs. on-drug conditions (Nowak *et al.*, 2005), although the forces remain at higher levels than in controls in the latter study. However, the on-stim vs. off-stim condition led to overshooting of grip force in the latter (Nowak *et al.*, 2005) but not in the former study (Wenzelburger *et al.*, 2002a). These differences may relate to different patients and positions of the active contacts of the electrodes. The pathophysiology of this desensitization for levodopa is not fully understood. The reduction of LID in patients with STN DBS has been ascribed to a reduction of dopaminergic dosage (Krack *et al.*, 1997; Bejjani *et al.*, 2000) but

GF$_{PEAK}$ (N)

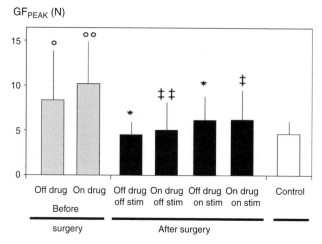

Figure 32.3. Mean peak grip force (GFPEAK+SD) before and 3 months after surgery. The overshooting of peak grip force was abolished in all conditions after surgery. °$P < 0.05$, °°$P < 0.01$ compared with controls. *$P < 0.05$ compared with off-drug state before surgery. ‡$P < 0.05$, ‡‡$P < 0.01$ compared with pre-surgical on-drug state. From Wenzelburger *et al.* (2002a).

there may also be a direct anti-dyskinetic effect of STN DBS. Substantial benefits on LID were seen regardless of whether the dopaminergic drugs after surgery could be reduced or not (Wenzelburger *et al.*, 2002a). Furthermore, the LID-suppressing effect of STN is not simply part of a general effect on all the Parkinsonian symptoms because switching off the stimulator led to a re-occurrence of severe akinesia and rigidity but left LID and force excess mainly unchanged. Therefore, the LID-suppressing effect may also reflect a desensitizing long-term effect of chronic STN DBS on LID and force regulation that outlasts even a temporary interruption of stimulation. The development of response fluctuations is believed to be due to the discontinuous pharmacological stimulation of the dopamine receptors by drug treatment. In contrast STN DBS is continuously stimulating the motor system. Such continuous stimulation may explain desensitization of both LID and grip-force excess.

The effect of therapy on grasping in tremor-dominant Parkinson's disease

The 'pill rolling' rest tremor of Parkinson's disease disappears almost completely as the hand is lifted off at the onset of a goal-directed movement. A low-amplitude action tremor with different features appears at the end of the movement just when the hand approaches the target (Wenzelburger *et al.*, 2000). While an isolated resting tremor has hardly any effect on the execution of voluntary movements, action tremor interferes with dexterous movements and can therefore be a source of significant disability (Raethjen *et al.*, 2005). Both types of tremors can be observed in Parkinson's disease (see also Chapter 27). The well-known low-frequency classical resting tremor (type I parkinsonian tremor) often re-emerges under postural conditions if the hand is held at a constant position for several seconds

(Deuschl *et al.*, 1998). The action tremor (types II and III parkinsonian tremor) that occurs at higher frequencies is present throughout the movement (Deuschl *et al.*, 1998). Quantitative studies have demonstrated that this higher-frequency tremor is activated during natural grasping movements even in patients who show the classical low-frequency re-emergent postural tremor (Forssberg *et al.*, 2000; Wenzelburger *et al.*, 2000). It has been postulated that this tremor may be an exaggeration of the central component of physiological tremor as it falls in the same frequency range (Forssberg *et al.*, 2000; Wenzelburger *et al.*, 2000) and that it may contribute to the loss of manual dexterity in PD (Gordon *et al.*, 1997; Ingvarsson *et al.*, 1997). The two main findings on action tremor in PD were a lack of response to levodopa and a correlation between the intensity of the action tremor and an impaired timing in the transition between reach-to-grasp and grasp-to-lift movements (Raethjen *et al.*, 2005). A strong correlation was observed between the resting tremor score, the postural/action tremor scores and a comparable levodopa response (see also Chapter 27). It is possible to clearly distinguish a separate higher frequency action (6.5–15 Hz band) from re-emergent tremor (3.5–6.5 Hz) by measuring the forces exerted on an object while lifting and holding it in the grip–lift task (Figure 32.4) (Raethjen *et al.*, 2005). The low-frequency oscillations during object holding were almost completely abolished under levodopa while the proportion of patients with higher frequency action

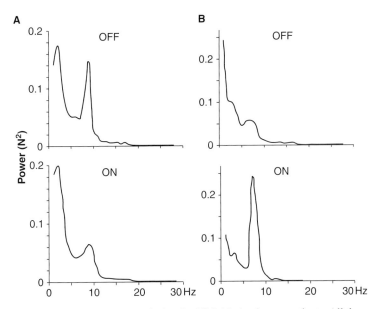

Figure 32.4. Examples of power spectra during the lift initiation in two patients. All the spectra show peaks in the 6–13 Hz range and high power in the very low frequency range because of the superimposed rise in lift force in this phase. Therefore, the exact peak frequency was sometimes difficult to determine in these spectra (see for example panel B upper spectrum). The higher-frequency tremor was therefore quantified by calculating the band power between 6.5 and 15 Hz which will always contain the higher-frequency tremor peak. While the tremor is reduced in one of the patients (A), it is even increased under levodopa in the other patient (B). From Raethjen *et al.* (2005).

tremor remained almost unchanged (Raethjen *et al.*, 2005). Thus, even if there may be some dampening in the amplitude of the high-frequency action tremor, these data show that it does not respond to levodopa to the same extent as the classical low-frequency parkinsonian resting tremor. This supports the notion that mechanisms other than the dopaminergic deficit seem to underlie the higher frequency (types II/III) action tremor. In patients with little, if any, clinical arm akinesia, action tremor can become very strong without any increase in the durations of movement transition. This is confirmed by the fact that the correlation disappears under levodopa when the majority of the patients have only mild clinical akinesia but a similarly strong action tremor (Raethjen *et al.*, 2005). This confirms that the basis of the impaired movement transition is the akinesia. However, the action tremor, which is largely independent of the dopaminergic deficit, can also affect movement timing. Although the higher frequency action tremor does not seem to be the main factor, it has the potential to further prolong movement transition times in combination with a marked overall akinesia, thereby impairing manual dexterity, and this symptom cannot be reversed by dopaminergic therapy.

The effect of cueing on grasping in Parkinson's disease

Problems with generation of repetitive movements are a key symptom of akinesia in Parkinsonism. Current physical therapy of parkinsonian patients is focused on this problem by providing external rhythms or environmental cues to facilitate gait and mobility (Nieuwboer *et al.*, 2007). Similar strategies could also be effective in training of the upper extremity. A study by Nowak *et al.* (2006) provides evidence that auditory cues may ameliorate the akinesia of grasping to lift an object. Such timing cues improved the akinesia of both the grasping and lifting components of the task in patients with UPDRS motor scores ranging from 20–55 in off-state. In on-state, with the STN DBS stimulation switched on, the effects of cueing vanished. Schenk *et al.* (2003) reported that reach-to-grasp movements improved in PD patients when an object was moved rapidly away from the subject or travelled along a defined path. These measurements underline clinical observations that even patients in deep off-state may show surprising capabilities of catching balls. A reason behind these observations could be that externally guided movements are less controlled by the basal ganglia than those which are determined by internal commands. The use of external cues to facilitate movements may provide a valuable supplementation of conventional physical therapies and devices are being developed that could apply valid cues for everyday life.

References

Bejjani, B. P., Arnulf, I., Demeret, S. *et al.* (2000). Levodopa-induced dyskinesias in Parkinson's disease: is sensitization reversible? *Ann Neurol*, **47**, 655–658.
Benice, T. S., Lou, J.-S., Eaton, R. & Nutt, J. (2007) Hand coordination as a quantitative measure of motor abnormality and therapeutic response in Parkinson's disease. *Clin Neurophysiol*, **118**, 1776–1784.

Brown, R. G., Dowsey, P. L., Brown, P. *et al.* (1999). Impact of deep brain stimulation on upper limb akinesia in Parkinson's disease. *Ann Neurol*, **45**, 473–488.

Deuschl, G., Bain, P. & Brin, M. (1998). Consensus statement of the Movement Disorder Society on Tremor. Ad Hoc Scientific Committee. *Mov Disord*, **13**, 2–23.

Fellows, S. J. & Noth, J. (2004). Grip force abnormalities in de novo Parkinson's disease. *Mov Disord*, **19**, 560–565.

Fellows, S. J., Noth, J. & Schwarz, M. (1998). Precision grip and Parkinson's disease. *Brain*, **121**, 1771–1784.

Forssberg, H., Ingvarsson, P. E., Iwasaki, N., Johansson, R. S. & Gordon, A. M. (2000). Action tremor during object manipulation in Parkinson's disease. *Mov Disord*, **15**, 244–254.

Gordon, A. M., Ingvarsson, P. E. & Forssberg, H. (1997). Anticipatory control of manipulative forces in Parkinson's disease. *Exp Neurol*, **145**, 477–488.

Guo, X., Hosseini, N., Hejdukova, B. *et al.* (2004). Load force during manual transport in Parkinson's disease. *Acta Neurol Scand*, **109**, 416–424.

Ingvarsson, P. E., Gordon, A. M. & Forssberg, H. (1997). Coordination of manipulative forces in Parkinson's disease. *Exp Neurol*, **145**, 489–501.

Krack, P., Limousin, P., Benabid, A. L. & Pollak, P. (1997). Chronic stimulation of subthalamic nucleus improves levodopa-induced dyskinesias in Parkinson's disease. *Lancet*, **350**, 1676.

Krack, P., Pollak, P., Limousin, P. *et al.* (1998). Subthalamic nucleus or internal pallidal stimulation in young onset Parkinson's disease. *Brain*, **121**, 451–457.

Krack, P., Pollak, P., Limousin, P. *et al.* (1999). From off-period dystonia to peak-dose chorea. The clinical spectrum of varying subthalamic nucleus activity. *Brain*, **122**, 1133–1146.

Lemon, R. N., Johansson, R. S. & Westling, G. (1995). Corticospinal control during reach, grasp, and precision lift in man. *J Neurosci*, **15**, 6145–6156.

Nieuwboer, A., Kwakkel, G., Rochester, L. *et al.* (2007). Cueing training in the home improves gait-related mobility in Parkinson's disease: the RESCUE trial. *J Neurol Neurosurg Psychiatry*, **78**, 134–140.

Nowak, D. A., Topka, H., Tisch, S. *et al.* (2005). The beneficial effects of subthalamic nucleus stimulation on manipulative finger force control in Parkinson's disease. *Exp Neurol*, **193**, 427–436.

Nowak, D. A., Tisch, S., Hariz, M. *et al.* (2006). Sensory timing cues improve akinesia of grasping movements in Parkinson's disease: a comparison to the effects of subthalamic nucleus stimulation. *Mov Disord*, **21**, 166–172.

Nutt, J. G. (2000). Clinical pharmacology of levodopa-induced dyskinesia. *Ann Neurol*, **47**, S160–S164; discussion S164–S166.

Raethjen, J., Pohle, S., Govindan, R. B. *et al.* (2005). Parkinsonian action tremor: interference with object manipulation and lacking levodopa response. *Exp Neurol*, **194**, 151–160.

Schenk, T., Baur, B., Steude, U. & Botzel, K. (2003). Effects of deep brain stimulation on prehensile movements in PD patients are less pronounced when external timing cues are provided. *Neuropsychologia*, **41**, 783–794.

Wenzelburger, R. (2005). Peak-dose dyskinesia; an acceptable price for mobility in late-stage Parkinson's disease? *Clin Neurophysiol*, **116**, 1997–1998.

Wenzelburger, R., Raethjen, J., Loffler, K. *et al.* (2000). Kinetic tremor in a reach-to-grasp movement in Parkinson's disease. *Mov Disord*, **15**, 1084–1094.

Wenzelburger, R., Zhang, B. R., Poepping, M. *et al.* (2002a). Dyskinesias and grip control in Parkinson's disease are normalized by chronic stimulation of the subthalamic nucleus. *Ann Neurol*, **52**, 240–243.

Wenzelburger, R., Zhang, B. R., Pohle, S. *et al.* (2002b). Force overflow and levodopa-induced dyskinesias in Parkinson's disease. *Brain*, **125**, 871–879.

Wenzelburger, R., Kopper, F., Zhang, B. R. *et al.* (2003). Subthalamic nucleus stimulation for Parkinson's disease preferentially improves akinesia of proximal arm movements compared to finger movements. *Mov Disord*, **18**, 1162–1169.

33

Therapy of focal hand dystonia

KIRSTEN E. ZEUNER, B. BAUR AND H. R. SIEBNER

Summary

Focal hand dystonia (FHD) is a disabling movement disorder. Affected patients show abnormal patterns of muscle activity of the forearm and hand while performing a specific task. This includes co-contractions of agonist and antagonist muscles and overflow of motor activity to muscles that are normally not involved in a given movement. Patients with writer's or musician's cramp may present with dystonic symptoms that only occur during a selective task (referred to as "simple" writer's cramp or musician's cramp) or may develop symptoms with multiple tasks (referred to as "dystonic" writer's or musician's cramp). Neurophysiological and neuroimaging studies in humans have identified several mechanisms that may be relevant to the pathophysiology of FDH. These mechanisms include impaired sensorimotor integration, maladaptive plasticity and deficient inhibition at various levels in the sensorimotor system. This work has been complemented by the successful establishment of a primate model in which excessive training of skilled finger movements induced a dystonia-like phenotype. Based on these lines of research, novel non-pharmacological interventions have been developed to improve dystonia in patients with FDH. In this chapter, we give an update on the range of therapeutic approaches that have been proposed for FDH.

Clinical presentation of focal hand dystonia

Task-specific dystonia of the hand often develops in individuals whose work involves skilled repetitive movements, requiring a high level of performance (Byl & Melnick, 1997; Frucht, 2004). In some patients, dystonia may show a progression and they may develop dystonia with other less specific motor actions of the involved limb.

The most common form of task-specific dystonia is writer's cramp. In this condition, dystonia affects handwriting. Dystonic symptoms may appear after prolonged writing, or even as soon as the pen is picked up. Patients show co-contractions and overflow of motor activity during handwriting which interferes with normal handwriting. Patients may develop an abnormal writing posture. Fingers, wrist and elbow can be inadequately flexed or extended. Patients produce excessive grip forces and exert an abnormally high vertical

Sensorimotor Control of Grasping: Physiology and Pathophysiology, ed. Dennis A. Nowak and Joachim Hermsdörfer. Published by Cambridge University Press. © Cambridge University Press 2009.

pressure on the tip of the pen. Handwriting is often painful and writing performance deeply disturbed. In up to 25% of patients with writer's cramp, the other hand also becomes affected if patients switch handwriting to the contralateral hand (Sheehy & Marsden, 1982).

Musicians such as pianists, guitarists and woodwind players are at particular risk for dystonia, because of the immense amount of sensorimotor training and the high demands on performance. Patients with musician's cramp experience deterioration of voluntary control of extensively practiced sensorimotor skills, manifesting as loss of control in fast passages, irregularity of trills, or involuntary flexion of one or more fingers (Altenmuller, 2003). Task-specific dystonia of the upper limb can be very disabling and may cause patients to give up their occupation (Lockwood, 1989).

Pathophysiology of focal hand dystonia

Abnormal sensory processing

Although patients show no sensory deficit on clinical examination, it is now evident that FHD is associated with impaired perception and sensorimotor integration (Hallett, 1995; Tinazzi *et al.*, 2003) (see Chapter 25). The performance in tactile spatial acuity compared with controls (Bara-Jimenez *et al.*, 2000; Zeuner *et al.*, 2002; Molloy *et al.*, 2003) is reduced. Spatial tactile discrimination seems to be disturbed in both hands, but the impairment is more pronounced in the affected compared with the unaffected hand (Sanger *et al.*, 2001; Molloy *et al.*, 2003). Recently, fMRI showed a widespread bilateral increase in regional neuronal activity in the basal ganglia while patients with writer's cramp performed a grating orientation task (Peller *et al.*, 2006).

Patients with FHD also show abnormal processing of sensory input from muscle spindle afferents (Hallett, 1995). Vibration of the dystonic limb in affected patients at rest can induce involuntary dystonic co-contractions of the muscles involved in dystonia (Tempel & Perlmutter, 1993; Kaji *et al.*, 1995b). The ability of vibration to provoke dystonic symptoms was reduced when muscle afferents were blocked with intramuscular injection of lidocaine (Kaji *et al.*, 1995a). Grunewald *et al.* (1997) demonstrated an impaired perception of movement-related proprioceptive input, but normal perception of static limb position.

Using magnetoencephalography (MEG), an abnormal homuncular organization of the finger representation in the primary somatosensory cortex in patients with focal hand dystonia has been described. A distorted cortical sensory representation was found in the somatosensory cortex contralateral (Elbert *et al.*, 1995, 1998; Bara-Jimenez *et al.*, 1998; Byl *et al.*, 2000; Meunier *et al.*, 2003) and ipsilateral to the affected limb (Meunier *et al.*, 2001). In musician's cramp, the cortical representation of the digits of the left hand of string players was shown to be larger compared with controls. In addition, the distances between the representations of individual digits in the somatosensory cortex were smaller in the affected hand relative to healthy controls (Elbert *et al.*, 1995, 1998). This further supports an abnormal sensory process in patients with dystonia.

Distorted cortical representations were also demonstrated in the motor cortex using transcranial magnetic stimulation (TMS) (Byrnes *et al.*, 1998) (see also Chapter 6). Patients with writer's cramp show displaced and distorted maps of corticomotor projections to the hand and forearm muscles. Interestingly, these map alterations were temporarily reversed with botulinum toxin treatment for the period of clinical improvement (Byrnes *et al.*, 1998).

It is difficult to incorporate all these clues about sensory dysfunction together into a clear paradigm, but since the sensory system is the major driving force behind the motor system, abnormalities of the sensory system could be relevant in causing motor dysfunction (Byl *et al.*, 1996a, 1996b; Bara-Jimenez *et al.*, 1998; Elbert *et al.*, 1998; Hallett, 1998a).

Maladaptive plasticity

The abnormal sensory and motor representations were interpreted as a result of maladaptive reorganization within the sensorimotor system. The concept of maladaptive sensorimotor plasticity is supported by a primate model of task-specific hand dystonia. Owl monkeys were trained to grasp a handgrip that repeatedly opened and closed. After months of training they developed a focal dystonia-like disorder (Byl *et al.*, 1996b). Electrophysiological mapping showed enlarged receptive fields with the breakdown of normally separated representations of different hand digits. It was presumed that stereotyped, repetitive inputs such as those found in musicians could cause this enlargement of the receptive fields and de-differentiation of representations in the sensory cortex in humans and lead to a motor disorder (Byl *et al.*, 1996a, 1996b, 1997; Byl & Melnick, 1997). Patients with writer's cramp also display an abnormal modifiability of sensorimotor circuits which may facilitate abnormal patterns of sensorimotor learning (Quartarone *et al.*, 2006).

Impaired activity of inhibitory circuits

Beside the sensory system the motor system has been investigated in detail in a number of studies using transcranial magnetic stimulation, electroencephalography and neuroimaging techniques (see Chapter 25). Transcranial magnetic stimulation of the primary motor hand area has revealed an increase in corticospinal excitability along with a reduced excitability of intracortical inhibitory circuits (Ikoma *et al.*, 1996; Berardelli *et al.*, 1998; Hallett, 1998b; Siebner *et al.*, 1999b). The inhibitory activity within subcortical (e.g. basal ganglia) and cortical (e.g. sensorimotor cortex) structures is thought to play a crucial role in the fine tuning of skilled movements and the inhibition of other competing sensorimotor representations (i.e. surround inhibition). Therefore, it has been proposed that in patients with FHD, deficient intracortical inhibition may lead to excessive cortico-muscular activation and abnormal movements (Ridding *et al.*, 1995; Berardelli *et al.*, 1998). The increased neuronal activity within the basal ganglia during a grating orientation task is also compatible with the concept that impaired center surround-inhibition within the basal ganglia–thalamic circuit leads to an overactivation of sensorimotor cortical areas during skilled movements.

Electroencephalogram studies in writer's cramp suggest a deficient inhibition in the primary sensorimotor cortex during movement initiation (Deuschl *et al.*, 1995; Hamano *et al.*, 1999; Toro *et al.*, 2000). Additionally, event-related EEG measurements reveal a deficient inhibitory control of motor programs in task-specific hand dystonia (Yazawa *et al.*, 1999; Berg *et al.*, 2001; Hummel *et al.*, 2002). Accordingly, several $H_2^{15}O$-PET activation studies have demonstrated a reduced movement-related activation of motor executive areas in idiopathic torsion dystonia and writer's cramp (Ceballos-Baumann *et al.*, 1995; Playford *et al.*, 1998; Ibanez *et al.*, 1999). This finding most probably reflects a primary dysfunction of cortical inhibitory interneurons in dystonia. This results in loss of finely tuned cortico-spinal motor output with subsequent abnormal involuntary posturing due to overflow of activity to extraneous muscles that are not involved in a selective task such as writing (Berardelli *et al.*, 1998).

In summary, it can be concluded that FHD is not only a disorder of movement execution, but also of movement preparation and sensorimotor processing. In addition, there is clear evidence for disorganization of the cortical representations for both affected and unaffected extremities in patients with task-specific dystonia, suggesting that focal dystonia is a generalized disorder manifesting with focal symptoms.

Treatment of task-specific dystonia

Pharmacological treatment

Task-specific dystonia affecting fine manual skills is a disabling neurological condition and the currently available treatments are limited in efficacy. Treatment options include non-pharmacological and pharmacological interventions, or a combination of both. Patients with mild symptoms are advised to reduce their task-specific activity or to consult an occupational therapist. Treatment with anticholinergics, baclofen and benzodiazepines usually produces little benefit. Intramuscular botulinum toxin (BTX) injection into affected forearm muscles is the current standard therapy for writer's and musician's cramp (Wissel *et al.*, 1996; Chen & Hallett, 1998; Jabusch *et al.*, 2005; Schuele *et al.*, 2005; Das *et al.*, 2006; Kruisdijk *et al.*, 2007), but improvement is far from perfect. Moreover, BTX injections carry the risk of inducing a disabling weakness of the hand, and regular injections are required to maintain a symptomatic therapeutic effect. Therefore, only about half the patients with writer's cramp injected with BTX injections remain on long-term treatment (Kruisdijk *et al.*, 2007). Thus, alternative treatment options are required for those patients showing no satisfactory response to BTX injections.

Neurostimulation

Based on the hypothesis that abnormal sensorimotor processing of muscle spindle afferent discharges may underlie the development of dystonia, transcutaneous electrical stimulation (TENS) has been studied as a novel treatment option. Ten patients with simple writer's cramp were treated with TENS stimulation of the forearm flexor muscles for 2 weeks. A beneficial effect was observed which lasted for 3 weeks (Tinazzi *et al.*, 2005a). The authors postulated that TENS induces reciprocal changes of corticospinal excitability in agonist and antagonist muscles (Tinazzi *et al.*, 2005b, 2006).

Transcutaneal magnetic stimulation has also been employed to transiently improve dystonia in writer's cramp. To reinforce the strength of intracortical inhibitory circuits, low-frequency (1 Hz) repetitive TMS (rTMS) was given to the primary cortex contralateral to the affected limb. A 30-min period of subthreshold 1 Hz rTMS enhanced the excitability of intracortical inhibitory circuits and improved handwriting in some patients (Siebner *et al.*, 1999a, 1999b). In a subsequent study, 1 Hz rTMS was combined with motor training. Patients received 1 Hz rTMS of the primary motor hand area for 10 minutes while they performed a "scribbling" task with the affected contralateral hand (Siebner *et al.*, 1999a). This intervention produced a transient improvement of dystonia in the majority of patients and improved the regularity of cyclic drawing movements. Cortical inhibition increased and led to some transient improvement (Siebner *et al.*, 1999b). A clinical benefit was also reported after 0.2 Hz rTMS of the dorsal premotor cortex contralateral to the affected limb (Murase *et al.*, 2005).

Immobilization

Another therapeutic approach that might utilize mechanisms of neuronal network plasticity is limb immobilization. This approach has been evaluated in patients with musician's and writer's cramp (Priori *et al.*, 2001; Pesenti *et al.*, 2004). In these studies, fingers and wrists in eight patients were immobilized for 4–5 weeks. After removal of the splint, patients were advised to slowly start with their daily activities including their specific task evaluated with the Arm Dystonia Disability Score and self-assessment. Substantial improvement lasted for at least 12 months. The positive clinical effect might be a result of a de-differentiation of the sensorimotor cortex during immobilization and an eradication of abnormal dystonic patterns.

Sensory and motor training programs

Based on the hypothesis that abnormal sensory processing could possibly cause a motor disorder, the use of sensory training has been suggested as a treatment for focal hand dystonia and was tested with Braille reading as a sensory training method to train spatial discrimination and localization. The extent of the change in somatosensory cortical representation of the digits as shown in owl monkey studies depends on the amount, simultaneity, rate and timing of sensory input (Byl *et al.*, 1997). Intense Braille reading can influence cortical representation (Sterr *et al.*, 1998a, 1998b, 1999) and produce superior tactile spatial acuity in blind Braille readers (Van Boven *et al.*, 2000). The purpose of sensory training was to reverse the abnormal somatosensory cortical dysfunction in patients with task-specific

A

B

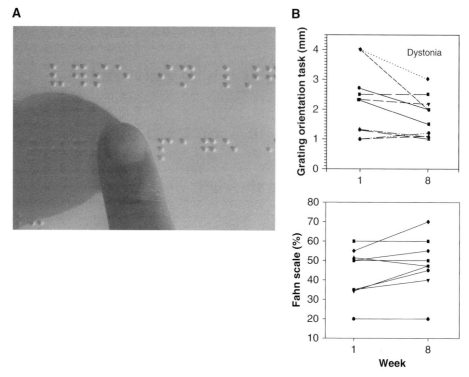

Figure 33.1. A. The left panel demonstrates the method of sensory training. Braille reading was performed with one finger at a time.

B. In the upper panel the thresholds of the grating orientation task at baseline and after 8 weeks of sensory training are given. After 8 weeks of sensory training patients improved significantly in their performance of the grating orientation task. The decrease in the threshold of the grating orientation task was paralleled with an improvement of hand dystonia when measured with the Arm Dystonia Disability Scale developed by Fahn (lower panel) (modified from Zeuner *et al.*, 2002).

dystonia. Sensory training was accomplished by training each finger individually to read Braille. Sensory discrimination, measured with the Johnson–van Boven–Phillips domes (Johnson & Phillips, 1981), improved in affected patients and controls and ameliorated motor function in writer's cramp patients (Figure 33.1) (Zeuner *et al.*, 2002; Zeuner & Hallett, 2003).

In contrast to training procedures which aim at changing sensory processing, another approach by Mai and colleagues focuses on direct training of handwriting movements. Patients with writer's cramp show co-contractions of antagonist arm and hand muscles which reduce the mobility of fingers and wrist. This typically leads to abnormalities in writing posture and inappropriate writing techniques like the exertion of excessive pressure on pen and desk or lacking arm transport. Concurrently, other aspects of the writing movement might be well preserved in writer's cramp patients (e.g. using proximal joints). The training approach by Mai *et al.* (Mai & Marquardt, 1999) is always tailored to the

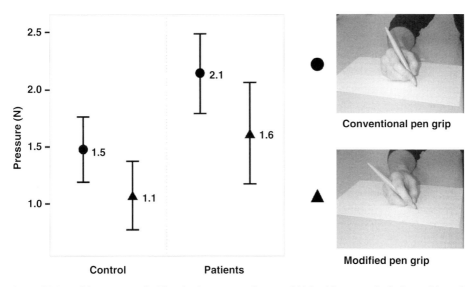

Figure 33.2. Writing pressure in 23 writer's cramp patients and 11 healthy controls during writing of a test sentence: means and 95% confidence intervals of means. Figure modified from Baur *et al.* (2006).

patient's individual pattern of preserved and dystonic handwriting aspects. It aims at the reduction of inadequate writing techniques and the re-training of efficient writing strategies. Therefore, various motor exercises are practiced under supervision: e.g. drawing quickly lines from left to right to improve arm transport or softly bending and stretching the fingers before or interspersed with writing to improve mobility of the finger joints. In addition, writing conditions (such as pen, pen grip, writing pad, etc.) might be altered temporarily during treatment to reduce the occurrence of dystonic movement patterns. The evaluation of 50 patients with writer's cramp revealed that the kinematic aspects of handwriting improved after training (Schenk *et al.*, 2004): writing speed and fluency, which were clearly disturbed in writer's cramp patients before training, were significantly ameliorated after training. This improvement was stable over a mean follow-up period of 20 months.

A modified pen grip (between the proximal and distal phalanges of index and middle finger, see Figure 33.2) may also be helpful in many patients. When the pen is held with the modified grip there is stronger contact between the pen and fingers already in the relaxed grip so that there is less need for active holding of the pen. Thus, the tasks of holding and moving the pen during handwriting are separated. Baur and colleagues (2006) showed that writer's cramp patients benefited immediately from the modified pen grip which led to a significant decrease of writing pressure exerted onto the desk. A further significant reduction of writing pressure and pain can be achieved by handwriting training with the modified pen grip (Baur *et al.*, in prep.).

To reverse motor abnormalities in focal hand dystonia, motor training programs have been discussed and tried further (Byl, 2003, 2004; Byl *et al.*, 2003). Experienced teachers have attempted to retrain musicians (Chamagne, 1993; Jabusch *et al.*, 2005). Byl *et al.* combined sensory discriminative training successfully with fitness exercises to improve sensory processing and motor control of the hand in patients with focal hand dystonia (Byl & McKenzie, 2000). They reported significant gains in motor control, motor accuracy, sensory discrimination and physical performance. All but one of their patients were able to return to work (Byl & McKenzie, 2000).

If maladaptive plasticity is one possible explanation for dystonic co-contraction, dystonia should improve with reorganization of the cortex. This concept has been employed by Candia *et al.* (1999) and Taub *et al.* (1999) who used constraint-induced movement therapy (see Chapter 29) in musician's cramp. In a method called sensorimotor retuning patients wore splints that immobilized one or more digits other than the dystonic fingers, while they performed a motor training. Improvement without the splint at the end of treatment lasted for several months and was very successful in piano players, but not in musicians who played wind instruments (Candia *et al.*, 1999, 2002, 2003; Taub *et al.*, 1999). Sensory motor retuning therapy led to subjective improvement evaluated by a dystonia evaluation scale, objective enhancement of movement smoothness and cortical reorganization measured by magnetoencephalography (Candia *et al.*, 2003).

A similar motor training model was designed for patients with writer's cramp. The patient's fingers and wrist were splinted, depending on the pattern of the dystonia, using thermoplastic material to reduce co-contraction. A plastic finger splint for the tip of the training finger with a pen attached to it was used for the training (Figure 33.3). Patients

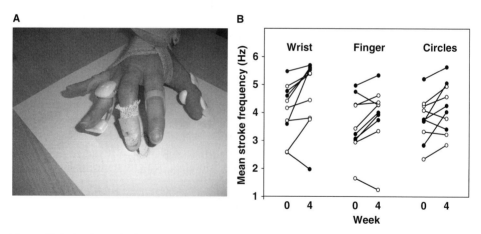

Figure 33.3. A. The method of task-specific motor training is shown. Patients trained each finger with a finger splint that had a pen attached to it and were advised to perform daily drawing and simple writing movements. The fingers that were not trained were splinted to avoid co-contraction in order to reduce abnormal synergism during the training.

B. The stroke frequency measured by the kinematic handwriting analysis is shown. Patients performed several different drawing and writing tasks (wrist movements, finger movements, circle drawing). Figure modified from Zeuner *et al.* (2005).

were instructed to train each finger individually by making clockwise and counter-clockwise circles on a piece of paper for 4–8 weeks daily (Zeuner *et al.*, 2005). The Arm Dystonia Disability scale (Fahn, 1989) revealed a positive clinical effect, whereas the kinematic writing analysis showed some improvement in simple drawing exercises. The finger that trained was not splinted, while the remaining four fingers were splinted to avoid co-contraction. The rationale behind this approach was to improve the dystonia by reducing the abnormal synergism and to reverse the reorganizational changes in the motor cortex.

In summary, a number of interesting therapeutic approaches for dystonia have emerged in recent years. However, most studies did not employ kinematic measurements and clinical scoring was done in a non-blinded fashion. This, however, would be desirable in order to ensure an objective assessment of clinical improvement. A reliable and objective measurement of motor performance is scale analysis to quantify dystonia in pianists (Jabusch *et al.*, 2004). Pianists with musician's cramp play on a digital piano that is connected to a computer. The principle of scale analysis is to measure irregularities, loudness and different timing parameters with a software program. In writer's cramp patients the kinematic analysis of writing movements using a digitizing tablet and the software CS (Marquardt & Mai, 1994) is a valuable objective method. Another open question is how long patients should continue with training. It has been observed that improvement in motor performance may wear off when training is discontinued (Zeuner & Hallett, 2003). Since it takes many years to develop the disorder, it may require many years of rehabilitation to achieve long-lasting improvement. It may even be necessary for patients to continue with training throughout life in order to sustain clinicial improvement.

References

Altenmuller, E. (2003). Focal dystonia: advances in brain imaging and understanding of fine motor control in musicians. *Hand Clin*, **19**, 523–538, xi.

Bara-Jimenez, W., Catalan, M. J., Hallett, M. & Gerloff, C. (1998). Abnormal somatosensory homunculus in dystonia of the hand. *Ann Neurol*, **44**, 828–831.

Bara-Jimenez, W., Shelton, P. & Hallett, M. (2000). Spatial discrimination is abnormal in focal hand dystonia. *Neurology*, **55**, 1869–1873.

Baur, B., Schenk, T., Fürholzer, W. *et al.* (2006). Modified pen grip in the treatment of writer's cramp. *Hum Mov Sci*, **25**, 464–473.

Baur, B., Fürholzer W., Jasper I., Marquardt C. & Hermsdörfer J. (in prep.). Effects of modified pen grip and handwriting training on writer's cramp.

Berardelli, A., Rothwell, J. C., Hallett, M. (1998). The pathophysiology of primary dystonia. *Brain*, **121**, 1195–1212.

Berg, D., Herrmann, M. J., Muller, T. J. *et al.* (2001). Cognitive response control in writer's cramp. *Eur J Neurol*, **8**, 587–594.

Byl, N. N. (2003). What can we learn from animal models of focal hand dystonia? *Rev Neurol (Paris)*, **159**, 857–873.

Byl, N. N. (2004). Focal hand dystonia may result from aberrant neuroplasticity. *Adv Neurol*, **94**, 19–28.

Byl, N. N. & Melnick, M. (1997). The neural consequences of repetition: clinical implications of a learning hypothesis. *J Hand Ther*, **10**, 160–174.

Byl, N. N. & McKenzie, A. (2000). Treatment effectiveness for patients with a history of repetitive hand use and focal hand dystonia: a planned, prospective follow-up study. *J Hand Ther*, **13**, 289–301.

Byl, N., Wilson, F., Merzenich, M. *et al.* (1996a). Sensory dysfunction associated with repetitive strain injuries of tendinitis and focal hand dystonia: a comparative study. *J Orthop Sports Phys Ther*, **23**, 234–244.

Byl, N. N., Merzenich, M. M. & Jenkins, W. M. (1996b). A primate genesis model of focal dystonia and repetitive strain injury: I. Learning-induced dedifferentiation of the representation of the hand in the primary somatosensory cortex in adult monkeys. *Neurology*, **47**, 508–520.

Byl, N. N., Merzenich, M. M., Cheung, S. *et al.* (1997). A primate model for studying focal dystonia and repetitive strain injury: effects on the primary somatosensory cortex. *Phys Ther*, **77**, 269–284.

Byl, N. N., McKenzie, A. & Nagarajan, S. S. (2000). Differences in somatosensory hand organization in a healthy flutist and a flutist with focal hand dystonia: a case report. *J Hand Ther*, **13**, 302–309.

Byl, N. N., Nagajaran, S. & Mckenzie, A. L. (2003). Effect of sensory discrimination training on structure and function in patients with focal hand dystonia: a case series. *Arch Phys Med Rehabil*, **84**, 1505–1514.

Byrnes, M. L., Thickbroom, G. W., Wilson, S. A. *et al.* (1998). The corticomotor representation of upper limb muscles in writer's cramp and changes following botulinum toxin injection. *Brain*, **121**, 977–988.

Candia, V., Elbert, T., Altenmuller, E. *et al.* (1999). Constraint-induced movement therapy for focal hand dystonia in musicians. *Lancet*, **353**, 42.

Candia, V., Schafer, T., Taub, E. *et al.* (2002). Sensory motor retuning: a behavioral treatment for focal hand dystonia of pianists and guitarists. *Arch Phys Med Rehabil*, **83**, 1342–1348.

Candia, V., Wienbruch, C., Elbert, T., Rockstroh, B. & Ray, W. (2003). Effective behavioral treatment of focal hand dystonia in musicians alters somatosensory cortical organization. *Proc Natl Acad Sci USA*, **100**, 7942–7946.

Ceballos-Baumann, A. O., Passingham, R. E., Warner, T. *et al.* (1995). Overactive prefrontal and underactive motor cortical areas in idiopathic dystonia. *Ann Neurol*, **37**, 363–372.

Chamagne, P. (1993). Functional dystonia in musicians: fundamental principles of the rehabilitation. *Ann Chir Main Memb Super*, **12**, 63–67.

Chen, R. & Hallett, M. (1998). Focal dystonia and repetitive motion disorders. *Clin Orthop*, **351**, 102–106.

Das, C. P., Dressler, D. & Hallett, M. (2006). Botulinum toxin therapy of writer's cramp. *Eur J Neurol*, **13 Suppl. 1**, 55–59.

Deuschl, G., Toro, C., Matsumoto, J. & Hallett, M. (1995). Movement-related cortical potentials in writer's cramp [see comments]. *Ann Neurol*, **38**, 862–868.

Elbert, T., Pantev, C., Wienbruch, C., Rockstroh, B. & Taub, E. (1995). Increased cortical representation of the fingers of the left hand in string players. *Science*, **270**, 305–307.

Elbert, T., Candia, V., Altenmuller, E. *et al.* (1998). Alteration of digital representations in somatosensory cortex in focal hand dystonia. *Neuroreport*, **9**, 3571–3575.

Fahn, S. (1989). Assessment of the primary dystonias. In T. Munsat (Ed.), *The Quantification of Neurologic Deficit* (pp. 241–270). Boston, MA: Butterworths.

Frucht, S. J. (2004). Focal task-specific dystonia in musicians. *Adv Neurol*, **94**, 225–230.

Grunewald, R. A., Yoneda, Y., Shipman, J. M. & Sagar, H. J. (1997). Idiopathic focal dystonia: a disorder of muscle spindle afferent processing? *Brain*, **120**, 2179–2185.

Hallett, M. (1995). Is dystonia a sensory disorder? *Ann Neurol*, **38**, 139–140.

Hallett, M. (1998a). The neurophysiology of dystonia. *Arch Neurol*, **55**, 601–603.

Hallett, M. (1998b). Physiology of dystonia. *Adv Neurol*, **78**, 11–18.

Hamano, T., Kaji, R., Katayama, M. *et al.* (1999). Abnormal contingent negative variation in writer's cramp. *Clin Neurophysiol*, **110**, 508–515.

Hummel, F., Andres, F., Altenmuller, E., Dichgans, J. & Gerloff, C. (2002). Inhibitory control of acquired motor programmes in the human brain. *Brain*, **125**, 404–420.

Ibanez, V., Sadato, N., Karp, B., Deiber, M. P. & Hallett, M. (1999). Deficient activation of the motor cortical network in patients with writer's cramp. *Neurology*, **53**, 96–105.

Ikoma, K., Samii, A., Mercuri, B., Wassermann, E. M. & Hallett, M. (1996). Abnormal cortical motor excitability in dystonia. *Neurology*, **46**, 1371–1376.

Jabusch, H. C., Vauth, H. & Altenmuller, E. (2004). Quantification of focal dystonia in pianists using scale analysis. *Mov Disord*, **19**, 171–180.

Jabusch, H. C., Zschucke, D., Schmidt, A., Schuele, S. & Altenmuller, E. (2005). Focal dystonia in musicians: Treatment strategies and long-term outcome in 144 patients. *Mov Disord*, **20**, 1623–1626.

Johnson, K. O. & Phillips, J. R. (1981). Tactile spatial resolution. I. Two-point discrimination, gap detection, grating resolution, and letter recognition. *J Neurophysiol*, **46**, 1177–1192.

Kaji, R., Kohara, N., Katayama, M. *et al.* (1995a). Muscle afferent block by intramuscular injection of lidocaine for the treatment of writer's cramp. *Muscle Nerve*, **18**, 234–235.

Kaji, R., Rothwell, J. C., Katayama, M. *et al.* (1995b). Tonic vibration reflex and muscle afferent block in writer's cramp. *Ann Neurol*, **38**, 155–162.

Kruisdijk, J. J., Koelman, J. H., Ongerboer de Visser, B. W., de Haan, R. J. & Speelman, J. D. (2007). Botulinum toxin for writer's cramp: a randomised, placebo-controlled trial and 1-year follow-up. *J Neurol Neurosurg Psychiatry*, **78**, 264–270.

Lockwood, A. H. (1989). Medical problems of musicians. *N Engl J Med*, **320**, 221–227.

Mai, N. & Marquardt, C. (1999). *Schreibtraining in der neurologischen Rehabilitation*. Dortmund, Germany: Borgmann Publishing.

Marquardt, C. & Mai, N. (1994). A computational procedure for movement analysis in handwriting. *J Neurosci Methods*, **52**, 39–45.

Meunier, S., Garnero, L., Ducorps, A. *et al.* (2001). Human brain mapping in dystonia reveals both endophenotypic traits and adaptive reorganization. *Ann Neurol*, **50**, 521–527.

Meunier, S., Lehericy, S., Garnero, L. & Vidailhet, M. (2003). Dystonia: lessons from brain mapping. *Neuroscientist*, **9**, 76–81.

Molloy, F. M., Carr, T. D., Zeuner, K. E., Dambrosia, J. M. & Hallett, M. (2003). Abnormalities of spatial discrimination in focal and generalized dystonia. *Brain*, **126**, 2175–2182.

Murase, N., Rothwell, J. C., Kaji, R. *et al.* (2005). Subthreshold low-frequency repetitive transcranial magnetic stimulation over the premotor cortex modulates writer's cramp. *Brain*, **128**, 104–115.

Peller, M., Zeuner, K. E., Munchau, A. *et al.* (2006). The basal ganglia are hyperactive during the discrimination of tactile stimuli in writer's cramp. *Brain*, **129**, 2697–2708.

Pesenti, A., Barbieri, S. & Priori, A. (2004). Limb immobilization for occupational dystonia: a possible alternative treatment for selected patients. *Adv Neurol*, **94**, 247–254.

Playford, E. D., Passingham, R. E., Marsden, C. D. & Brooks, D. J. (1998). Increased activation of frontal areas during arm movement in idiopathic torsion dystonia. *Mov Disord*, **13**, 309–318.

Priori, A., Pesenti, A., Cappellari, A., Scarlato, G. & Barbieri, S. (2001). Limb immobilization for the treatment of focal occupational dystonia. *Neurology*, **57**, 405–409.

Quartarone, A., Siebner, H. R. & Rothwell, J. C. (2006). Task-specific hand dystonia: can too much plasticity be bad for you? *Trends Neurosci*, **29**, 192–199.

Ridding, M. C., Sheean, G., Rothwell, J. C., Inzelberg, R. & Kujirai, T. (1995). Changes in the balance between motor cortical excitation and inhibition in focal, task specific dystonia. *J Neurol Neurosurg Psychiatry*, **59**, 493–498.

Sanger, T. D., Tarsy, D. & Pascual-Leone, A. (2001). Abnormalities of spatial and temporal sensory discrimination in writer's cramp. *Mov Disord*, **16**, 94–99.

Schenk, T., Bauer, B., Steidle, B. & Marquardt, C. (2004). Does training improve writer's cramp? An evaluation of a behavioral treatment approach using kinematic analysis. *J Hand Ther*, **17**, 349–363.

Schuele, S., Jabusch, H. C., Lederman, R. J. & Altenmuller, E. (2005). Botulinum toxin injections in the treatment of musician's dystonia. *Neurology*, **64**, 341–343.

Sheehy, M. P. & Marsden, C. D. (1982). Writers' cramp – a focal dystonia. *Brain*, **105**, 461–480.

Siebner, H. R., Auer, C., Ceballos-Baumann, A. & Conrad, B. (1999a). Has repetitive transcranial magnetic stimulation of the primary motor hand area a therapeutic application in writer's cramp? *Electroencephalogr Clin Neurophysiol Suppl*, **51**, 265–275.

Siebner, H. R., Tormos, J. M., Ceballos-Baumann, A. O. *et al.* (1999b). Low-frequency repetitive transcranial magnetic stimulation of the motor cortex in writer's cramp. *Neurology*, **52**, 529–537.

Sterr, A., Muller, M. M., Elbert, T., Rockstroh, B., Pantev, C. & Taub, E. (1998a). Changed perceptions in Braille readers. *Nature*, **391**, 134–135.

Sterr, A., Muller, M. M., Elbert, T. *et al.* (1998b). Perceptual correlates of changes in cortical representation of fingers in blind multifinger Braille readers. *J Neurosci*, **18**, 4417–4423.

Sterr, A., Muller, M., Elbert, T., Rockstroh, B. & Taub, E. (1999). Development of cortical reorganization in the somatosensory cortex of adult Braille students. *Electroencephalogr Clin Neurophysiol Suppl*, **49**, 292–298.

Taub, E., Uswatte, G. & Pidikiti, R. (1999). Constraint-induced movement therapy: a new family of techniques with broad application to physical rehabilitation – a clinical review. *J Rehabil Res Dev*, **36**, 237–251.

Tempel, L. W. & Perlmutter, J. S. (1993). Abnormal cortical responses in patients with writer's cramp [published erratum appears in *Neurology* (1994) **44**, 2411]. *Neurology*, **43**, 2252–2257.

Tinazzi, M., Rosso, T. & Fiaschi, A. (2003). Role of the somatosensory system in primary dystonia. *Mov Disord*, **18**, 605–622.

Tinazzi, M., Farina, S., Bhatia, K. *et al.* (2005a). TENS for the treatment of writer's cramp dystonia: a randomized, placebo-controlled study. *Neurology*, **64**, 1946–1948.

Tinazzi, M., Zarattini, S., Valeriani, M. *et al.* (2005b). Long-lasting modulation of human motor cortex following prolonged transcutaneous electrical nerve stimulation (TENS) of forearm muscles: evidence of reciprocal inhibition and facilitation. *Exp Brain Res*, **161**, 457–464.

Tinazzi, M., Zarattini, S., Valeriani, M. *et al.* (2006). Effects of transcutaneous electrical nerve stimulation on motor cortex excitability in writer's cramp: neurophysiological and clinical correlations. *Mov Disord*, **21**, 1908–1913.

Toro, C., Deuschl, G. & Hallett, M. (2000). Movement-related electroencephalographic desynchronization in patients with hand cramps: evidence for motor cortical involvement in focal dystonia. *Ann Neurol*, **47**, 456–461.

Van Boven, R. W., Hamilton, R. H., Kauffman, T., Keenan, J. P. & Pascual-Leone, A. (2000). Tactile spatial resolution in blind braille readers. *Neurology*, **54**, 2230–2236.

Wissel, J., Kabus, C., Wenzel, R. *et al.* (1996). Botulinum toxin in writer's cramp: objective response evaluation in 31 patients. *J Neurol Neurosurg Psychiatry*, **61**, 172–175.

Yazawa, S., Ikeda, A., Kaji, R. *et al.* (1999). Abnormal cortical processing of voluntary muscle relaxation in patients with focal hand dystonia studied by movement-related potentials. *Brain*, **122**, 1357–1366.

Zeuner, K. E. & Hallett, M. (2003). Sensory training as treatment for focal hand dystonia: a 1-year follow-up. *Mov Disord*, **18**, 1044–1047.

Zeuner, K. E., Bara-Jimenez, W., Noguchi, P. S. *et al.* (2002). Sensory training for patients with focal hand dystonia. *Ann Neurol*, **51**, 593–598.

Zeuner, K. E., Shill, H. A., Sohn, Y. H. *et al.* (2005). Motor training as treatment in focal hand dystonia. *Mov Disord*, **20**, 335–341.

34

Therapy of idiopathic normal pressure hydrocephalus

DENNIS A. NOWAK

Summary

The clinical spectrum of idiopathic normal pressure hydrocephalus (INPH) comprises gait impairment, cognitive decline and urinary incontinence, all associated with ventricular enlargement and normal cerebrospinal fluid (CSF) pressure on random spinal taps. There is significant variation in the clinical presentation and progression of the disorder and correct diagnosis frequently represents a challenge to the clinical neurologist. Several reports have suggested that the motor disability to be found in INPH may also involve the upper limbs and recent reports provide direct kinetic evidence for this suggestion. Grip-force analysis may also allow an objective evaluation of the beneficial effects of therapeutic strategies in this entity. This chapter reviews the pertinent literature upon the kinetic assessment of upper limb motor disability in the diagnosis and therapy of INPH.

The symptom complex of INPH

Idiopathic normal pressure hydrocephalus (INPH), first described by Hakim & Adams (1965) and Adams et al. (1965), is characterized by the clinical triad of gait disorder, dementia and urinary incontinence, all in the presence of ventriculomegaly and normal cerebrospinal fluid (CSF) pressure on random lumbar puncture. The cause of INPH is not known. When the clinical syndrome occurs as a result of other diseases, such as hemorrhage, traumatic brain injury, cerebral infarction or meningitis, it is referred to as *secondary normal pressure hydrocephalus* (Gallia et al., 2006). The incidence of INPH has been reported to be about two cases per 100,000 individuals (Vanneste et al., 1992; Krauss & Halve, 2004). The clinical symptom triad in INPH typically develops insidiously and affects people in the sixth to eighth decades of life (Fisher, 1982; McGirt et al., 2005; Marmarou et al., 2005; Wilson & Williams, 2006). Idiopathic normal pressure hydrocephalus appears to be a unique reversible form of neuronal injury, the mechanism of which is not well understood. The symptom complex of INPH has been explained by both mechanical (Hakim et al., 1970) and ischemic factors (Graff-Radford & Godersky, 1987; Krauss et al., 1996). It has been suggested that the widening of the ventricular CSF system causes stretching of the arterial vessels (Greitz et al., 1969), and that decreased compliance (Ekstedt & Friden, 1984) and high arterial pulse

Sensorimotor Control of Grasping: Physiology and Pathophysiology, ed. Dennis A. Nowak and Joachim Hermsdörfer. Published by Cambridge University Press. © Cambridge University Press 2009.

pressure both develop local vascular barotraumas and/or tangential shear stress to the surrounding white matter and its sensory-motor tracts (Hakim *et al.*, 1970). The clinical symptoms in INPH result from progressive lesioning of the sensory and motor tracts travelling through the cerebral white matter. Since the fibers of the corticospinal tract that supply motor function to the legs are located closest to the lateral ventricles, it appears no longer surprising that gait impairment is commonly the first symptom to be observed in INPH and the first to resolve with CSF evacuation (Hebb & Cusimano, 2001). Altogether hypokinetic gait disorder improves in 93% of INPH subjects after permanent CSF shunt implantation, while dementia and urinary incontinence are only half as likely to improve (Gallia *et al.*, 2006). Of the three cardinal symptoms of INPH, cognitive impairment is the least likely to improve promptly after CSF shunting (Hebb & Cusimano, 2001; Gallia *et al.*, 2006); however, in subjects who show improvement in gait disorder long-term improvement in memory has been documented (Duinkerke *et al.*, 2004).

Gait disturbance is frequently the first sign and widely held to be the most disabling symptom of the disorder (Adams *et al.*, 1965; Fisher, 1982; Wilson & Williams, 2006). The gait disorder in INPH is usually termed to as apraxic, hypokinetic, magnetic, parkinsonian and shuffling (Fisher, 1982; Stolze *et al.*, 2001). The semiology of the gait disorder in INPH is characterized by a slow, wide-based walking pattern with short shuffling steps and difficulty in turning and tandem walking, all in the absence of muscular weakness (Adams *et al.*, 1965; Fisher, 1982). Kinematic motion analysis of the gait pattern in INPH has demonstrated diminished gait velocity that is mainly due to reduced stride length, reduced step height during the swing phase of the gait cycle and enlarged balance-related gait measures, such as step width and foot rotation angles (Stolze *et al.*, 2001). After tapping of CSF the gait velocity has been found to increase by about 20% (Krauss *et al.*, 1996; Sudarsky *et al.*, 1997; Stolze *et al.*, 2001). The hypokinetic gait disorder of INPH shares several features in common with the gait disorder to be found in Parkinson's disease, such as reduced velocity due to diminished stride length, leg rigidity, a flexed posture and hampered postural reflexes (Sudarsky *et al.*, 1997; Stolze *et al.*, 2001). In INPH, however, both body sides are usually equally affected and the in-between distance of the feet is commonly larger compared with that seen in Parkinson's disease with V-shaped orientation of the feet.

Clinical experience, however, suggests that the clinical spectrum of INPH is commonly not limited to the classic clinical triad of symptoms with gait disorder representative of the hypokinetic motor disability (Soelberg Sorensen *et al.*, 1986; Blomsterwall *et al.*, 1995; Krauss *et al.*, 1997). In a series of 118 normal pressure hydrocephalus cases due to various etiologies, 75% of affected subjects exhibited signs of hypokinetic upper limb disability usually encountered in Parkinson's disease, such as akinesia, tremor and rigidity (Krauss *et al.*, 1997). The percentage of subjects who suffered from a hypokinetic motor disorder of the arm and hand was even more pronounced in the group of INPH (86%; Krauss *et al.*, 1997). Functional magnetic resonance imaging has shown reduced neural activity within the primary motor cortex in INPH for hand movements and an increase in neural activity after successful permanent CSF shunt procedure (Fukuhara *et al.*, 2001). Given the clinical recognition that gait disorder is not the only hypokinetic motor disorder in

INPH, it appears surprising that only a few kinematic studies have addressed the investigation of upper limb function in this entity (Matousek *et al.*, 1995; Nowak *et al.*, 2006).

The hypokinetic motor disorder in INPH also affects the upper limb

The control of isometric grip force is an essential feature of skillful manipulative tasks in everyday life (Flanagan & Johansson, 2002) and the investigation of grip forces has evolved a highly sensitive method to objectively evaluate impaired motor function of the hand (Nowak & Hermsdörfer, 2005). Recently (Nowak & Topka, 2006), the efficiency of grip-force scaling was comparatively studied in eight subjects with INPH (mean age 75 ± 8 years), eight subjects with Parkinson's disease (mean age 62 ± 14 years) and eight healthy control subjects (mean age 75 ± 7 years). All participants were right-handed. Diagnosis of INPH was based on clinical criteria for INPH (Hakim & Adams, 1965) consisting of gait disturbance with or without dementia and/or urinary incontinence. Computed tomography and/or magnetic resonance imaging of the brain had to disclose ventricular enlargement and the absence of cortical atrophy (Vanneste *et al.*, 1993; Krauss *et al.*, 1997). Functional disability was graded according to a modification (Krauss *et al.*, 1997) of the scale of Stein & Langfit (1974): six subjects needed some help or supervision and two subjects needed considerable help or supervision in activities of daily life. Marked clinical improvement of the gait disorder after removal of 30–50 ml of CSF on two occasions was considered confirmatory for the diagnosis. Parkinson's disease was diagnosed according to the UK brain bank criteria (Hughes *et al.*, 1992). The clinical symptom severity of Parkinson's disease was rated using the Hoehn & Yahr (1967) scale: three subjects had bilateral disease without impairment of balance and five subjects had mild to moderate bilateral disease with mild postural instability, but preserved independence in activities of daily life. Subjects with INPH were examined prior to CSF evacuation. Subjects with Parkinson's disease were examined after a 12-h overnight withdrawal of dopaminergic drugs.

Subjects were asked to grasp an instrumented object between the index finger and thumb of the right hand, lift it 5 cm above the support, hold it in that position for 7 seconds and then replace and release it. Ten such grip–lift trials were performed by each participant. The grip–lift task under discussion here comprises two different phases (see also Chapters 1 and 12): an early phase during which the grasping fingers establish a stable grasp and a late phase during which lift force increases until the object lifts from the support. The grip–lift task involves the subtle interaction between distal muscle groups, applying a stable grasp, and proximal muscle groups, involved in the lifting drive, of the arm. The following parameters were obtained for data analysis: (i) peak rate of grip force increase, (ii) peak grip force, (iii) peak acceleration of the lifting movement and (iv) static grip force established to hold the object stationary after the lift. The net lift force was calculated from the product of the object mass and the vectorial summation of accelerations along the object's X-, Y- and Z-axes including gravity (see Figure 34.1 and also see Chapter 1).

Figure 34.2 shows average traces ($\pm$ 1 SD) of grip force, rate of grip-force increase and acceleration due to lift over time for the groups of healthy subjects, subjects with

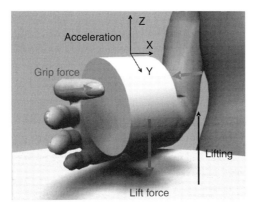

Figure 34.1. Illustration of the grip–lift task. The instrumented object was grasped between the index finger and thumb in opposition, lifted 5 cm above the support, held for 7 seconds and then replaced and released. The object incorporated a force sensor to register grip force and linear acceleration sensors to measure accelerations in three dimensions. The mass of the object was 350 g. Grip surfaces were sandpaper at a medium grain (No. 240).

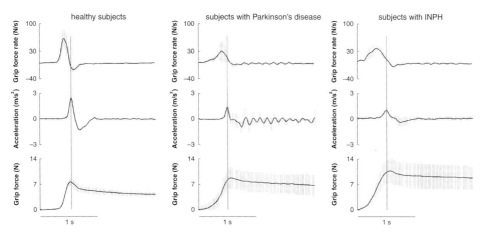

Figure 34.2. Average traces (± 1 SD) of grip-force rate, acceleration and grip force obtained from grip–lift trials performed by healthy subjects, subjects with Parkinson's disease and subjects with INPH. The rate of grip force increase and the lift acceleration appear to be smaller and the grip force signal appears to be greater and more irregular in patients when compared to healthy controls. The dotted vertical lines indicate time of peak acceleration.

Parkinson's disease and INPH subjects. Healthy subjects increased grip force rapidly around 150 ms prior to lift-off. Subjects with INPH and subjects with Parkinson's disease, however, started to increase grip force 500 ms prior to lift-off and the amount of grip-force increase over time was small. Patients with INPH and Parkinson's disease generated significantly greater and more variable peak and static grip forces than healthy subjects when grasping and lifting the object. In addition, the peak rate of grip-force increase and the peak lifting accelerations were significantly smaller for subjects with INPH and Parkinson's disease, compared with healthy subjects. Interestingly, additional oscillations due to re-emergent resting tremor were evident within the traces of acceleration, rate of grip-force increase and grip force for grip–lift trials performed by Parkinsonian subjects. A slowing of the rate of grip-force increase and abnormally high grip forces in relation to the lift forces are typical characteristics of the hypokinetic movement disorder in Parkinson's disease (Ingvarsson *et al.*, 1997; Fellows *et al.*, 1998; Benice *et al.*, 2007). All these features of impaired manual performance were observed with similar severity in both subjects with Parkinson's disease and INPH subjects.

In conclusion, this study showed clear evidence for a hypokinetic deficit of upper limb motor performance in INPH. The hypokinetic motor disability shares several features in common with that found in Parkinson's disease. There were, however, also clear differences in performance between INPH subjects and those with Parkinson's disease. Figure 34.2 illustrates oscillations within the grip-force and acceleration traces obtained from grip–lift trials of subjects with Parkinson's disease that arise from tremor. Frequency analysis detected oscillations only within the grip force signals (minimum amplitude of 0.5 N, minimum duration of 50 ms) of Parkinsonian subjects, but not of INPH subjects, during a 1-second period of stationary holding of the object. These oscillations had an average frequency of 4.6 ± 1 Hz (range: 3.3–6.4 Hz) and were attributed to re-emergent resting tremor.

Cerebrospinal fluid drainage improves the hypokinetic motor disorder of the upper limb in INPH

To answer the question whether the hypokinetic motor disability of grasping improves with CSF evacuation and could be used as a clinical marker for the effectiveness of such treatment, eight subjects with INPH (mean age: 74 ± 7 years) were tested prior to and following drainage of CSF by a spinal tap (Nowak *et al.*, 2006). Eight healthy subjects (mean age 75 ± 7 years) served as a control group. The diagnosis of INPH was by clinical criteria (Hakim & Adams, 1965) and computed tomography or magnetic resonance imaging evidence of ventricular enlargement without evidence of cortical atrophy. Functional disability was graded using a modification of the Stein & Langfitt (1974) scale (Krauss *et al.*, 1997): four INPH subjects needed some help or supervision in activities of daily life and four INPH subjects needed considerable help or supervision. Gait disturbance was rated according to Krauss *et al.* (1997): four subjects showed a considerably unstable gait with tendency to fall and four subjects were unable to walk unassisted. Cognitive impairment was graded using the Mini Mental Status examination (Folstein *et al.*, 1975). Clinical improvement after lumbar puncture and removal of 40 ml CSF on two occasions was assumed to be confirmatory of the diagnosis of INPH.

Prior to and after the second confirmatory CSF evacuation subjects performed ten grip–lift trials with an instrumented object as described above (see Figure 34.1). The following parameters were used for data analysis: (i) peak rate of grip-force increase, (ii) peak grip

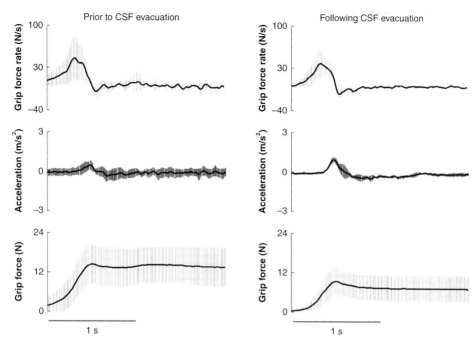

Figure 34.3. Average data (± 1 SD) of grip-force rate, acceleration and grip force obtained from grip–lift trials performed prior to and following CSF drainage.

force, (iii) peak acceleration of the lift and (iv) static grip force established to hold the object stationary. Figure 34.3 shows average traces (± 1 SD) of grip-force rate, acceleration and grip force obtained from grip–lift trials performed by INPH subjects prior to and following CSF tapping. Subjects with INPH generated significantly smaller lift accelerations and significantly greater grip forces compared with healthy control subjects, regardless of whether they were tested prior to or after CSF evacuation. The peak rates of grip-force increase were not significantly different between healthy subjects and INH subjects both prior to and after the lumbar tap. Cerebrospinal fluid evacuation did not change the rate of grip-force increase when establishing a stable grasp and the acceleration of the lifting movement. However, peak grip forces, static grip forces (established when holding the object stationary) and the ratio between peak grip and lift forces all significantly decreased after CSF drainage. Thus, spinal tapping improves the overshoot in grip force to be found in INPH subjects.

Importantly, the severity of the hypokinetic gait disorder as assessed by the score of Krauss *et al.* (1997) and the severity of cognitive impairment as assessed by the Mini Mental Status examination (Folstein *et al.*, 1975) were both significantly correlated with the amount of grip-force overflow as indexed by the ratio between peak grip and lift forces prior to and following CSF evacuation. This is an essential finding suggesting that the measurement of grip force may serve as a clinical adjunct to assess the severity of the hypokinetic motor disability in INPH. Cerebrospinal fluid drainage significantly improved force overflow, gait disturbance and cognitive function. The average force ratio was reduced from 4.8 prior to CSF evacuation to 3.5 following CSF evacuation (27%). The average Mini Mental status improved from 19 points prior to CSF evacuation to 24 points following CSF evacuation (21% improvement). The average grade of gait disturbance improved from 2.4 before CSF evacuation to 1.3 following CSF evacuation (90%). These data indicate that the analysis of grip force adds an additional clinical test for the quantification of the effectiveness of CSF drainage on the hypokinetic motor disorder in INPH.

References

Adams, R. D., Fisher, C. & Hakim, S. (1965). Symptomatic occult hydrocephalus with "normal" cerebrospinal fluid pressure. *N Engl J Med*, **273**, 117–126.

Benice, T. S., Lou, J. S., Eaton, R. & Nutt, J. (2007). Hand coordination as a quantitative measure of motor abnormality and therapeutic response in Parkinson's disease. *Clin Neurophysiol*, **118**, 1776–1784.

Blomsterwall, E., Bilting, M., Stephensen, H. & Wikkelso, C. (1995). Gait abnormality is not the only motor disturbance in normal pressure hydrocephalus. *Scand J Rehabil Med*, **27**, 205–209.

Duinkerke, A., Williams, M. A., Rigamonti, D. & Hillis, A. E. (2004). Cognitive recovery in idiopathic normal pressure hydrocephalus after shunt. *Cogn Behav Neurol*, **17**, 179–184.

Ekstedt, J. & Friden, H. (1984). CSF hydrodynamics for the study of the adult hydrocephalus syndrome. In K. Shapiro, A. Mamarou & H. Portnoy (Eds.), *Hydrocephalus* (pp. 363–384). New York, NY: Raven Press.

Fellows, S. J., Schwarz, M. & Noth, J. (1998). Precision grip in Parkinson's disease. *Brain*, **121**, 1771–1784.

Fisher, C. M. (1982). Hydrocephalus as a cause of disturbances of gait in the elderly. *Neurology*, **32**, 1358–1363.

Flanagan, J. R., Johansson, R. S. (2002). Hand movements. In V. S. Ramachandran (Ed.), *Encyclopedia of the Human Brain*, Vol. 2 (pp. 399–414). San Diego, CA: Academic Press.

Folstein, M. F., Folstein, S. E. & McHugh, P. R. (1975). Mini-Mental State. A practical method for grading the cognitive state of patients for the clinician. *J Psychiatr Res*, **12**, 189–198.

Fukuhara, T., Luciano, M. G., Liu, J. Z. & Yue, G. H. (2001). Functional magnetic resonance imaging before and after ventriculoperitoneal shunting for hydrocephalus – case report. *Neurol Med Chir (Tokyo)*, **41**, 626–630.

Gallia, G. L., Rigamonti, D. & Williams, M. A. (2006). The diagnosis and treatment of idiopathic normal pressure hydrocephalus. *Nat Clin Pract Neurol*, **2**, 375–381.

Graff-Radford, N. R. & Godersky, J. C. (1987). Idiopathic normal pressure hydrocephalus and systemic hypertension. *Neurology*, **37**, 868–871.

Greitz, T. V., Grepe, A. O., Kalmer, M. S. & Lopez, J. (1969). Pre- and postoperative evaluation of cerebral blood flow in low pressure hydrocephalus. *J Neurosurg*, **31**, 644–651.

Hakim, S. & Adams, R. D. (1965). The special clinical problem of symptomatic hydrocephalus with normal CSF pressure. *J Neurol Sci*, **2**, 307–327.

Hakim, S., Vengas, J. G. & Burton, J. D. (1970). The physics of the cranial cavity, hydrocephalus and normal pressure hydrocephalus: mechanical interpretation and mathematical model. *Surg Neurol*, **5**, 187.

Hebb, A. O. & Cusimano, M. D. (2001). Idiopathic normal pressure hydrocephalus: a systematic review of diagnosis and outcome. *Neurosurgery*, **49**, 1166–1184.

Hoehn, M. M. & Yahr, M. D. (1967). Parkinsonism. Onset, progression and mortality. *Neurology*, **17**, 427–442.

Hughes, A. J., Daniel, S. E., Kilford, L. & Lees, A. J. (1992). Accuracy of clinical diagnosis of idiopathic Parkinson's disease: a clinico-pathological study of 100 cases. *J Neurol Neurosurg Psychiatry*, **55**, 181–184.

Ingvarsson, P. E., Gordon, A. M. & Forssberg, H. (1997). Coordination of manipulative forces in Parkinson's disease. *Exp Neurol*, **145**, 489–501.

Krauss, J. K. & Halve, B. (2004). Normal pressure hydrocephalus: survey on contemporary diagnostic algorithms and therapeutic decision-making in clinical practice. *Acta Neurochir*, **146**, 379–388.

Krauss, J. K., Regel, J. P., Vach, W. *et al.* (1996). Vascular risk factors and arteriosclerotic disease in idiopathic normal pressure hydrocephalus of the elderly. *Stroke*, **27**, 24–29.

Krauss, J. K., Regel, J. P., Droste, D. W. *et al.* (1997). Movement disorders in adult hydrocephalus. *Mov Disord*, **12**, 53–60.

Marmarou, A., Young, H. F., Aygok, G. A. *et al.* (2005). Diagnosis and management of idiopathic normal-pressure hydrocephalus: a prospective study in 151 patients. *J Neurosurg*, **102**, 987–997.

Matousek, M., Wikkelso, C., Blomsterwall, E., Johnels, B. & Steg, G. (1995). Motor performance in normal pressure hydrocephalus assessed with an optoelectronic measurement technique. *Acta Neurol Scand*, **91**, 500–505.

McGirt, M. J., Woodworth, G., Coon, A. L. *et al.* (2005). Diagnosis, treatment, and analysis of long-term outcomes in idiopathic normal-pressure hydrocephalus. *Neurosurgery*, **57**, 699–705.

Nowak, D. A. & Hermsdörfer, J. (2005). Grip force behavior in neurological disorders: toward an objective evaluation of manual performance deficits. *Mov Disord*, **20**, 11–25.

Nowak, D. A. & Topka, H. (2006). Broadening a classic clinical triad: the hypokinetic motor disorder of normal pressure hydrocephalus also affects the hand. *Exp Neurol*, **198**, 81–87.

Nowak, D. A., Gumprecht, H. & Topka, H. (2006). CSF drainage ameliorates the motor deficit in normal pressure hydrocephalus. Evidence from the analysis of grasping movements. *J Neurol*, **253**, 640–647.

Sand, T., Bovim, G., Grimse, R. *et al.* (1994). Idiopathic normal pressure hydrocephalus: the CSF tap-test may predict the clinical response to shunting. *Acta Neurol Scand*, **89**, 311–316.

Soelberg Sorensen, P., Jansen, E. C. & Gjerris, F. (1986). Motor disturbances in normal-pressure hydrocephalus. Special reference to stance and gait. *Arch Neurol*, **43**, 34–38.

Stein, S. C. & Langfitt, T. W. (1974). Normal-pressure hydrocephalus: predicting the results of cerebrospinal fluid shunting. *J Neurosurg*, **41**, 463–470.

Stolze, H., Kuhtz-Buschbeck, J. P., Drücke, H. *et al.* (2001). Comparative analysis of gait disorder of normal pressure hydrocephalus and Parkinson's disease. *J Neurol Neurosurg Psychiatry*, **70**, 289–297.

Sudarsky, L. (1997). *Clinical Approach to Gait Disorders of Aging: An Overview.* Philadelphia, PA: Lippincott-Raven.

Vanneste, J., Augustijn, P., Dirven, C., Tan, W. F. & Goedhart, Z. D. (1992). Shunting normal-pressure hydrocephalus: do the benefits outweigh the risks? A multicenter study and literature review. *Neurology*, **42**, 54–59.

Vanneste, J., Augustijn, P. & Tan, W. F. (1993). Shunting normal pressure hydrocephalus: the predictive value of combined clinical and CT data. *J Neurol Neurosurg Psychiatry*, **56**, 251–256.

Wilson, R. K. & Williams, M. A. (2006). Normal pressure hydrocephalus. *Clin Geriatr Med*, **22**, 935–951.

Index